W9-BIB-969

COMPOSITION AND PROPERTIES OF CONCRETE

McGRAW-HILL CIVIL ENGINEERING SERIES

Harmer E. Davis, *Consulting Editor*

BABBITT, DOLAND, AND CLEASBY Water Supply Engineering
BENJAMIN Statistically Indeterminate Structures
CHOW Open-channel Hydraulics
DAVIS, TROXELL, AND WISKOCIL The Testing and Inspection of Engineering Materials
DUNHAM Advanced Reinforced Concrete
DUNHAM Foundations of Structures
DUNHAM The Theory and Practice of Reinforced Concrete
DUNHAM AND YOUNG Contracts, Specifications, and Law for Engineers
HALLERT Photogrammetry
HENNES AND EKSE Fundamentals of Transportation Engineering
KRYNINE AND JUDD Principles of Engineering Geology and Geotechnics
LEONARDS Foundation Engineering
LINSLEY, KOHLER, AND PAULHUS Applied Hydrology
LINSLEY, KOHLER, AND PAULHUS Hydrology for Engineers
LUEDER Aerial Photographic Interpretation
MATSON, SMITH, AND HURD Traffic Engineering
MEAD, MEAD, AND AKERMAN Contracts, Specifications, and Engineering Relations
NORRIS, HANSEN, HOLLEY, BIGGS, NAMYET, AND MINAMI Structural Design for Dynamic Loads
PEURIFOY Construction Planning, Equipment, and Methods
PEURIFOY Estimating Construction Costs
TROXELL, DAVIS, AND KELLY Composition and Properties of Concrete
TSCHEBOTARIOFF Soil Mechanics, Foundations, and Earth Structures
WANG AND ECKEL Elementary Theory of Structures
WINTER, URQUHART, O'ROURKE, AND NILSON Design of Concrete Structures

COMPOSITION AND PROPERTIES OF CONCRETE

SECOND EDITION

GEORGE EARL TROXELL
Professor of Civil Engineering, Emeritus
University of California, Berkeley

HARMER E. DAVIS
Professor of Civil Engineering
University of California, Berkeley

JOE W. KELLY
Professor of Civil Engineering, Emeritus
University of California, Berkeley

McGRAW-HILL BOOK COMPANY New York St. Louis
San Francisco Toronto London Sydney

COMPOSITION AND PROPERTIES OF CONCRETE

Library of Congress Catalog Card Number 68-13104

65286

234567890 MAMM 754321069

PREFACE

Today, more than ever before, the civil engineer is required to give thought and time to the problems of concrete making and utilization. The results accomplished in the field by the construction engineer and the concrete inspector depend upon their knowledge of concrete and of the materials from which it is made. Satisfactory designs of structures are dependent to a considerable extent upon the familiarity of the design engineer with the desirable and the undesirable characteristics of concrete.

The beginner may feel somewhat perturbed on undertaking his study of concrete because seemingly indefinite factors in the manufacturing process apparently tend to yield a product of somewhat indefinite properties. Be it said, however, that improved methods of testing and inspection are resulting in the control of the qualities of concrete within more and more well-defined limits.

Although the rapid advance during the last few decades in the use and knowledge of concrete has made a detailed study of this material quite extensive, there are certain simple principles which have been developed and which can be set down for the guidance of the beginner.

This text and manual is designed as a guide to the student in a comprehensive course in the study of plain concrete. Part I is a descriptive text in which sufficient information is provided so that he can intelligently understand the many factors having a bearing on the proportioning, production, testing, and control of plain concrete. Much of this information has appeared in publications of various technical societies and associations but has been selected and condensed here to make it more useful to the student. It may well serve as a guide to the practicing engineer in selecting and using the cement, fine aggregate, coarse aggregate, and admixtures for a given structure. It covers the proportioning and mixing of these materials, as well as the placing and curing of the concrete, to produce a finished product of suitable and predictable quality and economy.

Several chapters are devoted to the properties of concrete, their significance, and how they are affected by the many steps involved in the fabrication of the product. Attention is given not only to ordinary concrete as used in buildings and highways made of ordinary aggregates but also to the problems involved in mass concrete and other special concretes.

Competent inspection is essential for the attainment of the best results. Hence the problems of the inspector and information on the usual records that he must keep are included in this book.

During the period since publication of the first edition, there have been many new developments in the field of admixtures and other materials for use with concrete. Also much additional information is available concerning cements and the various properties of concrete. As much as is practicable of this new material has been included in this second edition.

Lightweight concrete is now more commonly used for concrete structures, and heavyweight concrete for shielding radiation is required for many special structures; thus additional information on such concretes is presented. Various new types of mechanical equipment are discussed, as they have an appreciable impact on modern concreting operations. The chapters on cement and on forms for concrete have been completely rewritten.

Two new tests have been introduced in Part II which comprises instructions for tests that have been selected to illustrate as efficiently as possible the most important facts and principles connected with the use of cement, aggregate, and concrete.

It is intended that by his laboratory work the student will become familiar with the nature and properties of concrete as well as with the methods of testing cement, aggregate, mortar, and concrete. By following the instructions given herein and by referring to published standards, he should gain familiarity with current specifications for cement and concrete. Through a study of his test results, supplemented by a study of data given in the text, he should come to recognize the properties of these materials and to distinguish between satisfactory and defective samples. It is hoped that by the combination of these exercises he will acquire a thorough understanding of the factors which contribute to the production and control of quality concrete. By summarizing his work in written form, the student will obtain practice in the formulation of engineering reports.

Appendix J is a list of references, classified by subject, which has been augmented in this edition. Brief comments indicating the scope and content of a work are given under some of the items. Throughout the text, references to sources of data or suggestions as to material for further study are made by use of numbers in brackets, which numbers refer to the corresponding bibliographical item listed in Appendix J. Reference to ASTM Specifications and methods of testing are made by indication of the ASTM Serial Designation in brackets.

This work includes information from many sources, and the authors have endeavored to give credit where it is due. Special acknowledgment is given to the Portland Cement Association, the American Concrete Institute, the American Society for Testing and Materials, and the U.S. Bureau of Reclamation for permission to use material from their publications.

The authors also desire to acknowledge the helpful association with Alexander Klein, who reviewed the new chapter on cement, and to other reviewers who proffered various suggestions.

George Earl Troxell
Harmer E. Davis
Joe W. Kelly

CONTENTS

Preface v

PART I COMPOSITION AND PROPERTIES OF CONCRETE

Chapter 1 The Nature of the Problem 3

1.1 Composition of Concrete 3
1.2 Functions of the Paste and Aggregate 4
1.3 General Proportions of Ordinary Concretes 5
1.4 Influence of Quality of Paste upon Properties of Concrete 6
1.5 Concrete Making 7

Chapter 2 Concrete-making Materials—Portland Cement 10

2.1 Cementing Materials 10
2.2 Portland Cement 10
2.3 Influence of Cement on Durability of Concrete 11
2.4 Types of Portland Cement 12
2.5 Basic Constituents of Cements 13
2.6 Chemical Formulas and Processes 16
2.7 Manufacture of Portland Cement 18
2.8 Chemical Analysis of Portland Cement 21
2.9 Major Compounds in Portland Cement 23
2.10 Influence of Composition upon Characteristics of Portland Cement 28
2.11 Fineness of Cement 29
2.12 Hydration Reactions in Cement Paste 31
2.13 Heat of Hydration 35
2.14 Physical Aspects of the Setting and Hardening Process 36
2.15 Structure of Cement Paste 40
2.16 Role of Paste Structure and Composition in the Behavior of Concrete 46
2.17 Specification Requirements for Portland Cements 49
2.18 High-early-strength Cements 54
2.19 Portland-pozzolan Cements 55
2.20 Slag Cements 56
2.21 Portland Blast-furnace-slag Cement 56
2.22 Masonry Cements 56
2.23 Expansive Cements 57
2.24 High-alumina Cement 58

Chapter 3 Aggregates 60

3.1 Preliminary Remarks 60
3.2 General Characteristics 61
3.3 Data Needed for Proportioning Mixtures 62

3.4 Sampling Aggregates 65
3.5 Specific Gravity 66
3.6 Unit Weight and Voids 68
3.7 Moisture and Absorption 74
3.8 Gradation 76
3.9 Sieve Analyses 77
3.10 Grading Charts 81
3.11 Maximum Size of Aggregate 82
3.12 Grading Requirements 84
3.13 Quality Requirements 86
3.14 Deleterious Substances 87
3.15 Reactive Aggregates 90
3.16 Handling and Storing Aggregates 93

Chapter 4 Water, Admixtures, and Miscellaneous Materials 94

4.1 Mixing Water 94
4.2 Water for Washing Aggregates 95
4.3 Water for Curing Concrete 95
4.4 Types of Admixtures 95
4.5 Workability Admixtures 97
4.6 Water-reducing and Set-retarding Admixtures 97
4.7 Air-entraining Admixtures 98
4.8 Gas-forming Admixtures 101
4.9 Accelerators 102
4.10 Expansion-producing Admixtures 103
4.11 Pozzolanic Materials 103
4.12 Bonding Admixtures 105
4.13 Curing Aids 105
4.14 Miscellaneous Materials 106
4.15 Steel Reinforcement 106

Chapter 5 Properties of Fresh Concrete 108

5.1 General 108
WORKABILITY; CONSISTENCY 108
5.2 Workability 108
5.3 Consistency 110
5.4 Slump, Flow, and Ball Tests for Consistency 112
5.5 Powers Remolding Test 115
5.6 Other Tests 115
BLEEDING; STIFFENING; SETTING 116
5.7 Bleeding 116
5.8 Plastic Shrinkage and Settlement 117
5.9 Stiffening and Setting 118
5.10 Pressure on Forms 119
AIR ENTRAINMENT 120
5.11 Entrained Air 120
5.12 Measurement of Entrained Air 121
COMPOSITION; PROPERTIES 123
5.13 Unit Weight; Yield; Cement Factor 123
5.14 Proportions of Components 125

5.15 Uniformity 126
5.16 Temperature of Components and Mix 127

Chapter 6 Proportioning of Concrete Mixes 130

6.1 General 130
6.2 Methods of Expressing Proportions 131
6.3 Aggregate-Paste Relationships 133
6.4 Variables in Proportioning 136
6.5 Trial Method of Proportioning 137
6.6 Mix Adjustments 141
6.7 ACI Method of Proportioning 142
6.8 ACI Method for Small Jobs 150
6.9 ACI Method for Structural Lightweight Concrete 151
6.10 ACI Method for No-slump Concrete 154
6.11 Arbitrary Proportions 155
6.12 Proportioning by Maximum Density of Aggregate 155
6.13 Proportioning by Surface Area of Aggregate 156
6.14 Proportioning by Fineness Modulus of Aggregate 157
6.15 Proportioning by Voids-Cement Ratio and Mortar Voids 158
6.16 Proportioning by Void Content of Coarse Aggregate 160

Chapter 7 Manufacture of Concrete 162

BATCHING 162
7.1 Batching 162
7.2 Weight-batching Equipment 163
7.3 Checking Weighing Equipment 164
7.4 Volumetric Batching Equipment 166
7.5 Batching Cement 167
7.6 Irregularities in Batching 167
7.7 Water- and Admixture-measuring Equipment 167
MIXING 169
7.8 Mixing 169
7.9 Types of Mixers 169
7.10 Time of Mixing 171
7.11 Mixer Efficiency 171
7.12 Hand Mixing; Retempering 173
7.13 Ready-mixed Concrete 173
CONVEYING 175
7.14 Conveying 175
7.15 Batch Containers 175
7.16 Pump and Pipeline 178
7.17 Pump Sizes 181
7.18 Cleaning the Concrete Pump 182
7.19 Pneumatic Method 183
7.20 Chutes and Belts 183

Chapter 8 Placing and Curing Concrete 185

8.1 Preparations for Placing 185
8.2 Placing 187

COMPACTION 192
8.3 Compaction 192
8.4 Hand Tamping 192
8.5 Vibrators 192
8.6 Vibrator Efficiency 193
8.7 Concrete Mix for Vibratory Compaction 193
8.8 Proper Use of Vibration 194
CURING 195
8.9 The Curing Period 195
8.10 Curing Methods 196
8.11 Curing of Pavements and Other Structures 197
8.12 Curing Temperatures 199
8.13 Steam Curing 201
8.14 Concrete Work during Cold Weather 202
8.15 Calcium Chloride in Concrete during Cold Weather 205
8.16 Concrete Work during Hot Weather 207
8.17 Curing in the Laboratory 209
OPERATIONS AFTER CURING 209
8.18 Removal of Forms 209
8.19 Patching 210
8.20 Prevention of Damage 211

Chapter 9 Forms for Concrete 213

9.1 Requirements of Forms 213
9.2 Factors Affecting Form Pressures 213
9.3 Lateral Pressure Values 214
9.4 Uplift Due to Lateral Pressure 216
9.5 Other Loads 217
9.6 Formwork Materials 217
9.7 Form Design 219
9.8 Wall Forms 220
9.9 Beam Forms 220
9.10 Column Forms 221
9.11 Construction of Forms 222
9.12 Slip-form Construction 222
9.13 Coating of Forms 224
9.14 Failure of Forms 225

Chapter 10 Strength of Concrete 227

10.1 Properties of Hardened Concrete 227
SIGNIFICANCE OF STRENGTH 227
10.2 Resistance to Applied Forces 227
10.3 Strength as a Measure of General Quality 228
10.4 Nature of Strength 228
KINDS OF STRENGTH 230
10.5 Compressive Strength 230
10.6 Tensile Strength 231
10.7 Flexural Strength 233
10.8 Shear Strength 234
10.9 Bond with Reinforcement 234

10.10 Nondestructive Indications of Strength 236
FACTORS AFFECTING STRENGTH 238
10.11 Effect of Component Materials 238
10.12 Effect of Proportions and Uniformity 240
10.13 Effect of Curing Conditions 242
10.14 Effect of Loading Conditions 247
10.15 Effect of Exposure to High Temperatures 248
FACTORS AFFECTING RESULTS OF STRENGTH TESTS 250
10.16 Specimens vs. Structures 250
10.17 Effect of Size and Shape of Specimen 251
10.18 Effect of Conditions of Casting 254
10.19 Effect of Moisture Content of Specimen 254
10.20 Effect of Temperature of Specimen 255
10.21 Effect of Bearing Conditions 255
10.22 Effect of Rate of Loading 257

Chapter 11 Permeability and Durability 261

PERMEABILITY 261
11.1 Pore Structure of Concrete 261
11.2 Significance of Permeability 262
11.3 Permeability Tests 262
11.4 Factors Affecting Watertightness 263
11.5 Effect of Water and Cement 263
11.6 Effect of Aggregates 265
11.7 Effect of Curing 265
11.8 Effect of Admixtures and Coatings 265
11.9 Uniformity of Concrete 266
11.10 Absorption 267
DURABILITY 268
11.11 Deterioration of Concrete 268
11.12 Weathering 268
11.13 Weathering Resistance as Affected by Aggregate, Cement, and Water 269
11.14 Air-entrained Concrete 272
11.15 Freeze-Thaw Tests 273
11.16 Reactive Aggregates 275
11.17 Sulfate Waters 278
11.18 Leaching and Efflorescence 281
11.19 Chemical Attack 281
11.20 Corrosion of Embedded Metal 281
11.21 Wear 282
11.22 Restoration of Disintegrated Concrete 284

Chapter 12 Volume Changes and Creep 290

12.1 Types of Volume Change in Concrete 290
12.2 Significance of Volume Changes and Creep 291
12.3 The Gel Structure as Related to Volume Changes 292
12.4 Shrinkage of Fresh Concrete 232
12.5 Autogenous Volume Changes 232
12.6 Expansive Cements 294
SHRINKAGE AND EXPANSION DUE TO MOISTURE CHANGES 294

12.7 Factors Affecting Shrinkage and Expansion 294
12.8 Effect of Cement and Water Contents 295
12.9 Effect of Composition and Fineness of Cement 297
12.10 Effect of Type and Gradation of Aggregate 298
12.11 Effect of Admixtures 302
12.12 Effect of Age at First Observation 302
12.13 Effect of Moisture and Temperature Conditions 303
12.14 Effect of Duration of Tests 304
12.15 Effect of Size and Shape of Specimen 304
12.16 Effect of Absorptiveness of Forms 305
12.17 Effect of Carbonation 306
12.18 Effect of Reinforcement 307
12.19 Prepacked Concrete 308
12.20 Thermal Volume Changes 308
CREEP OF CONCRETE 309
12.21 Factors Affecting Creep 309
12.22 Effect of Stress and Age When First Loaded 311
12.23 Effect of Water-Cement Ratio and Mix 312
12.24 Effect of Composition and Fineness of Cement 313
12.25 Effect of Character and Grading of Aggregate 314
12.26 Effect of Moisture Conditions of Storage 315
12.27 Effect of Size of Mass 315
12.28 Creep in Axial Tension and Compression 316
12.29 Estimation of Creep 316
12.30 Creep Recovery 317
12.31 Creep of Reinforced Concrete 318
12.32 Creep of Prestressed Concrete 319

Chapter 13 Other Properties 321

ELASTIC PROPERTIES OF CONCRETE 321
13.1 Modulus of Elasticity 321
13.2 Methods for Determining Moduli of Elasticity 321
13.3 Effect of Method of Test on Modulus of Elasticity 325
13.4 Effect of Characteristics of Concrete on Modulus of Elasticity 326
13.5 Relationship of Modulus of Elasticity to Strength 329
13.6 Effect of Type of Loading on the Modulus of Elasticity 330
13.7 Sustained Modulus of Elasticity 330
13.8 Significance of Poisson's Ratio 330
13.9 Factors Affecting Poisson's Ratio 331
THERMAL PROPERTIES 331
13.10 Thermal Conductivity 331
13.11 Condensation as Related to Thermal Conductivity 332
13.12 Thermal Properties and Their Relationships 334
13.13 Temperature Rise in Mass Concrete 335
EXTENSIBILITY AND CRACKING 339
13.14 Cracking of Concrete 339
13.15 Extensibility and Cracking 342
13.16 Thermal Stress and Cracking 343
MISCELLANEOUS PROPERTIES 345
13.17 Fire Resistance 345

13.18 Fatigue Strength 348
13.19 Unit Weight 350

Chapter 14 Special Types of Concrete 352

14.1 Architectural Concrete 352
FLOOR SURFACES 353
14.2 Types and Requirements 353
14.3 Preparation of Base 353
14.4 Concrete Mix 354
14.5 Placing and Finishing 354
14.6 Curing and Protection 355
14.7 Surface Hardeners 356
SPRAYED MORTAR AND CONCRETE: GUNITE AND SHOTCRETE 356
14.8 Use and Limitations 356
14.9 Equipment 357
14.10 Preparation of Base 358
14.11 Aggregate 358
14.12 Rebound 359
14.13 Mix 359
14.14 Mixing and Placing 359
14.15 Curing 360
MASS CONCRETE 360
14.16 Characteristics of Mass Concrete 360
14.17 Special Treatment of Mass Concrete 361
14.18 Effect of Temperature and Other Variables on Properties of Mass Concrete 362
MISCELLANEOUS CONCRETES 365
14.19 Concrete Placed under Water 365
14.20 Vacuum Concrete 366
14.21 Concrete for Radiation Shielding 367
14.22 Lightweight Concrete 368
14.23 Grouting without Pressure 371
14.24 Pressure Grouting 372
14.25 Grouted Concrete 373
14.26 Chemical Prestressing Using Expansive Cements 373

Chapter 15. Inspection 375

15.1 Need for and Scope of Inspection 375
15.2 Inspection Organization 376
15.3 Qualifications of the Inspector 377
15.4 Responsibility 378
15.5 Inspector Training 378
15.6 Relations with Superior Officers 379
15.7 Relations with the Contractor 379
15.8 Authority of the Inspector 382
15.9 Specification Is Inspector's Guide 382
15.10 Inspection before Concreting 383
15.11 Inspection of Concreting 383
15.12 Inspection after Concreting 384
15.13 Concrete Samples for Tests 384
15.14 Molding Specimens 384

15.15 Storing and Shipping Specimens 385
15.16 Required Strength of Test Specimens 385
15.17 The Field Laboratory 386

Chapter 16 Inspection Records and Reports 389

16.1 General Comments 389
16.2 Batching and Mixing Record 390
16.3 Record of Materials 390
16.4 Record of Placing and Curing 391
16.5 Daily Reports 391
16.6 Diary 392
16.7 Photographs 392
16.8 Summary Report 393

Chapter 17 Analysis and Presentation of Data 395

17.1 The Problem of Transmission of Information 395
17.2 Variations in Data 396
17.3 Grouping of Data 396
17.4 Central Tendency 398
17.5 Dispersion 398
17.6 Probable Error 402
17.7 Limits of Uncertainty of an Observed Average 402
17.8 Number of Tests to Obtain a Desired Accuracy 404
17.9 Significant Figures to Retain in Presenting Test Results 405
17.10 Required Average Strength 406
17.11 Statistical Summaries 406
17.12 Tables 407
17.13 Figures 407

PART II INSTRUCTIONS FOR LABORATORY WORK

General Instructions 413

Test 1 Normal Consistency and Time of Set of Portland Cement 415
2 Strength of Type I Portland Cement and Type III High-early-strength Cement Mortars at Various Ages 418
3 Effect of Curing Conditions upon Compressive Strength of Portland Cement Mortars 422
4 Sieve Analysis of Concrete Aggregates 424
5 Specific Gravity, Unit Weight, Moisture Content and Absorption of Concrete Aggregates 426
6 Characteristics of Fresh Concrete 432
7 Effect of Water-Cement Ratio upon Compressive Strength and Consistency of Concrete of Uniform Mix 436
8 Effect of Water-Cement Ratio upon Compressive Strength, Cement Factor, and Cost of Concrete of Uniform Consistency 440
9 Trial-mix Proportioning of Concrete 443
10 Concrete-mix Proportioning by ACI Calculation Method 445
11 Adjustment of Concrete Mix to Give Desired Cement Factor or Water-cement Ratio at Constant Consistency 446

12 Adjustment of Concrete Mix to Produce a Given Change in Consistency 448
13 Effect of Capping Materials and End Conditions before Capping upon Compressive Strength of Concrete Cylinders 450
14 Effect of Shape of Test Specimen upon Indicated Compressive Strength of Concrete 452
15 Physical and Mechanical Properties of Concrete 454
16 Splitting Tensile Strength of Concrete Cylinders 456
17 Demonstration of Entrained Air in Concrete 458

APPENDIXES

A Summary of Useful Values 461
B Instructions on Operation of Testing Machines 462
C Procedure for Making the Slump Test 465
D Procedure for Making the Ball-penetration Test 466
E Procedure for Making the Flow Test 467
F Procedure for Making the Remolding Test 468
G Procedure for Batching and Mixing Concrete and Molding Compression-test Cylinders 470
H Procedure for Capping Compression Cylinders with Gypsum Compounds 472
I Procedure for Capping Compression Cylinders with Sulfur Compound 473
J Selected References and Specifications Pertaining to Plain Concrete 475

Index 515

COMPOSITION AND PROPERTIES OF CONCRETE

PART ONE

COMPOSITION AND PROPERTIES OF CONCRETE

ONE
THE NATURE OF THE PROBLEM

1.1 Composition of concrete. Concrete is a composite material which consists essentially of a binding medium within which are embedded particles or fragments of a relatively inert mineral filler. In portland cement concrete the binder or matrix, either in the plastic or in the hardened state, is a combination of portland cement and water; it is commonly called the "cement paste." The filler material, called "aggregate," is generally graded in size from a fine sand to pebbles or fragments of stone which, in some concretes, may be several inches in diameter.

In practical concrete mixtures, the overall proportions of these principal components, the binder and the aggregate, are controlled by the requirements that, (1) when freshly mixed, the mass be workable or placeable, (2) when the mass has hardened, it possess strength and durability adequate to the purpose for which it is intended, and (3) cost of the final product be a minimum consistent with acceptable quality. A diagrammatic representation of the composition of concrete of the proportions used in construction is shown in Fig. 1.1.

The aggregate occupies roughly three-quarters of the space within a given mass. For convenience, particles smaller than about 3⁄16 in. in diameter are designated as fine aggregate or sand. Natural coarse aggregates may consist of gravel or crushed stone. Other materials employed as aggregates include slag, cinders, and artificial lightweight aggregates made of burned clay or shale.

The space not occupied by aggregate, roughly one-quarter of the entire volume of an average concrete, is filled with cement paste and air voids. After concrete has been placed, even though it has been compacted with considerable thoroughness, some entrapped air remains within the mass. In a freshly made and compacted concrete of suitable proportions, the volume of unavoidable entrapped air is comparatively small, usually not over 1 or 2 percent. For particular purposes, however, there has developed in recent years the practice of incorporating in the mixture special

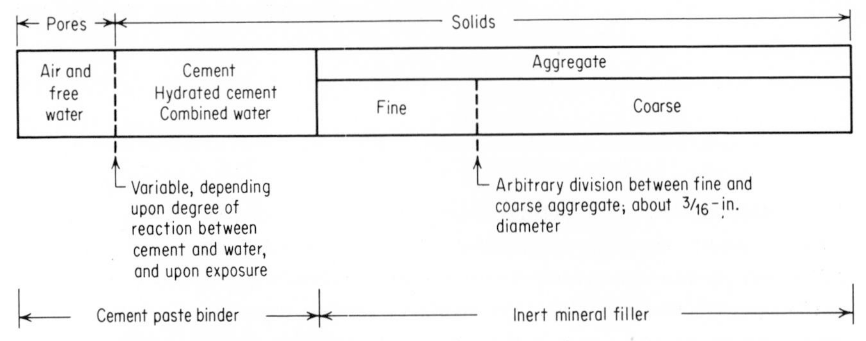

Fig. 1.1. Composition of concrete.

air-entraining agents, with the result that small air voids, amounting sometimes to several percent of the volume of the mass, are distributed throughout the paste.

The solid portion of hardened concrete is composed of the mineral aggregate and the hardened cement paste, which may include some of the original cement, and a new product formed by combination of the remainder of the cement with some of the water. After any period, the amount of free water left depends upon the extent of combination of cement and water, and upon possible loss of water from the mass due to evaporation under drying conditions.

1.2 Functions of the paste and aggregate. The binder material, the cement-water paste, is the active component of concrete and has two main functions: (1) to fill the voids between the particles of the inert aggregates, providing lubrication of the fresh, plastic mass and water tightness in the hardened product, and (2) to give strength to the concrete in its hardened state. The properties of the hardened paste depend upon (1) the characteristics of the cement, (2) the relative proportions of cement and water, and (3) the completeness of chemical combination between the cement and water. This chemical process is often referred to as "hydration," although other processes are undoubtedly involved. Hydration of the cement requires time, favorable temperatures, and the presence of moisture. The period during which concrete is definitely subjected to favorable temperature and moisture conditions is known as the "curing" period. Curing periods varying from 3 to 14 days are commonly used on construction work. In the laboratory, a common curing period is 28 days. Adequate curing is essential for the production of quality concrete.

The aggregate has three principal functions: (1) to provide a relatively cheap filler for the cementing material; (2) to provide a mass of particles which are suitable for resisting the action of applied loads, abrasion, the percolation of moisture, and the action of weather; and (3) to reduce the volume changes resulting from the setting and hardening process and from moisture changes in the cement-water paste. The properties of concretes resulting from the use of particular aggregates depend upon (1) the mineral character of the aggregate particles, particularly as related to strength, elasticity, and durability; (2) the surface characteristics of the particles, particularly as related to workability of the fresh concrete, and bond within the hardened mass; (3) the grading of the aggregates, particularly as related to the workability, density, and economy of the mix; and (4) the amount of aggregate in unit volume of concrete, particularly as related to cost and to volume changes due to drying.

1.3 General proportions of ordinary concretes. The properties of both freshly mixed and hardened concrete are intimately associated with the characteristics and relative proportions of the component ingredients.

In fresh concrete the aggregate is, in effect, suspended in the cement paste; there must be, then, sufficient paste not only to coat the aggregates but also to fill the voids between them. The consistency, or degree of wetness, of the mass is controlled by the fluidity of the paste, by the quantity of aggregate per unit volume of paste, and by the gradation and shape of the aggregate particles. For most work a plastic consistency is desirable—with either too dry, too wet, or too harsh a mixture, segregation tends to occur, and a defective product may result.

In hardened concrete, properties such as strength are functions of the density of the paste, which in turn is controlled by the ratio of water to cement in the original mixture. Hence, there are practical limits to the proportions of cement, water, and aggregate in normal concretes.

The general proportions of a group of practical mixtures are illustrated in Fig. 1.2. All mixes are of the same consistency. The diagram shows, for a given volume of fresh concrete, the relative volume occupied by the aggregate particles, the cement, and the water (the small volume of air voids is neglected here). Even over the wide range in mixes shown, it is noteworthy that the aggregate occupies between 65 and 75 percent of the volume of the concrete—a major portion of the mass. Over this range the volume of the cement varies from about 16 to 8 percent of the volume of the total mix; that is, there is about half as much cement in the lean mix as in the rich mix.

The total water varies relatively little over the range, being about 19 percent of the total volume of concrete for the richest mix shown and about 17 percent for the leanest mix shown. Advantage may be taken of this important fact in adjusting mixes to a required cement content. By exchanging fine aggregate for cement on a solid-volume basis, while the total

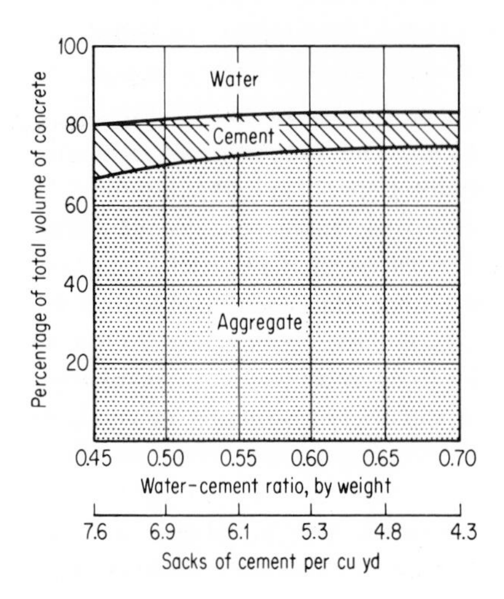

Fig. 1.2. Composition of concrete mixtures of uniform consistency (3 to 4-in. slump).

water content is kept constant, approximately the same consistency is maintained.

From the fact that the water content is practically constant while the cement content varies considerably, it is apparent that the dilution of the paste, expressed, say, as the ratio of water to cement, must be greater for the lean mixes than for the rich ones. Stated in practical terms, it means that as the amount of cement in a mix is reduced, more water must be used per sack of cement, if constant consistency is to be maintained. It is apparent that an intimate relation exists between richness of mix and water-cement ratio in mixes of the same consistency.

In the hardening process, a unit quantity of a particular cement can potentially combine with a specific quantity of water. Thus for a given quantity of cement paste, a paste having a higher water-cement ratio (small percentage of cement) will have a larger volume of potentially uncombined water than will a paste of lower water-cement ratio. Since capillary-pore space derives from uncombined water, the paste in the leaner mixes has a more porous structure than that in the richer mixes.

1.4 Influence of quality of paste upon properties of concrete. The cement paste has been characterized as the active element in concrete. The performance of concrete is largely influenced by the properties of the paste, provided the mineral aggregates are of satisfactory quality. Ordinarily, sound strong aggregates are not difficult to secure.

For a cement of given chemical composition, the strength and porosity of the paste-structure are dependent almost entirely upon the water-cement ratio (see Art. 10.12). So long as plastic workable mixes are used, the lower the water-cement ratio, the greater the strength and water-tightness.

This is true regardless of the richness of mix. Whatever measures of strength are used (compressive, tensile, or flexural strength) the relationships between them and the water-cement ratio are similar [100, 101].[1]

Durability, or resistance to weathering, is a function of both strength and watertightness. The quality of the paste thus has a direct influence upon this important property. This has been shown not only by tests but also by studies of the condition of structures in the field [101].

It is apparent that other properties which are affected by the character of the paste structure will be influenced by the water-cement ratio. For example, the higher the water-cement ratio, the greater the shrinkage of paste due to drying. The volume change of concrete, however, is dependent upon the quantity of paste as well as upon quality of the paste; in some cases the shrinkage due to the effect of increase in water-cement ratio may be offset by the effect of increased quantity of aggregate.

A basis that has been used in the past for selecting a water-cement ratio for a particular job is the compressive strength of concrete. However, even with fairly high water-cement ratios, ample strengths are obtained with most present-day normal portland cements. Hence any limitations that are placed upon the water-cement ratio should also be governed by considerations of resistance to weathering. Recommendations on this basis, for various types of structures and degrees of exposures, have been made available (see Table 6.5 and Ref. 100). In any case, fixing the water-cement ratio gives a measure of control over the potential quality of the concrete.

1.5 Concrete making. The making of concrete is a manufacturing process, even though the plant may often be temporary and the product is made in the field. In many areas, however, concrete is now produced in a central mixing plant and is hauled to the construction site. In any case, the problems of material procurement, of personnel organization, of quality control and economics parallel those of any other manufacturing process. When the plant is temporary, the problems of proper plant layout and widely varying sources of material must be recognized as a part of each job. Furthermore, from a general point of view, because the process of securing a concrete of desired quality is not completed until the concrete structure is finished, each job presents its own peculiar problems of forming, placement, and curing. These considerations call for special experience and judgment on the part of the manufacturer.

An engineer or engineering staff is usually responsible for the technical phases of the process, whether the operating organization is provided by a contractor or by the owner. The engineer must be familiar with the principles of concrete making as well as the art of producing it. He must

[1] Numbers in brackets indicate references listed in Appendix J.

know how to prepare the specifications to select the raw materials, to control the batching, mixing, and placing, and to inspect and test the product in any stage of completion.

The essential problem for the engineer is to obtain a satisfactory product at a reasonable cost. A satisfactory concrete is one which has the necessary and desired properties, such as *workability* in the plastic material, and *uniformity, strength, watertightness, durability,* and *volume constancy* in the hardened product. A concrete in which these several factors are properly interrelated may be said to possess "balance."

Control implies that concrete is produced in such a manner that there is deposited in place an adequate supply of uniform, workable concrete which, when adequately cured, will have the desired service qualities. It involves carrying out the various steps in the process in an economical manner, the making of preliminary tests to ensure suitable materials, and such follow-up tests as may be necessary to confirm the quality.

It is not enough to select a water-cement ratio and other proportions presumed to give, on paper, a desired strength or similar property, because unless the mix is workable, it cannot be placed properly. A mixture that is too dry or too harsh for given placing conditions and handling equipment will inevitably result in rock pockets. Too wet a mixture gives rise to segregation, honeycomb, and an accumulation of laitance, or scum, at the top of each lift. No matter how watertight the paste may be, a concrete is not watertight if it contains rock pockets and cracks. No matter how great the potentialities of a mix may be with regard to strength and durability, these qualities cannot be realized unless they are built into every cubic yard of concrete in the field, through adequate placing, compacting, and

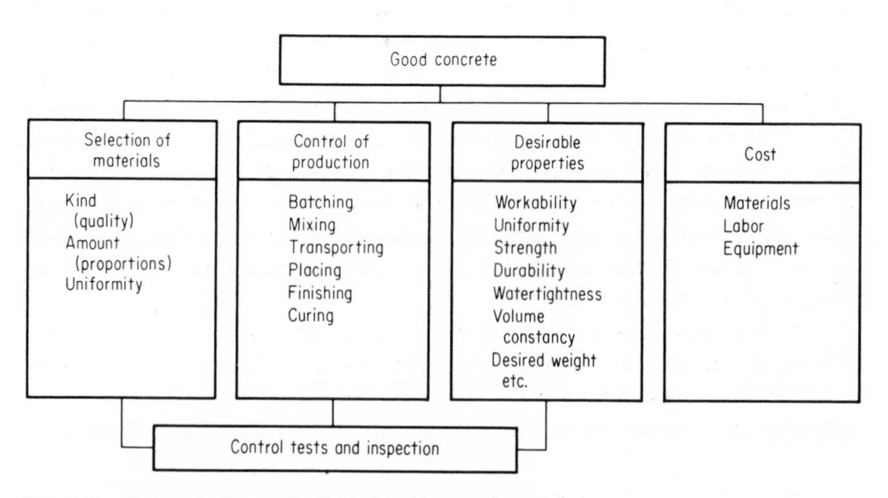

Fig. 1.3. Factors in the production of good concrete.

curing conditions. Good judgment, foresight, and competent inspection are necessary to secure a balance among all the factors in the manufacture of concrete. Too much emphasis cannot be placed upon the necessity for good workmanship if the final product taken as a whole—the structure—is to be of high quality.

The factors to be considered in producing good concrete are summarized diagrammatically in Fig. 1.3.

QUESTIONS

1. What factors control the proportions of a concrete mix?

2. What percentage of a concrete mass generally is occupied by the aggregate?

3. What factors control the amount of free water in a hardened concrete mass?

4. Discuss the functions of the cement-water paste (*a*) in fresh concrete and (*b*) in hardened concrete.

5. The properties of the hardened cement-water paste depend upon what conditions?

6. Hydration of a cement is affected by what factors?

7. List the principal functions of the aggregate in concrete.

8. What properties of the aggregate influence the properties of the resulting concrete, and what is their effect?

9. What factors control the consistency of concrete?

10. Discuss the variation in aggregate and water contents of normal, workable concretes as the mix varies from lean to rich in cement.

11. Is ample water included in the usual concrete mix to hydrate the cement? Is it ever necessary to provide additional water? Explain.

12. What effect does the water-cement ratio have on the strength, porosity, and durability of a concrete mass?

13. Describe the characteristics of some concrete mixes which cause the mix to be undesirable for good work.

14. List the factors to be considered in producing good concrete.

TWO
CONCRETE-MAKING MATERIALS—PORTLAND CEMENT

2.1 Cementing materials. For constructional purposes a cement is a material capable of developing, after appropriate reactions have taken place, those adhesive and cohesive properties that make it possible to bond together mineral fragments to produce a hard continuous compact mass of masonry. There are a variety of cementing materials in use for this purpose; two important classes of such materials are calcareous cements and bituminous binders. The cementing or bonding action of the calcareous cements is attained through a chemical reaction involving lime or lime compounds.

For structures which must sustain and transmit appreciable load and which must offer appreciable resistance to disintegration under a range of adverse conditions of exposure, by far the most important of the cementing materials (except for use in pavements) are the calcareous hydraulic cements. Hydraulic cements have the special property of setting and hardening under water. Key and essential components of these cements are lime and silica. In the presence of water, the lime and silica or lime-silica compounds, react to form ultimately a hardened product containing hydrated calcium silicates.

2.2 Portland cement. For the purposes of this book, the discussion is directed primarily to portland cement, which, of all the hydraulic cements, is in the most widespread use for constructional purposes at the present time. The origin of the name "portland" cement is usually attributed to Joseph Aspdin, a brick mason in England, who in 1824 took out a patent for making artificial stone; he called his product portland cement because the mortar made with it resembled in color the stone which was quarried on the Isle of Portland. The production of material which would approximate the composition and properties of present-day portland cements, however, did not begin in appreciable quantity until the latter part of the nineteenth century.

As manufactured for use in construction work, portland

cement is a finely powdered substance, usually gray or brownish gray, composed largely of artificial crystalline minerals, the most important of which are calcium and aluminum silicates. The calcium silicate compounds, upon reaction with water, produce the new compounds capable of imparting the stonelike quality to the mixture. These and the other compounds characteristic of portland cement are described in more detail in subsequent articles.

While the term "cement" normally refers to the dry, unreacted powdered material, it is also sometimes applied to the reaction product that serves as the binder in mortar or concrete; hydrated cement or hardened cement paste would be a more specific term for the latter state.

The particles range in size from about 0.5 micron (about 0.00002 in.) in diameter to perhaps 80 microns (0.0032 in.). Of most present-day portland cements, the major part passes a No. 200 sieve (74-micron openings).

The specific gravity of the particles of portland cement generally ranges from about 3.12 to 3.16. The unit weight of bulk cement varies with the degree of compactness; in the United States, 1 cu ft bulk of cement is taken to be 94 lb, which is the net weight of a commercial sack, or bag of cement. In large quantity, the unit of measure is the barrel, which is the equivalent of four sacks or 376 lb.

The rate of hardening, strength attained, durability, and other properties are controlled to a considerable degree by the relative proportions of the essential compounds and by the fineness to which the cement is ground. The development of the desirable properties is also influenced by the ambient conditions (temperature and the availability of water) during the hardening process.

2.3 Influence of cement on durability of concrete. With the widespread use of portland cement concrete under varying conditions of exposure and service, it is natural that shortcomings should have become apparent. The gradual development of checks, cracks, spalls, or a state of general disintegration in an important concrete structure is a matter of real concern. It is only fair to state that perhaps many more failures of concrete structures have been due to poorly made concrete than to defective cement. However, as the result of intensive research, there has developed substantial understanding of the problem of durability. The general problem of concrete durability is discussed in Chap. 11, but a few of the influencing factors which may be partly controlled by proper choice of cement are mentioned here.

The most important general agency of distintegration is weathering action: wetting, freezing, thawing, and drying. Except for severe exposure conditions, a well-made concrete is the best insurance for good service, but even some types of well-built structures, such as concrete pavements, may be subject to disintegration, as the weathering cycle may be severe not only in intensity but also in frequency. An important contribution was made by the discovery that the presence of minute air voids, distributed throughout the cement paste, greatly enhanced the resistance of concrete to disintegration under the action of repeated freezing and thawing. As a result, chemicals are now available which cause air to be entrained in a concrete during mixing; these chemicals may be added at the mixer, but they are also incorporated in a finished cement. However, for successful use in the field, conditions of batching, mixing, and placing must be carefully controlled.

An important field of use of concrete is for substructures, port structures, pavements, and canal linings; in any of these cases the concrete may be exposed to waters containing chlorides and sulfates, especially of magnesium and sodium. The sulfates are the more aggressive in their action, and if the exposure is accompanied by wetting and drying, the disintegration is greatly accelerated. It has been found that cements having a relatively low content of tricalcium aluminate are much more resistant to the action of these aggressive waters than are cements of normal composition. Hence, for use in important structures subjected to exposure of this kind, it has become the practice to specify cements of modified composition.

2.4 Types of portland cement. The types of portland cement which are in current use in the United States are listed in Table 2.1. Requirements for these cements are given in the American Society for Testing Materials Specification C150.[1]

[1] Citations preceded by a letter symbol refer to ASTM Standards as designated and published in current books of ASTM Standards (see Appendix J).

Table 2.1 Current types of portland cement*

General description	Designation	Use
General-purpose or normal	Type I	For general concrete construction where special properties are not required
Modified general-purpose	Type II	For general concrete construction exposed to moderate sulfate action or where it is required that the heat of hydration be somewhat lower than for normal cement
High-early-strength	Type III	For use when rapid hardening is required
Low-heat	Type IV	For use where it is required that the heat of hydration be a practicable minimum
Sulfate-resisting	Type V	For use where a high resistance to the action of sulfates is required

* From Ref. C150.

The properties and performance of these several types of cement are largely governed by the compound composition although rapid hardening may be influenced by fineness of grinding as well as appropriate compound composition. See Arts. 2.9 and 2.10 for discussion of composition.

Type I, the general purpose or "normal" cement, is by far the most widely used and is the type most likely to be available from any mill; its composition may vary considerably among mills. The composition of type II is such (lower tricalcium aluminate content) that concretes made with this cement provide a somewhat greater resistance than those made with type I to disintegration by attack of aggressive chemicals, notably the sulfates, found in some soils and water; this type of cement also gains strength somewhat more slowly than type I; it has rather extensive use in some parts of the country. The high-early-strength cement, type III, by virtue of composition or fineness or both hardens more rapidly than the type I cement and finds use in concreting in cold weather and in making repairs where the structure must be replaced in use as soon as possible; the type III cement is generally available. The type IV cement has relatively low generation of heat during setting and hardening, and the type V cement has relatively much greater resistance to the action of sulfates than the other cements; because of their rather special nature, these cements are not usually carried in stock and usually must be made on special order.

In addition to the five basic types of cement, three variations of these types are produced by incorporating in them, during grinding, chemical additives which cause the entrainment of air during the concrete-mixing process. The air-entraining cements are designated as types IA, IIA, and IIIA, and correspond to types I, II, and III; requirements for these cements are given in ASTM Specification C175. The air-entraining cements are being more widely used in structures where there is expectation that the concrete will be exposed to severe frost action. Air entrainment can also be obtained by adding a suitable admixture to the concrete during the mixing process.

2.5 Basic constituents of cements. A knowledge of a number of the common chemical consituents of cements is important to an understanding of the nature and behavior of the cements. Also, in the field of cement technology, a number of special usages of terms and particular ways of representing chemical compounds have grown up. Some of these terms and some of the concepts pertinent to the chemistry of cement are discussed in this and the following articles.

Lime. Chemically, lime is calcium oxide, CaO. The term is often loosely used, however, to refer to the presence of other compounds in which calcium is a key element (e.g. a "high-lime" cement is one in which the content of the characteristic calcium silicates is relatively high).

As a commercial material, "lime" may be furnished in the form of quicklime, CaO, or hydrated lime, $Ca(OH)_2$. Slaked lime, as used in construction, is generally lime hydrated with an excess of water and is in the form of a wet paste or putty.

Materials in which calcium is the key element are called "calcareous" materials. The commonly occurring calcareous mineral is calcite, whose chemical formula is $CaCO_3$ (or $CaO \cdot CO_2$). Quicklime is usually made by heating limestone, or some form of calcium carbonate other than calcite, to a temperature sufficiently high (generally about 1800°F or 1000°C) to drive off the carbon dioxide:

$$CaCO_3 + \text{heat} \rightarrow CaO + CO_2\uparrow$$

Quicklime reacts rapidly with water, accompanied by considerable evolution of heat:

$$CaO + H_2O \rightarrow Ca(OH)_2 + \text{heat}\uparrow$$

When lime is used as the cementing agent in mortar for masonry or plaster, it is used in the hydrated form. Lime mortar ultimately hardens by reaction with carbon dioxide from the atmosphere, reverting to calcium carbonate:

$$Ca(OH)_2 + CO_2 \rightarrow CaCO_3 + H_2O\uparrow$$

This process is very slow, however, and straight lime mortars are now seldom used for construction.

In portland cement, the lime is largely combined in the form of calcium silicates and calcium aluminates. Some small amounts of uncombined lime (CaO) and the hydrate $Ca(OH)_2$ are generally present in finished cement, for one reason or another.

Silica. Chemically, silica is silicon dioxide, SiO_2. Silica occurs in nature in crystalline form as quartz and in a variety of amorphous (and hydrated) forms, such as opal. While quartz is a highly stable form of silica, some other forms under appropriate conditions can react fairly readily, although slowly at normal atmospheric temperatures, with lime and alkali oxides.

A large number of the rocky or earthy materials on the earth's surface are, or contain, "silicates." The basic structural arrangement of atoms in the crystal lattice of the silicates is a tetrahedral grouping of four oxygen atoms around a silicon atom joined to form chains, sheets, or other crystalline structures which characterize a great variety of siliceous minerals. The more important synthetic minerals in portland cement are silicates, as are the corresponding hydrated reaction products in the hydraulic cements.

Alumina. Chemically, alumina is Al_2O_3 (sometimes called aluminum sesquioxide[1]). It can occur naturally as the mineral corundum, but this is not important in cement technology. The association of aluminum with oxygen (along with other elements) in characteristic crystal lattice arrangements, gives rise to the "aluminates."

Important natural aluminum-bearing minerals are the clay minerals; substances containing clay or claylike materials are also called "argillaceous" materials. In the manufacture of cement, the argillaceous materials also furnish the silica. For example, the chemical composition of one of the simpler clay minerals, kaolinite, is $H_4Al_2Si_2O_9$ ($Al_2O_3 \cdot 2SiO_2 \cdot 2H_2O$). Portland cement and some other hydraulic cements characteristically contain certain calcium aluminates.

Iron. Depending on its valence or oxidation number, iron may combine with oxygen to form "ferrous" oxide, FeO, or "ferric" oxide, Fe_2O_3. The most important ores of iron are hematite, Fe_2O_3, and its hydrate limonite, $Fe_2O_3 \cdot H_2O$.

Because most of the natural raw materials from which portland cement is made contain some iron, characteristic iron compounds occur in most portland cements. There are numerous complex minerals containing ferric iron and oxygen, often referred to for some purposes as "ferrites" (not to be confused with the α form of pure iron.

Gypsum. Chemically, gypsum is a hydrated calcium sulfate, $CaSO_4 \cdot 2H_2O$, and is a fairly common natural mineral. When completely dehydrated at a temperature above 400°F, it changes crystalline form and becomes a compound known as anhydrite, which also occurs as a natural mineral. If finely ground gypsum is heated at a somewhat lower temperature (between 250 and 350°F), only a part of the "water of crystallization" is driven off, and the product known as plaster of paris is produced:

$$2(CaSO_4 \cdot 2H_2O) + \text{heat} \rightarrow (CaSO_4)_2 \cdot H_2O + 3H_2O\uparrow$$

Both anhydrite and plaster of paris compose, or form the basis for, a variety of relatively quick-setting "gypsum plasters"; with the addition of water to such plasters the compound gypsum again forms as a hardened mass of interlocking crystals.

In addition to serving as the source for a type of cement, gypsum plays an important role in the manufacture of portland cement. During the grinding process, an appropriate small quantity of gypsum or anhydrite is added to control the time of set of the cement.

In the chemical analysis of cement, the presence of the gypsum is indicated by SO_3, called "sulfuric anhydride."

[1] When the combining ratio of oxygen to another element is 1½ it is called a sesquioxide. Another important sesquioxide in cement is that of iron, Fe_2O_3. Collectively they are referred to by the abbreviation R_2O_3. (R is an abbreviation for radical.)

Magnesia. Chemically, magnesia is magnesium oxide, MgO. As a mineral, it is called "periclase." The magnesium-bearing compounds found in hydraulic cements derive largely from the magnesium carbonates in the raw materials. While the mineral magnesite, $MgCO_3$, occurs in nature, the mineral dolomite. $CaMg(CO_3)_2$ or $CaCO_3 \cdot MgCO_3$, is more common. Most natural limestones contain some magnesite or dolomite, and many so-called deposits of dolomite would more properly be described as dolomitic limestones.

Magnesian or dolomitic limes, composed of CaO + MgO in varying proportions, are made commercially for use in mortars for masonry work. They tend to give a smoother-working mortar than high-calcium lime. It is necessary, however, that high-magnesian limes be completely hydrated before use. In the manufacture of portland cement the presence of magnesium-bearing compounds is undesirable, and the materials are selected or blended to keep the magnesia content of the finished cement below a specified amount (see Arts. 2.12 and 2.17).

Alkalies. The alkali oxides, or the "alkalies," Na_2O and K_2O, are often present in portland cement in very small quantity coming from sodium or potassium-containing compounds occurring as impurities in the natural raw materials. Knowledge is limited concerning all the detailed reactions of the alkali compounds in the cement-making and cement-hydration process. With some types of aggregates a reaction can occur between the alkalies and certain compounds in the aggregate producing a disruptive expansion after the concrete has hardened.

2.6 Chemical formulas and processes. In cement chemistry it is customary to report the results of a chemical analysis in terms of the oxides of the elements present (see Art. 2.8 for an example). Most of the compounds present in cement, however, do not occur as oxides, nor do the oxides, as such, form the unit cell of the crystalline structure of the major cement compounds. Nevertheless, for convenience or through custom, the chemical formulas for many of the compounds are expressed in terms of combinations of oxides. Based on the custom of expressing the chemical formulas in terms of oxides, there also developed the practice of representing the oxides by abbreviations. Abbreviations for the common oxides are given in Table 2.2. Although most of these abbreviations are the same as some of the standard symbols for chemical elements, the context in which they are used will usually avoid confusion.

Abbreviations for characteristic compounds which can be found in portland cement are shown in Table 2.3. In commercial cements, these compounds do not necessarily occur in pure form, as will be discussed subsequently.

A major reaction during the process of hardening of hydraulic cements is the chemical combination with water—a process termed "hydration." With some substances the hydration process involves a complete change of form and internal structure and often a substantial change in volume; with other substances the forces which bond the water (or its ions) to the structure are weak and the process is easily reversible; with still others, the number of molecules of water taken up in the hydration process may be variable. In some substances the hydrate may be the more stable of possible forms of the substances. Some simple illustrations of the hydration process were mentioned in the discussion of lime and gypsum (Art. 2.5).

Table 2.2 Names and abbreviations of common oxides

Oxide	Common name	Abbreviation
CaO	Lime	C
SiO_2	Silica	S
Al_2O_3	Alumina	A
Fe_2O_3	Iron	F
H_2O	Water	H
SO_3	Sulfuric anhydride	$\bar{S}$
MgO	Magnesia	M
Na_2O	Soda	N
K_2O	Potassa	K

The chemical formula for a hydrated compound may be expressed in alternate ways, for example, hydrated lime could be represented by the formula $CaO \cdot H_2O$ just as well as by $Ca(OH)_2$ or by the abbreviated form, CH. In the hydration of the more complex compounds in portland cement,

Table 2.3 Names and abbreviations of major compounds

Name	Chemical composition	Abbreviation
Tricalcium silicate	$3CaO \cdot SiO_2$	C_3S
Dicalcium silicate	$2CaO \cdot SiO_2$	C_2S
Tricalcium aluminate	$3CaO \cdot Al_2O_3$	C_3A
Tetracalcium alluminoferrite*	$4CaO \cdot Al_2O_3 \cdot Fe_2O_3$	C_4AF

* The iron compounds may occur in a solid-solution system. C_4AF is only one of several possible compounds (see Art. 2.7).

some constituents may be thrown out; for example, in the hydration of tricalcium silicate, lime hydrate is released:

$$2C_3S + 6H \rightarrow C_3S_2H_3 + 3CH$$

2.7 Manufacture of portland cement. The raw materials used to produce a portland cement must provide, in suitable form and proportions, compounds containing lime, silica, and alumina. Natural calcareous deposits, such as limestone and shell beds, are common sources of lime. Natural argillaceous deposits, such as clay, shale, and slate deposits, supply both silica and alumina. Natural deposits of argillaceous limestone or of marl (a calcareous clay) can supply all three basic ingredients although not necessarily in the right proportions for a desired portland cement composition.[1]

In some situations, the principal raw materials may have a deficiency or an excess of one or other of the essential ingredients; in such event, supplemental materials of suitable composition may be used to adjust the raw mix to the desired proportions of the ingredients.

Most natural deposits also contain compounds other than lime, silica, and alumina. For example, nearly all clays contain some iron as mentioned in Art. 2.5. The presence of some iron is useful in the manufacturing process because it combines with some of the lime and/or alumina to form separate lime-alumina-iron compounds, thus reducing the amount of C_3A that would otherwise be formed. Along with the alumina, it also serves as a flux (i.e., promotes fusion) and thus aids in the formation of the calcium silicates at a lower temperature than would otherwise be necessary. If the iron content of the raw materials is very low, iron may be added to the raw mix to control the proportion of the C_3A.

The grayish color of cement is caused by iron. White portland cement is produced in some mills by using specially selected raw materials containing so little iron that the finished cement will have less than ½ percent of iron.

Other compounds which often occur in natural deposits used as raw material for cement making include magnesia, alkalies, phosphates, etc. Some of these have deleterious effects if present in appreciable quantity in the finished cement. Additions may have to be made or some special preliminary processing may have to be undertaken to maintain the presence of such material below a harmful level.

In the cement plant the raw materials are first subjected, if need be, to preliminary treatment (such as crushing, screening, etc.) and are stock-

[1] In a few localities in the United States and elsewhere, there occur natural deposits of argillaceous limestone, called "cement rock," containing lime, silica, and alumina in such proportions that a hydraulic cement can be produced simply by heating the raw material to a sufficiently high temperature and grinding it. While this natural cement resembles portland cement, it is no longer produced in large quantity because of its variable composition.

piled. Based on chemical analyses, the proportions of the several raw materials which would result in a finished cement of desired composition are calculated. Various rules and indexes have been developed to control the chemical proportions of the ingredients. The "raw mix" is prepared by grinding and blending the chosen parent materials, sometimes in a dry form and sometimes in the form of a slurry. The wet process (using slurry) is favored by many modern cement plants. A finely ground, intimate mixture of the raw materials is important to attain as complete a reaction as possible during the subsequent burning.

The prepared mixture is fed into the upper end of a rotary kiln, a long cylindrical furnace capable of rotating slowly on its axis and slightly inclined to the horizontal. Present-day commercial cement kilns vary from about 300 to 700 ft in length and about 12 to 25 ft in diameter. As the kiln rotates, the materials pass slowly from the upper end to the lower end at a rate controlled by the slope and speed of rotation of the kiln. This is a continuous process, as contrasted with the batch process used in the vertical kilns in the early history of cement making. The kiln may be fired by gas, oil, or pulverized coal as fuel, which is sprayed into the kiln at its lower end. As the material passes through the kiln, its temperature is raised to a level at which the essential reactions can take place.

The process of heating the ingredients to produce the desired product is known as "burning" the cement. The term "calcining" is also used; this usage derives from the early manufacture of limes and meant heating limestone to a temperature sufficient to drive off the CO_2 and thus produce lime. Heating a mixture of materials to the point where a more or less coherent mass is attained without melting the entire mass is called "sintering." This state is also referred to as "incipient fusion." The temperature (range) at which this state of incipient or partial fusion takes place is called the "clinkering" temperature because the output of the kiln after this burning process occurs is in the form of rough-textured lumps or pellets, ranging in size from perhaps 1/16 to 1 or 2 in., called "clinker."

By control of the rate of travel through the kiln, the material is held within a critical temperature range for a sufficient period of time so that a recombination of the lime-silica-alumina mixture takes place to form the characteristic minerals of portland cement. Depending on the raw materials to be processed, the maximum temperatures which are developed near the lower end of a kiln are generally in the range 2500 to 2900°F (about 1400 to 1600°C).

Various reactions take place in succession with increasing temperature as the materials move from the upper end through the kiln [203, p. 117] At 212°F (100°C) free water is evaporated. A considerable amount of loosely bound water is lost from the clay in the range 300 to 650°F (about 150 to 350°C), and the more firmly bound water begins to be driven from the clay beginning at about 930°F (500°C). The $MgCO_3$ begins to decompose and lose its CO_2 beginning at about 1110°F (600°C). The $CaCO_3$ begins to decompose at about 1650°F (900°C), at which temperature some reaction also begins to take place between lime and clay. Some liquid begins to form at about 2280°F (1250°C) and the major compound formations begin to take place above about 2330°F (1280°C). It has been estimated [201, p. 52] that some 20 to 30 percent of the mix is liquefied in the hotter part of the kiln, and that the major reactions which produce the characteristic cement compounds take place largely in the liquid state although solid state reactions also occur. Temperatures that are too high may induce the formation of undesired compounds and also may result in severe manufacturing difficulties. A balance has to be struck between the rate of reaction and fuel costs.

After burning, the clinker is cooled and stored. Sometimes the cooling of the hot clinker is accelerated by spraying it with water.

The clinker is ground with a small amount of calcium sulfate mineral, usually gypsum, to the desired degree of fineness. The principal function of the gypsum is to control the time of set of the cement when it is mixed with water on the job. Sometimes small amounts of other materials are added in the grinding process, either to facilitate grinding or to impart special properties to the finished cement, such as air entrainment.

A schematic diagram or flow sheet, showing the several steps in the wet and in the dry processes, is shown in Fig. 2.1. For descriptions of

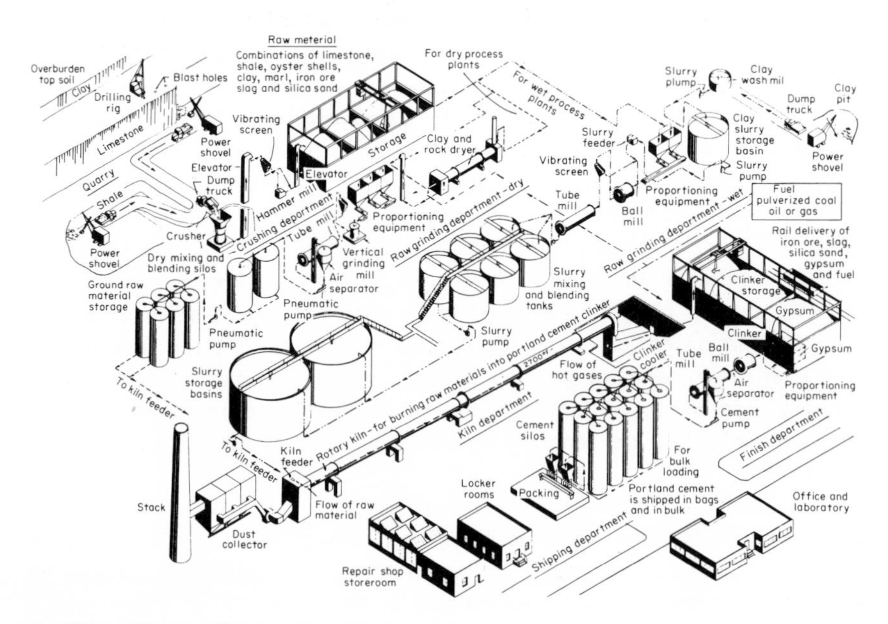

Fig. 2.1. The manufacture of portland cement. Isometric flow diagram. (*Portland Cement Association.*)

the details of cement manufacture, see Refs. 201, 203, 206. For some historical aspects of cement production, see Refs. 210 to 216.

2.8 Chemical analysis of portland cement. The results of a routine chemical analysis of portland cement are reported in terms of the oxides of the constituent elements [C114]. Table 2.4 indicates the general proportions of the major constituents in the clinker, expressed as oxides, for a range of ordinary commercially made portland cements. It may be noted that these four major constituents account for over 90 percent of the total. Also, it may be noted that the oxide composition varies among the cements represented over a fairly narrow range. However, a relatively small change in composition as indicated by the oxide analysis may result in an appreciable change in the proportions of the actual compounds that make up a cement.

A typical oxide analysis of a type I portland cement is shown in Table 2.5, which includes constituents other than the major ones given in Table 2.4.

The role of the major constituents will be discussed in greater detail in subsequent articles. At this point however, it is pertinent to note the significance of certain other items reported in the usual chemical analysis.

Table 2.4 Major constituents of normal portland cement clinker, in terms of the oxides

Major constituents in terms of the oxides	Percent, by wt	
	Representative average	Approximate range
Lime, CaO	63	60–65
Silica, SiO_2	21	20–24
Alumina, Al_2O_3	6	4–8
Iron, Fe_2O_3	3	2–5

Table 2.5 Typical chemical analysis of a type I portland cement

Principal oxides	Percent, by wt	Other determinations	Percent, by wt
SiO_2	20.67	MgO	2.58
Al_2O_3	5.96	Na_2O	0.12
Fe_2O_3	2.35	K_2O	0.94
CaO	63.62	Loss on ignition	1.37
SO_3	2.13	Insoluble residue	0.26
		Free CaO	1.43

The sulfuric anhydride or sulfur trioxide, SO_3, which appears in the results of a chemical analysis of a finished cement, comes largely from the gypsum, which is added in an amount normally ranging from 2 to 6 percent during the grinding of the clinker. The amount of $CaSO_4$ in the cement can be approximated by multiplying the amount of SO_3 by 1.7, which is the molar ratio of $CaSO_4$ to SO_3. Some small amount of sulfur is often present as sulfides in the clinker, deriving from impurities in the raw materials or from the fuel in the burning process. A.S.T.M. cement Specifications [C150] place limits of between 2.3 and 4.0 percent on the SO_3, depending on the alumina content of the cement and the type of cement (see Art. 2.17).

The percentage of magnesia, MgO, usually ranges from 1 to 4 percent, deriving from magnesium compounds in the raw materials. Most specifications place a limit of 5 percent on MgO in order to control potential detrimental expansion due to hydration of this compound in hardened concrete. However, not all the MgO reported in the oxide analysis is necessarily in the oxide form in the cement; some magnesia may be in solid solution in other compounds.

The "loss on ignition" (loss in weight of a sample after heating to a full red heat, 1832°F or 1000°C) is determined to give an indication of possible prehydration or carbonation due to improper or prolonged storage of the clinker or finished cement. Some fraction of the ignition loss derives from loss of the chemically bound water in the gypsum. Normally, in a satisfactory cement the ignition losses do not exceed about 2 percent although a maximum somewhat greater than this is permitted under existing specifications.

The "insoluble residue" is that fraction of the cement sample which is insoluble in hydrochloric acid and derives principally from the silica which has not reacted to form the cement compounds which are soluble in this acid. The amount of insoluble residue thus can serve to indicate the completeness of the reactions in the kiln.

The amounts of the alkalies, Na_2O and K_2O, are now often determined in order to give warning of possible difficulties in the use of the cement. The total alkalies, generally expressed in terms of Na_2O, may range in amount from 0.4 to 1.3 percent in ordinary cements. These alkalies can react with certain opaline silicates or other materials found in some aggregates. If the alkalies are present above some small but critical amount, about 0.6 percent, disintegration of concretes containing such aggregates can occur due to disruptive expansion of the reaction products after the concrete has hardened. A limiting value of the combined alkalies is often specified in cements which are to be used with reactive aggregates. It has been found, also, that the potentially destructive effect of alkali-aggregate reaction may be offset by incorporating in the concrete mix a finely divided siliceous material which will react with the alkalies before the concrete hardens. Too large an alkali content can also affect the setting time of cement.

The amount of CaO given under the listing of the principal oxides accounts for all calcium in the sample of cement analyzed. It includes not only the calcium in combination in the cement compounds, but also the calcium in the gypsum and any "free" lime, e.g., uncombined lime or lime hydrate. It is now not uncommon to include a determination of free lime in the chemical analyses of cement although a special analysis is required to determine it. In computing the potential compound composition, it is sometimes desired to correct the amount of total CaO by subtracting the amount of free lime present. Further, it may be desirable to estimate the amount of lime that is present as the "hard-burned' form of CaO, although the control of unsoundness due to the presence of uncombined lime in cement is accomplished by an expansion test rather than chemical analysis (see Arts. 2.12 and 2.17). The total free lime (including hydrated lime) in freshly ground cement may be as much as 3 percent; the amount of uncombined CaO is probably usually less than 1 percent in most modern commercial cements.

There may be varying and very small amounts of compounds of such other elements as phosphorus, manganese, and titanium in a cement. They are generally present in amounts too small to warrant inclusion in ordinary chemical analyses; they normally will total less than 1 percent.

2.9 Major compounds in portland cement. The story of the identification and description of the compounds which make up portland cement is a long one, filled at some points with controversy; it is a story for which the final chapter has not yet been written [201, 203]. The kind and number of compounds that may develop in the clinker depend on the complex variety of reactions that can occur among the several constituents. Complicating the story is the fact that the nature of the original materials utilized to furnish the lime, silica, and alumina, as well as the methods of preparation and burning of the raw mix, have a significant bearing on the composition of the clinker produced.

Studies of the constitution of portland cement by a variety of means, both theoretical and experimental, have shown, however, that four major characteristic artificial "minerals" or "phases" can result from the thermal processing of appropriate combinations of the raw constituents. The term "phase" is commonly used by cement chemists in the sense it is used in physical chemistry: a homogeneous, physically distinct portion of matter in a heterogeneous system, separated from other phases by a distinct physical boundary. There are a number of difficulties in being able to give a complete and unequivocal description of the actual compound composition of the cement; thus the term phase is probably preferable to compound.

A small change in the proportions of some of the constituents can result in a substantial change in the amount of some of the phases produced; the presence and amount of minor constituents can also affect some of the phases. The degree to which the potential reactions can proceed to completion or "equilibrium" depends (1) in part on the intimacy with which the basic constituents can be brought together (a function of fineness and intermixing of the raw materials) and (2) in part on the temperature and period of time that the mix is held in the critical temperature range. Further, the internal structure of some of the phases can be affected by the rate at which the clinker is cooled.

Nevertheless the major phases are usually designated in terms of the chemical compounds which they approximate: a tricalcium silicate (C_3S) phase; a dicalcium silicate (C_2S) phase; a tricalcium aluminate (C_3A) phase; and a solid solution sometimes called the "ferrite phase." The abbreviation F_{ss} is sometimes used for the ferrite solid-solution phase.

In the early days of cement research, investigators were able to distinguish several separate phases or "minerals" by use of the petrographic microscope. For want of a better designation and since their composition was then unknown, four of these minerals were called "alite," "belite," "celite," and "felite" [203]. Subsequently, from phase-equilibrium studies, x-ray diffraction analyses, and other studies, alite was found to be essentially C_3S; belite and felite were different forms of C_2S; and celite was the ferrite phase.

In commercial portland cement the compounds probably only rarely occur in pure form. Also, localized inclusions and solid solutions may occur. For example in the C_3S phase, ions of Mg and Al may substitute for Ca at random points. Thus there is considerable merit in calling the phase that is mainly C_3S by its mineral name of alite; this usage is becoming more common in the current literature.

Dicalcium silicate, C_2S, can occur in several polymorphic[1] forms; the form mainly produced in modern portland cements—and the only form having cementing value—is the β-dicalcium silicate, β-C_2S. This β-phase in commercial cement is the mineral that is termed belite.

The ferrite phase is a solid solution of variable composition believed to be in the range C_2F and C_6A_2F (i.e., the ratio of A and F ranges from 0 to 2).[2] In earlier literature on portland cement chemistry this phase was designated as tetracalcium aluminoferrite (C_4AF), which composition the ferrite phase is believed by many investigators to approximate [203, 204].

[1] Polymorphs are substances having the same chemical composition but different arrangements of atoms or molecules in the crystal lattice.

[2] Compounds which are considered to be potential components of the F_{ss} solid-solution system are: C_2F, C_6AF_2, C_4AF, C_6A_2F.

The principal phases, at least C_3S and β-C_2S, as confirmed by x-ray diffraction studies tend to be microcrystalline; i.e., the atoms have a definite ordered arrangement or lattice structure even though some of the discrete units of the crystalline material may be of submicroscopic dimensions. In commercial clinker, however, there may also be some amorphous material ("glass"), formed as the result of rapid cooling of material that had become liquefied during the burning process. Some investigators now believe that the amount of glass is probably very small and that some material that appears to be glass actually has a crystalline structure on a submicroscopic scale. It is considered that the glassy phase in portland cement clinker consists for the most part of calcium aluminoferrites and calcium aluminates. A reproduction of a photomicrograph indicating the appearance of the several phases is shown in Fig. 2.2.

Fig. 2.2. Photomicrograph of a polished and water-etched section of portland cement clinker magnified 750 times. The large grey angular crystals that often contain small spherical inclusions are C_3S; the ragged grey patchy crystals are beta C_3S; the darkest crystals are C_3A; immediately to the left of center are two small periclase (M_gO) crystals; and the bright interstitial material is the ferrite phase. *(Portland Cement Association.)*

In spite of the various uncertainties concerning the actual phase composition of a given cement, a calculation of the "potential" compound composition, for assumed idealized conditions, provides a useful means of distinguishing among several types of portland cement and a general guide for selection of proportions of the raw constituents.

The Bogue method [201] of calculating the potential compound composition from the oxide analysis assumes that the chemical reactions involved in the formation of the major phases have proceeded to equilibrium, that the conditions of cooling have not altered that phase equilibrium, that no glass has been formed, that the ferrite phase may be calculated as C_4AF, and that the presence of minor constituents may be ignored. In making the calculation, the total CaO given by the oxide analysis is often reduced by the amount of uncombined lime which is determined separately[1] and by the amount of CaO in the gypsum ($CaO \cdot SO_3 \cdot 2H_2O$) present as indicated by the SO_3 content. The Fe_2O_3 is assigned to the C_4AF and the remaining Al_2O_3 is assigned to the C_3A. The remaining CaO is apportioned to the C_3S and β-C_2S. This procedure was systematized in equation form as follows:

$$\begin{aligned}
\%C_3S &= (4.071 \times \%CaO) - (7.600 \times \%SiO_2) \\
&\quad - (6.718 \times \%Al_2O_3) - (1.430 \times \%Fe_2O_3) - (2.852 \times \%SO_3) \\
\%C_2S &= (2.876 \times \%SiO_2) - (0.7544 \times \%C_3S) \\
\%C_3A &= (2.650 \times \%Al_2O_3) - (1.692 \times \%Fe_2O_3) \\
\%C_4AF &= (3.043 \times \%Fe_2O_3)
\end{aligned}$$

Later, various investigations proposed modifications to the Bogue calculation, taking into account deviations from complete equilibrium (of reactions) and variations in the aluminate and ferrite phases [203]. The ASTM Specification for portland cement [C150] provides that the above equations shall be used for calculating the potential compound composition only when the ratio of the percentages of alumina to ferric oxide is 0.64 or more. When the alumina–ferric oxide ratio is less than 0.64, the ASTM Specification recognizes that a calcium aluminoferrite solid solution, also abbreviated as $ss(C_4AF + C_2F)$, is formed and provides that the amount of solid solution of ferrites and of C_3S shall be calculated as follows:

$$\begin{aligned}
\%ss(C_4AF + C_2F) &= (2.100 \times \%Al_2O_3) + (1.072 \times \%Fe_2O_3) \\
\%C_3S &= (4.071 \times \%CaO) - (7.600 \times \%SiO_2) - (4.479 \times Al_2O_3) \\
&\quad - (2.859 \times \%Fe_2O_3) - (2.852 \times \%SO_3)
\end{aligned}$$

[1] In recent practice, because modern well-burned clinker usually contains relatively little free lime, and because the calculations give only an indication of the compound composition, it is common in routine calculations to ignore the uncombined lime. Also, the insoluble residue, which was formerly taken into account and subtracted from the silica, is now often ignored.

and notes that no C_3A will be present in cement having such composition. The C_2S is calculated as shown for cements with the higher A/F ratio.

In recent years, methods have been developed for the quantitative determination of the amount and composition of the major phases in samples of clinker by x-ray diffraction [235]. The calculated potential compound composition remains, however, a valuable index for judging the general characteristics of portland cements.

Representative average compound compositions for the five types of cement recognized by the ASTM [C150] are shown in Table 2.6. The percentages of the compounds shown are averages for a number of commercially made cements of each type, in use during the 1940s and 1950s. In 1960, both the ASTM and Federal Specifications removed the maximum limitations that had been placed on C_3S for the type II and V cements prior to that time. This change may result in a trend toward higher C_3S contents (and hence higher early strength and heat of hydration) for those types of cement [106].

It should be noted that the calcium silicate compounds constitute some 70 to 80 percent of the cement; for most cements they constitute close to 75 percent. In general, a small increase in the proportion of lime in the raw materials appreciably increases the percentage of C_3S and decreases the C_2S. Cements with high percentages of C_3S are sometimes referred to as "high-lime" cements. An increase in the iron content of the raw materials decreases the percentage of C_3A by causing formation of more of the ferrite phase. Practical considerations relating to the raw materials available and to the conditions of manufacture and economic considerations impose certain limits on composition, however.

Using the method of calculation described above, the cement whose chemical analysis is shown in Table 2.4 may be found to have the following

Table 2.6 Representative average compound compositions for five types of portland cement*

Type of cement	General description	Potential compound composition, percent			
		C_3S	C_2S	C_3A	C_4AF
I	General-purpose	49	25	12	8
II	Modified general-purpose	46	29	6	12
III	High-early-strength	56	15	12	8
IV	Low-heat	30	46	5	13
V	Sulfate-resistant	43	36	4	12

* From Ref. 106.

potential composition: C_3S, 52.4 percent; C_2S, 19.7 percent; C_3A, 11.8 percent; and C_4AF, 7.2 percent. It would be classed as a type I cement.

2.10 Influence of composition upon characteristics of portland cement. Some of the characteristics of the principal compounds are given in Table 2.7. From a consideration of the properties of the compounds it may be expected that, by altering their proportions, the properties of the cement may be controlled. For example, to secure a cement of low heat generation, such as would be used in massive dam construction, a relatively high C_2S content and low C_3S and C_3A contents are desirable.

In general, the early strength of a portland cement will be higher with higher percentages of C_3S, but if moist curing is continuous, the later-age strengths, after about 6 months, will be greater for the higher percentages of C_2S. Although the total percentage of silicates does not vary much in commercial cements, yet an increase in the summation tends to increase strengths at all ages. C_3A contributes to strength developed during the first day after wetting, because it is the earliest of the four principal compounds to hydrate; however, it is the least desirable compound in cement. A low C_3A cement generates less heat, develops higher ultimate strength, and exhibits greater resistance to destructive elements than a cement containing larger amounts of this compound [206, 263].

From examination of Table 2.6 in relation to Table 2.7, it may be noted that type I and type III cements are similar in their content of the aluminate and ferrite phases but that the type III (high-early-strength) cements are higher in C_3S and lower in C_2S than the type I (general-purpose) cements. In the type II (modified general-purpose) cements and the type V (sulfate-resistant) cements, the C_3A is reduced (by increasing the ferrite phase) in order to promote sulfate resistance. In these cements the C_3S is also reduced in relation to the C_2S, giving cements that can be expected

Table 2.7 Characteristics of major compounds which occur in portland cement*

Property	Relative behavior of each compound			
	C_3S	C_2S	C_3A	C_4AF
Rate of reaction	Medium	Slow	Fast	Slow
Heat liberated, per unit of compound	Medium	Small	Large	Small
Cementing value, per unit of compound:				
Early	Good	Poor	Good	Poor
Ultimate	Good	Good	Poor	Poor

* From Ref. 263.

to harden more slowly than the type I, general-purpose cements. The relatively lower heat generation of the type IV (low-heat) cements is derived from holding both the C_3S and the C_3A to considerably lower limits than is characteristic of the other types; it may be expected that the early rate of strength development of the type IV cement will be considerably less than that of the general-purpose cements.

2.11 Fineness of cement. Regardless of composition, an important factor which influences the rate of reaction of cement with water is the size of the particles. For a given weight of a finely ground cement, the surface area of the particles is greater than for a coarsely ground cement, hence the rate of reaction with water is greater and the hardening process can proceed more rapidly. If, however, a cement is ground too finely, the extremely fine particles may be prehydrated by moisture present as vapor in the grinding mills, or during storage, so that no cementing value can be derived from such particles.

Furthermore, the completeness with which a cement can react with water is influenced by the particle size. The cores of the coarser particles may take years to hydrate under practical conditions, and there is evidence that very coarse particles may never be completely hydrated.

Of interest in connection with fineness of cement is the particle-size *distribution*. For routine testing and for transmitting information on fineness, however, a single factor, if it is a representative index, is a more practical way of indicating particle size. Such factors are the percentage passing or retained on some given size of sieve or set of sieves, and a measure of the surface area of the particles [266, 267].

For many years, specifications for portland cement required that the minimum residue on a No. 200 sieve (200 meshes per inch, clear openings = 0.0029 in., or 74 microns) should not exceed 22 percent. Since, however, more than 90 to 95 percent of most modern cements pass the No. 200 sieve, and since the small percentage retained on this sieve does not always correlate well with the particle-size distribution of fine cements, this requirement has been dropped from a number of specifications for portland cement. In some testing laboratories a method of wet sieving using the No. 325 sieve (openings = 0.0017 in. or 44 microns) has been retained for some special studies to provide additional information, but a number of difficulties are inherent in the use of extremely fine-mesh sieves.

In recent years, with the development of suitable methods which can be reduced to a routine procedure, it has become customary to state the fineness of cement in terms of the *specific surface*, which is the calculated surface area of the particles, in square centimeters, per gram of cement.

Although the percentage of material passing a No. 200 sieve cannot be taken as a criterion of the specific surface of a particular cement, a

Table 2.8 Comparison of various measures of fineness of cement*

Specific surface, sq cm per g		Approximate range in percent passing the No. 200 sieve, by wt
Wagner turbidimeter	**Blaine air permeability**	
1,250	2,500	76–83
1,500	3,000	82–90
1,750	3,300	87–93
2,000	3,600	92–96
2,250	3,800	95–98
2,500	4,000	96–99

* Values shown are for general information only. Range in specific surface corresponding to a given percentage passing the No. 200 sieve is very wide.

rough comparison between sieve fineness and specific surface is given in Table 2.8. The specific surface of commercial standard cements produced today ranges from perhaps 2,800 to 3,600 sq cm per g by the air permeability method. For high-early-strength cements the range is probably about 3,600 to 4,500 sq cm per g.

The finer the cement, the higher the specific surface. While the calculated specific surface is only an approximation of the true surface area, it appears to serve very satisfactorily as a measure of fineness; good correlations are obtained between specific surface and the properties of cement which are influenced by fineness of particles. The most active part of a cement is the material finer than 10 or 15 microns. Since surface area varies as the square of the diameter of a particle, an increase in the percentage of material in this size range is much more effective in increasing specific surface, and hence the activity, than is a corresponding reduction in some of the coarser fractions.

In some methods in current use, in which the principle of sedimentation in a liquid is employed, the specific surface is, in effect, computed from the particle-size distribution. Observations of the quantity of cement remaining in suspension (determined by the concentration, or density) at given elevations in the fluid column after various intervals of time yield data from which the particle-size distribution can be calculated. The calculations are based upon the assumption that the particles are spheres and that Stokes' law, which relates particle size to velocity of settlement of a particle in the fluid, may be applied; assumptions are also made regarding size distribution in the very fine sizes, since a detailed analysis in this range requires extended periods of time.

In one ASTM method for determining specific surface, the Wagner turbidimeter is employed. With this apparatus, a sample of cement is dispersed in kerosene in a tall glass container. A beam of light is passed horizontally through the column of suspended cement at given elevations on a stated time schedule, and the turbidity (a function of the concentration, or density, of the cement still in suspension) is measured by a photoelectric cell. The specific surface is calculated from the data obtained, but the particle-size distribution may also be determined [268, C115].

A hydrometer has also been used by some laboratories to determine the density, or concentration, of a suspension of cement in kerosene or similar liquid [270, 271].

Related to the sedimentation methods is the elutriation process, such as is employed in the Roller air analyzer. In the Roller apparatus, a stream of air is passed upward through a sample of cement at various increasing velocities, so that several fractions varying successively from fine to coarse are lifted and carried out of the sample chamber [272].

The turbidity principle is employed in a different way in the Klein turbidimeter. Here a sample of cement is suspended in castor oil in a shallow dish, and the turbidity is measured photoelectrically by passing a beam of light vertically through the suspension. The viscosity of the liquid is such that the particles do not settle appreciably during the course of the test, and, by calibration, the specific surface is calculated directly [226]. The results obtained with the Klein method are about 20 percent lower than those obtained with the Wagner method because of differences in assumptions as to particle-size distribution in the very fine sizes.

Another principle that is employed for determination of surface area is the air permeability of an aggregation of cement particles. The number and size of pores in a sample of given density is a function of the size of the particles and their size distribution, and determines the rate of air flow through the sample. The Blaine air-permeability apparatus consists of a means of drawing a definite quantity of air through a prepared sample of definite density [273, 274, C204]. Comparative tests on various samples of cement have shown that the specific surface as determined by use of the Blaine apparatus is about double that determined by other methods because of fundamental differences in the theories involved. However, the specific surface obtained by any one of these methods serves to evaluate the fineness of any given sample.

2.12 Hydration reactions in cement paste. When portland cement[1] is mixed with water, a series of chemical reactions begins to take place. The reac-

[1] In this chapter the term "cement" is used specifically to refer to the dry, powdered aggregation of the several characteristic artificial anhydrous phases or minerals together with an amount of gypsum ($CaSO_4 \cdot 2H_2O$) sufficient to induce "normal" setting in a particular cement.

tions of cement with water are generally described as the "hydration" process. The hydration process involves much more, however, than the simple attachment of water molecules (or OH ions) to the original cement compounds.[1] Although some simple hydrates are formed for example, $Ca(OH)_2$, the complex processes of dissolution and precipitation result in a reorganization of the constituents of the original compounds to form new, however hydrated, compounds. Further, the rates at which the reactions proceed differ for the several compounds; even under favorable conditions of temperature and the continued presence of water, in ordinary concrete mixtures, some of the reactions may require a considerable period—sometimes several years—to reach completion. Although it should be recognized that all reactive phases continue to react with water (if it is available) throughout the hydration process, the reactions of the aluminates in particular, and the aluminoferrites to a lesser degree, significantly affect the course of events during the very early stages of the process while the silicates play a dominant role in the late (hardening) stages.

The ultimate strength-giving structure of the hardened cement paste[2] owes its properties primarily to the hydrated calcium silicates. From the fact that about 75 percent of an ordinary portland cement is composed of calcium silicate compounds, it may be expected that these compounds and/or their hydrated derivatives will comprise the major fraction of a cement paste at any stage of hydration. The hydration of the bulk of the calcium silicate compounds takes place, however, after the "setting" period has occurred.

As water is brought into contact with the cement particles, reactions immediately begin at the surfaces of the particles, and ions from all the reactive phases begin to go into solution. Of particular importance is the early and continued presence of lime in solution; in the early stages of the reaction process, an appreciable amount of the lime in solution derives from the gypsum, but some comes from the other compounds. Ions from the alkalies and other minor constituents are also present, but discussion herein will be confined to the reactions involving the major compounds. Although the final setting of a normally behaving cement paste is influenced by the reaction of the C_3S phase [242, 244], it is generally thought that the surfaces of the calcium silicate phases are soon covered with their

[1] For a discussion of hydrolysis, hydration, and other processes involved in the cement-water reaction, see Ref. 201, chap. 22.

[2] The term "cement paste" is used to designate the material which results from the mixture of cement and water. A "fresh" cement paste is one in which the cement-water reaction has not proceeded far enough to cause appreciable stiffening and the mass behaves as a viscous suspension. The term "hardened" cement paste refers to the condition of the paste after setting has occurred and the mass has acquired a degree of rigidity.

own hydrated reaction products and that during most of the setting period the activity of the silicates is relatively small as compared with the activity involving the aluminate phases. The process of setting (see Art. 2.14) is importantly conditioned by the reactions involving calcium sulfate, most often present in the form of gypsum, and the aluminate compounds.

The reaction of C_3A alone with water is immediate; solution and the formation of a crystalline hydrate occur rapidly with liberation of a large amount of heat. In portland cement, unless this violent reaction of C_3A is moderated by some means, "flash set" occurs (see Art. 2.14). With gypsum present as a retarder, the gypsum and C_3A in solution react to form a relatively insoluable sulfoaluminate coating on the C_3A phases, which slows down the reaction.

Either one or both of two calcium aluminosulfate hydrates (often called the "sulfoaluminates") are formed in hydrating portland cements as the result of the interaction of the C_3A, gypsum, and water. One, called the "high-sulfate" form, usually designated by its mineral name of "ettringite," can be represented as

$$3CaO \cdot Al_2O_3 \cdot 3CaSO_4 \cdot 30\text{–}32H_2O \qquad \text{or} \qquad C_6A\bar{S}_3H_{30-32}$$

The "low-sulfate" form can be represented as

$$3CaO \cdot Al_2O_3 \cdot CaSO_4 \cdot 12H_2O \qquad \text{or} \qquad C_4A\bar{S}H_{12}$$

The details of formation of the sulfoaluminates are not clear: in part because of the complexity of the reactions that are taking place during the early hydration period of cements; in part because in cements of different compositions, variations in the relative amounts of the constituents (both initially and with time) induce varying sequences of reactions; and in part because of difficulties of identfying and determining quantitatively the amount of these compounds in the presence of the several other compounds present in the hydrating cement.

It would appear that with cements having low C_3A, the liquid phase of a setting paste can have a relatively low concentration of aluminate ions and relatively high concentration of SO_3 (as well as calcium ions), which condition would favor the formation of ettringite rather than the low-sulfate form of the sulfoaluminate. Under other conditions, the relative concentrations may be such that they favor early formation of the low-sulfate form. Some investigators have observed that if the SO_3 is depleted early in the course of the reaction, ettringite may be converted to the low sulfate form by further reaction with C_3A [224, 233, 234, 237, 241, 242, 244]. It appears that an excess of calcium ion must be available for at least some of the reactions; it has also been observed that calcium hydroxide may play some independent role in the retardation of the setting reactions.

A clarification of this aspect of the chemistry of cement hydration is much to be desired. Although calcium sulfate is added to cement to serve as a retarder, what has been termed the "optimum" gypsum content (see Art. 2.14) is sometimes selected on the basis of inducing favorable performance of the hardened cement as regards strength and minimal volume change [246, 247, 249]; a calcium sulfate content based on this criterion is not necessarily that which may be the most desirable as regards the products of the reaction between C_3A and $CaSO_4$.

Subsequently, during the hardening stage, it is believed that a solid solution of calcium aluminate hydrates and low-sulfate sulfo-aluminate may be produced together with calcium aluminate hydrates in crystalline form [204; 237; 223, p. 205].

The ferrite phase reacts much more slowly than the aluminate phase. Although even less well understood than those of the aluminate phase, the reactions appear to be analogous to them [232]. In addition to possible sulfate compounds (sulfoferrites), other important reaction products appear to include compounds or solid solutions of compounds within a series that ranges from hydrated calcium ferrites and calcium aluminates to calcium aluminosilicates and calcium ferrosilicates; because a group of natural minerals called "garnets" are silicates of this latter type, hydrated reaction products of the ferrite phase having compositions similar thereto have been referred to as a hydrogarnet phase [204, p. 344; 223, pp. 205, 359; 224; 237].

The stoichiometric relations for the reactions of the calcium silicates are usually given for the C_3S as

$$2Ca_3SiO_5 + 6H_2O \rightarrow Ca_3Si_2O_7 \cdot 3H_2O + 3Ca(OH)_2$$

or $$2(3CaO \cdot SiO_2) + 6H_2O \rightarrow 3CaO \cdot 2SiO_2 \cdot 3H_2O + 3(CaO \cdot H_2O)$$

or $$2C_3S + 6H \rightarrow C_3S_2H_3 + 3CH$$

and for the C_2S as

$$2Ca_2SiO_4 + 4H_2O \rightarrow Ca_3Si_2O_7 \cdot 3H_2O + Ca(OH)_2$$

or $$2(2CaO \cdot SiO_2) + 4H_2O \rightarrow 3CaO \cdot 2SiO_2 \cdot 3H_2O + CaO \cdot H_2O$$

or $$2C_2S + 4H \rightarrow C_3S_2H_3 + CH$$

Several points should be noted about the reactions of the silicate phases. First, the reaction of the C_2S proceeds at a much slower rate than that of the C_3S. Second, the reaction of either C_3S or C_2S with water results in the same (actually closely similar) hydrated calcium silicate compound. It is this hydrate that forms the basic gel structure that characterizes hardened portland cement paste.

Third, the reactions of both silicates produce appreciable lime hydrate (the C_3S reaction produces about 39 percent lime hydrate in the reaction products while the C_2S reaction produces about 18 percent lime hydrate in the reaction products). This lime becomes an integral part of the microstructure of hardened cement paste.

Fourth, studies by a variety of methods [237, 238] have shown that the ratio of the CaO to SiO_2 in the hydrated calcium silicates may actually vary somewhat, and also the water of hydration may vary. In other words, the hydration of the calcium silicates produces a series or family of hydrated silicates in which the water of hydration may vary (within limits).

Finally, in commercially produced portland cement, the actual calcium-silicate phases, alite and belite, usually contain some Al and/or Mg atoms within their structures, which can be carried over into the reaction products.

The hydrated calcium silicate that is formed has a molecular arrangement that resembles a natural, although rare, mineral called tobermorite.[1] Because of this and because of the size and other properties of the particles, the hydrated material is often called tobermorite gel although tobermorite-*like* gel might be a more pertinent description [204, 225, 226, 243].

The delayed hydration of uncombined lime (CaO) and periclase (MgO) which may be present in the original cement can be a source of difficulty. The hydrated reaction product resulting from the combination of CaO and H_2O occupies considerably greater volume than the original uncombined CaO called quicklime. If uncombined CaO is confined within the structure of a hardened mass of cement paste and hydration begins to take place, stresses can be induced in the structure of the paste due to the tendency of the hydrating lime to expand. If a sufficient amount of uncombined lime is contained in the hardened mass, the stresses induced as a result of the delayed hydration may be sufficiently large to cause disruption of the mass. The term "unsoundness" applied to cements refers to this kind of action. Cement specifications call for a soundness test, which essentially determines whether a sample of paste will undergo undue expansion under accelerated curing conditions (see Art. 2.17).

Periclase (MgO) hydrates more slowly than lime, but when it hydrates it undergoes a volumetric expansion even greater than that which occurs when lime combines with water. If a sufficient amount of this MgO hydrates within a hardened mass of paste, mortar, or concrete, it can cause severe cracking and disintegration. Cement specifications usually place a maximum limit on the MgO that may be present in a cement (see Art. 2.17).

2.13 Heat of hydration. It has been pointed out that the setting and hardening process is a chemical reaction. In common with many chemical reactions, the hydration of cement is accompanied by liberation of heat. Concrete is in itself a fair insulator, and in a large mass such as a dam,

[1] After the town of Tobermorey, Scotland. Actually the term characterizes a series or family of minerals.

the heat generated in setting and hardening is not readily dissipated. This results in a rise in temperature and expansion of the mass, so that later cooling and contraction of the hardened concrete may result in the development of serious cracks (see Arts. 13.14 to 13.16 and 14.16 to 14.18).

While in massive concrete structures the actual temperature rise is controlled, in part, by rate of placement of the concrete, and in some large structures such as Hoover Dam by a system of embedded cooling pipes, the amount of heat which is generated can be materially reduced by modifying the cement composition. This is accomplished by limiting the amounts of tricalcium silicate and tricalcium aluminate. Specification requirements for a low-heat cement are given in Tables 2.9 and 2.10. For a summary of the early developments of special cements for mass concrete, see Ref. 218.

The heat of hydration of normal portland cement is of the order of 85 to 100 cal per g; that for a low-heat cement such as was used in the construction of Hoover Dam is about 60 to 70 cal per g. A number of procedures have been used by different investigators for determining the heat of hydration. The ASTM method is based on the measurements of the heat of solution in hydrofluoric acid of hydrated and unhydrated samples of the cement [C186].

2.14 Physical aspects of the setting and hardening process. For concrete to be molded or cast in the many forms that have made it such a versatile and useful product, it is necessary that the freshly mixed mass remain in a plastic condition for a sufficient period to permit transporting, placing and compacting under a variety of practical, and sometimes adverse, conditions. On the other hand, for a number of reasons including that of economy, it is desirable that the mass should harden and develop strength within a reasonable time after it has been finally cast in place. An understanding of the nature of the setting and hardening process of the cement paste is important to the control of the quality of the concrete product under the conditions that are encountered in practice.

Although a paste (or the concrete which contains it) can be readily manipulated for periods of perhaps 1 to 2 or more hours after mixing, it gradually stiffens. As the hydration reactions proceed, not only do the reaction products take up what was originally "free" water, but the gel and other reaction products begin to occupy more space, and the mobility of the paste is decreased. Finally, increasing numbers of particles of gel and other products make sufficiently close contact and develop bonds of increasing strength, and if the mass is left undisturbed, it begins to develop rigidity. At some point, the mass can sustain more or less arbitrary load without flowing, and the paste is said to have set.

In the early days of cement making, in order to determine whether or not a cement would probably behave satisfactorily on the job, tests were developed to determine its setting characteristics. However, although based on keen observation of the changes in condition of the paste with time, the measures of the state of the stiffening process may be regarded as more or less arbitrary. In order to avoid the complications of having a mineral aggregate present, tests for time of set are performed on a "neat" cement paste, that is, a mixture of only cement and water.

The water content of a paste has a marked effect upon the time of set. In acceptance tests of cement, the water content is regulated by bringing the paste to a standard condition of wetness, called "normal consistency." Normal consistency is that condition for which the penetration of a standard weighted needle into a paste is 10 mm in 30 sec [C187]. An example of the variation in consistency of a neat cement paste with changes in water content (as indicated by needle penetration) is shown in Fig. 2.3; in this example, normal consistency is obtained with 25.7 percent water. The water content at normal consistency for neat cement pastes made from available commercial portland cements ranges from about 20 to 26 percent. The paste at normal consistency is fairly stiff and is used only for making specimens for determinations of time of set and soundness. In contrast, the water contents of pastes in ordinary concretes range from about 40 percent upward. Since the manipulation of the paste, the temperature, and other conditions have important effects on normal consistency and time of set, detailed standard procedures are prescribed [Part II, Test 1; C191, C266].

The time of set of the neat paste at normal consistency is determined by the ability of a specimen of the paste to sustain the weight of specified small rods or needles. There is determined what is called "initial" set and "final" set. Initial set indicates the beginning of noticeable stiffening. Final set is taken to indicate the beginning of what may be regarded as the hardening period in pastes of this consistency. The standard specifications for cements of types I through V require that initial set shall not

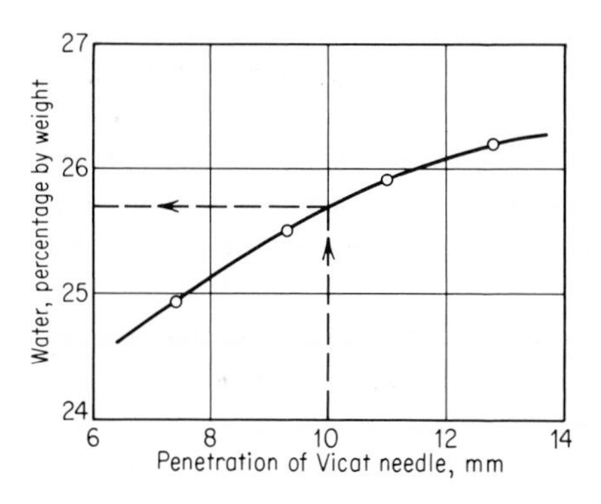

Fig. 2.3. Typical relationship between penetration of neat cement paste and water content.

occur in less than one hour as determined by the Gillmore needle (or 45 min by the Vicat needle), and that final set shall occur within 10 hr. Most portland cements meet these requirements by a wide margin. In general they exhibit initial set in 2 to 4 hr and final set in 5 to 8 hr. In normal concretes, with considerably higher water contents, set occurs later than in neat paste of standard consistency.

The setting times of pastes made with properly retarded cements are affected somewhat by cement composition, but fineness of cement and temperature, in addition to water content, are important factors. Within the common range of fineness, the finer cements tend to set more rapidly because of greater chemical activity of the finer particles. Low temperatures retard and high temperatures accelerate set.

Based on studies of reaction rates, the main events that take place during the setting process of a normal cement paste are visualized as proceeding through four stages [247, 242]. In the first stage, which lasts only for several minutes after contact of cement with water, there is a relatively high rate of heat generation as the cement grains are wetted and initial dissolution of the compounds and reaction of the constituents begin to take place, after which the rate of heat generation drops to a relatively low value. In the second stage, sometimes called the "dormant" period, lasting from perhaps 1 to 3 or 4 hr, activity proceeds at a low rate; it is during this period that the cement grains slowly build up the initial coating of reaction products, and sedimentation or bleeding, if any, takes place. The third stage begins as heat evolution begins to rise again; it has been suggested that in a properly retarded cement, this renewed activity is caused by rupture of the initial (and still weak) gel coating on the crystals of the C_3S phase, so water can again have access to fresh reactive surfaces of the C_3S; a progressive process of breakup and resealing is postulated, resulting finally in the production of calcium silicate gel of sufficient amount and strength that the mass begins to exhibit increasing rigidity; the activity reaches a peak at about 6 hr in pastes of normal consistency and later in pastes of higher water-cement ratios. It is in this third stage that the degrees of rigidity described as initial and final set are reached. As the activity of the third stage subsides, the fourth stage or period of hardening is begun [247, 242].

With a cement having more than some small amount of C_3A, if there is insufficient gypsum present, an extremely rapid reaction (within a few minutes) of water with the C_3A results in the formation of a sufficient mass of a crystalline C_3A hydrate, which causes what is called "flash set." Flash or quick set is defined as the rapid development of rigidity in a freshly mixed cement paste or concrete usually with the evolution of considerable heat, which rigidity cannot be overcome nor can plasticity be regained by further mixing without the addition of water [C451]. As explained in Art. 2.12, the presence of gypsum can moderate the C_3A reaction through the formation of relatively insoluble calcium sulfoaluminates.

The amount of gypsum should bear some relation to the amount of alumina-bearing compounds present. It is desirable that the gypsum be used up by, or not too long after, final set, in order to avoid the rapid production of ettringite after the paste has gotten well into the hardening stage. In fresh paste, the increase in volume resulting from the formation of ettringite can be accommodated without damage to the paste structure.

Too much gypsum results in hardened paste of lowered strength and undesirable expansion and increase in volume. There is thus an "optimum" gypsum content for a cement, which depends upon its composition [246, 247, 249]. The amount of gypsum added to the clinker at the time of grinding is ordinarily about 4 to 6 percent by weight. The maximum amount is controlled in current specifications by placing a limit in terms of the SO_3 as determined in the chemical analysis. For example, ASTM C150, for a type I cement, calls for a maximum limit of 2.5 percent SO_3 when the C_3A is 8 percent or less and 3.0 percent SO_3 when the C_3A is more than 8 percent.

Another phenomemon known as "false set" sometimes occurs. False set is the rapid development of rigidity in a cement paste without the generation of much heat, which rigidity can be overcome and plasticity regained by further mixing without the addition of water [C451]. During the grinding process (or for other reasons), some of the gypsum ($CaSO_4 \cdot 2H_2O$) may be dehydrated to the soluble anhydrite or the hemihydrate form. When the cement is mixed with water, under some conditions the lower hydrates of $CaSO_4$ rapidly hydrate again to gypsum, forming a mass of fine crystals, and in so doing cause stiffening. Continued mixing or agitation can break up this crystalline mass and the mixture can be used without damage to the ultimate product. ASTM C451 prescribes a procedure for investigating false set.

A distinction should be made between rapid setting and rapid hardening of cement. A rapid *setting* cement is one in which the reactions take place at such a rate that the mass becomes rigid and begins to acquire some degree of strength at ages earlier than those for a normally retarded portland cement. Too rapid setting of cements intended for normal use may cause difficulties in handling. On the other hand, there may be conditions where it is important for the cement to set very rapidly. To stop the flow of water in joints in tunnelling operations, special, very rapid setting cements may be used. To obtain setting in a reasonable length of time under cold weather conditions where the normal setting process would be unduly delayed by the low temperatures, accelerators, such as $CaCl_2$, may be added to the mixture. A rapid *hardening* cement is one in which the reactions and gain in strength, *after* setting has occurred, take place at a rate greater than that in an ordinary general-purpose (type I) cement. Such cements are ordinarily called high-early-strength cements. As discussed in Art. 2.18, accelerated hardening can be obtained not only by using a cement having rapid hardening properties, but also by using admixtures which accelerate the reactions, by inducing temperatures which favor more rapid hardening, or by using richer mixtures.

2.15 Structure of cement paste. A mass of cement paste, beginning as a mixture of discrete particles with water, until it becomes a mature hardened substance undergoes continuous transformation. It is desirable to formulate adequate concepts of the structure of the paste at several stages in its development in order to have a basis for understanding the behavior of mortars and concretes and for adopting measures for control of properties.

Fresh paste. Immediately after cement has been mixed with water, the resulting paste may be regarded as a suspension in which solid phase is dispersed in a liquid phase even though the suspension is by no means a dilute one. For an ordinary cement, the particles of ground clinker and gypsum, ranging in size from about 0.5 micron to perhaps 80 microns, have a specific surface of the order of 3,000 to 4,000 sq cm per g as measured by air-permeability methods. The surface-mean-diameter[1] for such a cement would lie perhaps between 5 and 10 microns. In the range of practical water contents, the particles would be separated by average distances of the order of 5 to 10 microns. The particles are small enough and close enough together so that the paste acts as a plastic mass [242]. This plastic cement-water system, truly a *paste,* forms the basic matrix in which aggregate particles and air bubbles[2] may be embedded (or dispersed) to form mortar or concrete.

Over the range of water contents found in ordinary concrete mixes, corresponding to water-cement ratios ranging from, say, 0.4 to 0.8 by weight, the proportions of the cement solids by absolute volume in a mass of fresh paste ranges from about 40 to 28 percent.

In a suspension of fine particles, such as portland cement particles in water, it is conceivable that there is some characteristic average separation of the particles which is a function of gravitational force, buoyant force, interparticle attractive forces (van der Waals forces), and repulsive electrostatic forces deriving from the ionic charges on the surface of the particles. If the initial concentration of solid particles in the suspension (as governed by the water-cement ratio) is less than some critical value and the mass

[1] The diameter of one-size spheres in an aggregation having the same specific surface and weight as the aggregation of cement particles.

[2] For the purposes of this article, air bubbles formed during the mixing process or other air voids resulting from less-than-complete compaction or from spaces left by water after bleeding has occurred are not considered as a part of the cement paste.

is let to stand without agitation, it would be expected that the mass of solids would settle and consolidate until the particles are separated by that distance at which the forces tending to cause consolidation and those of repulsion are in equilibrium.

In the range of water-cement ratios occurring in ordinary concretes, "bleeding" of free water to the surface of the paste[1] (or conversely, the settlement of particles in the suspension) does in fact occur during the first few hours after mixing [236]. Because of the initial closeness of the particles and because of the interparticle attractive forces, the mass tends to consolidate as a single floc (that is, the particles of different size do not settle at different rates as they would in a dilute suspension).

The plasticity characteristics of a fresh paste also are governed by the distance between the particles and the interparticle attractive and repulsive forces. Fineness of particles (affecting surface area) and plasticizing agents can affect the magnitude of the interparticle forces and thus change the mobility of the suspension.

In a mass of paste, after bleeding has taken place, it is the initial arrangement and spacing of the cement particles, which establishes the basis for the ultimate structure of the paste that develops as the subsequent products of chemical reaction form [242]. Within the boundaries of the mass of cement paste, the volume of the space between and surrounding the original particles of cement determines the density of the porous mass that will eventually evolve. The volume of water which initially fills this space in relation to the volume of the cement solids defines the effective water-cement ratio for the purposes of the discussion which follows.

Cement gel. A substantial part of the hydration product is a "gel." Technically, a gel of the kind here involved is a two-component disperse system in which the solid phase is of colloidal dimensions (see Fig. 2.4). Colloidal particles, which range in size from about the wave length of light down to the dimensions of large molecules, have surface areas of such magnitude that surface forces are large relative to the weight of a particle.

[1] If pastes of relatively high water-cement ratio are worked or agitated too much in place, fine particles may be brought to the top surface forming what is called "laitance."

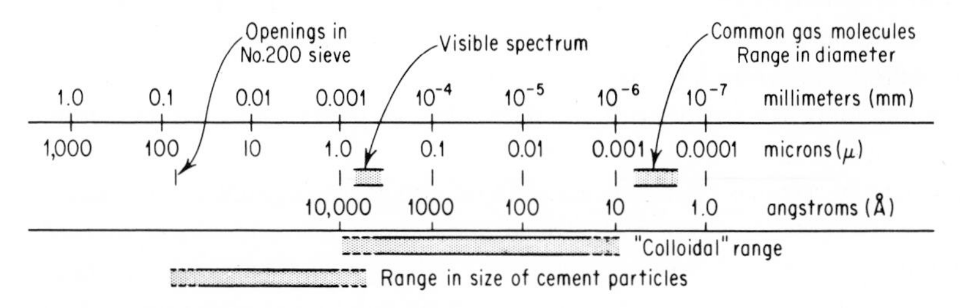

Fig. 2.4. Small-dimension relationships.

If these surface forces and the arrangement of particles are such to provide strong attraction between the particles and to induce mechanical properties characteristic of the solid state, the gel is said to be a rigid gel. In a rigid gel, both the dispersed component (solid) and the dispersion medium are, in effect, continuous throughout the system [204, p. 19].

The particles of the solid (dispersed) phase can be either crystalline or amorphous (in various shapes and forms—bulky, platy, fibrous, or even long-chain macromolecules). They may be bonded in various ways, such as by ionic or covalent bonds, electrostatic attractions, and/or van der Waals forces. The essential requirement is that they adhere together in such a way to form an open framework, no matter how irregular the structure.

The hydrated calcium silicate of gel-like nature that is formed as a result of the reaction of cement and water was once thought by some to be an amorphous material. However, an increasing body of evidence, from x-ray diffraction and other studies, is providing a much improved concept of the molecular structure. Although the "particles" of gel are of extremely small size, the arrangement of their constituent molecules appears to be such to justify regarding the gel as a microcrystalline substance. Some evidence supports the view that the molecular arrangement forms a kind of layered structure somewhat analogous to the structure found in some clay minerals although possibly not as regularly ordered.

The structure that is formed is evidently only a few molecules thick although the other two dimensions may be relatively much greater. The resulting "sheets", however, seem to be rolled or folded in various ways; they appear to be in the form of needles, fibers, slivers, and elongated, crumpled sheets in micrographs made with the electron microscope. This material forms the colloidal gel which in a cement paste is called tobermorite gel [225, 226, 243]. Figure 2.5 indicates the form of particles of calcium silicate hydrate occurring in hydrated pastes, designated as tobermorite (G) by Brunauer [225]. To obtain the electron micrographs, samples hydrated for 8 years were suspended in an organic liquid and the particles dispersed by ultrasonic vibration. The calcium silicate hydrate, or tobermorite gel, has a specific surface, as estimated by several approaches, of something in excess of 2×10^6 sq cm per g, which is about 1,000 times the specific surface of the cement from which it is derived [236, 236a]. A reproduction of a photomicrograph of a fractured surface of a sample of partially hydrated cement paste is shown in Fig. 2.6.

The reaction of cement with water results, of course, in products other than tobermorite gel. Crystalline $Ca(OH)_2$ is produced along with the gel from the silicate phases, and the aluminates and ferrites produce crystalline or amorphous reaction products. These substances are evidently dispersed throughout the mass of tobermorite gel. Although the particles of these substances apparently are in the low-micron or submicron-size range, their

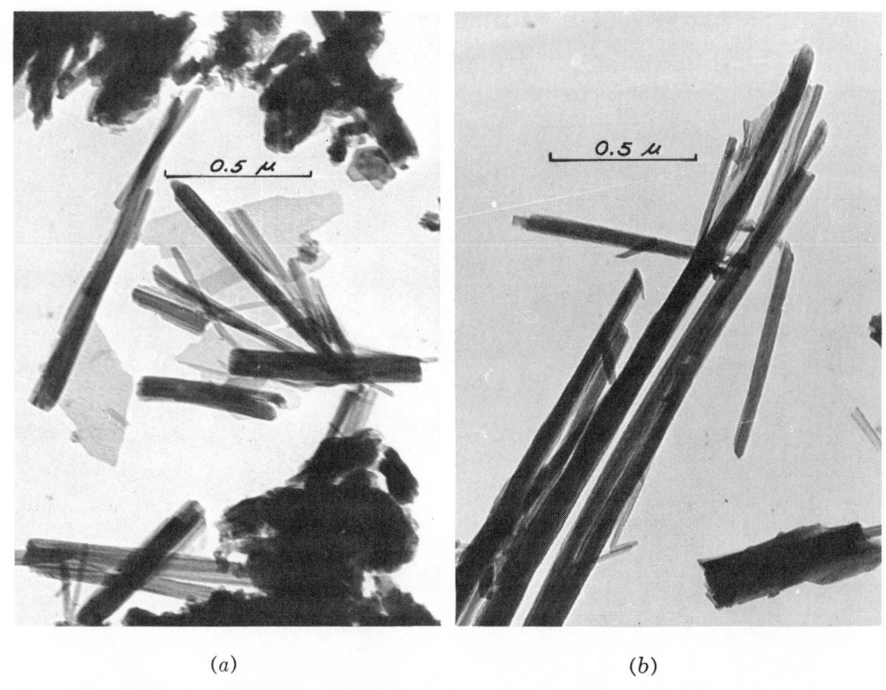

(*a*) (*b*)

Fig. 2.5. Electron micrographs of dispersed tobermorite (G) particles (*a*) **from tricalcium silicate paste;** (*b*) **from dicalcium silicate paste. Magnification: $10^4 \times 8$.** (*Portland Cement Association.*)

surface area is small compared with that of the tobermorite gel. Nevertheless, the amount of tobermorite gel produced and the enormous surface area it provides are such as to overwhelmingly condition the behavior of hardened cement paste. The term "cement gel" is often applied, loosely but with some justification, to the total mass of reaction products [236].

Gel formation. As the surfaces of the cement particles are attacked by water, solution of the original constituents, hydrolysis, and the formation of molecules of the reaction products takes place. After the formation of an initial layer of gel, some reaction products precipitate in the space made available by dissolution of the cement particles, and some are evidently formed outside of the growing layer of gel as constituent molecules migrate through the porous gel structure. The stage called "final set" probably occurs when filaments of gel growing outward from the surfaces of the dispersed particles extend far enough to develop the bond needed to impart to the mass the characteristics of a rigid gel structure. Following the establishment of such a framework, reaction products formed outside the gel layer on the particles must take place within the confines of this framework [242].

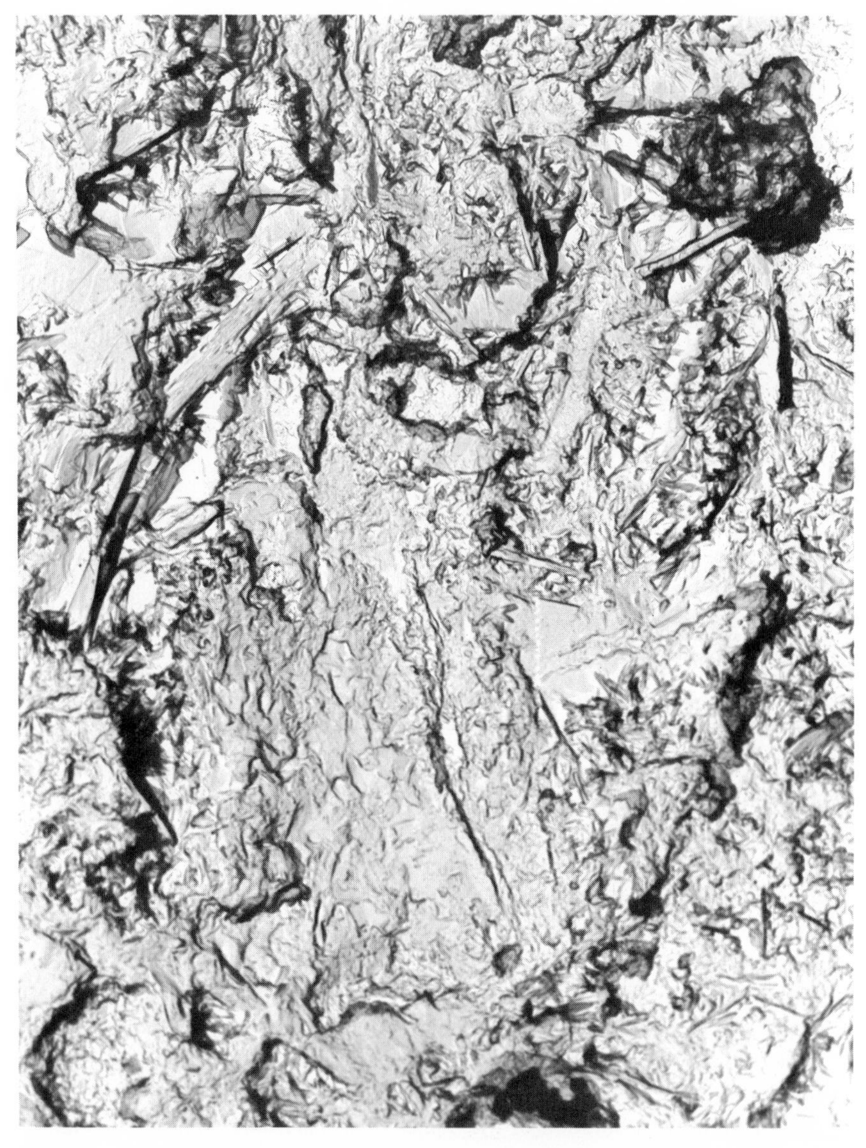

Fig. 2.6. Photomicrograph of a fractured surface of portland cement paste, $W/C = 0.45$ and cured 14 days. Growths of the calcium silicate hydrate can be seen extending from the surface—probably into large pore space. These kinds of growths are characteristic of young pastes and very porous pastes. *(Portland Cement Association.)*

Structure of hardened paste. The cement gel is conceived to be made up of a collection of interwoven filaments of tobermorite gel interspersed with other reaction products. The space between the various particles is called the "gel-pore" space. The gel pores are estimated to have average representative dimensions something of the order of 15 to 20 Å [236]. Water in pores of such small size, under the influence of surface forces of the gel particles, evidently does not exhibit the same properties as free water.

If the reaction products have not completely filled the space that existed between the original cement particles, there will be space which can have dimensions considerably larger than those characteristic of the gel pores. This space is called "capillary-pore" space. During the early hardening period the capillary pores may be more or less interconnected throughout the mass of paste, and the permeability of such pastes is observed to be relatively high. With pastes of moderate-to-low water-cement ratios, the capillary pores tend to become completely filled with cement gel as hydration proceeds toward completion. As the capillary pores become filled with gel, they become discontinuous. Obviously, the dimensions of both the gel pores and the capillary pores must range over some spectrum of size, and the distinction between gel pores and capillary pores serves as a convenient concept rather than a rigid differentiation on the basis of size [230].

It has been determined that if 1 cu cm (absolute volume) of cement undergoes complete reaction, a cement gel (containing all reaction products) having a volume (solid particles plus gel pores) of about 2.1 to 2.2 cu cm is produced [231, 245]. Some of this gel occupies the space formerly occupied by the original cement. Thus 1.1 to 1.2 cu cm of space outside the boundaries of the original cement grain is needed. It has been found that space of this magnitude within the mass of cement paste must be available if the reaction is to proceed to completion; if something less than this is available, then some of the cement remains unhydrated [236].

These space relationships at four stages of hydration, calculated for paste having a range in water-cement ratios, are illustrated in Fig. 2.7 assuming that the cement has a specific gravity of 3.15 and that the volume of the reaction product (solid plus gel pores) is 2.1 times the volume of the original cement. For example, for a water-cement ratio by weight of 0.5, the volume of water corresponding to 1 cu cm (absolute volume) of cement is 1.58 cu cm; thus the volume of the fresh paste is 2.58 cu cm as shown in Fig 2.7*a*. The percentage of water-filled voids in this paste is 61. After one-third of the original cement is hydrated, the cement gel has a volume of $0.333 \times 2.1 = 0.70$, and the volume of unreacted cement plus cement gel is 1.37 cu cm (Fig. 2.7*b*). When the cement is completely hydrated, the volume of the cement gel is 2.1 cu cm, and the capillary space is $2.58 - 2.10 = 0.48$ cu cm (Fig. 2.7*d*).

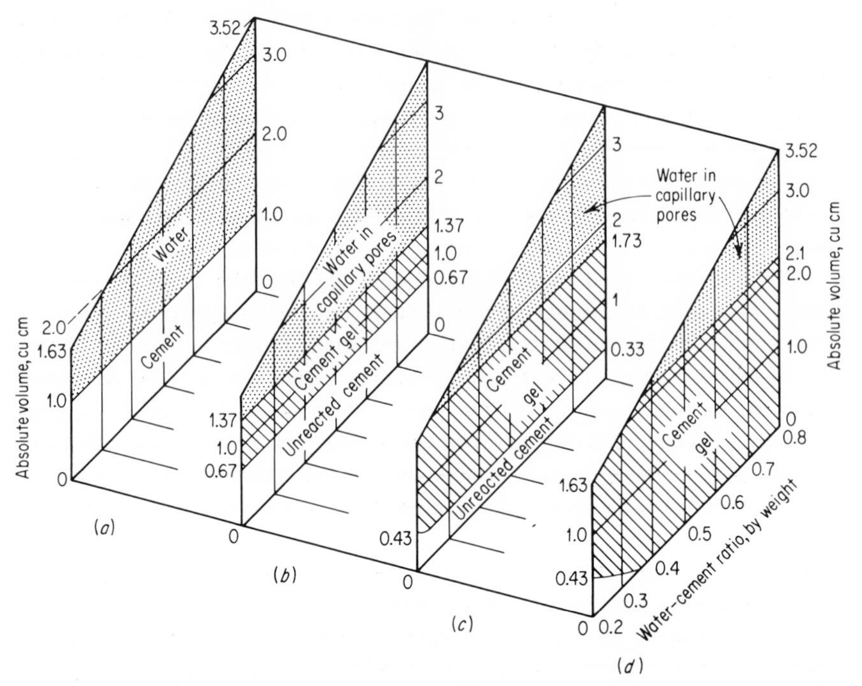

Fig. 2.7. Space relationships for paste components at various stages of hydration. (*a*) Fresh paste; (*b*) reaction one-third complete; (*c*) reaction two-thirds complete; (*d*) reaction completed. Based on concepts in Refs. 230, 236, 245. Assumptions: specific gravity of cement = 3.15. When completely hydrated 1 cu cm of cement produces 2.1 cu cm of cement gel.

Under the assumptions made here, with a paste having a water-cement ratio of 0.35 by weight, the original water-filled space would be completely filled with cement gel at 100 percent hydration, and with pastes having water-cement ratios lower than this, all of the original cement could not be hydrated. There does not appear to be firm evidence as to the exact magnitude of this critical water-cement ratio to permit complete hydration.

It has been estimated that the percent of gel-pore space in cement gel is about 26 to 28 percent [236, 245]. In the range of intermediate water-cement ratios used in ordinary concretes, in which some capillary space would also exist in the cement paste, the porosity of the hardened, fully hydrated cement paste may be 40 to 55 percent.

2.16 Role of paste structure and composition in behavior of concrete. A number of aspects of the behavior of concrete may be predicated from the structure of the hardened cement paste and from the chemical nature of its constituents. Because the basic structure of the hardened paste is a rigid gel with a considerable porosity and a large specific surface, it is capable of holding a substantial quantity of water under the influence of attractive forces of varying degrees of magnitude. It may be expected that the amount of water held within the overall gel structure will vary with humidity (vapor pressure) of the surrounding atmosphere. Some water occupies the capillary-pore space, but a significant amount occupies the gel-pore space. The gain or loss of gel water, by altering the forces within the gel structure, results in expansion or contraction of the paste (and of the concrete in which it may be used). While a part of this volumetric change is reversible, complete drying of the paste results in some volumetric shrinkage that is not recovered by subsequent wetting.

Because of the fine network of interconnected gel pores (and usually also some capillary-pore space, not always interconnected), hardened cement paste is permeable to water. It is to be expected that as hydration proceeds and more of the space between the originally discrete cement grains is filled with cement gel, the permeability will decrease. The permeability of a fully hydrated paste may be of the order of 10^6 times lower than that of a paste only a day or so old. The permeability of a mature paste is comparable to that of fairly dense rocks, such as marble or diorite [236, 245]. Because of the forces induced in water contained in pores of the very small size that exist in cement gel, the flow characteristics of water in such pores are not the same as those of free water [254]. Voids and channels in a concrete mass outside the gel mass per se may, of course, also affect the permeability of concrete.

Because water in the gel pores does not freeze at the same temperature as free water, it may be expected that a paste having a fully developed, dense gel structure will exhibit greater resistance to freezing and thawing than one having a higher volume of water-filled capillary pores. On the other hand, small, non-water-filled air bubbles distributed throughout the mass enhance resistance to freezing and thawing by providing a source of relief to pressures built up by freezing of water in the capillary spaces [252].

The strength of a paste derives in large degree from the bonds formed between the very small particles that compose the cement gel. Generally, the greater the number of such particles and the denser the gel structure, the stronger the gel mass [245]. This is why the ratio of water to cement in the original mix serves as a useful index of potential strength. The amount of pore space (and hence density of paste) that can finally be developed is a function of the initial water content; and the amount of gel that can potentially develop is a function of the relative amount of cement present. Increase in amount of gel, as hydration proceeds, results in increase in strength.

If water is removed from the paste structure, not only is the gel volume decreased, but some of the interparticle attractive forces are increased; thus pastes or concretes tested after drying exhibit greater strength than if they were tested in a fully saturated condition.

Because of the preponderance of lime and lime compounds in cement paste, the paste may be regarded as essentially chemically basic in nature. Hence, exposure to acidic solutions can be expected to cause dissolution and damage to concrete. Even long exposure to chemically pure water can cause lime to be leached, but in a dense concrete, other than superficial reaction is negligible over long periods in pure water.

When certain alumina-bearing compounds are present in the cement of a hardened concrete, its exposure to water containing sulfate ions results in the formation of ettringite, accompanied by a volumetric expansion within the fabric of the hardened paste which can result in disruption of the gel structure. Hence, for concretes that will be exposed to sulfate-containing soils or waters, low C_3A cements (type II or V) are often specified.

The disruption of hardened cement paste that can be caused by delayed hydration of hard burned CaO and MgO has been mentioned in Art. 2.8.

Because CO_2 is present in the atmosphere, under some conditions, it may react with the $Ca(OH)_2$ or other lime-bearing compounds in the hardened cement paste. It has been observed that when hardened pastes are exposed to air containing CO_2 at relative humidities, ranging from perhaps 35 percent to something less than 100 percent, a decrease in volume occurs that is called "carbonation shrinkage" [203, p. 473]. This shrinkage takes place even though the reaction of CO_2 with $Ca(OH)_2$ gives a product having a volume greater than that of the $Ca(OH)_2$ consumed. The mechanism of this phenomenon is not fully understood; one view that has been postulated is that as the reaction takes place, $Ca(OH)_2$ is dissolved, some of the resulting $CaCO_3$ in solution is deposited elsewhere in the paste structure, and the existing tensile stresses in the gel water cause a decrease in the space previously occupied by the $Ca(OH)_2$ [255]. While surface crazing has been observed to result from carbonation shrinkage, such action is confined to only a very thin layer near the surface and apparently is not harmful to the remaining mass of concrete.

When CO_2 is dissolved in water it produces a weak acid; thus under some circumstances of exposure of paste to CO_2-bearing water, some mild dissolution may be caused.

Compounds of sodium and potassium are usually present in portland cement as minor constituents. These alkalis can react rather strongly with certain silicates in some aggregates, with resultant volumetric expansion and disintegration of a hardened concrete mass. For this reason special steps should be taken to offset this potential reaction if the aggregates to be used contain such silicates. One procedure is to limit the alkali content of the cement to some small amount which experiment and experience has shown has seldom resulted in deterioration. Another procedure is to use a siliceous (pozzolanic) admixture in the concrete mixture that will readily react with both the alkalies and calcium hydroxide during the early stages of the hardening process.

2.17 Specifications requirements for portland cements. Specifications for cement generally include a statement of chemical and physical requirements, methods of testing, limitations on additions, and provisions concerning packaging, marking, storage, and inspection and rejection. Of interest here are the nature of, and reasons for, the chemical and physical requirements.

The principal chemical requirements for the several types of portland cement manufactured under ASTM Specifications are outlined in Table 2.9. The background and implications of these requirements have been discussed in Arts. 2.5, 2.8, and 2.9.

The principal physical requirements of the several types of portland cement manufactured under ASTM Specifications are given in Table 2.10. The usual physical requirements prescribed for acceptance of a cement are concerned with (1) fineness, (2) time of set, (3) soundness, and (4) strength. In addition, for some types of cement, there may be requirements having to do with heat of hydration, false set, and air content of mortar made with the cement. A discussion of the determination of fineness is given in Art. 2.11. Background on the meaning of time-of-set requirements and on false set is given in Art. 2.14. Comments on heat of hydration are given in Art. 2.13. While a basis has been laid for the significance of the soundness test in the discussions of free lime and magnesia in several preceding articles, some summary comments on soundness are pertinent here. Also, some brief comments relative to the strength tests made for acceptance of cement will be made.

Soundness. Unsoundness in (set) cement is caused by undue expansion of some of the constituents, which expansion is manifested by cracking, disruption, and disintegration of the mass. One source of unsoundness in cement is the delayed hydration of free lime incased within the cement particles (see Art. 2.12). A protective film prevents the immediate hydration of the free lime. However, moisture may finally reach the lime after the cement has set, and since lime expands with considerable force when hydrated under restraint, its delayed hydration may disrupt the mass. One advantage of a slow setting cement is that more time is given to hydrate the lime before the mass becomes rigid.

Another possible cause of unsoundness is the presence of too high a magnesia content (see Art. 2.12). Standard specifications limit the magnesia content to 5 percent.

Fine grinding of the raw materials brings them into more intimate contact when burned, so that there is less chance of free lime existing

Table 2.9 Chemical requirements for portland cements*

	Type I	Type II	Type III	Type IV	Type V
Silicon dioxide (SiO_2), min, percent	...	21.0			
Aluminum oxide (Al_2O_3), max, percent	...	6.0	...	...	†
Ferric oxide (Fe_2O_3), max, percent	...	6.0	...	6.5	†
Magnesium oxide (MgO), max, percent	5.0	5.0	5.0	5.0	5.0
Sulfur trioxide (SO_3), max, percent:					
When $3CaO \cdot Al_2O_3$ is 8 percent or less	2.5	2.5	3.0	2.3	2.3
When $3CaO \cdot Al_2O_3$ is more than 8 percent	3.0	...	4.0		
Loss on ignition, max, percent	3.0	3.0	3.0	2.5	3.0
Insoluble residue, max, percent	0.75	0.75	0.75	0.75	0.75
Tricalcium silicate ($3CaO \cdot SiO_2$,)‡ max, percent	...	...	...	35	
Dicalcium silicate ($2CaO \cdot SiO_2$),‡ min, percent	...	...	...	40	
Tricalcium aluminate ($3CaO \cdot Al_2O_3$),‡ max, percent	...	8	15§	7	5
Sum of tricalcium silicate and tricalcium aluminate, max, percent	...	58¶			
Tetracalcium aluminoferrite plus twice the tricalcium aluminate‡ ($4CaO \cdot Al_2O_3 \cdot Fe_2O_3$) + 2($3CaO \cdot Al_2O_3$), or solid solution ($4CaO \cdot Al_2O_3 \cdot Fe_2O_3 + 2CaO \cdot Fe_2O_3$), as applicable, max, percent	...	...	...	...	20.0

* From Refs. C150, C175.

† The tricalcium aluminate shall not exceed 5 percent, and the tetracalcium aluminoferrite ($4CaO \cdot Al_2O_3 \cdot Fe_2O_3$) plus twice the amount of tricalcium aluminate shall not exceed 20 percent.

‡ The expressing of chemical limitations by means of calculated assumed compounds does not necessarily mean that the oxides are actually or entirely present as such compounds. Refer to Art. 2.9 for the formulas used to calculate the percentages of the major compounds in a cement.

§ When moderate sulfate resistance is required for type III cement, tricalcium aluminate may be limited to 8 percent. When high sulfate resistance is required, the tricalcium aluminate may be limited to 5 percent.

¶ This limit applies when moderate heat of hydration is required and tests for heat of hydration are not requested.

in the clinker. Thorough burning of the raw materials further reduces the potential amount of free lime. Finally, fine grinding of the clinker tends to expose the free lime so that it will hydrate quickly before the mass of concrete hardens appreciably.

By allowing unsound cement to aerate for several days, the free lime is given an opportunity to hydrate and the cement may become sound. As aeration is harmful to the strength of cement, it should not be "seasoned" except when unsound. Owing to the possible beneficial effects of aging unsound cement, specifications [C150] allow the acceptance of the cement if it passes a retest using a new sample at any time within 28 days after the first test.

One method for detecting unsound cements is the "autoclave" test [C151]. An autoclave is essentially a high-pressure steam boiler. The test specimens are bars made of neat cement paste of normal consistency, 1 by 1 in. in cross section and having an effective gage length of 10 in. Beginning at the age of 24 hr after molding, the test specimens are placed in the autoclave and subjected to an atmosphere of high pressure steam at 295 psi (corresponding to a temperature of 420°F) for 3 hr. After cooling, the length of the bars is measured and compared with the original length before treatment. Portland cements which exhibit an expansion of not more than 0.80 percent are considered to be sound [C150]. The high temperatures used in these tests accelerate the hydration of the cement so that within the short period of the test, conditions are developed which would require much longer periods under normal curing. Although the accelerated soundness tests are made under abnormal conditions, yet the results of many such tests, in comparison with long-time normal curing, bear witness to their value.

Strength. The strength of cement is usually determined from tests on mortars made with the cement. Both tension and compression tests have been standardized by the ASTM [C150, C109, C190].

Tensile-strength tests are made on mortar briquets having a net cross-sectional area of 1 sq in. and composed of 1 part cement to 3 parts sand, by weight. For making a mortar a special pure silica sand (standard Ottawa sand), sized between the No. 20 and No. 30 sieves, is used. The water requirement for this "standard mortar" is computed from the water requirement for normal consistency of neat-cement paste. The consistency of this standard mortar is fairly stiff. Based upon correlation between tests of a large number of cements and mortars the following empirical relationship was found by the ASTM Committee on Cement:

$$\text{Water for standard mortar, in \% of dry cement + sand} = 6.5 + \frac{\text{\% water for normal consistency}}{6}$$

A more general formula for water requirement of a mortar is given in Ref. C190. In order to obtain reliable strength results all operations, such as molding, curing, and testing, must be carried out under specified conditions. For normal cements tests are commonly made at the ages of 3, 7, and 28 days.

Since the tensile strength of cement in a concrete is not ordinarily a major factor in controlling the strength of structural members, the value of the briquet test has often been questioned. This and the fact that struc-

Table 2.10 Physical requirements for portland cements[a]

	Type I	Type II	Type III	Type IV	Type V
Fineness, specific surface, sq cm per g (alternate methods):[b]					
Turbidimeter test:					
Average value, min	1,600	1,600	...	1,600	1,600
Minimum value, any one sample	1,500	1,500	...	1,500	1,500
Air permeability test:					
Average value, min	2,800	2,800	...	2,800	2,800
Minimum value, any one sample	2,600	2,600	...	2,600	2,600
Soundness:					
Autoclave expansion, max, percent	0.80	0.80	0.80	0.80	0.80
Time of setting (alternate methods):[c]					
Gillmore test:					
Initial set, min, not less than	60	60	60	60	60
Final set, hr, not more than	10	10	10	10	10
Vicat test (Method C191):					
Set, min, not less than	45	45	45	45	45
Air content of mortar, prepared and tested in accordance with Method C185, max, percent by volume, less than	12.0	12.0	12.0	12.0	12.0
Compressive strength, psi:[d]					
The compressive strength of mortar cubes, composed of 1 part cement and 2.75 parts graded standard sand, by weight, prepared and tested in accordance with Method C109, shall be equal to or higher than the values specified for the ages indicated below:					
1 day in moist air	...	...	1,700		
1 day in moist air, 2 days in water	1,200	1,000	3,000		
1 day in moist air, 6 days in water	2,100	1,800	...	800	1,500
1 day in moist air, 27 days in water	3,500	3,500	...	2,000	3,000

Table 2.10 Physical requirements for portland cements[a] (Continued)

	Type I	Type II	Type III	Type IV	Type V
Tensile strength, psi:[d]					
The tensile strength of mortar briquets composed of 1 part cement and 3 parts standard sand, by weight, prepared and tested in accordance with Method C190, shall be equal to or higher than the values specified for the ages indicated below:					
1 day in moist air	...	...	275		
1 day in moist air, 2 days in water	150	125	375		
1 day in moist air, 6 days in water	275	250	...	175	250
1 day in moist air, 27 days in water	350	325	*d*	300	325
Heat of hydration:[e]					
7 days, max, cal per g	...	70			
28 days, max, cal per g	...	80			
False set, final penetration, min, percent[f]	50	50	50	50	50

[a] From Ref. C150, C175.

[b] Either of the two alternate fineness methods may be used at the option of the testing laboratory. However, in case of dispute, or when the sample fails to meet the requirements of the Blaine meter, the Wagner turbidimeter shall be used, and the requirements shown for this method shall govern.

[c] The purchaser should specify the type of setting time test required. In case he does not so specify, or in case of dispute, the requirement of the Vicat test only shall govern.

[d] The purchaser should specify the type of strength test required. In case he does not so specify, the requirements of the compressive-strength test only shall govern. The strength at any age shall be higher than the strength at the next preceding age. Unless otherwise specified, the compressive- and tensile-strength tests for types I and II cement will be made only at 3 and 7 days. If, at the option of the purchaser, a 7-day test is required on type III cement, the strength at 7 days shall be higher than at 3 days.

[e] These requirements apply only when specifically requested; when the heat of hydration requirements are specified, the strength requirements for type II shall be 80 percent of the values listed.

[f] This requirement applies only when specifically requested.

tural concretes ordinarily contain a plastic mortar composed of a graded sand have led to the development of a compressive test of plastic mortar as an alternate acceptance test for cements. In the standard compressive-strength test, the specimen is a 2-in. mortar cube composed of 1 part cement of 2.75 parts sand, by weight. The sand is a graded Ottawa silica sand, and the water requirement is determined experimentally. A molded sample of the mortar is jigged on a special plate called a flow table (Art. 5.2 and Appendix E), and if the *increase* in diameter of the pile is between 100 and 115 percent of the original diameter, it is considered to be of the desired plastic consistency. As a guide in preparing trial batches, a water content of about 47 to 49 percent, by weight, is suggested [C109].

2.18 High-early-strength cements. In many instances in construction it is highly desirable that sufficient strength be developed within, say, 3 days, so that, for example, forms may be removed and construction resumed, or a roadway may be opened to traffic promptly, rather than after the week or longer period that is required with a normal cement.

During the last four decades, portland cement has constantly been improved to produce higher-strength concrete. Present-day standard cements are at least a third stronger than those of 40 years ago. However, various high-early-strength cements have been developed which are capable of producing as great strength in 1 to 3 days as normal portland cement does in 28 days. Because of their greater cost, they are used only where high early strength is a special requirement.

Three factors are of importance in the production of a high-early-strength cement (type III): (1) chemical composition; (2) degree of chemical combination of the raw materials as influenced by the fineness of grinding, degree of blending, and completeness of burning of the raw material; and (3) the fineness of grinding of the clinker.

The effect of chemical composition upon rate of hardening has been mentioned in Art. 2.10. The development of more effective grinding equipment has made possible the finer grinding of raw materials, which in turn has permitted a more intimate association of these components in the burning operation. The use of better blending equipment in the dry process and the introduction of the wet process of manufacture combined with longer kilns have all contributed to the production of a superior clinker. Finally, a high-early-strength cement is often ground more finely than normal cement. In fact a high-early-strength cement may be produced by grinding some cements of normal composition to a high degree of fineness.

High-early-strength cements are not necessarily superior to normal portland cements in all respects. It has been observed that the use of high-early-strength cements has not infrequently resulted in greater checking and cracking of concrete in structures subjected to drying conditions.

Owing to the relatively high percentage of C_3S and C_3A in most high-early-strength cements, the heat of hydration of such cements is higher than for normal cements. In massive structures this chemical heat is not dissipated readily and causes an appreciable rise in temperature of the concrete mass. Over a period of time, after the concrete has hardened, the structure cools and contracts. Owing to the restraint of the foundations or to differential restraints within the mass, tensile stresses are developed which, in long structures, exceed the tensile strength of the concrete and cause cracking. The higher the temperature resulting from chemical activity, the greater the later temperature drop and so the greater the possibility of cracking.

During cold weather the use of a high-heat cement in ordinary construction may prove advantageous, in that there is slightly less likelihood of damage due to freezing of the concrete and the concrete tends to develop adequate strength before freezing occurs.

It is not essential that high-early-strength cement be used in making high-early-strength concrete. According to the water-cement-ratio "law" (Arts. 6.3 and 10.12) the use of a sufficiently low water content with normal cement will produce high-strength concrete. However, the use of rich mixes, made necessary by a low water-cement ratio, may give rise to such disadvantages as large heat generation and slightly greater shrinkage upon drying.

Calcium chloride is sometimes used as an admixture in concrete to accelerate the hardening, particularly in cold-weather construction. The use of this is discussed in Art. 8.15.

Further, protection from loss of heat or the use of heating devices to maintain adequate ambient temperatures can serve to promote more rapid hardening of concretes in cold weather.

2.19 Portland-pozzolan cements. When portland cement combines with water during setting and hardening, lime is liberated from some of the compounds. The amount of lime that can be liberated appears to be as much as 20 to 30 percent by weight of the cement [203, 251]. Under unfavorable circumstances, a concrete structure may be subject to disintegration owing to the leaching of this lime from the mass. One way in which this difficulty can be overcome is to mix or grind some pozzolanic material with the cement. A pozzolan is a siliceous material which reacts with lime in the presence of moisture at ordinary temperatures to give a relatively stable strength-producing calcium silicate. Pumicite and diatomaceous earth, for example, have pozzolanic properties. A number of portland-pozzolan cements which have been commercially produced in recent years contain about 25 percent of pozzolan, although percentages ranging perhaps from 10 to 50 have been considered. Compared with portland cements of similar characteristics, portland-pozzolan cements gain strength relatively slowly, although the ultimate strengths of concretes made with portland-pozzolan cements may equal or exceed those made with similar portland cement. Extended periods of moist curing are necessary to develop the benefits of pozzolanic action [203, 264, 265].

Specifications for portland-pozzolan cement are given by ASTM C340. A type I cement having appreciable pozzolanic additions is designated as type IP. If it also contains an air-entraining agent it is designated as type IPA.

2.20 Slag cements. Slag cements are produced by intergrinding a suitable blast-furnace slag with hydrated lime. Hot slag, which is a waste product from the blast furnace used in smelting iron ores, is usually first granulated by a stream of cold water. It is then dried, ground, mixed with a high-calcium hydrated lime, and pulverized. Depending on the composition of the slag, the proportions of the two components may vary from 25 to 45 lb of lime for 100 lb of slag.

Slag cements have a low specific gravity ranging between 2.7 and 2.85, and are usually light-colored. As the strengths obtained with slag cements are generally low and variable, their use is limited to unimportant structures or to those massive structures whose strength is not of great importance. Some slag cements are used as a masonry cement.

2.21 Portland blast-furnace-slag cement. Portland blast-furnace-slag cement is an intergrind of portland cement clinker and granulated blast-furnace slag with the addition of a small amount of gypsum to control the set. The slag component should be between 25 and 65 percent of the total cement, to satisfy the requirements of ASTM C205, and the slag should conform to the following composition:

$$\frac{CaO + MgO + \frac{1}{3}Al_2O_3}{SiO_2 + \frac{2}{3}Al_2O_3} \geqq 1$$

If ground with a type I clinker, it is designated as type IS.

The early strengths obtained with this cement are almost as high as for a type I portland cement, and at later ages the strength may be even higher. Its cost is a little lower.

2.22 Masonry cements. Masonry mortars, for use in brick masonry and block masonry, require masonry cements having greater plasticity than is obtainable with ordinary portland cements. These masonry cements consist of blends of cementing materials with other materials. The cementing materials may be portland cement, portland-pozzolan cement, natural cement, slag cement, hydraulic lime, or hydrated lime. Other materials which may be blended with the cementing materials are limestone dust, chalk, talc, pozzolans, clay, and gypsum; some of these act as plasticizers to increase the plastic qualities of a masonry mortar.

ASTM Specifications [C91] cover two types of masonry cement: type I for general-purpose non-load-bearing masonry, and type II for masonry where high strength is required. The required compressive strengths of

2-in. cubes of a 1:3 mix using a 50:50 blend of standard and graded Ottawa sand mortar are as follows:

Age, days	Compressive strength, minimum, psi	
	Type I	Type II
7	250	500
28	500	900

2.23 Expansive cements. Because of the inherent tendency of ordinary portland cement concretes to shrink on drying, with consequent tendency to crack when the concrete mass is restrained from shrinking, many efforts have been made to eliminate or reduce drying shrinkage. One approach is to prepare a mixture containing compounds that will expand sufficiently to offset the shrinkage tendency of the normal hydrated cement compounds. Interest in the possibility of "expansive" cements has been augmented by the increasing use of prestressed concrete, in which it would be desirable in some applications to utilize expansive properties of the concrete to induce prestress in the steel. Considerable experimental and developmental work has been done in recent decades [256 to 261].

A number of reactions that will result in expansion of the cement paste have been mentioned in Arts. 2.5, 2.8, and 2.12. It was pointed out that if undue expansion of some components takes place in hardened concrete, disruption and disintegration takes place. Thus, attention has been focused on developing compounds and/or processes by which useful expansion can take place at an early enough period so that the paste can increase in volume without permanent damage to the paste structure.

Most of the efforts to develop usefully expansive cements have utilized the formation of calcium sulfoaluminates for the expansive mechanism. The mixtures which comprise the expansive cement generally contain normal portland cement together with a mixture or compound that serves as the expansive agent; in some compositions a stabilizer or moderator is also included. The expansive agents that have been tried contain ingredients which essentially supply lime sulfates and aluminates. Gypsum, bauxite, and some lime-bearing compound, such as calcite, are typical raw materials which are used in a number of ways. In the preparation of some expansive cements, the expansive agent is prepared by burning the above-mentioned materials to form a clinker, which is then interground with portland cement clinker. In the preparation of others, mixtures of the several ingredients are simply blended with the portland cement.

2.24 High-alumina cement. Although not of the same composition as portland cement, aluminous, or high-alumina cement, is a hydraulic cement that is used to make concrete in the same manner as portland cement. It has certain useful special properties such as rapid hardening and high resistance to the action of sulfate waters. With special aggregates such as firebrick, it can be used to make refractory concrete that can stand high furnace temperatures.

High-alumina cement is made by grinding a compound formed by the fusion of limestone and bauxite. While the composition varies, chemical analyses of representative cements shows the principal oxides to be approximately as follows: CaO, 35 to 44 percent; Al_2O_3, 33 to 44 percent; SiO_2, 3 to 11 percent; and Fe_2O_3, 4 to 12 percent. The principal products of hydration of high-alumina cements at normal atmospheric temperatures are calcium-aluminate hydrates and some hydrated colloidal alumina [203, 204].

Hydration is practically complete in about 24 hr, and the strengths attained in 1 day equal those attained by portland cement in about 1 month. Very considerable heat is liberated by this rapid reaction. At reaction temperatures above about 35°C in the presence of moisture, the resulting strength is markedly lowered, so that dissipation of the heat of hydration becomes a serious problem. It should not be used in massive sections.

In refractory concrete, when the temperature is increased, the hydration products are first dehydrated, and then at sufficiently high temperatures the dehydrated compounds react with the aggregates to form a ceramic bond.

QUESTIONS

1. What is the volume and weight of one sack of cement?

2. In general, portland cement is made from what raw materials?

3. Describe the general methods of manufacture of cement by both the wet and dry processes.

4. What changes occur within a rotary kiln in the manufacture of portland cement?

5. Why is gypsum added to the clinker in the production of cement?

6. State the typical oxide composition of normal portland cement.

7. Name the four principal compounds in portland cement and the approximate percentage of each in a normal type I cement.

8. Tabulate the characteristics of the principal compounds which occur in portland cement as to (*a*) rate of reaction, (*b*) heat liberated, and (*c*) cementing value.

9. What is a pozzolan, and why is it sometimes used with portland cement?

10. Describe certain common conditions which may cause the disintegration of concrete structures.

11. How do high-early-strength type III cements differ from normal type I cements?

12. Does high-early-strength cement have any disadvantages for certain classes of concrete? Explain.

13. Define specific surface of a cement. How is it determined?

14. What is meant by "normal consistency" and how is it determined?

15. Discuss the influence of various factors upon the time of set of cement.

16. What causes the set of cement?

17. What is meant by "unsoundness" and how is it caused?

18. How may unsoundness be determined?

19. How is the strength quality of a given cement determined?

20. Define "heat of hydration." What is its significance?

21. List the current types of portland cement and the general use made of each type.

22. Following is a chemical analysis of a portland cement. Compute the potential compound composition and state what type of cement it is.

SiO_2	21.73%	SO_3	1.72%
Fe_2O_3	2.63	Free lime	1.02
Al_2O_3	5.73	Loss on ignition	1.27
CaO	63.75	Insoluble residue	0.15
MgO	2.42		

23. Consider a hardened cement paste having an original effective water-cement ratio of 0.55 by weight. Find the amount of unreacted cement, the amount of cement gel, and the amount of water-filled capillary-pore space when the hydration process is 75 percent complete. Express these in terms of volumes relative to an initial 1 cu cm of cement (by absolute volume). Assume that the specific gravity of the cement is 3.12, that complete hydration of 1 cu cm of cement produces 2.06 cu cm of cement gel, and that water is continuously available to the hydrating paste. Find the water-cement ratio at which, if the hydration proceeds to completion, there would be no capillary-pore space and no unreacted cement.

THREE
AGGREGATES

3.1 Preliminary remarks. The mineral aggregate comprises the relatively inert filler material in a portland-cement concrete. However, inasmuch as the aggregate usually occupies from about 70 to 75 percent of the total volume of a mass of concrete, its selection and proportioning should be given careful attention in order to control the quality of the concrete structure. In choosing aggregate for use in a particular concrete, attention should be given to three general requirements: economy of the mixture, potential strength of the hardened mass, and probable durability of the concrete structure.

One of the important properties of aggregate for concrete is the gradation of the particles. In order to provide a dense packing, or arrangement of particles in place, in a concrete mass, the aggregate must be suitably graded from fine to coarse. Gradation of particles is also an important factor in controlling workability of the fresh plastic concrete, although the grading to produce optimum workability produces a packing of particles somewhat less dense than that which, theoretically, gives the maximum density of aggregate in the concrete mass. Gradation influences both economy and strength. In general, the more aggregate that can be crowded into a given volume of concrete, the more economical is the resulting product. On the other hand, with a given amount of aggregate per unit volume of concrete, the more workable a mix is—as the result of suitable gradation—the lower the water requirement and the greater the strength.

In order that a concrete be durable, in so far as the influence of the aggregate is concerned, it is important (1) that the aggregate be resistant to weathering action, (2) that no unfavorable reaction take place between the aggregate minerals and components of the cement, and (3) that the aggregate contain no impurities which affect the strength and soundness of the cement paste. With some aggregates, over a long period of time there may occur a slight interaction between the paste and the aggregate at the surface of the particles; this action is beneficial in that it promotes good bond within the concrete mass, but in general the extent of such reaction appears to be small if not negligible. If, however, either weathering action or chemical reactions affect the aggregate particles to such an extent that appreciable expansive forces are developed in the concrete mass, partial or complete disintegration of the concrete may ensue. One of the problems in the selection of aggregates is to avoid or to minimize these sources of difficulty [309, 310].

In addition to the above-mentioned general considerations special aggregates may be used for specific purposes, such as for making heavy-weight, radiation-shielding concrete (Art. 14.21), lightweight concrete (Art. 14.22), or for concrete for areas subject to abrasion (Art. 11.21). For such and similar cases, special qualities may be required in the aggregates in order that the service requirements be met.

It is the purpose of this chapter to discuss some of the properties of aggregates which have a bearing on the production of serviceable concretes. Three types of properties should be recognized: One group comprises those physical characteristics on which it is desirable to have information for use in the calculation of proportions of mixtures. A second group includes properties which affect the durability of concrete. The third group of properties has to do with special requirements.

3.2 General characteristics. Aggregates may be generally classified as to source, as to mineralogical composition, as to mode of preparation, and as to size. Such classifications serve principally as an aid in becoming familiar with types of aggregate or in identifying particular lots of aggregate. The acceptance of aggregates for use on the job should, in the final analysis, be governed by specific information regarding their quality.

With reference to source, aggregates may be natural or artificial, i.e., the substance of which the particles are composed may be the result of natural processes, or it may have been produced by some industrial process. The natural sands and gravels are the product of weathering and the action of running water, while the "stone" sands and crushed stones are reduced from natural rock by the crushing and screening of quarried material. Natural aggregates may be derived from any or all the rock types: igneous, sedimentary, or metamorphic; however, it should be noted that not all members of these geologic groups make satisfactory aggregates for concrete. Some natural aggregates are preponderantly siliceous in composition, while others are calcareous, but the form in which the principal minerals occur, and the presence of secondary, or accessory, minerals are more important than the average composition.

Artificial aggregates are usually produced for some special purpose, such as, for example, burned, expanded clay aggregates for making lightweight concrete. Some artificial aggregates are a by-product of an unrelated industrial process, such as cinders and blast-furnace slag. For making heavy-weight concrete, steel rivet punchings have been used. It may be noted, however, that natural aggregates may be used for special concretes, such as pumice for lightweight concrete, and magnetite (iron ore) for heavy-weight or radiation-shielding concrete.

In some localities and for some types of work, pit-run aggregates are used, but on important jobs where it is desired to control the quality of the concrete, the aggregates are processed by washing or cleaning and by screening. Crushing may also be involved in the case of quarried stone and oversize gravel.

Aggregate smaller than about ¼ in. in diameter is classified as fine aggregate or sand [C33]. However, some commercially used sands have maximum sizes considerably smaller than this size. There are frequently available for concrete work two or three grades of sand and several sizes (size groups) of coarse aggregate, e.g., No. 4 to ¾-in., ¾ to 1½-in., 1½ to 2½-in., etc. ASTM Specifications C33 and D448 list size groups to which many producers conform, but not all these groups may be available in any one locality.

The principal qualifications of aggregates for concrete are that they be clean, hard, tough, strong, durable, and of the proper gradation [C33]. Under cleanness comes the requirement of freedom from excess of silt, soft or coated grains, mica, harmful alkali and organic matter. Tests have been developed for detecting or measuring these undesirable constituents. Hardness is determined by an abrasion test, toughness by an impact test. The strength may be determined by crushing tests of rock cores, or may be evaluated by comparative tests of mortars or concretes made of the aggregate in question and aggregates of known or standard quality. The durability or soundness may be determined by a freezing-thawing test or a test involving the alternate soaking in a sodium or magnesium sulfate solution and drying. The gradation of the particles may be determined by a "sieve" analysis. Fortunately, in most regions of the country, there is usually little difficulty in securing materials which meet the necessary requirements. The question of more or less satisfactory materials is largely a matter of cost.

See Art. 12.10 for the effect of type and grading of aggregates upon shrinkage and expansion of concrete due to moisture changes, Art. 12.25 for effects of similar conditions upon creep of concrete with time, and Art. 12.20 for effect of kind of aggregate upon thermal volume changes.

3.3 Data needed for proportioning mixtures. For the calculation of batch quantities, for making mix adjustments, for computing effective water-cement ratios and yields and for making estimates of quantities required for jobs, some of or all the following information regarding the aggregates is needed:

1. Specific gravity of aggregate particles
2. Unit weight of aggregate in bulk
3. Free moisture and absorption
4. Gradation of aggregates

While the basic nature of the above-listed properties is familiar to the engineer from studies of physics and from job experience, special usages have developed in regard to their determination in concrete construction practice. It appears desirable, therefore, for the engineer to have a clear concept of the basic reasons for the particular usages.

Figure 3.1 represents, in idealized form, an aggregation of mineral particles. If the void spaces between the particles were filled with cement paste, the resulting combination would be concrete. The total volume of concrete, then, equals the volume of the paste[1] plus the space occupied by the aggregate *particles*. The space occupied by the particles, designated herein as the "solid volume" of the aggregate, is conveniently determined from the weight and the specific gravity of the particles. However, a particle itself may contain voids or pores, some entirely enclosed, and some connected to the surface as illustrated diagrammatically in Fig. 3.2. Since the cement paste is fairly viscous and cannot penetrate the smaller surface pores, and in no case can it enter the isolated pores, the calculation of the displacement of the particles in a mass of cement paste must take this condition into account. In effect, the solid volume of a particle that is desired is that which would be enclosed within a membrane stretched over the particle and conforming to the main outlines of the periphery as shown in Fig. 3.2. The solid volume of a given mass of aggregate is the

[1] In this discussion, it is assumed that the paste comprises cement + water + *entrapped air*.

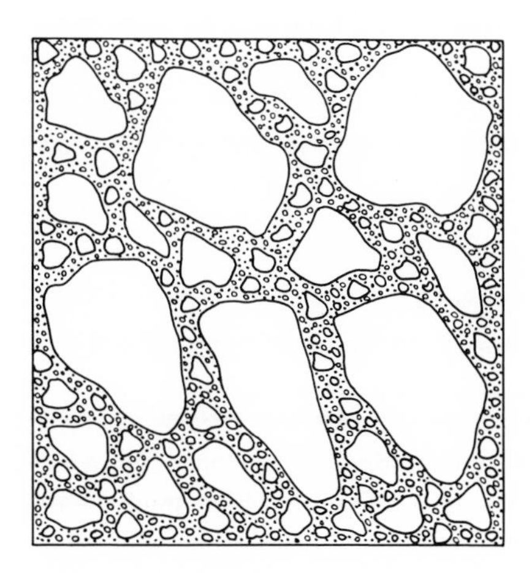

Fig. 3.1. An aggregation of mineral particles such as used in a concrete mix.

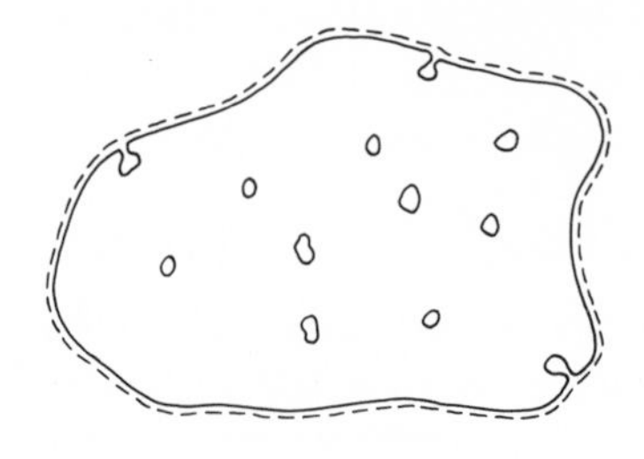

Fig. 3.2. Solid volume of aggregate. All voids within the dotted line are included in solid volume.

sum of the solid volumes of all the particles. Obviously it is less than the overall or "bulk volume" of the mass.

Quantities of aggregate, for the purpose of batching or estimating, may be indicated by either weight or bulk volume. When accurate control of batch quantities is desired, the weight basis is generally accepted as preferable. To convert from weight to volume, it is necessary to know the *unit weight,* normally expressed in pounds per bulk cubic foot.

The aggregate which is batched at the concrete mixer may carry more or less water. Some of this water may be present in such form as to contribute to the dilution of the cement paste, in which case it is called "free," or "surface," water. Water that is not available to become part of the mixing water is designated "absorbed" water. At the other extreme the aggregate may be so dry that it absorbs some of the mixing water. The neutral state in which water is neither contributed to nor withdrawn from the mixing water is taken to be that when the aggregate particles are saturated but surface-dry. Hence in stating moisture contents, and in making calculations involving moisture carried by the aggregate, recognition should be given to these conditions.

Aggregates are usually supplied for use on the job in two or more size groups or "sizes." Methods that have been or are used to choose the relative proportions of the various sizes are as follows (see also Chap. 6):

1. Trial mixes of concrete to obtain maximum economy with good workability
2. Empirical criteria based upon
   a. Unit weights or void contents
   b. Sieve analyses and grading diagrams
3. Trial mixtures of dry aggregates to obtain maximum density
4. Rule-of-thumb ratios

The first method is probably the most satisfactory in the long run, since the final criterion of optimum proportions is a concrete which most nearly possesses the necessary economy and workability. However, with more than two sizes of aggregate the number of trials may become large, so that familiarity with the significance of unit weights and gradations (method 2) serves as an excellent guide to the most desirable combinations. The third method may also serve as a guide, except that, as has been pointed out, maximum density of dry aggregate does not give optimum workability in a concrete mixture, so that either trial mixes or empirical modification of proportions is still necessary. Rule-of-thumb ratios should be used only on small jobs where quality is not paramount and it is not economically feasible to undertake a study of the proper proportioning of the aggregates.

3.4 Sampling aggregates. Before determining the properties of an aggregate, it is essential that a sample be obtained which is truly representative of the material to be used. As the procurement of the sample is complicated by segregation which occurs when it is handled, it is essential that the method of sampling compensate for segregation. Reference D75 describes some recommended methods.

Sand often contains some free moisture which tends to gravitate to the bottom of the mass if allowed to stand for some time. This moisture tends to prevent segregation of sand. However, the coarse particles of any pile of *dry* aggregate, having a sloping top surface, tend to roll toward the bottom of the slope.

For sampling sands and some coarse aggregates stored in a pile, a tube sampler should be used. It is usually a steel pipe about 2 in. in diameter and 6 ft long, having a point at one end and a handle at the other. It has a series of openings along one side, with a slight projection or ear at each hole. After forcing the pipe horizontally into the aggregate, it is rotated so that the ears scoop some aggregate into the tube which is then withdrawn while keeping the openings on top. Samples should be taken at several locations. If a tube is not available, use a shovel to obtain several samples from well beneath the surface. The use of a board held just above the point of sampling will serve to exclude unwanted surface material. For sampling aggregates on a conveyor belt, it is best to stop the belt, if possible, and to remove all material from a short length of the belt at various points. If it is not possible to stop the belt, sample the discharged material at selected intervals. For aggregates in a bin, select a series of samples at intervals at the discharge chute. In all cases the number and size of samples will depend on the quantity and the uniformity of the aggregate. It should be collected in clean, sturdy bags, or other containers, and adequately marked.

Samples of sand should be reduced to test size by use of a sand splitter or by the quartering method, but only the latter should be used for coarse aggregate. The quartering method, shown in Fig. 3.3, involves flattening the sample, dividing into quarters, discarding two diametrically

Fig. 3.3. The quartering method. It is used to reduce large sample of aggregates to representative smaller-sized samples for testing. *(Portland Cement Association.)*

opposite quarters, and combining the two remaining quarters, taking care to include the dust and fines with each quarter. This process is repeated until the size of sample is reduced sufficiently.

3.5 Specific gravity. The specific gravity of a substance is, in effect, the ratio of the unit weight of the substance to the unit weight of water. In the metric system, the unit weight of water (under standard conditions) may be taken as unity; hence the unit weight of a substance, in grams per cubic centimeter, is numerically equal to the specific gravity. In the English system, the unit weight of a substance (in pounds per cubic foot) is divided by 62.4 (the uint weight of water) to obtain the specific gravity.

Applied to concrete aggregates, the term specific gravity customarily refers to the density of the individual particles, not to the aggregated mass as a whole. As mentioned in the preceding article, in some aggregates the particles are dense and impermeable, while in others they may be porous. In connection with concrete mix calculations it is desired to know the space occupied by the aggregate particles within the relatively thick cement paste regardless of whether or not pores or internal voids exist within the particles. Hence, there is determined what is called the bulk specific gravity of the particles.

The bulk specific gravity is defined as the ratio of the weight in air of a given volume of a material (including both the permeable and impermeable voids normal to the material) at the standard temperature to the weight in air of an equal volume of distilled water at the standard temperature.[1]

In concrete testing 20°C (68°F) is taken as the standard reference temperature. Fluctuations of ±10°F will not cause an error greater than that normally tolerable in concrete testing. Within the limits of accuracy feasible in the testing of aggregates, the use of tap water is usually acceptable. To determine the true (or "absolute") specific gravity of the mineral matter composing particles which contain internal voids, it would be necessary to reduce it to a powder sufficiently fine to eliminate the effect of any impermeable voids; however the specific gravity on this basis serves no useful purpose in connection with concrete manufacture.

Standard procedures for the determination of the bulk specific gravity of coarse and fine aggregate are given in ASTM Specifications C127 and C128. The procedures described in Part II, Test 5, are essentially in accord with these specifications.

For use in the computation of concrete mixes the bulk specific gravity is always determined for saturated surface-dry aggregates. The specific gravities of a number of commonly used aggregates fall within the range 2.6 to 2.7, although there are satisfactory materials for which the specific gravity falls outside this range. The specific gravities of a few types of aggregates are given in Table 3.1.

The specific gravity of a material multiplied by 62.4 gives the weight of 1 cu ft of that substance; this is sometimes called the "solid-unit weight." The weight of a given quantity of particles divided by the solid-unit weight gives the "solid volume" of the particles (inclusive of internal pores in the particles):

$$\text{Solid volume, cu ft} = \frac{\text{wt, lb}}{\text{solid-unit wt, pcf}} = \frac{\text{wt, lb}}{\text{sp gr} \times 62.4} \tag{3.1}$$

For example, if a gross cubic foot of aggregate weighs 105.0 lb, and the aggregate particles have a specific gravity of 2.65, the space occupied by the aggregate particles alone is:

$$\text{Solid volume} = \frac{105.0}{2.65 \times 62.4} = \frac{105.0}{165.4} = 0.635 \text{ cu ft}$$

[1] Other definitions of specific gravity adopted by the ASTM [E12] are as follows:

1. Absolute specific gravity is the ratio of the weight referred to vacuum of a given volume of the material (without pores) at a stated temperature to the weight referred to vacuum of an equal volume of gas-free distilled water at a stated temperature.

2. Apparent specific gravity is the ratio of the weight in air of a given volume of the impermeable portion of a permeable material (that is, the solid matter including its impermeable pores or voids) at a stated temperature to the weight in air of an equal volume of distilled water at a stated temperature.

In a batch of concrete the sum of the solid volumes of the cement, aggregates, and water gives the nominal volume (exclusive of any air voids) of concrete produced per batch. Equation (3.1) is the basis for making this calculation.

Table 3.1 Specific gravities of various types of stone used for aggregates

	Bulk specific gravity	
Material	**Average**	**Range**
Sandstone	2.50	2.0–2.6
Sand and gravel*	2.65	2.5–2.8
Limestone	2.65	2.6–2.7
Granite	2.65	2.6–2.7
Trap rock	2.90	2.7–3.0

* Sands and gravels are usually a mixture of several kinds of rock materials, so the specific gravity will depend upon the preponderant type.

3.6 Unit weight and voids. Unit weight is the weight of a unit volume of aggregate, usually stated in pounds per cubic foot. In estimating quantities of materials, and in mix computations when batching is done on a volumetric basis, it is necessary to know the conditions under which the aggregate volume is to be measured: (1) loose or compact, and (2) dry, damp, or inundated. For general information and for comparison of different aggregates, the standard conditions are dry and compact; for scheduling volumetric batch quantities, the unit weight in the loose, damp state should be known. The unit weight for any given conditions may be determined by weighing the aggregates required to fill an appropriate container of known volume. The procedures for determining unit weight as given in Part II, Test 5, follow essentialy ASTM Specification C29.

With respect to a mass of aggregate, the term "voids" refers to the spaces *between* the aggregate particles; numerically, this void space is the difference between the gross, or over-all, volume of the aggregate mass and the space occupied by the particles alone. As indicated in the following paragraph, the void space can ordinarily be determined indirectly from the unit weight of the aggregate and the specific gravity of the particles [C30], although sometimes it is found by measuring the quantity of water required to inundate a sample of aggregate which has been compacted into a container.

The percentage of voids *between* the particles in a given gross or overall volume of aggregate can be determined by the expression

$$\text{Voids, \%} = \frac{\text{solid-unit wt} - \text{unit wt}}{\text{solid-unit wt}} \times 100$$
$$= \frac{62.4 \times \text{sp gr} - \text{unit wt}}{62.4 \times \text{sp gr}} \times 100$$

For example, if the unit weight of an aggregate is 105.0 pcf and the specific gravity of the particles is 2.65, the percentage of voids is

$$\text{Voids, \%} = \frac{62.4 \times 2.65 - 105.0}{62.4 \times 2.65} \times 100 = 36.5$$

or, from the previous example, in Art. 3.5, where the solid volume was 0.635 cu ft,

$$\text{Voids, \%} = 100 - 63.5 = 36.5$$

For a given specific gravity, the greater the unit weight, the smaller the percentage of voids, and hence the better the gradation of the particles.

A criterion that has been used for the selection of aggregate combinations for concrete mixtures is that the density (percentage of solids) should be as great as possible. The unit weight and the void content are functions of the density of an aggregate mass. Since the percentage of solids (or percentage of voids) are controlled to a considerable extent by the grading, shape, and surface texture of the particles, the unit weight and void content then serve as a rough index of the efficiency of the grading. However, since the workability of a concrete mix is also influenced by grading, and since gradings which produce maximum density of aggregates tend to produce harsh mixes, density of the aggregates alone cannot be taken as a final criterion.

In addition to the gradation, shape, and surface texture of particles, the unit weight is influenced by the specific gravity of the particles, the moisture condition of the aggregate, and the compactness of the mass. As a result of variations in these four factors, a rather wide range in values of unit weight is to be expected. An indication of the general range in unit weight for common natural aggregates is given in Table 3.2.

For a mass of one-sized spheres packed so that the centers of the spheres have a cubical arrangement (the loosest state), the void content is 48 percent; if the spheres are so packed that their centers lie at the apexes of imaginary tetrahedrons (densest state for one-sized particles), the void content is 26 percent. If the particles happen to range in size, the void content will then always be less than that of correspondingly arranged particles of uniform size. An indication of the range in void contents of ordinary sands and gravels is given in Fig. 3.4. In the experiments

Table 3.2 General range in unit weight of common natural aggregates

Material	Moisture condition	Unit wt, pcf Loose	Unit wt, pcf Compact
Sand	Dry	90–100	95–115
	Damp	85–95	
Gravel, No. 4–¾-in	Dry or damp	92–98	99–107
Gravel, No. 4–1½-in	Dry or damp	95–103	104–112
Mixed sand and gravel, 1½-in, max	Dry	. . .	110–125
	Damp	100–115	
Crushed stone, No. 4–¾-in.	Dry or damp	88–94	95–103
Crushed stone, No. 4–1½-in	Dry or damp	91–99	100–108

on which these data are based, the sand and the gravel were each segregated into three component size groups and then recombined. The results are plotted in the form of contours of equal void content on a trilinear diagram.[1] For the sand, minimum voids amounting to 32 percent occurred with about 30 to 50 percent of the finest component. For the gravel, minimum voids of 30 percent occurred with about 20 to 40 percent of the finer component. An indication of the void contents of various combinations of sand and coarse aggregate is given in Fig. 3.5. The minimum void content occurs with mixtures containing roughly about 40 to 60 percent sand, but the amount of sand to give a minimum varies with the maximum size of aggregate and with the kind of coarse aggregate, gravel, or crushed stone. The minimum void content ranges from about 22 to 29 percent. For the dry loose state, the void content is generally about 5 to 7 percent greater with sand and 3 to 4 percent greater with gravel, than for the dry compact state. For aggregates of normal range in specific gravity (2.55 to 2.65), a unit weight of 100 pcf corresponds to a void content of about 37 to 39 percent, and an increase in unit weight of. 5 pcf corresponds to a decrease in voids of about 3 percent.

[1] A point on a trilinear chart shows the percentage of each component of a three-component system. Values increase from 0 to 100 from the base line for each component to the opposite apex. For convenience, the scale of values is shown along one of the edges, as indicated in the diagram at the right.

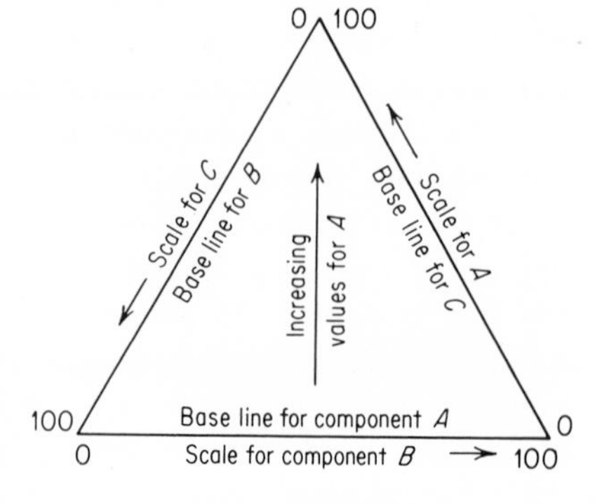

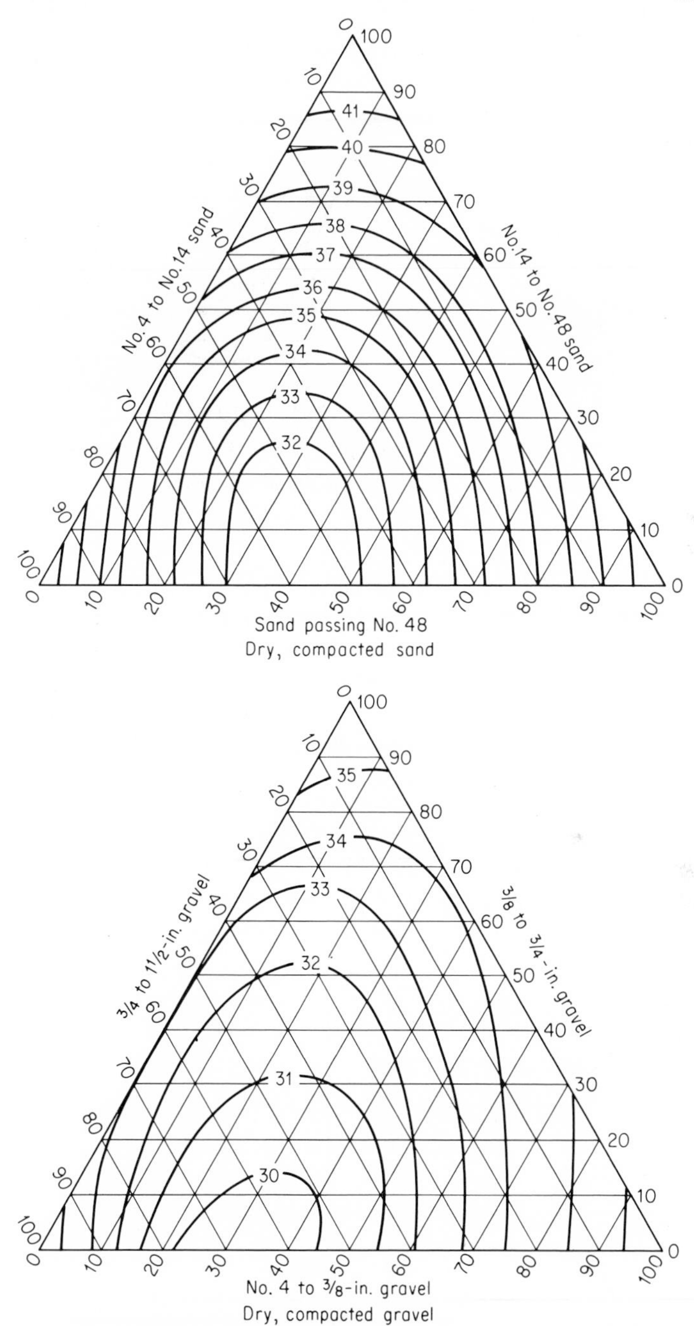

Fig. 3.4. Void contents of sands and gravels [302].

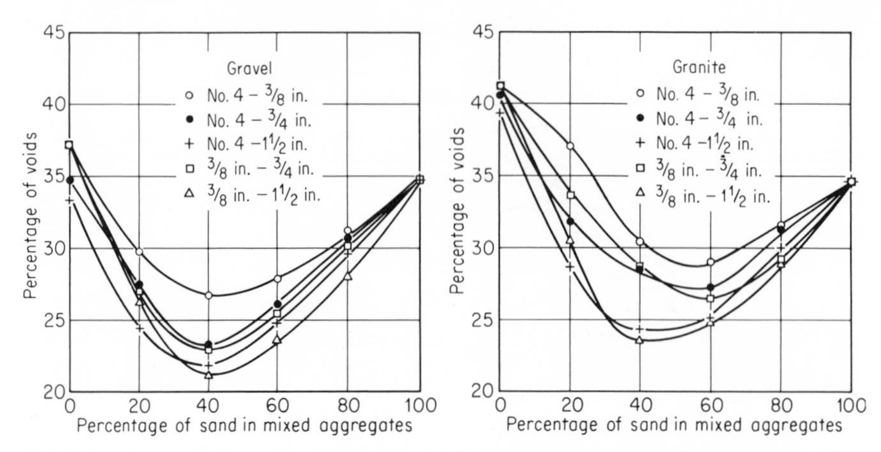

Fig. 3.5. Void contents of combinations of dry, compacted fine, and coarse aggregates [302].

In fine aggregate, free or surface moisture holds the particles apart and prevents them from adjusting themselves to occupy a minimum volume; hence, there may result a marked decrease in the weight of aggregate in a given measured volume and a very material increase in the percentage of voids. This phenomenon is known as "bulking." Up to a free-water content of about 4 percent for some coarse sands and 8 percent for some fine sands, the more free water, the greater the bulking. The finer the

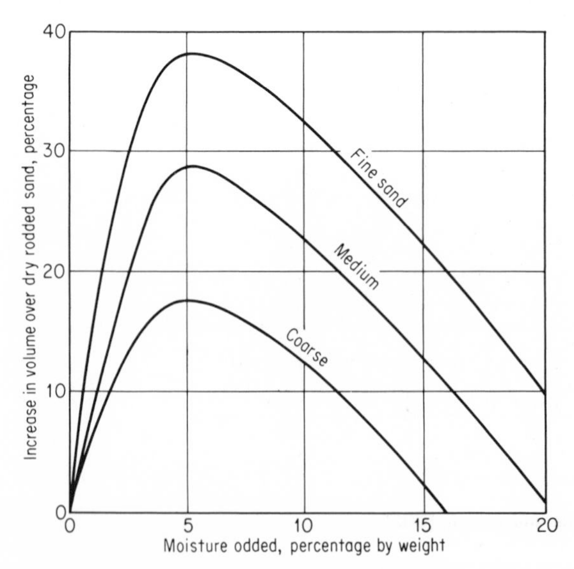

Fig. 3.6 Effect of moisture on bulking of sand. *(Portland Cement Association.)*

aggregate, the more pronounced the bulking. The maximum increase in volume may amount to as much as 40 percent for a fine sand or 25 percent for a coarse sand. The corresponding reduction in unit weight may be about 25 percent for a fine sand or 15 percent for a coarse sand. Coarse aggregates exhibit negligible bulking due to moisture. The result of some tests which demonstrate the bulking of fine, medium, and coarse sands are shown in Fig. 3.6.

To compute the number of cubic feet of damp, loose aggregate corresponding to 1 cu ft of dry, rodded aggregate, use is made of a ratio called the "bulking factor." After determining the unit weight under both conditions as well as the moisture content of the damp aggregate, then

$$\text{Bulking factor} = \frac{\text{unit weight of surface-dry, rodded aggregate}}{\begin{pmatrix}\text{unit weight of damp,}\\ \text{loose aggregate}\end{pmatrix} - \begin{pmatrix}\text{weight of surface}\\ \text{moisture in unit}\\ \text{volume of damp,}\\ \text{loose aggregate}\end{pmatrix}}$$

Table 3.3 shows the method of computing the bulking factor for a damp, loose sample of a fine aggregate and of a coarse aggregate.

Sand which is completely submerged or "inundated" shows no bulking. The volume of inundated sand is practically the same as that of surface-dry

Table 3.3 Bulking of aggregate in damp, loose condition

Item	Sand	Coarse aggregate
1. Wt of damp sample, oz	35.0	32.5
2. Wt of oven-dried sample, oz	33.0	31.7
3. Wt of water in damp sample (item 1 — item 2)	2.0	0.8
4. % of total moisture in terms of dry aggregate $\left(\frac{\text{item 3}}{\text{item 2}} \times 100\right)$	6.0	2.5
5. % of absorption (assumed)	1.0	1.0
6. % of surface moisture (item 4 — item 5)	5.0	1.5
7. Wt per cu ft damp, loose, lb (by test)	97.2	94.4
8. Wt of surface-dry aggregate in 1 cu ft of damp, loose material, lb $\left(\text{item 7} \times \frac{100}{100 + \text{item 6}}\right)$	92.6	93.0
9. Wt of water in 1 cu ft of damp, loose material, lb (item 7 — item 8)	4.6	1.4
10. Wt per cu ft of surface-dry compact aggregate, lb (by test)	112.0	99.0
11. Bulking factor $\left(\frac{\text{item 10}}{\text{item 8}}\right)$	1.21	1.06

material. Inundated sands have a higher unit weight than dry aggregates because of the 15 to 25 percent of water required for inundation.

The unit weight of lightweight aggregates may vary over a considerable range, depending upon the type, grading, and source of the material. Aggregates made of ordinary blast-furnace slag may range in unit weight from 60 to 85 pcf. Experience indicates that concrete made with slag aggregate which weighs less than about 70 pcf compacted is rather difficult to place; also, concrete made with slag aggregate having a unit weight less than about this value tends to be somewhat lower in strength than concrete made with heavier material. Standard specifications place a minimum limit of 70 pcf on slag for general use in concrete [C33]. Specifications for lightweight aggregates for structural purposes call for a maximum unit weight (dry, loose) of 55 pcf for the coarse, 70 pcf for the fine, and 65 pcf for the combined aggregate [C330]. For insulating concretes the unit weights should be appreciably lighter [C332], but their strength is reduced correspondingly. The unit weight for coarse expanded blast-furnace slag, clay, shale, slate or diatomite aggregates sized from No. 4 to ½ in. usually varies within the range 40 to 50 pcf, and for fine aggregates of similar materials it may vary from 55 to 65 pcf. For coarse natural pumice of size No. 4 to ¾ in. the unit weight may range from 30 to 40 pcf, and for fine pumice passing the No. 8 sieve it may range from about 50 to 65 pcf.

3.7 Moisture and absorption. As regards moisture content, the various states in which an aggregate may exist are:

1. Oven-dry—all moisture, external and internal, driven off, usually by heating at 100 to 110°C
2. Air-dry—no surface moisture on the particles, some internal moisture, but particles not saturated
3. Saturated surface-dry—no free or surface moisture on the particles, but all voids within the particles filled with water
4. Damp, or wet—saturated and with free or surface moisture on the particles

The various moisture conditions are represented diagrammatically in Fig. 3.7.

In a concrete mix, if the aggregates are not fully saturated, some of the mixing water is absorbed; on the other hand, free moisture on the surfaces of the aggregate particles becomes a part of the mixing water. Hence in computations of the net or effective ratio of water to cement, and in calculations relating to proportions of mixes by weight, the saturated surface-dry condition is used as the basis.

On the job, the fine aggregates usually carry some free moisture.

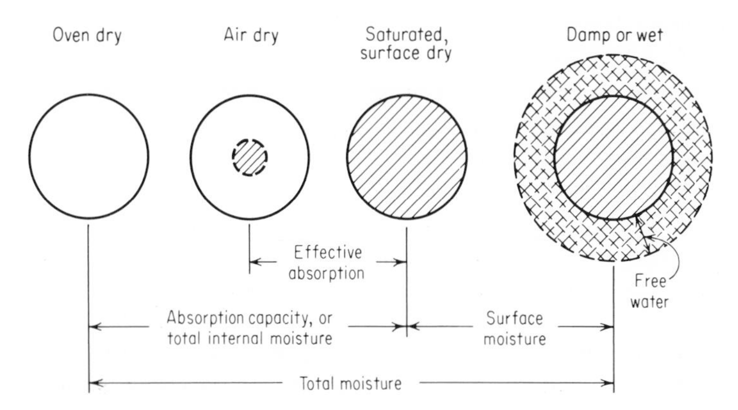

Fig. 3.7. States of moisture in aggregate. Heavy circle represents the aggregate; cross-hatching represents moisture.

Freshly washed coarse aggregates contain free water, but since they dry quickly, they are ordinarily partially dry when used.

The total internal moisture content of an aggregate in the saturated surface-dry condition may be termed the "absorption capacity," although it is sometimes referred to simply as the "absorption" [C127, C128]. The amount of water required to bring an aggregate from the *air-dry* condition (see Fig. 3.7) to the saturated surface-dry condition is herein termed the "effective absorption."

The absorption capacity is determined by finding the weight of a surface-dried sample after it has been soaked for 24 hr and again finding the weight after the sample has been dried in an oven; the difference in weights, expressed as a percentage of the *dry* sample weight, is the absorption capacity [C127, C128]. Coarse aggregates are considered to be surface-dry when they have been wiped free of visible moisture films with a cloth. The saturated surface-dry condition of fine aggregate is usually taken as that at which a previously wet sample just becomes free-flowing. As drying proceeds, an indication that this point is reached is given when the material in a pile, formed by a small conical mold under specified conditions, slumps upon removal of the mold (see Part II, Test 5; 306; C128).

The absorption capacity is a measure of the porosity of an aggregate; it is also used as a correction factor in determinations of free moisture by the oven-drying method. Approximate values of the absorption capacity of several types of aggregate are given in Table 3.4.

The effective absorption of an air-dry aggregate can be determined by either of two methods: By one method there is determined the gain in weight between the air-dry and the saturated surface-dry conditions after water soaking for, say, ½ hr. By the second method there is found the difference between the absorption capacity and the moisture content of the

Table 3.4 Approximate absorption capacities of various types of stone used for aggregates

Material	Absorption capacity, % by wt
Average concrete sand	0–2
Average gravel; crushed limestone	½–1
Trap rock; granite	0–½
Sandstone	2–7
Very light porous materials	Up to 25

air-dry sample, as determined by oven-drying. These two methods give slightly different results owing to the difference in soaking periods employed.

The free moisture is commonly determined by drying or by displacement methods, although other methods have been employed [304, 305]. When drying is employed, the sample is completely dried out, and the absorption capacity is subtracted from the total percentage of moisture in the aggregate to obtain the free moisture. In the laboratory, oven-drying is most commonly employed; in the field, drying is sometimes accomplished by mixing the sample with some volatile solvent such as alcohol and burning off the solvent, or by use of a Speedy moisture tester in which a chemical is used to react with the surface moisture and develop a pressure which is a measure of that moisture.

In obtaining the free moisture by displacement methods, the bulk specific gravity on a saturated surface-dry basis is employed. Determinations are made by the use of a specially constructed flask [C70], by the use of a pycnometer (see Part II, Test 5), or simply by weighing the sample under water [536]. The ASTM flask method is used only for fine aggregate. The pycnometer, also, is ordinarily used only for fine aggregate.

The finer the aggregate, the more free water it can carry. Approximate moisture contents of average job materials are shown in Table 3.5.

3.8 Gradation. A suitable gradation of the combined aggregate in a concrete mix is desirable in order to secure workability and to secure economy in the use of cement. For mixes of given consistency and cement content a well-graded mixture produces a stronger concrete than a harsh or poorly graded one, since less water is required to give suitable workability. Fortunately, however, provided proper adjustments are made in the mix, a fairly wide latitude in grading may be tolerated without seriously affecting the properties of the resulting concrete.

For some types of work, if suitably sized commercial materials are available, the aggregates may be combined in arbitrary proportions, based upon experience. On important jobs, however, it has been found advantageous to make special studies to determine the desirable combination of materials.

Table 3.5 Approximate moisture contents of common aggregates

Aggregate	Approximate surface moisture	
	Wt, %	Gal/cu ft
Moist gravel or crushed stone	½–2	1/16–¼
Moist sand	1–3	⅛–⅓
Moderately wet sand	3–5	⅓–½
Very wet sand	5–10	½–1

Most attempts to secure optimum gradings have been directed toward attaining a combination of materials to produce maximum density consistent with good workability of concrete, and minimum cement requirement for concrete of given consistency. Three general means may be employed to approach this result: (1) calculated combinations based upon filling the voids in the aggregate with successively smaller sizes of material, (2) combinations calculated from sieve analyses, and (3) trial combinations. In general it has been found that a "continuous" grading including all size groups is the most satisfactory, although it is not desirable that equal amounts be retained on each sieve. Some claim has been made as to the merits of "gap" gradings, i.e., gradings in which one or more than one intermediate size group is omitted. In any event, a grading in which one particular size is present in excess has always been found to be harsh, giving rise to what has been called "particle interference" [330]. Excessively fine sands are to be avoided, as they are not as economical as well-graded coarse sands. The general appearance of a fine sand and a well-graded coarse sand is shown in Figs. 3.8 and 3.9, which also show how these sands look when separated into several sizes.

3.9 Sieve analyses. The particle-size distribution of an aggregate is determined by a "mechanical analysis," which is usually made by shaking the material through a series of sieves, nested in order, with the smallest on the bottom. These sieves have square openings and are usually constructed of wire mesh. General specifications for sieves for testing purposes are given in ASTM E11.

In the testing of concrete aggregates, there is generally employed a series of sieves in which any sieve in the series has twice the clear opening of the next smaller size in the series. Two series of sieves conforming to this relation are the U.S. Series and the Tyler Series. Commonly used sieves, with both the U.S. and Tyler designations are given in Table 3.6. Sometimes closer sizing than is given by these sieve series is desired, in which case intermediate or "half" sizes are employed; the ½-in. and 1-in.

Fig. 3.8. Sample of poorly graded sand. Lower part of figure shows sand separated into four sizes finer than $\frac{1}{16}$ in. Ordinarily such a sand is not as economical as a coarser sand. *(Portland Cement Association.)*

Fig. 3.9. Sample of well-graded sand. Lower part of figure shows sand separated into 6 sizes from fine up to $\frac{1}{4}$ in. This is a good sand for concrete work. For good workability, at least 10 percent should pass a No. 50 sieve. *(Portland Cement Association.)*

Table 3.6 Sieve series commonly used for analysis of aggregates for concrete

U.S. Series* sieve size	Clear opening, in.	Tyler Series† sieve size	Clear opening, in.
No. 100	0.0059	No. 100	0.0058
No. 50	0.0117	No. 48	0.0116
No. 30	0.0232	No. 28	0.0232
No. 16	0.0469	No. 14	0.046
No. 8	0.0937	No. 8	0.093
No. 4	0.187	No. 4	0.185
3/8-in.	0.375	3/8-in.	0.371
(1/2in.)‡	0.500	(1/2-in.)‡	0.525
3/4-in.	0.750	3/4-in.	0.742
(1-in.)‡	1.00	(1-in.)‡	1.05
1 1/2-in.	1.50	1 1/2-in.	1.50

* From Ref. E11.
† Except for half sizes, openings are within tolerances permitted by E11.
‡ Half sizes.

sieves shown in Table 3.6 are half sizes. The sieve analysis procedure outlined in Part II, Test 4, employs either the U.S. or Tyler series of sieves and follows the major requirements of ASTM C136.

The data from sieve tests are recorded and reduced in tabular form; the quantities usually shown in successive columns are (1) weight retained on each sieve, (2) percent retained on each sieve, and (3) total (cumulative) percent *coarser* than each sieve or total percent *passing* each sieve. The results of the tests are commonly reported in terms of item 3 above, as illustrated in Table 3.7.

From the results of a sieve analysis, a factor called the "fineness modulus" is sometimes computed. The fineness modulus is the sum of the percentages in the sieve analysis divided by 100, when the sieve analysis is expressed as total percentages (by weight or absolute volume) coarser than the square mesh sieves (not including half sizes) in the Tyler or U.S. series. Examples of calculation of the fineness modulus are given in Table 3.7 for aggregates up to 1½-in. size, but the 3 and 6-in. sieves would be used for aggregates up to the 6-in. size.

An interpretation of the fineness modulus might be that it represents the (weighted) average sieve of the group upon which the material is retained, No. 100 being the first, No. 50 the second, etc. Thus, for a sand with a fineness modulus of 3.00, sieve No. 30 (the third sieve) would be the average sieve size. It should also be noted, in this connection, that there are an infinite number of gradings which will produce a given fineness modulus. This is a particular disadvantage of the use of this factor, and

Table 3.7 Examples of sieve analyses and calculation of fineness modulus of aggregates

	Cumulative percent coarser than given sieve				
(1) Standard sieve size	(2) Sand, medium fine	(3) Sand, medium coarse	(4) Gravel, nominal No. 4 to ¾-in.	(5) Gravel, nominal No. 4 to 1½-in.	(6) Combined aggregate,† 0 to 1½-in.
1½-in.	...	...	...	0	0
¾-in.	...	...	2	49	29
⅜-in.	...	0	59	81	49
No. 4	0	4	99	100	62
No. 8	9	15	100	100	66
No. 16	28	37	100	100	75
No. 30	49	62	100	100	85
No. 50	79	85	100	100	94
No. 100	96	98	100	100	99
Fineness modulus	2.61	3.01	6.60	7.30	5.59

* Suggested as an "all-service" grading for sand for concrete.
† 40 percent sand (col. 3) + 60 percent gravel (col. 5). Values in col. 6 equal 40 percent of col. 3 plus 60 percent of col. 5.

therefore the fineness modulus should be used only in comparing materials whose gradings are similar.

In combining separate aggregates on the basis of sieve analyses, two methods have been used to compute the relative proportions of the various size groups: By one method the proportions are selected so that the gradation of the combined aggregates falls within desired limits or approximates some "ideal" grading. Such an ideal grading is one which has been found by experience or by test to represent, for the given conditions (kind and maximum size of aggregate, cement content, and consistency), a combination producing concrete of optimum quality [607]. By the second method, the proportions of the component sizes of aggregate are selected so that the fineness modulus of the combined aggregate lies within limits which have been found desirable by test [600].

Since the cement content of a mixture will affect the plasticity or workability thereof, it follows that the richer the mix, the harsher may be the gradation of the aggregates, taken by themselves. For a given type of work, however, (e.g., building construction) the richness of mix usually varies over a relatively small range, so that ordinarily grading limits may be established for such conditions without regard to cement content.

It should be noted in passing that since the particles are three-dimensional and the sieve openings are two-dimensional, the method of separation by sieving is at best an approximation (though in many cases a good one if representative samples are obtained). For this reason it is desirable to check calculated proportions by making up a trial mix. It should also be noted that in practice, since commercially available stock sizes which are not sized between individual sieves of any series must be employed, it is impossible to conform absolutely to any ideal grading.

3.10 Grading charts. A grading chart (see Fig. 3.10) is a useful method of showing the grading of individual and combined aggregates. The ordinates are commonly constructed to read total percentage coarser (or passing), 0 to 100, from bottom to top. In this text values coarser than each sieve are shown as they are used for computing the fineness modulus discussed previously. The abscissas are the sieve openings plotted to some scale. Although arithmetic scales and even warped scales are sometimes used, a "logarithmic" plot is very convenient, since in the regular sieve series, with the consecutive sizes of opening related by a constant ratio, logarithmic spacing is equal spacing. Hence from left to right along the chart, lines at equal intervals represent the successive sieves in the series. Figure 3.10 is constructed in this way. It is recommended that the spacing between the sieve sizes be approximately equal to 10 percent on the vertical scale. Sieve sizes intermediate between those in a given series may be located by the ratio of the differences of the logarithms of the openings concerned. Thus, the nominal 1-in. sieve would be represented by a line halfway between the lines representing the ¾-in. and 1½-in. sieves.

The points representing the results of an analysis are connected to form the "grading curve" of the given aggregate. This, if plotted as indicated above, is sometimes called "the cumulative logarithmic plot." The data given in Table 3.7, cols. 3 and 5, are so plotted in Fig. 3.10.

A combined grading curve may be constructed from the data of the individual curves as follows: (1) Find the percentage coarser than a particular sieve size contributed by each individual aggregate by multiplying its corresponding ordinate by the percentage of the given aggregate to be used; (2) add the partial percentages so obtained for each sieve size to get the ordinate to the curve for the combined aggregate; (3) remember that if none of an aggregate is coarser than a given sieve, the ordinates for that and all larger sieves are zero percent. Suppose, for example, that a sand and gravel, whose grading curves are shown in Fig. 3.10, are to be combined in the ratio 40:60. The ordinate to the combined curve at the No. 8 sieve would be $0.40 \times 15 + 0.60 \times 100 = 66$. At the ⅜-in. sieve, the ordinate would be $0.60 \times 81 = 49$.

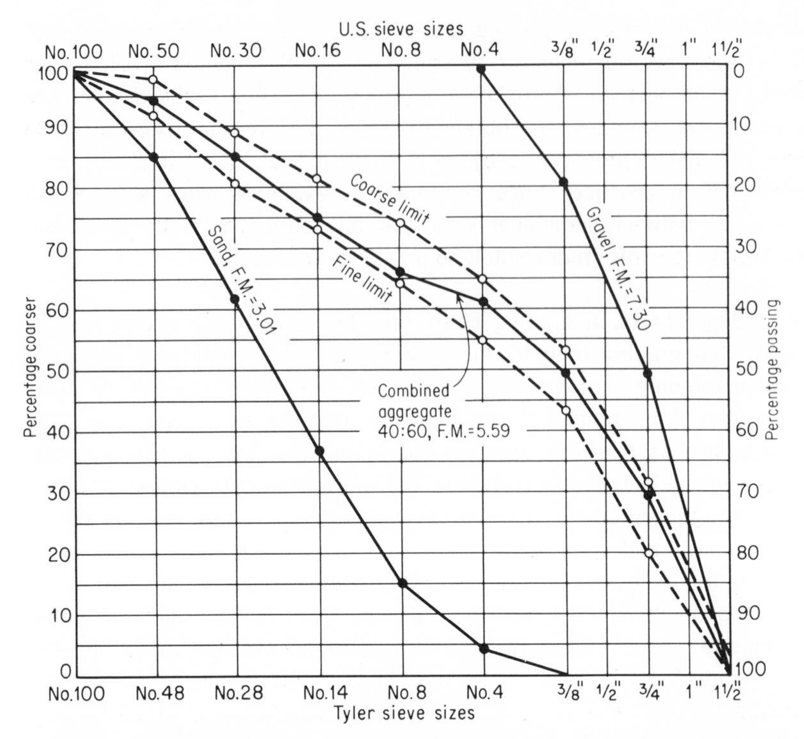

Fig. 3.10. Aggregate gradation chart. Data shown are for illustrative purposes only. Fine and coarse limits are for use in Test 4 (Part II). See Table 3.8 for computation of fineness modulus of combined aggregate.

If an ideal grading curve has been established and plotted, the curve for the combined aggregates may be made to approach the ideal curve by using trial percentages of the individual gradings involved. If the idea of permissible fine and coarse limits is employed, a useful device is to represent these by dotted lines upon the chart, the lower line being the fine limit and the upper the coarse limit. The combined grading curve is then made to fall within these boundary lines. Suitable grading limits for a 1½-in. maximum size gravel aggregate for use in mixes of about 1:5 to 1:6 by weight are shown in Fig. 3.10 and in Table 3.8, col. 7.

3.11 Maximum size of aggregate. The maximum size of an aggregate is determined from its sieve analysis and is generally designated by the commercial sieve size next coarser than the largest size on which 15 percent or more is retained.

The maximum size of aggregate to be used depends upon where the concrete is to be placed. In building construction the maximum size should not be larger than one-fifth of the narrowest dimension of the form in which the concrete is to be used, nor larger than three-quarters of the minimum clear distance between reinforcing bars [1801]. In making concrete test cylinders, the maximum size of aggregate preferably should not be greater than one-quarter of the diameter of the mold, although one-third is allowed [C31]. The range in maximum size of aggregate for particular classes of work is indicated in Table 6.3.

Where no physical restrictions are placed upon size, the use of larger maximum sizes usually gives greater economy in the use of cement. Gonnerman [111] says on this point:

For a given maximum size there is a limit to the amount of aggregate that can be used with a given quantity and quality of paste and still maintain a given degree of plasticity. In order to increase the amount of aggregate that can be carried in suspension by a given cement paste it is necessary to increase the maximum size and thereby extend the range over which the particles are graded. [Figure 3.11], based on laboratory and field data, illustrates the influence of size of aggregate on the cement content of concrete of a given slump and quality of paste. Since this curve is based on carefully designed mixes using well-rounded aggregates, it represents the best results that have been obtained with the maximum size of aggregate indicated when incorporated in mixes of a consistency suitable for hand placing in large sections. This figure shows a marked economy in cement as the maximum size of aggregate is increased up to about 3 in., but beyond this point the use of the larger sizes would be justified only if the cost of obtaining them would be offset by the saving in cement. In the range of maximum sizes of 2 to 4 in., a careful study should

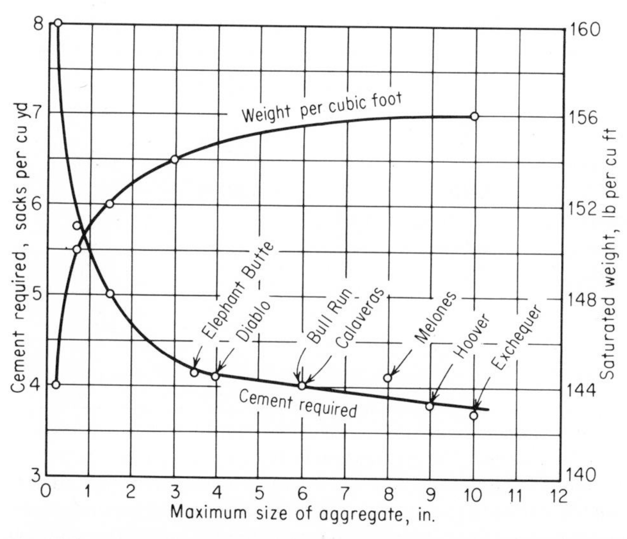

Fig. 3.11. Effect of size of aggregate upon cement requirement and unit weight of concrete of given water-cement ratio and consistency. Water-cement ratio = 6.5 gal per sack of cement. Slump 3 to 5 in. (*Portland Cement Association.*)

be made to determine whether it would be economical to introduce into the mix some of the larger sizes of particles that might be available. In this connection it is of interest to note that the U.S. Bureau of Reclamation has adopted 6 in. as the maximum allowable size of aggregate for its more recent work.

For additional information see Art. 14.16.

The use of a larger maximum size of aggregate reduces the voids in the coarse aggregate and results in a lower sand requirement in the combined aggregate, but sizes larger than 6 in. give rise to difficulties in handling and placing.

3.12 Grading requirements. The range in grading permitted for a concrete sand [C33] is shown in Table 3.8, col. 2. These are wide limits, and so far as many jobs are concerned, a closer grading might be written. In the interests of control of production of concrete, the gradation of sand

Table 3.8 Grading limits for aggregates

	Percent passing given sieve					
(1) Standard sieve size	(2) Fine aggregate, C33*	(3) Coarse aggregate,† No. 4 to ¾ in.	(4) Coarse aggregate,† No. 4 to 1 in.	(5) Coarse aggregate,† No. 4 to 1½ in.	(6) Coarse aggregate,† No. 4 to 2 in.	(7) Combined aggregate,‡ 0 to 1½ in.
2 in.				100	95–100	
1½-in.			100	95–100		98–100
1 in.		100	95–100		35–70	
¾ in.		90–100		35–70		68–80
½ in.			25–60		10–30	
⅜ in.	100	20–55		10–30		47–57
No. 4	95–100	0–10	0–10	0–5	0–5	35–45
No. 8	80–100	0–5	0–5			26–36
No. 16	50–85					18–27
No. 30	25–60					11–19
No. 50	10–30					2–8
No. 100	2–10					1–2

* The minimum percentages shown for material passing the No. 50 and No. 100 sieves may be reduced to 5 and 0, respectively, if the aggregate is to be used in air-entrained concrete containing more than 4½ bags of cement per cu yd, or in non-air-entrained concrete containing more than 5½ bags of cement per cu yd, or if an approved mineral admixture is used to supply the deficiency in percentages passing these sieves. The fineness modulus should not be less than 2.3 nor more than 3.1.

† From Ref. C33, but not all sizes covered by Ref. C33 are shown in this table.

‡ For use in Part II, Test 4; corresponds to grading for certain types of building construction

furnished to any particular job should remain reasonably uniform; a requirement with regard to uniformity is that variation in fineness modulus greater than 0.20 either way from that of the sand as originally accepted for a job shall be cause for rejection [C33].

Experience has shown that for mixes of given water-cement ratio the fine aggregate should contain a sufficient amount of material passing the No. 50 sieve if satisfactory workability is to be secured. The minimum amount of fines passing the No. 50 sieve varies with the composition of the mix. Mixes rich in cement require relatively small amounts, while leaner mixes require a considerable quantity of the fine material. A 6½-gal-per-sack mix requires about 10 percent fine aggregate passing the No. 50 sieve for satisfactory workability. When sand is deficient in fines, increasing the sand percentage, the common method of securing smoother working concrete, will not produce real workability or prevent segregation and bleeding except in rich mixes.

A desirable minimum limit for the material passing the No. 100 sieve is from 3 to 5 percent. Fines in sand must not be confused with silt, loam, or other impurities which are undesirable.

When sand having a large percentage of coarse particles is used, the coarse aggregate should contain but little material of the size of the maximum sand particles in order to avoid a so-called "popcorn," "grainy" concrete, difficult to finish. This may occur when screenings or pea gravel are used.

In general, for concrete work, the fineness modulus of a sand should lie between 2.6 and 3.1. There is no ideal grading for sand, since what is ideal for one set of conditions is not ideal for another. Subject to these limitations a suggested gradation for an "all-service" concrete sand is shown in Table 3.7, col. 3.

The U.S. Bureau of Reclamation specifications usually provide that the sieve analysis of a sand be within the following limits:

Sieve size	Percentage retained (individual)
No. 4	0–5
No. 8	5–15*
No. 16	10–25*
No. 30	10–30
No. 50	15–35
No. 100	12–20
Pan	3–7

* If the individual percentage retained on the No. 16 sieve is 20 percent or less, the maximum limit for the percentage retained on the No. 8 sieve may be increased to 20 percent.

Table 3.9 Suggested gradings of natural coarse aggregates of various maximum sizes*

Max size aggregate, in.	Percent of coarse aggregate in each fraction				
	No. 4 to 3/8 in.	3/8 to 3/4 in.	3/4 to 1 1/2 in.	1 1/2 to 3 in.	3 to 6 in.
3/4	27–45	55–73			
1 1/2	15–25	30–35	40–55		
3	10–15	15–25	20–40	20–40	
6	8–15	12–20	20–30	23–32	20–35

* From Ref. 106.

Permissible grading limits for coarse aggregates [C33] are given in Table 3.8, cols. 3 to 6. Desirable gradings for coarse aggregates, as recommended by the U.S. Bureau of Reclamation, are indicated in Table 3.9.

It may be noted that the specification limits for gradings are broad. While, as has been indicated, some variation in the gradation of aggregates does not seriously affect the quality of concrete, the selection of desirable and practicable gradings and of aggregate sizes presents a problem of some difficulty to those who are new in the field, particularly when questions of mix design are involved. For this reason discussions of some length have been given in these pages regarding grading requirements; the relative amount of space given should not be construed, however, as placing undue importance on grading as compared with requirements for strong, durable aggregates.

3.13 Quality requirements. Apart from those properties which contribute to the making of a good concrete mix, to render satisfactory service in concrete over a period of time an aggregate must be clean, sound, hard, and strong. The implications of these requirements as to quality are large. Essentially the problem is one of durability, but the factors which may promote disintegration are numerous and varied.

The principal factors which cause disintegration are weathering action (wetting and drying, freezing, and thawing—leading to mechanical disintegration); adverse chemical reactions between components of the cement and components of the aggregate, or between water and some of the components of the aggregate; and secondary effects of service loadings. For convenience, these factors will be discussed in relation to (1) deleterious substances, usually present in relatively small amount, which may be regarded as foreign inclusions in the main body of the aggregate, and (2) reactive aggregate minerals which are characteristic of the aggregate and remain a source of difficulty regardless of processing.

Numerous tests have been devised in attempts to detect the presence of the undesirable components of aggregates. Some are excellent indicators, but the fact remains that a service record is the best clue to aggregate quality. If a record of service is lacking, the next most valuable indices are accelerated durability tests on concrete. However, in the development of new sources of supply, and in the day-by-day control of supplies in which the quality may fluctuate, the aggregate tests are an important aid, especially if the engineer is familiar with their significance and limitations.

3.14 Deleterious substances. Harmful substances which may be present in aggregates may be classified as shown in Fig. 3.12 and as follows [110]:

1. Substances causing an adverse chemical reaction
2. Substances which undergo disruptive expansion
3. Clay and surface coatings
4. Particles having an unduly flat or elongated shape
5. Structurally soft and/or weak particles

In regard to adverse chemical reactions, one of the principal offenders is organic material, often in the form of humus, or organic loam. Contamination of this sort usually is found only in a fine aggregate. In some unusual cases, organic contamination has occurred through shipment in railroad cars previously loaded with sugar, which, in addition to certain other organic substances, inhibits or prevents portland cement from harden-

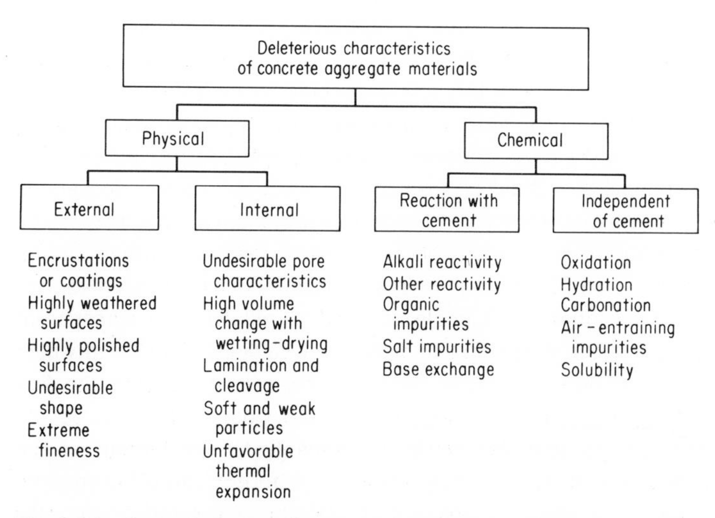

Fig. 3.12. Suggested classification of deleterious aggregates. (*Based on Swenson and Chaly* [310].)

ing. Coal particles and lignite may sometimes cause difficulty; the determination of the amount of coal, lignite, and wood or fiber present in an aggregate is made by separation in a liquid having a specific gravity of about 2.0 [C123]. For detection of organic impurities containing humic acids, a colorimetric test is used [C40]; however, in case of doubt, a strength test should be made on a mortar containing the suspected sand and the results compared with those of a standard mortar [C87]. In some aggregate deposits, sulfide nodules may occur, and in some regions the water used for washing the aggregates may contain harmful chemicals.

Some materials, partly because of weathering action, and partly because of reaction with chemical elements in concrete, may undergo volumetric expansion sufficient to cause disruption of a concrete. Some shales, some cherts (colloidal silica), particles built up of layers of iron oxide ("chocolate bars"), and many types of clayey limestones are offenders in this class. While visual inspection may serve to indicate their presence in many cases, their effect is determined by accelerated weathering tests. The currently used soundness test for aggregates [C88] involves repetitions of soaking in sodium or magnesium sulfate and oven-drying and the determination of the amount and kind of disintegration resulting from this cyclic treatment. In case of doubt an accelerated freezing-thawing test on a mortar or concrete made of the suspected aggregate may be preferable. Some engineers consider a simple absorption test to be an adequate indicator, but there appears to be no general agreement on the limits which should be adopted.

Very fine particles in an aggregate may be present simply in the form of dust, or may form a coating on the larger aggregate particles. Material of this sort, if present in sufficient quantity, may unduly increase the water requirement for a mix, or if it varies from batch to batch, which is likely if it is present at all, undesirable fluctuations in the consistency may occur. Probably the worst offender in this class is clay, especially in the form of surface coatings, in which case it prevents good bond between cement paste and aggregate and causes concrete to expand and contract excessively, thus leading to cracking. Crusher dust in crushed stone is not so serious and may usually be readily removed at the processing plant by washing. The undesirable fine material in an aggregate is, in effect, defined as the material passing a No. 200 sieve [C117] but is best determined by the sand-equivalent test [D2419] as follows: a sample of sand is placed in a transparent cylinder and shaken up in a solution of water and calcium chloride to which glycerin and formaldehyde have been added. This solution causes the fine particles to flocculate and settle more rapidly than would otherwise be the case. The coarse sand particles settle immediately to the bottom of the graduate, and the fine dust and clay settle at different rates but remain in suspension for some time. A reading at 20 min of the top of the clay suspension and another based on the top level of the sand are made. The sand reading divided by the clay reading, times 100, gives the sand-equivalent value. A small amount of clay gives a high sand-equivalent value. The California Division of Highways, which developed this test, requires a value of about 75 to 80. Materials which do not meet this requirement can be made to comply by additional washing.

Very flat or elongated particles, if present in substantial amount, affect the workability of a concrete mix, and appreciable numbers of flat pieces tend to affect durability if they are oriented in such a way as to cause accumulations of water and laitance underneath their bottom surfaces. Apparently, about 10 or 15 percent of such shaped particles can be tolerated in an aggregate, but current specifications do not cover, or do not place quantitative limits on, inclusions of this sort.

Weak particles tend to lower the strength of concrete, although they must be present in some appreciable quantity before there is a noticeable effect on compressive strength. The effect is more marked upon flexural or tensile strength. The direct and most satisfactory way to evaluate the effect of such particles in a particular aggregate is through strength tests of the concrete in comparison with strength tests of concrete in which an aggregate of proven quality is used. For fine aggregates, the details of such a comparison test are specified by the ASTM [C87].

Soft particles may be objectionable not only from the standpoint of strength and durability, but especially if the surface of the concrete is to be subject to wear or abrasion (see Art. 11.21). The abrasive resistance of the concrete is a function of the wear resistance of the aggregate, as cement paste has little resistance to abrasive conditions.

The wear resistance of an aggregate is determined by use of a Los Angeles abrasion machine [C131], which consists of a cylindrical drum 28 in. in diameter and 20 in. long mounted on a horizontal shaft. A shelf 3½ in. wide runs the length of the drum on the inside. The test sample usually consists of 5,000 g of oven-dried aggregate of a specified grading. An abrasive charge, usually of twelve 1⅞-in. steel or cast-iron balls, is placed in the drum which is then rotated 500 revolutions at 30 to 33 rpm. The wear caused by the tumbling and dropping of the aggregate and the balls in the drum is determined from the material which, after test, will pass a No. 12 sieve.

The results of various tests using this equipment showed that the Los Angeles abrasion values for the aggregates investigated agreed with their service behavior in concrete, i.e., for low wear values their service performance was better. Also, these tests indicate that the lower the percentage of wear, the higher the concrete strength in flexure and in compression [110]. General practice dictates that coarse aggregate should lose not more than 10 percent after 100 revolutions, nor more than about 40 to 50 percent after 500 revolutions.

Suggested limits on deleterious substances in aggregates are given in Table 3.10.

Table 3.10 Recommended limits for deleterious substances in concrete aggregates*

	Recommended maximum limit, wt, %	
	Fine aggregate	**Coarse aggregate**
Clay lumps	1.0	0.25
Soft fragments	...	5.0
Chert that disintegrates in five soundness cycles:		
1. Severe exposure	...	1.0
2. Mild exposure	...	5.0
Material passing No. 200 sieve:		
1. Concrete subject to surface abrasion	3.0	1.0
2. All other classes of concrete	5.0	1.0
Coal and lignite:		
1. Where surface appearance of concrete is of importance	0.5	0.5
2. All other concrete	1.0	1.0

* From Ref. C33.

3.15 Reactive aggregates. In recent years there has been recognized a source of disruption caused by reaction between the alkalies, Na_2O and K_2O, in some cements, and certain minerals present in some aggregates. Numerous tests have shown opaline silica to be a serious offender, but other forms of silica and possibly other minerals may be involved. Known reactive substances in addition to opal are chalcedony, tridymite, cristobalite, zeolite, huelandite (and probably ptilolite); glassy to cryptocrystalline rhyolites, dacites, and andesites, and their tuffs; and certain phyllites. The formation of the products of the reaction between the alkalies and the aggregate causes abnormal expansions, which, however, sometimes do not take place until 2 or more years after the concrete has been placed. In a number of cases in recent years the integrity of important structures has been impaired by these expansions. The obvious steps that may be taken to eliminate this difficulty are two: to detect and eliminate the reactive aggregates and to limit the alkali content of the cement below some critical value. Some specifications for special cements limit the alkalies calculated as Na_2O equivalent to 0.6 percent. Another possible remedy may be the inclusion of some material in the mixture which will nullify the reaction or cause the reaction to spend itself while the concrete is still relatively soft and plastic (see Art. 11.16).

Petrographic analyses [C295] may be helpful but are not satisfactory for positive determinations of the degree of reactivity of an aggregate, as too little is known about the specific mineral constituents of various rock types which contain silica in forms susceptible to reactivity with alkalies in cement. However, aggregate which will cause this trouble in concrete may be recognized if it produces expansion within a short time in test specimens made with high-alkali cement [C227]. This determination is carried out using a 1:2.25 mortar mix, the aggregate having a specified gradation. Coarse aggregates can be tested by crushing to sand size. After 24 hr of curing at 73°F the specimens should be stored at 100°F in sealed containers, which contain free water, not in contact with the specimens, to maintain the desired humidity. Expansion is generally considered to be excessive if it exceeds 0.05 percent at 3 months or 0.10 percent at 6 months [C33].

Lack of expansion in a short period is no guarantee, however, against a delayed reaction, for which no reliable accelerated test has been developed. In general, aggregates known to have, or suspected of, injurious activity should be used only with cement of low-alkali content or not at all.

To speed up the determination of the potential deleterious reactivity of an aggregate, the U.S. Bureau of Reclamation developed a chemical test [C289] in which a sample of the aggregate is pulverized and treated with a sodium hydroxide solution. The degree of reactivity is determined from the amount of silica dissolved by the solution, and the reduction in alkalinity of the solution. This method has been used successfully in rating many aggregates for which service records and mortar-bar expansion data are available. The test can be made quickly and, while not completely reliable in all cases, provides helpful information especially where results of the more time-consuming tests are not available.

Another damaging reaction has been identified, in which certain rare argillaceous dolomitic limestones rather than siliceous aggregates react with cement alkalies to produce aggregate expansion and consequent cracking of the concrete. Their identification involves determining the calcite and dolomite contents of the aggregate. If they have approximately equal proportions of these materials, they should be tested further to determine the expansion of aggregate prisms [1120].

Some sand-gravel aggregates have been involved in concrete deterioration that has been attributed to a cement-aggregate reaction other than the common alkali-aggregate reaction. The following special tests have been developed to detect this phenomenon [1120].:

1. Concrete test involving heating, drying, cooling, and soaking cycles. An expansion of 0.07 percent or more at 1 year (285 cycles) indicates a potentially deleterious cement-aggregate combination.

2. Accelerated concrete test involving heating and cooling with constant water spray.

3. Mortar test in accordance with ASTM C342. Expansions of 0.200 percent or more at 1 year are considered unsatisfactory [C33].

Incorrect methods of stockpiling aggregates cause segregation and breakage.

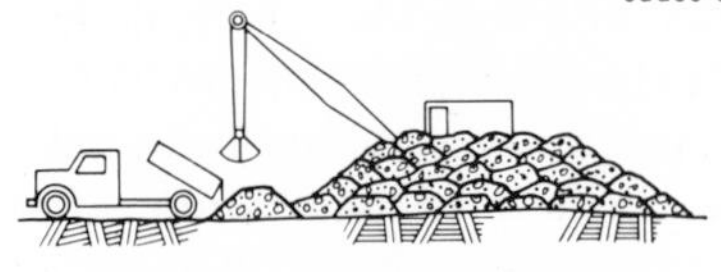

Preferable

Crane or other means of placing material in pile in units not larger than a truck load which remain where placed and do not run down slopes.

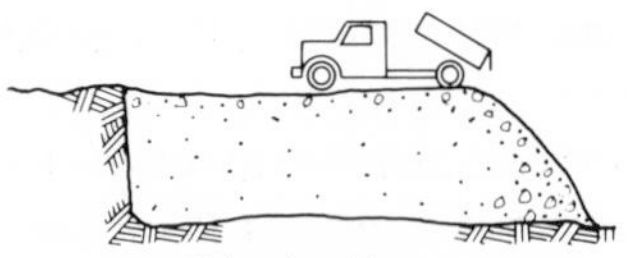

Objectionable

Methods which permit the aggregate to roll down the slope as it is added to the pile, or permit hauling equipment to operate over the same level repeatedly.

Limited acceptability – generally objectionable

Pile built radially in horizontal layers by bulldozer working from materials as dropped from conveyor belt. A rock ladder may be needed in this setup.

Bulldozer stacking progressive layers on slope not flatter than 3:1. Unless materials strongly resist breakage, these methods are also objectionable.

Stockpiling of coarse aggregate when permitted (Stockpiled aggregate should be finish screened at batch plant; when this is done no restrictions on stockpiling are required.)

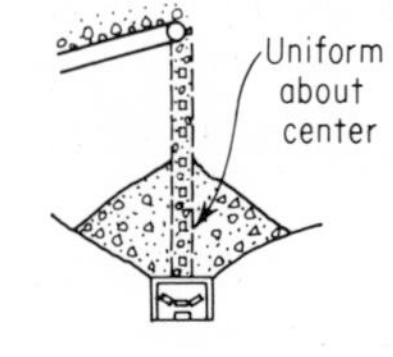

Correct

Chimney surrounding material falling from end of conveyor belt to prevent wind from separating fine and coarse materials. Openings provided as required to discharge materials at various elevations on the pile.

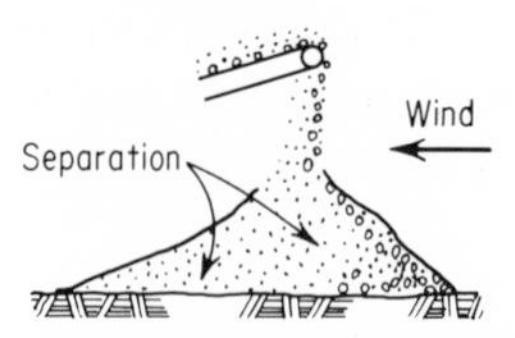

Incorrect

Free fall of material from high end of stacker permitting wind to separate fine from coarse material.

Unfinished or fine aggregate storage (dry material)

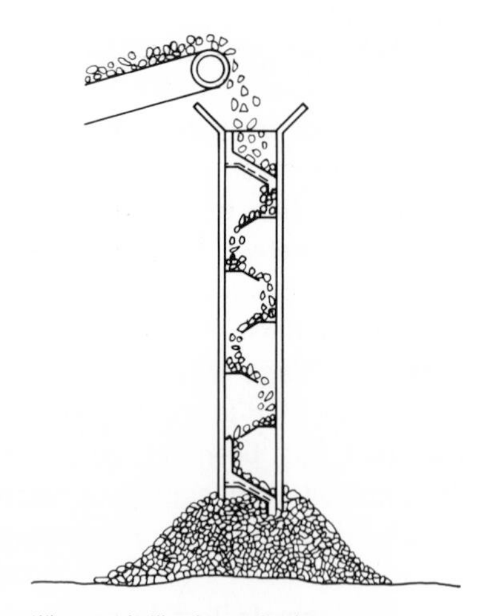

When stockpiling large-sized aggregates from elevated conveyors, breakage is minimized by use of a rock ladder.

Finished aggregate storage

Fig. 3.13. Correct and incorrect handling of aggregates. (*From ACI Committee* 614 [700].)

3.16 Handling and storing aggregates. Unless care is taken in the handling and storing of aggregates, there is a marked tendency for segregation of the fine and coarse particles to occur. Figure 3.13 shows some correct and incorrect methods of handling and storing aggregates. In general, there is less danger of segregation in coarse aggregates if they are screened into two or more size ranges and handled separately.

QUESTIONS

1. List the principal requirements of aggregates for concrete.

2. What data on aggregates are required for the design of a concrete mix?

3. Define "solid volume" of an aggregate for concrete.

4. What methods may be used to select the relative proportions of the various aggregates for a concrete mix? Discuss the applications and limitations of each method.

5. How does the specific gravity of an aggregate, as used in concrete technology, differ from the usual definition?

6. If a given lot of aggregate weighs 80 lb and its bulk specific gravity is 2.70, what is its solid volume?

7. If the unit weight of an aggregate is 102 pcf and its bulk specific gravity is 2.65, what is the percentage of voids in the aggregate?

8. List the various factors which influence the unit weight of an aggregate.

9. Define "bulking" of aggregates and discuss its significance.

10. Discuss the relative bulking tendencies of fine and coarse sands.

11. As regards moisture content, discuss the various states in which an aggregate may exist.

12. What various methods may be used to determine the free moisture in a sand?

13. What is the significance of the grading of a combined aggregate?

14. Describe the process of quartering an aggregate. Why is it done?

15. Define "fineness modulus" of an aggregate. What is its significance?

16. If the fine and coarse aggregates shown in Fig. 3.10 are to be combined in the ratio 35:65, compute the ordinate to the combined curve at the No. 14 sieve.

17. Discuss the effect of a maximum size of aggregate on the placeability and economy of a concrete mix.

18. What is the effect of the fineness of the sand and the richness of mix on the optimum percentage of sand?

19. The fineness modulus of a concrete sand should lie within what range of values?

20. Discuss the quality requirements of an aggregate.

21. How may the harmful expansions of reactive aggregates be prevented?

FOUR
WATER, ADMIXTURES, AND MISCELLANEOUS MATERIALS

4.1 Mixing water. Water used for concrete mixtures should contain no substance which can have an appreciably harmful effect upon strength (i.e., upon the process of hydration of the cement) or upon durability of the concrete in service. Water that is acceptable for drinking purposes is satisfactory for use as mixing water. If clear water from streams, not subject to contamination by domestic wastes, does not have a brackish or salty taste, it is acceptable. Water from a supply not approved by public health authorities for domestic consumption should be examined for the nature and extent of contamination. Small amounts of impurities can be tolerated with no apparent detrimental effect [400], and generally the procurement of acceptable water does not present a serious problem.

Substances which, if present in sufficient amounts in mixing water, may have an injurious effect upon concrete are silt, oil, acids, alkalies and salts of alkalies, organic matter, and sewage. Sources of supply which should be regarded with suspicion are streams carrying large concentrations of suspended solids, streams carrying industrial and domestic wastes, small streams and wells in mining country (acid mine waters), and wells, small lakes, and small streams in arid alkali country.

No general limits of tolerance for degree of contamination have been developed. The U.S. Bureau of Reclamation suggests limits on suspended silt, in terms of a turbidity test, of 2,000 parts per million [106]. The American Association of State Highway Officials Specification T26 suggests tests to determine acidity, alkalinity, suspended solids and chemical composition but does not give tolerance limits.

A practcal recourse for evaluating the effect of using a water of questionable quality is to make comparative tests for time of set and soundness of the cement and for strength and durability of mortars in which there is employed the contaminated water and water of proven quality [C87]. There should be no appreciable change in time of set and soundness through the use of the water in question. A tolerable limit for reduction in strength through the use of the water in question is given as 10 percent by the AASHO T26 and was suggested as 15 percent by Abrams [400]. In passing, it is of interest to note that under the criterion just mentioned, Abrams found sea water (having a salinity of about 3.5 percent) to be acceptable for use in making plain concrete. The effect of saline, acid, or alkali waters upon the corrosion of reinforcing bars or chemical alteration of certain types of mineral aggregates should receive consideration where possibilities of such reactions exist; however, there appears to be no substantial evidence that the use of sea water for mixing has alone caused appreciable corrosion of reinforcing steel [1188], but all researchers do not agree [1189]. Tests show that use of sulfate water in concrete produces little ill effect until an SO_4 concentration of about 1 percent is reached [106].

In situations where it is extremely difficult to secure good water, such as in desert areas, the final decision as to source of supply or as to processing readily available water may be governed by the importance of appreciable loss in concrete quality in comparison with the cost of securing good water.

4.2 Water for washing aggregates. The most important effect of the use of impure water for washing aggregates is the deposition of coatings of silt, salts, or organic material on the surfaces of the particles. The concentration of impurities in water to cause deleterious coatings is much greater than that which is harmful in mixing water, but no definite limits can be stated, and again recourse must be made to comparative tests to judge potentially harmful effects.

4.3 Water for curing concrete. Except for possible discoloration, the presence of silt, oil, or moderate amounts of salt in curing water does not appear to have harmful effects. However, waters containing appreciable concentrations of acid or organic substances should be regarded with suspicion and subjected to investigation.

A detailed discussion of curing as a process in the manufacture of concrete is given in Chap. 8.

4.4 Types of admixtures. Admixtures are substances used in cements, mortars, and concretes for the purpose of improving or imparting particular properties. Among the purposes for which admixtures are most commonly used are [405]:

1. To improve workability of fresh concrete
2. To reduce the water required
3. To improve durability by entrainment of air

4. To accelerate setting and/or hardening and thus to produce high early strength
5. To promote a pozzolanic reaction with lime liberated by cement during hydration
6. To aid in curing
7. To impart water-repellant or waterproofing properties

Admixtures are sometimes also used for the following purposes:

8. To cause dispersion of the cement particles when mixed with water
9. To retard setting
10. To improve wear resistance (hardness)
11. To reduce or offset shrinkage during setting and hardening
12. To cause expansion of concrete and automatic prestressing of steel
13. To aerate mortar or concrete to produce a lightweight product
14. To impart color
15. To offset or reduce some adverse chemical reaction
16. To reduce bleeding
17. To reduce the evolution of heat

Many admixtures can be useful and effective in accomplishing their intended purposes. On the other hand for some admixtures there have been made extravagant claims which are not fully justified. In making a decision as to whether or not an admixture is desirable or necessary, the following factors should receive consideration: (1) the possibility of accomplishing the desired result by a small modification in the basic mix, (2) the additional cost of using the admixture as against the additional cost of a modified basic mixture, (3) possible adverse effects on properties other than that which the admixture is intended to improve.

Some admixtures accomplish more than one purpose, e.g., calcium chloride under some conditions improves workability, accelerates setting and hardening, and tends to aid curing because of its hygroscopic property. Commercial admixtures may contain materials that separately would belong in two or more groups. For example, a water-reducing admixture may be combined with an air-entraining admixture. Some admixtures, especially finely powdered substances, while they do improve workability, increase the shrinkage of concrete upon drying. An admixture should be judged in relation to its overall effect upon the serviceability of a concrete [405].

The amount of any admixture used is of great importance because it may affect simultaneously such properties as water requirement, air content, rate of hardening, bleeding, and strength of the concrete. Since relatively small quantities of some admixtures are used, it is essential that suitable and accurately adjusted dispensing equipment be used.

Before any admixture is approved for use it should be tested with the actual concreting materials to be used on the job to determine that it has no adverse effects on any of the many important properties of the fresh or hardened concrete. As the specific effects of an admixture vary with different cements, water-cement ratios, mixing temperature, ambient temperature, and other job conditions, it is generally recommended that the proportions of the admixture used be adjusted to meet job conditions [405].

The number of admixtures which has been proposed for use in concrete is large. Further, the popularity of particular admixtures tends to change with time. For the purposes of this text, attempt is made only to discuss briefly the principal classes of admixtures, as the use of admixtures does not alter the basic principles of concrete making. For more detailed consideration, reference is made to the technical literature (see list of references for Chap. 4). The specifications for some chemical admixtures for concrete are given in ASTM C494.

4.5 Workability admixtures. A number of powdered admixtures, such as pumicite, hydrated lime, diatomaceous earth, bentonite, and fly ash, have been used to improve workability. Their use may become desirable when an aggregate is deficient in grading (lacks sufficient fines) or the cement has a marked tendency to bleed. Powdered admixtures are often added to the concrete batch at the mixer, although sometimes they are premixed with the dry cement. The use of large quantities of a finely powdered admixture generally tends to require a higher water-cement ratio, so that unless appropriate adjustments are made in the mix, some adverse effect may be expected upon strength, durability, and drying shrinkage. Possible exceptions to this are those materials which exhibit a high degree of pozzolanic activity; in this case the admixture is in effect a part of the cement, if the curing conditions are such that the pozzolanic reaction can proceed.

Water-reducing agents discussed in the next article can serve to increase workability of a mix. Also, air entrained in the form of minute bubbles improves workability, sometimes markedly. Air entrainment is discussed in Art. 4.7.

4.6 Water-reducing and set-retarding admixtures. Water-reducing admixtures are organic materials used to improve the quality of concrete, obtain required strength with less cement, or increase the slump of a given mixture without increasing the water content. They also may improve the workability of concrete containing aggregates that are harsh, or poorly graded, and may serve effectively in concrete that must be placed by means of a pump or under difficult conditions.

Set-retarding admixtures delay the initial setting and hardening of the concrete, with accompanying reduction in the mixing water required. They are used primarily to offset the accelerating and damaging effect of high temperature and to keep concrete workable during the entire placing period.

The admixtures most commonly used are: lignosulfonic acids and their salts, hydroxylated carboxylic acids and their salts, or modifications or derivatives of these chemicals.

Lignosulfonates may be used to extend the setting time of concrete 30 to 60 percent at temperatures of 65 to 100°F. In the amounts normally used they may entrain 2 to 6 percent or even more air in the concrete. Concrete containing a lignosulfonate retarder usually requires 5 to 10 percent less water than does similar concrete without the admixture. Compressive strengths at 2 or 3 days are usually equal to or higher than for comparable concrete without the admixture, and the strength at 28 days or later may be 10 to 20 percent higher. These materials may cause some reduction of bleeding and settlement of fresh concrete, depending on the amount of air which they entrain.

Hydroxylated carboxylic acid salts serve as water-reducing, non-air-entraining retarders. Used in the proportion required to retard the set by 30 percent, the water may be reduced 5 to 8 percent in either air-entrained or non-air-entrained concrete. The rate of bleeding and bleeding capacity are increased. Compressive strengths during the first 24 hr are lower but after 3 days are higher by 10 to 20 percent.

All water-reducing agents enhance the air-entraining properties of air-entraining cements as well as the amount of air entrained by a given amount of an air-entraining admixture.

Both types of these admixtures are available in either powder or liquid form. Powders may be added with the cement or the aggregate but preferably with the latter. If entirely soluble they may be dissolved in water and added as a solution. Liquids may be added with the mixing water or added after the other constituents of the concrete have been partially mixed; they should not come in contact with the cement prior to addition of the mixing water.

Some materials are variable in their action, retarding the set of some cements and accelerating the set of others. Also, some materials act as retarders when used in certain amounts and as accelerators when used in other quantities. Therefore, great care must be taken when employing such materials, and advance experiments should be conducted before using them on any job.

4.7 Air-entraining admixtures. Within recent years there has developed the practice of using admixtures which cause the entrainment of air or formation of gas in concrete for the purpose of improving the resistance to freezing and thawing (see Art. 11.14) and for other purposes. These developments are considered by many to be one of the most important advances in the art of concrete making in the last three decades. The principal application has been to pavement concretes; entrained air increases the resistance to surface scaling resulting from the application of salts for ice removal, but structural concretes now commonly include these agents. Considerable literature is now available on this subject [e.g., 420–431].

Aeration, in the form of finely divided bubbles, uniformly distributed throughout the concrete mass, may be produced in two ways: by use of a foaming agent which causes air to be entrained during mixing, or by the use of an agent which by reaction (with constituents usually present in the cement) produces a gas. For producing air entrainment during mixing, a number of organic compounds such as natural resins, tallows, and sulfonated soaps or oils have been tried; several proprietary agents such as Darex have been approved for use under ASTM Specification C175. Gas-forming agents are discussed in the following article.

The general effects of air entrainment are to increase workability, decrease density (unit weight), decrease strength, reduce bleeding and segregation, and increase durability. Air entrainment makes possible a reduction in the sand content of the mix in an amount approximately equal to the volume of the entrained air. Also, each percent of entrained air permits a reduction in mixing water of about 3 percent, with no loss in slump and even some gain in workability. By redesign of the mix to maintain constant workability, the tendency to decrease strength can be partially offset, so that the reduction in strength will rarely exceed 15 percent [405]. For most types of exposed concrete a slight reduction in strength is far less significant than the improved resistance to frost action. The decision concerning whether to use air entrainment, or how much air to entrain, generally depends upon to what degree strength can be sacrificed in the interest of improved durability.

For average mixes, each percent of air entrained causes a reduction in strength of about 3 or 4 percent. Experiments indicate that marked improvement in resistance to freezing and thawing does not occur unless the air content is greater than about 3 percent. Depending on the maximum size of aggregate, the content of entrained air should be as follows [1120]:

Maximum size of aggregate, in.	**Recommended air content of concrete, percent by volume**	
	Desired average	**Minimum**
½	8	7
¾	7	6
1	6	5
1½	5½	4½
2	5	4
3	4½	3½
6	4	3

The finer the aggregate, the greater the required air content, so that for mortar the optimum air content is about 10 to 13 percent.

The air-entraining agent may be introduced by (1) using a "treated" cement, with which the agent has been interground, or (2) by adding the agent at the mixer. Since the agents are used in only very small amounts, it is claimed by some that intergrinding with the cement gives more uniform dispersion and hence more consistent results. On the other hand, variations in type of cement, mix proportions, mixer action, temperature, and other variables influence the amount of air entrained, so that it is claimed by others that the addition of the agent at the mixer permits adjustment to a desired quanity of air under particular conditions.

A number of factors affect the amount of air entrained by a given quantity of agent, and thus they affect the properties of the resulting concrete. The effect of the amount of Vinsol resin on the air content of concrete mixes of constant cement content is shown in Fig. 4.1. It appears that the air content increased progressively with the amount of Vinsol resin in the concrete. Also, for the richer concrete mixes, more Vinsol resin was required to produce a given air content. These general effects have been noted for other agents as well. If any finely powdered admixtures, such as fly ash, are added to the mix, they act like cement in reducing the air content for a given amount of agent.

The effect of time of mixing upon the air content varies with the type of air-entraining agent and whether it is interground with the cement or

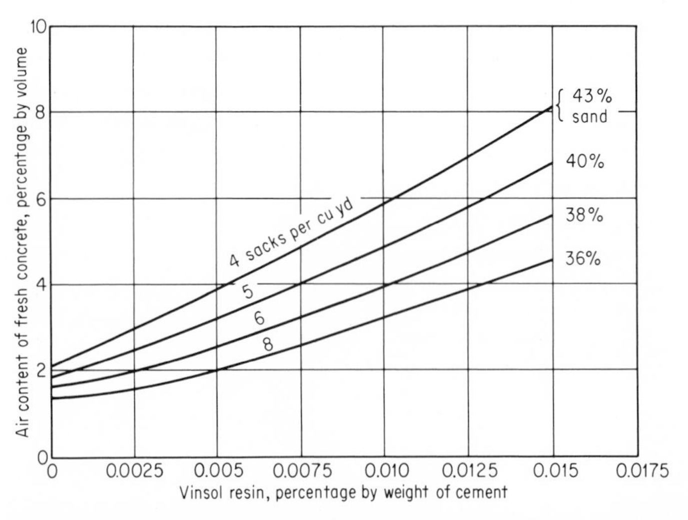

Fig. 4.1. Effect of amount of Vinsol resin on air content of concrete mixes of constant cement content. Maximum aggregate 1 in.; 3 to 4-in. slump. (*From Gonnerman* [1147].)

added at the mixer. In general, the change with time is not large for usual mixing periods, but for extended mixing periods of 5 or 10 min the air content may be somewhat higher or lower depending upon its activity. A peak value of the air content usually occurs between 3 to 15 min of mixing.

As the water-cement ratio is increased, the average size of the air voids, the distance between the air voids, and the freezable water content of the cement paste increase for given conditions, resulting in a reduced resistance of the concrete to freezing and thawing.

Less air will be entrained at 100°F than at 70°F, and more will be entrained at 40°F, so that with other things held constant, the air content varies inversely with the temperature. Air contents, after placing and vibration, may be about one-fifth less than when first mixed, but resistance to laboratory freezing and thawing has not been found to be affected adversely by loss of air caused by vibration, provided the concrete originally contained a satisfactory air system. Presumably this is also true for field conditions.

Methods of conducting comparative tests for judging the acceptability of new air-entraining agents have been developed by the ASTM [C233].

Some lightweight mixes (see Art. 13.10 and 14.22) are made by using proteins to produce foam which is then mixed into a cement slurry. Aggregates may be added for the heavier, stronger mixes. The proprietary foaming agents such as Elasticel can produce unit weights from 30 to 115 pcf and strengths from almost nothing up to 3,000 psi.

4.8 Gas-forming admixtures. During the period while setting takes place in concrete, the solids tend to "subside" in the forms, principally as a result of the bleeding of free water from the mass. In many mixtures, as the result of this subsidence void spaces of appreciable magnitude may occur under reinforcing bars and under the larger pieces of coarse aggregate. To offset this tendency, there are sometimes used agents which produce gas (in the form of minute, dispersed bubbles) in an amount intended to be approximately equal to the effective change in volume resulting from subsidence. The gas dispersed in a concrete by the use of generating agents improves resistance to frost action in the same way as entrained air, but to a lesser degree, as the resulting voids are usually larger and farther apart. Additionally a gas-forming agent may be used in sufficiently large quantity to produce very lightweight cellular concrete or mortar (see Art. 14.22).

Agents which have been used to produce gas bubbles in concrete are aluminum or zinc powder and hydrogen peroxide. Aluminum powder, which is the most commonly used, produces hydrogen gas as the result of a reaction with the alkaline hydroxides in the cement paste [430, 431]. The amounts added are usually in the range of 0.005 to 0.02 percent by weight of cement although larger amounts may be used for the production of low-strength cellular concrete. Hydrogen peroxide liberates oxygen, but the action may be too fast to be highly effective.

4.9 Accelerators. Accelerators are used to shorten the time of set and to increase the rate of hardening for the purposes of (1) permitting earlier removal of forms, (2) shortening the curing period, (3) advancing the time when a structure can be placed in service, (4) offsetting the retarding effects of low temperatures, or (5) compensating for the retarding effects of some other admixture [405]. Their use requires that no delays occur in placing the concrete, as otherwise early stiffening and loss of slump will occur.

Agents which accelerate setting and hardening include calcium chloride [D98], triethanolamine, and some soluble carbonates, silicates (such as water glass), and fluosilicates. A more rapidly hardening concrete than normal may also be obtained by the use of high-early-strength portland cements and of aluminous cements.[1] Some accelerators, such as triethanolamine, may produce a flash set, but the ultimate strength may be lowered markedly.

The most commonly used accelerating admixture is calcium chloride, either directly as such, or as an ingredient in a number of proprietary compounds. Calcium chloride may be used in amounts up to 2 percent by weight of the cement. The maximum amounts in relation to temperature which have been suggested when calcium chloride is used as an accelerator [pp. 47–48 of 109] are shown in the following tabulation:

Prevailing temperature, °F	Maximum wt of calcium chloride, lb per sack of cement
70 or less	2
70 to 90	1½
90 or greater	1

During freezing weather, with cold materials, amounts up to 4 percent have been used although 2 percent is a conservative maximum. Calcium chloride increases the rate of early heat development, accelerates the set, but lowers the freezing point of the concrete only slightly; thus other protection against freezing will be required.

Where a considerable period may elapse before the concrete is placed, as in the delivery of ready-mixed concrete, it may be advisable to introduce the calcium chloride a few minutes before the concrete is to be discharged

[1] It is of interest to note in passing that the time of set for most portland cements would be very short unless controlled by the action of gypsum used as a *retarder*.

from the drum. It is generally purchased in flake form, but is added to the mix in *complete* solution in the mixing water. Its use generally increases the 1- and 3-day strengths, but the effect at later ages becomes progressively less (see Art. 8.15). The accelerating effect is not the same with all brands of cement [405, 440–442]. Calcium chloride causes some reduction in bleeding and lowers the resistance of concrete to sulfate attack, but it usually increases the drying shrinkage, increases expansion under moist curing, and increases the expansion caused by any alkali-aggregate reaction [405]. Calcium chloride does not promote corrosion of the usual reinforcement in concrete where adequate cover is provided for the steel, but it should not be used in prestressed concrete as it may cause stress corrosion of the prestressing steel. Further details concerning the use of calcium chloride are given in Arts 8.15 and 10.11.

The use of accelerators in concrete made during hot weather should be considered with caution, because of the possibility of premature stiffening of the mix.

Some chemicals, such as sodium silicate (water glass), which cause very rapid setting result in relatively low ultimate strengths and probably poor durability, and should be used only for emergency work such as stopping leaks.

4.10 Expansion-producing admixtures. Expansion-producing materials which are incorporated in "expanding" or "drying-shrinkage-compensating" cements have been discussed in Art. 2.24. These materials, and others, may also be used as admixtures.

Expansion-producing admixtures either expand themselves during the hydration period of the concrete, or they react with other constituents of the concrete with resulting expansion. The expansion may offset the later drying shrinkage or it may be greater, even to produce self-stressing concrete.

In addition to the sulfoaluminate cement covered in Art. 2.23 granulated iron and chemicals to cause its oxidation and expansion in the presence of air and moisture are used in grouting operations. Control of the amount of oxidizing catalyst is required to procure the desired amount of oxidation and expansion. After the desired expansion has occurred, all exposed surfaces should be sealed to prevent wetting and later expansion.

4.11 Pozzolanic materials. Certain minerals, predominantly siliceous in nature, when in a finely divided state can combine with lime in the presence of water to form cementitious compounds (see Art. 2.19). A number of naturally occuring materials such as pumicite, opaline shales and cherts, diatomaceous earth, certain shales and clays, and some materials resulting from manufacturing processes, such as fly ash, have been employed successfully as pozzolans [456]. Usually the clayey materials and diatomaceous earth must be calcined at 1000 to 1700°F to make them suitable.

Pozzolanic materials are used in combination with portland cement, often as a replacement for a part of the cement. The pozzolanic reaction, which occurs with the lime liberated by the hydrating portland cement, requires the continued presence of water. For successful use of portland-pozzolan cements at normal temperatures, sustained moist curing is important.

Portland-pozzolan cements (the combination of portland cement with an appreciable amount of a pozzolanic material, say, in excess of 15 percent by weight of the portland cement) have been used for concrete in massive structures such as dams and large piers and for structures exposed to sea water or sulfate-bearing waters. In massive structures an important advantage which accrues by its use, in addition to the obvious saving in cement cost, is the reduction in amount of total heat liberated within the mass with a resulting smaller temperature rise and thermal volume change. Use of a pozzolanic material with other than sulfate-resisting portland cements generally increases resistance of the concrete to attack by sea water and other aggressive waters. The relative improvement is greater for concrete of low-cement content. The use of pozzolans with sulfate-resisting portland cements does not increase sulfate resistance. Pozzolans tend to reduce permeability, especially in lean mixes. Some pozzolanic materials prevent the harmful action of certain reactive aggregates (Art. 11.16), improve the workability of the lean mixes used in mass concrete, and reduce bleeding and segregation. The effect of pozzolanic substitutions on strength vary with the particular materials, cements, and mixes. With lean mixes the strength is often increased, while the strength of rich mixes is generally decreased. Disadvantages lie in the relatively slow rate of strength development and the possibility that in exposed sections, if special attention is not given to maintaining moist conditions and temperatures sufficiently above freezing, serious deficiency in strength and low durability may result. Some pozzolans require more water than portland cement. This increases the drying shrinkage and may cause increased cracking.

Pozzolans vary considerably in their effect on concrete mixes. Some not only may cause the undesirable effects noted above, such as excessive drying shrinkage and reduced strength and durability, but if used in insufficient amounts, may even cause deleterious reaction with cement alkalies. Before a pozzolan is approved for use on a job, it should be tested in combination with the cement and aggregate for that job to ascertain its suitability.

In recent years, where portland-pozzolan cements have been employed, pozzolanic material has generally been used in amounts ranging from 10 to 30 percent of the cement by weight. The specific gravities of pozzolanic materials are lower than that of portland cement, so that the percent substitution stated on a volume basis is greater than that stated on a weight basis.

For additional information on pozzolanic cements, see Art. 2.20, Refs. 405, 450–459, C340, and U.S. Federal Specification SS-C-208c.

4.12 Bonding admixtures. Water emulsions of various organic materials are mixed with portland cement or mortar grout for application to a hardened concrete surface, just prior to the placement of fresh concrete over it, to increase the bond strength between the old and new concrete. These materials have been found useful in patching concrete, for bonding cement plaster, and in the formulation of cement paints.

The commonly used bonding admixtures are made from natural or synthetic rubbers or any of many organic polymers or copolymers. The polymers include polyvinyl chloride, polyvinyl acetate, and acrylics. Other bonding agents are two-component, modified epoxy resin compounds. All of these bonding agents are effective only on clean, sound surfaces since the strength of the bond is only as good as the strength of the material to which it is attached.

4.13 Curing aids. Curing involves the retention of sufficient free water in the cement paste so that the process of cement hydration may readily proceed. (Time and favorable temperatures are also involved.) A common means of assuring water retention for the desired curing period is to keep the surface of the cement mass moist or wet. Water may be applied directly in the field by ponding, or by intermittent or continuous spraying; in the laboratory submersion was formerly used, but most modern laboratories have curing rooms in which test specimens are stored in an atmosphere maintained at or near 100 percent relative humidity.

A moist environment is sometimes maintained by form retention, or by placing upon, or banking against, the concrete a layer of moist earth or damp straw; strips of continuously wet burlap or cotton mats are also used. However, the use of earth, straw, or burlap may cause discoloration of the concrete surface.

When sufficient water is difficult to provide, other means of aiding the curing process can be employed, such as admixtures and surface coatings. As an integral (mixed with the concrete) curing aid, calcium chloride or some product containing calcium chloride is used. The uses of this chemical for curing is covered by ASTM Specification D98.

Commonly used surface coatings for curing purposes function as membranes to prevent evaporation. Waterproof paper, bituminous and paraffin emulsions (the Hunt process), and coal tar cut-backs are among the products which have been used successfully. (See ASTM Specification C309.) Since the regions where water is scarce often have a high degree of sunlight exposure, light rather than dark coatings are preferable in order to minimize heat absorption by the concrete. On construction of aqueducts in hot desert areas, whitewash has been used over the membrane curing coating.

A summary of a comprehensive study of the effectiveness of a number of curing aids is given in Ref. 460. Details concerning the use of the various curing aids in the manufacture of concrete and also methods for hot- and cold-weather concreting are given in Chap. 8.

4.14 Miscellaneous materials. Various special and miscellaneous materials such as water-repelling agents (so called "waterproofers"), coloring agents, paints for concrete surfaces, surface hardeners, to name a few, are sometimes used in concrete construction work. A few sources of information are given in Refs. 470–499.

For oiling wood forms, a heavy mineral oil is commonly used.

4.15 Steel reinforcement. While a discussion of steel reinforcement is beyond the scope of this elementary text on plain concrete, the selection, specification, acceptance, placing, and inspection of steel reinforcement are matters with which the engineer in practice may be concerned. For information, see standard texts on reinforced concrete. Also see ASTM Specifications A15, A16, A82, A160, A184, A185, corresponding AASHO Specifications, and Ref. 499.

QUESTIONS

1. Discuss requirements for mixing water for concrete.

2. Discuss possible deleterious effects of the following in mixing water: silt, oil, acids, alkalies, organic matter.

3. If positive information on the suitability of a water for mixing purposes is lacking, and existing specifications do not cover limits for the existing contaminating substance, how can decision be made concerning the use of the water?

4. What is the effect of using sea water for mixing concrete?

5. What can be done when all readily available water is contaminated?

6. Discuss requirements for water for washing aggregates; for curing concrete.

7. What is an "admixture"?

8. List the principal purposes for which admixtures are most commonly used. Name also some special purposes.

9. Mention some of the effects of calcium chloride when it is used as an admixture in concrete.

10. Describe some of the principal types of workability admixtures, and what supplementary effects they have on concrete properties other than workability.

11. What is the purpose of air entrainment, when it is known that an appreciable air content tends to lower the strength of concrete?

12. What materials are used for air entrainment?

13. Under what conditions would it be desirable to use a gas-producing agent rather than an air-entraining agent, even if the final air contents were the same?

14. How are air-entraining agents introduced into a concrete mix?

15. How does the optimum amount of entrained air vary with the maximum size of aggregate in the concrete?

16. What materials may serve as water-reducing admixtures?

17. What benefits may result from the use of water-reducing admixtures?

18. What materials may be used as bonding agents for concrete, and what precautions must be taken in their use?

19. Discuss briefly the advantages and disadvantages of "accelerators."

20. What is "pozzolanic action"? What is the predominant mineral compound in a pozzolanic material? Mention some pozzolanic materials.

21. Discuss the pros and cons of the use of portland-pozzolan cements. Indicate, especially, precautions that must be taken.

22. Describe several commonly used curing aids. What is the underlying objective in the use of curing aids?

FIVE
PROPERTIES OF FRESH CONCRETE

5.1 General. Beginning as soon as water is brought into contact with the other ingredients, freshly mixed concrete gradually undergoes several changes until (normally within a few hours) it becomes rigid, or "hardened," concrete. In this chapter are discussed various elements of the stiffening process, some influencing factors, and pertinent tests and computations.

WORKABILITY; CONSISTENCY

5.2 Workability. It is desirable that freshly mixed concrete be relatively easy to transport, deposit, consolidate, and finish and that it remain free from segregation during these operations. The composite quality sought, involving ease of placement and resistance to segregation, is termed "workability." The workability of concrete depends on a number of properties which cannot be satisfactorily measured; there is, in fact, no general agreement what all these properties are. Further, workability is a *relative* property; a concrete that is workable under some conditions may not be workable under some other conditions. The necessary workability may vary with the equipment for mixing, transporting, or consolidating or with the size and shape of the mass to be formed. For example, a rather stiff concrete suitable for massive construction could not be placed in narrow, deep forms filled with intricate reinforcement. Usually a workable concrete is plastic although under certain conditions of placement stiff concretes are usable and are therefore considered "workable."

The basic influence of materials and proportions on workability is summarized by F. R. McMillan [101] in the following statements regarding plasticity:

The factors governing the plasticity of a concrete mixture . . . [are]:

1. Relative quantities of paste and aggregates
2. Plasticity of the paste itself
3. Grading of aggregates
4. Shape and surface characteristics of aggregate particles

For any given paste, that is, a quantity of cement with its definite proportion of water, decreasing the amount of paste with respect to the quantity of aggregate stiffens the mixture, and increasing the amount of paste renders the mix more fluid. If the quantity of paste is reduced to the point where there is not enough to fill the spaces and actually float the aggregate particles, the mix will become granular or harsh and will be impossible of proper placement.

Similarly, for a given quantity of paste and aggregate the plasticity of the mix will depend upon the relative quantities of cement and water in the paste. A paste that is high in cement and low in water content will itself be stiff and cannot carry much aggregate without becoming so stiff as to be wholly unplaceable. On the other hand, if the cement content of the paste is low and the water content high, the paste may be so thin and watery that it will be unable to hold the aggregates in that cohesive mass which is the very embodiment of plasticity.

The grading of the aggregates affects the plasticity of the concrete: (a) by affecting the quantity of paste necessary to fill the spaces thoroughly and surround the aggregate particles completely, and (b) by affecting the resistance which is offered to the mobility of the mass through the varying combinations of sizes.

As in the case of grading, the shape and surface characteristics of the particles affect the plasticity of the mix through their effect on the amount of paste required and on the friction between the particles as the concrete is molded. Angular particles or those with rough surfaces require a greater amount of paste for the same mobility of mass than is necessary for well-rounded particles or those with smooth and slippery faces, other conditions remaining the same.

The essential properties (somewhat idealized) which influence workability are (1) the shear resistance or force required to start flow; (2) the mobility of the mass after flow has started (harshness or the resistance offered to the movement of the mass is the converse expression of this property); (3) cohesiveness or resistance to segregation; and (4) stickiness, which, although it is related to cohesiveness, may exist in varying degrees in a mass which has little tendency to segregate and may affect the ease of manipulation and finishing.

In practice, properties involved in workability include consistency, mobility in the conveying and handling devices, and segregation (including bleeding).

Factors affecting workability include quantity and characteristics of cement, gradation and shape of aggregates, quantity of water, quantity of entrained air, and type and quantity of finely divided or chemical admixtures.

Workability should be distinguished from "consistency," which term, as used in concrete practice, relates to the degree of *wetness* of concrete. Consistency has to do with the force-flow relationships alone (the shear resistance and the mobility, mentioned above). The ordinary consistency tests (slump, flow, and penetration) do not uniquely define consistency since they indicate the flow under one particular force only. Within limits, wet concretes are more workable than dry (stiff) concretes, but concretes of the same wetness (consistency) may differ in workability.

Both workability tests and consistency tests are relative measures only because they do not determine fundamental quantities.

Inasmuch as true measures of the qualities which make up workability are difficult if not impossible to obtain, it seems desirable here to point out the value of visual inspection. By systematic inspection of a mix for significant quantities (cohesiveness, sandiness, and troweling workability) which are readily observable, together with a measurement of the consistency, a reasonable estimate can be made of workability [502]. This examination is particularly of value to the beginner in developing judgment concerning desirable and undesirable characteristics of a mix and is employed in Test 6, Part II.

5.3 Consistency. Consistency is a practical consideration in securing a workable concrete. Usually it is taken to denote the fluidity or wetness as indicated by the slump or corresponding tests. It is useful as a major factor of workability.

Table 5.1 Range in slump and flow of concrete for various degrees of consistency

Consistency	Slump, in.	Flow, %	Remarks
Dry	0–1	0–20	Crumbles and falls apart unless carefully handled; can be consolidated into rigid mass under vigorous ramming, heavy pressure, or vibration but exhibits voids or honeycomb unless special care is used
Stiff	½–2½	15–60	Pile tends to stand upright; holds together fairly well but crumbles if chuted; with care and effort can be tamped into solid dense mass; satisfactory for vibratory consolidation
Medium	2–5½	50–100	Alternative terms are plastic, mushy, quaking; easily molded although some care is required to secure complete consolidation
Wet	5–8	90–120	Pile flattens readily when dumped; can be poured into place
Sloppy	7–10	110–150	Grout or mortar tends to run out of pile, leaving coarser material behind

Table 5.2 Consistency, cement content, and aggregate size employed on various types of concrete construction

Type of construction	Typical structures	Consistency	Cement factor, sacks per cu yd	Maximum size of aggregate, in.
Massive	Dams, heavy piers, large open foundations	Stiff	2½–5	3–6
Semimassive	Piers, heavy walls, foundations, heavy arches, girders	Stiff; medium	4–6	2–3
Heavy building	Large structural members, small piers, medium footings; wide to moderately wide spacing of reinforcement	Medium wet	5–7	1–2
Light	Small structural members, thin slabs, small columns, heavily reinforced sections, closely spaced reinforcement	Wet	5½–7	½–1

The effect of increase in wetness on concrete having given proportions (dry) of cement to aggregate may be seen by the following step-by-step analysis: with the first increments of water, the dry mass bulks. As more increments are added, this bulking increases to a maximum which depends on the nature of the materials used; then as more water is added, the bulking gradually decreases until the maximum density of the mass is reached. At that point the mass becomes plastic. The addition of more water causes the mass again to increase in volume and to become more fluid, until a point is reached at which the paste is so thin that it can no longer hold the aggregate in suspension. Then segregation results because of the dilution of the mixture. The minimum amount of water which should be used is that which will overcome the bulking of the mass, and the maximum is that which will produce concrete of the required strength and workability and which will not start segregation.

For convenience, various degrees of wetness of a mix may be roughly classified and described as dry, stiff, medium, wet, and sloppy. A concrete is said to be of "medium" or "plastic" consistency when it is just wet enough to flow sluggishly—not so dry that it crumbles or so wet that water or paste runs from the mass. The range in values of slump and flow (see Art. 5.4) corresponding to these arbitrary degrees of consistency is indicated in Table 5.1; the values for each class are not to be taken as absolute. Consistencies employed on various types of construction are shown in Table 5.2.

5.4 Slump, flow, and ball tests for consistency. The "slump test" is only an approximate measure of consistency, but it defines ranges of consistency well enough for most practical work. It is the consistency test most widely used both in the field and in the laboratory. It is performed by measuring the subsidence, in inches, of a pile of concrete formed in a mold which has the shape of a truncated cone of base diameter 8 in., top diameter 4 in., and height 12 in. (see Fig. 5.1). Details of the procedure [C143] are given in Appendix C herein. A rough indication of workability can be obtained by tapping the side of the slumped pile with the tamping rod; a cohesive concrete will not break apart or crumble. Also, slumped concrete can be worked and its surface smoothed with a trowel. As shown in Fig. 5.2, the observed slump is affected considerably by the temperature of the concrete; the higher the temperature the less the slump. This effect should be recognized when making any consistency test at temperatures much above or below normal.

The "flow test" is performed by jigging a pile of concrete on a metal table [C230] and noting the spread of the pile as a percentage of the original

Fig. 5.1. Measuring slump of concrete. *(Portland Cement Association.)*

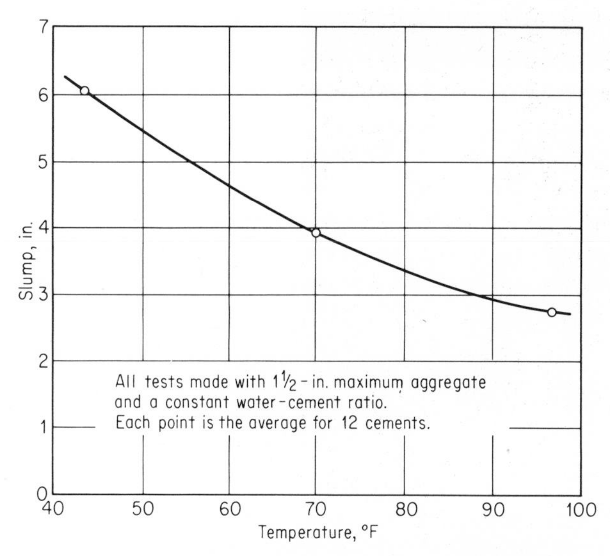

Fig. 5.2. The slump of concrete decreases as the temperature increases. (*From U.S. Bureau of Reclamation* [106].)

formed diameter. At the start of the test the concrete has a base diameter of 10 in., a top diameter of 6¾ in., and a height of 5 in. Details of the procedure [C124] are given in Appendix E herein. Although the results are more reproducible than those of the slump test, the flow test is largely limited to laboratory use because of the unwieldiness of the apparatus. It has been pointed out that the weakness of the flow test lies in the facts that the flow is unrestricted, that some of the aggregate rides along only partly embedded in the mortar, and that at the end of the test the mass is scattered instead of being homogeneous.

The "ball test," commonly referred to as the "Kelly ball" test, is a convenient and useful means of measuring the consistency of concrete as it rests in a container or in the forms [505, C360]. It is performed by measuring the penetration, in inches, of a 6-in. hemispherical surface under a weight of 30 lb (see Fig. 5.3). Roughly, 1 in. of penetration corresponds to 2 in. of slump although as shown in Fig. 5.4 the ratio is not constant over a range, even under a given set of conditions. When it is desired to use ball penetration as an indication of specified slump, simultaneous tests should be made under the job conditions and a proportionality factor established.

The size, shape, and weight of the "ball" penetrator are such that the concrete is positively displaced over an area large enough to avoid effects of individual pieces of aggregate and that the area and volume of displacement increase in greater proportion than the depth of penetration. The

Fig. 5.3. A Kelly ball being used in measuring the consistency of concrete. The penetration is read directly from the scale on the vertical stem.

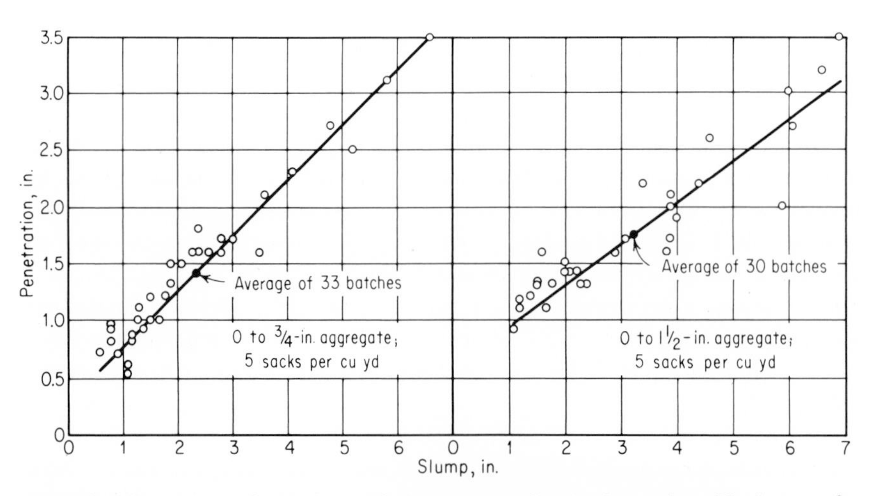

Fig. 5.4. Comparison of Kelly-ball and slump tests. Penetration varies with slump under given conditions, but the ratio is not necessarily constant. (*From Kelly and Polivka* [505].)

test covers the range from very stiff consistencies to consistencies as wet as should ever be used in construction. The ball test is relatively free from personal errors whereas the slump test is subject to several personal errors. A disadvantage of the ball test for use in the laboratory, where test batches are usually small, is that the minimum sample required (to avoid boundary conditions) is several times that required for a slump test.

5.5 Powers remolding test. The Powers test to determine "remolding effort" as an indication of workability is performed in the apparatus diagrammed in Fig. F.1 of Appendix F [511]. In a cylindrical container 12 in. in diameter and 8 in. high there is suspended concentrically a ring 8¼ in. in diameter, with its lower end 2¾ in. above the bottom of the container. The container is clamped onto a flow table which is set for ¼-in. drops. A standard slump cone is set in the ring and is filled with concrete as in the standard slump test; the cone is then removed. A flat plate, which with its guide rod weighs 4.30 lb, is set on the slumped pile of concrete. The apparatus is then jigged until the concrete is at the same level on both sides of the ring; the number of drops required is noted as the remolding effort. Details of the test are given in Appendix F.

In general, the less the remolding effort the greater the slump; however, the relation between remolding effort and slump varies with richness of mix, grading of aggregate, and other factors. The remolding test seems well able to distinguish between the plastic characteristics of mixes even of the same consistency (slump). It is widely used in laboratory work in the United States.

5.6 Other tests. Many attempts have been made to determine the workability of concrete by measurement of consistency, penetration, flow, compaction, or effort required to produce deformation. However, apparently none of the methods devised so far take account of all the inherent factors. The methods now more or less in use range from practically static consistency tests to measurement of mechanical effort or electrical power required to make freshly mixed concrete change its shape or position under given conditions. Space limitations herein do not permit a description or discussion, but some references for further study are given in the following list of apparatus and methods. In the list, no distinction is attempted among workability, consistency, degree of compaction, segregation, flow, or other factors of workability. Brief descriptions and diagrams of several of the tests are given in Ref. 514.

Thaulow concrete tester [513]: Slump specimen on drop table; remolding effort indicated by number of drops to reshape specimen.

Vebe consistometer [500, 514]: Slump specimen on vibrating table; remolding effort indicated by time of vibration to reshape specimen.

Wigmore consistometer [514, 515]: Cylindrical container on drop table; metal ball on vertical guide rod penetrates into concrete; number of drops required to penetrate a fixed distance determined.

Compacting-factor apparatus [503]: Concrete dropped successively through two inverted truncated cones into and overflowing a cylindrical container which is then struck off; ratio of concrete in cylinder to concrete from same batch thoroughly compacted in same container taken as compacting factor. As a supplement or alternative [506], weight of heaped specimen (before strike-off) is determined. Method considered suitable for no-slump concrete [603].

Penetration of various rods, cones, or balls—some under static conditions and some dropped or vibrated.

Burmister flow trough [501]: Flow of sample of concrete down inclined trough, the upper end of which is raised a fixed distance and dropped a fixed number of times; distance of front of flow measured.

Slope (flow) test [509]: Sloping channel with lower end constricted and with vibrator attached; weighed quantity of concrete introduced at upper end, and time (in seconds) for half the concrete to flow out determined.

For fluid grout: Flow cone; torque meter; viscosimeter [514].

For concrete in mixer during mixing [507, 510, 512]: Resistance to movement of pivoted arm with plate, truncated cone, or cylinder in stream of falling concrete; location of center of gravity of batch by observing balance of tilting mixer; power or torque required to rotate mixer drum.

BLEEDING: STIFFENING: SETTING

5.7 Bleeding. The tendency for water to rise in freshly placed concrete is known as "bleeding," or "water gain." It results from the inability of the constituent materials to hold all the mixing water dispersed as the relatively heavy solids settle. Some of the bleeding water reaches the surface, some is trapped under horizontal reinforcing bars and larger pieces of aggregate, and some from the lower portion dilutes the cement-water paste in the upper portion.

If the loss of water by bleeding were uniform throughout, it would improve the quality of concrete by reducing its water-cement ratio and allowing it to become more compact. In fact, this desirable effect is obtained to a considerable degree in the vacuum process of extracting water from concrete (see Art. 14.20) and in the casting of ornamental "cast stone" in absorbent molds. In most concrete, however, the extraction is not uniform and the effects are undesirable. Usually the upper portion of the concrete is weaker than the average of the full depth. Also, water pockets formed under reinforcing bars and aggregate reduce bond and facilitate percolation of water. Further, the rising water tends to carry with it many fine soft particles which weaken the top portion and in extreme cases form a scum, called "laitance," over the surface. This laitance tends to prevent the bonding of a subsequent layer to the concrete below; it should be thoroughly removed by some method such as brushing or jetting. In floors and pavements, accumulation of bleeding water at the surface delays the finishing operations and dilutes the paste at the all-important finished surface.

Factors tending to minimize bleeding include favorable gradings and proportions of aggregate, low water contents, relatively high cement contents, finely ground cements, natural (rounded) sands with an adequate percentage of fines, certain finely divided mineral admixtures, and shallow placements or lifts. Air entrainment (see Art. 5.11) is very effective. In some cases the potential contraction due to bleeding has been offset through the use of special expansive-type cements or, in grouts, of aluminum powder which forms minute bubbles of gas.

The ASTM test for bleeding of concrete [C232] employs a cylindrical container of capacity approximately ½ cu ft (diameter 10 in., height 11 in.). The sample of concrete is placed in the container, filling it to a depth of 8 in., and is consolidated and smoothed off. The container is covered. At 10-min intervals during the first 40 min and thereafter at 30-min intervals, the water accumulated on the surface is withdrawn and measured. The amount of bleeding water is expressed as the percentage of net mixing water contained in the sample. The criterion may be taken as either the total at the end of bleeding or the bleeding at the end of a stated interval. The progress of bleeding can be observed by plotting a diagram of bleeding against time.

5.8 Plastic shrinkage and settlement. After concrete has been placed in a deep form, such as that for a building column or a wall, it can be observed that the top surface subsides. Also, when forms are stripped from a column or a wall, or even from a test cylinder, sometimes there are observed short, approximately horizontal cracks which indicate that there has been a tendency toward subsidence. The reduction in volume of the concrete is called plastic shrinkage, or setting shrinkage. It is largely due to bleeding [518] as water rises to the top, but it may result also from leakage and absorption of water by forms, absorption of water by aggregate, and/or combination of water with cement. A contributing cause of settlement, aside from reduction in volume of concrete, may be bulging or settlement of the forms.

Rapid drying of fresh concrete in slabs, even within minutes of placement, may cause evaporation of water which exceeds the rate of bleeding of water to the surface. When this condition occurs, according to Lerch

[517], the concrete near the surface has attained some initial rigidity but cannot flow plastically to accommodate the rapid volume change of plastic shrinkage, nor has it attained sufficient strength to withstand tensile stress; thus plastic-shrinkage cracks may develop. He recommends such corrective measures as dampening the subgrade, forms, and aggregates; avoiding extremes of temperature; starting the curing very soon; using temporary coverings or fog spray between placing and finishing; and using windbreaks and sunshades to reduce evaporation.

Even if actual cracking does not occur while the concrete is still plastic, it may develop later from potential cracks resulting from early surface drying.

5.9 Stiffening and setting. Fresh concrete, if left undisturbed, gradually stiffens until it may be said to have "set." However, there is no well-defined point at which concrete sets or passes from the plastic to the rigid condition. For practical purposes concrete should remain sufficiently plastic over a period of at least ½ hr, and preferably 1 hour or so, to permit being transported and consolidated without unusual measures and adverse effects; on the other hand, it should harden within a reasonable time for construction purposes. Under favorable conditions, up to several hours after first being mixed, concrete which has stiffened considerably may be rendered plastic and reconsolidated by vibration or remixing [520].

Practical interest in determining setting time of concrete lies in evaluating need for, and effectiveness of, set-controlling admixtures, regulating maximum times of mixing and/or transit, and providing protection from adverse weather conditions. The setting time of cement has been found to be no index to the setting time of concrete, but the setting time of mortar correlates fairly well with that of concrete. As in the case of cement, the setting time of concrete is necessarily an arbitrary value taken at some point in the gradual process of stiffening and defined in terms of a particular test method and apparatus [516]. Test methods which have been proposed include measurements of electrical resistance, consistency (see Arts. 5.3 and 5.4), wave velocity, bleeding characteristics, heat of hydration, volume change, deformability, compressive strength, flexural strength, strength of bond with smooth pull-out pins, time limit of replasticizing by vibration, and resistance to penetration of a plunger.

Of the foregoing methods, a penetration-resistance method similar to that developed by Tuthill and Cordon [520] has been adopted as a tentative ASTM Standard [C403] for concrete with slump greater than zero. In brief, a sample of mortar is sieved from the concrete, is placed in a container, and is subjected at intervals over a period of hours to penetration to a depth of 1 in. by a flat-ended plunger. The plungers employed range in face area from 1 to 1/40 sq in. The resistance per square inch of face is then determined. A curve is plotted, from which the time of initial set

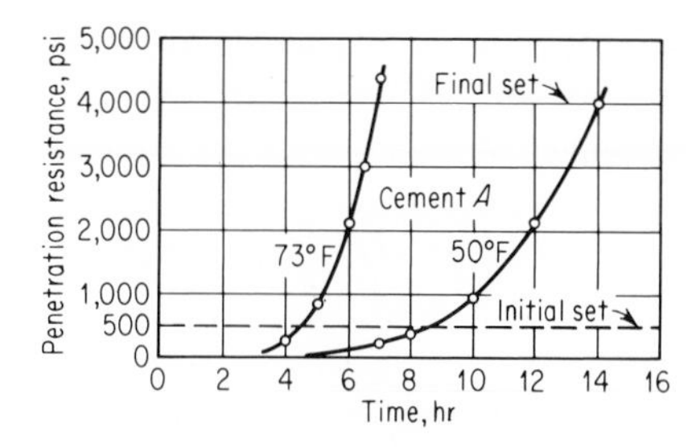

Fig. 5.5. Example of effect of temperature on setting time of concrete. (*From T. M. Kelly* [516].)

is taken at a penetration resistance of 500 psi, and the time of final set is taken at 4,000 psi. These points, although arbitrarily chosen, are useful in comparative studies of various factors and in specifying limits on the rate of hardening of concrete in construction. The penetration limit of 500 psi corresponds to the "vibration limit," beyond which concrete can no longer be made plastic again by vibration. The penetration limit of 4,000 psi corresponds to a mortar compressive strength of approximately 100 psi. An example is shown in Fig. 5.5.

The rate at which concrete stiffens is greatly affected by temperature, as indicated by Fig. 5.5; at low temperatures the setting may be unduly delayed for purposes of construction, and at high temperatures it may be unduly accelerated. Temperature and/or other factors such as absorption of mixing water by aggregates, grinding of aggregates in the mixer, and combination of water with cement may call for special control of job conditions. As one instance, the U.S. Bureau of Reclamation [106] limits the allowable loss in slump between the mixer and the forms to 1 in. Undesirably rapid increases in stiffness of freshly mixed concrete are "false" set, or the rapid development of rigidity without the evolution of much heat, and "flash" set, during which considerable heat is evolved. False set can be dispelled by further mixing without addition of water, but flash set cannot. The occurrence of either type of abnormal setting can usually be traced to conditions of manufacture of the cement, perhaps aggravated by job conditions such as high temperature.

5.10 Pressure on forms. For some time after placement, fresh concrete exerts pressure on the forms that confine it laterally. As the top surface of concrete being placed is brought higher above a given elevation, the lateral pressure at that elevation rises to a maximum which is a consideration of consequence in the design and construction of the formwork. The maximum, which is less than that for a fluid of the same density at the same elevation, depends on many factors including the rate and depth of placement; the consistency, temperature, and rate of stiffening of the concrete; the smoothness and permeability of the forms; and the method of consolidation of the concrete (for example, if vibrators are used, the pressures are relatively high). In general, the more rapid the rate of placement and the less rapid the rate of stiffening, the greater is the maximum pressure developed. See Art. 9.3 for quantitative values of lateral pressures under a variety of conditions.

AIR ENTRAINMENT

5.11 Entrained air. By "entrained" air is meant that which is intentionally introduced into concrete in the form of uniformly distributed minute bubbles (of the order of 1/100 to 1/1,000 in.) in appropriate amounts to produce a desired interbubble spacing and resultant desirable effects. "Entrapped" air is unintentionally included during the batching and mixing operations and is random in amount, size, and shape of the various inclusions; usually it amounts to a percent or so of the volume. Air-entraining agents and their effects on durability and strength of *hardened* concrete are discussed in Arts. 4.7 and 11.14. With regard to *fresh* concrete, air entrainment reduces mixing-water requirement, increases plasticity, reduces bleeding and segregation, and reduces unit weight.

Workability is improved in part because the air bubbles increase the volume of mortar; further, they have flexibility of shape and thus aid the movement of the rigid aggregates. They are most effective with regard to the finer range of sand particles; in fact, experiments have shown that mixtures with sufficient sand of favorable gradation of the portion between the No. 100 and No. 14 sieves but with no cement at all can be made plastic by air entrainment [522, 523]. In practice, it is desirable to reduce the volume of sand by approximately the amount of entrained air; when this reduction is made, the consistency remains about the same and the water requirement is less. Considerations in proportioning of air-entrained concrete are given in Chap. 6.

Many factors affect the amount of air entrained. Other conditions being equal and within practical limits, the air entrainment at the conclusion of mixing is greater for (1) greater amounts of air-entraining agent, (2) leaner mixes, (3) more of the fine aggregate, (4) more of the No. 30 to No. 50 size fraction in the sand, (5) smaller maximum sizes of aggregate (from 1½ in. down), (6) less of finely divided mineral admixtures, (7) wetter consistencies, (8) lower temperatures of concrete, (9) stronger mixing action, and (10) longer mixing. In general, the operations of conveying (especially pumping), placing, and consolidating tend to reduce the air content. A normal amount of vibration, say up to 15 sec, does not materially affect the amount of entrained air. One study revealed that 3 min of vibration reduced the percentage of air by 1 to 3 units over a range of consistencies. It is important that uniformity be maintained not only in the materials and proportions but also throughout the operations of batching, mixing, and placing.

5.12 Measurement of entrained air. Of various methods of determining the air content of fresh concrete, three (pressure, volumetric, and gravimetric) have been standardized by ASTM; these are briefly described in the following paragraphs.

Pressure method [C231]. This widely used method is based on the principle of Boyle's law that increase in pressure on a gas decreases its volume in proportion; the method makes use of the fact that in concrete only the air is compressible. Pressure is applied to a concrete sample and the reduction in volume is observed; the amount of air (entrained and entrapped) is then calculated or is indicated by a calibrated gage. The method is applicable for concretes except for those containing highly porous aggregates; in some cases, however, a correction for absorption and porosity can be determined by making a pressure test on a sample of the aggregate alone.

The air meter, one form of which is shown in Fig. 5.6, consists of a bowl and a cover capable of being attached in order to form a rigid,

Fig. 5.6. Measuring air content of fresh concrete by the pressure method. (*Portland Cement Association.*)

pressure-tight assembly. The cover is fitted with a graduated transparent tube, an air pump, and a pressure gage. The sample of concrete is placed in the bowl, consolidated, and struck off. The cover is attached. Water is added over the concrete up to a given mark near the top of the tube. A known air pressure indicated by the gage is applied, and the change in level of the water surface is observed. From the pressure and the corresponding change in volume, the volume of air in the concrete is computed.

A laboratory exercise in the determination of air content by the pressure method is given as Test 17 (Part II).

Volumetric method [C173]. In the volumetric method, the volume of air is determined by displacement as the difference between the volume of a sample of concrete under water and the volume of the same sample after the air has been washed out by agitation.

The air meter shown in Fig. 5.7 consists of a bowl and a cover section of capacity approximately the same as that of the bowl, capable of being attached to form a rigid, tight assembly. The cover is fitted with a graduated transparent tubular neck. The sample of concrete is placed in the bowl, consolidated, and struck off. The top section is clamped on. Water is added up to a given mark on the tube, and the top of the tube is closed. The apparatus is then agitated by rocking, shaking, and/or rolling until the water level in the tube becomes constant. The change in level of the water indicates the volume of air (entrained and entrapped) driven out by the washing action. The method is particularly useful for concretes containing lightweight or highly porous aggregates.

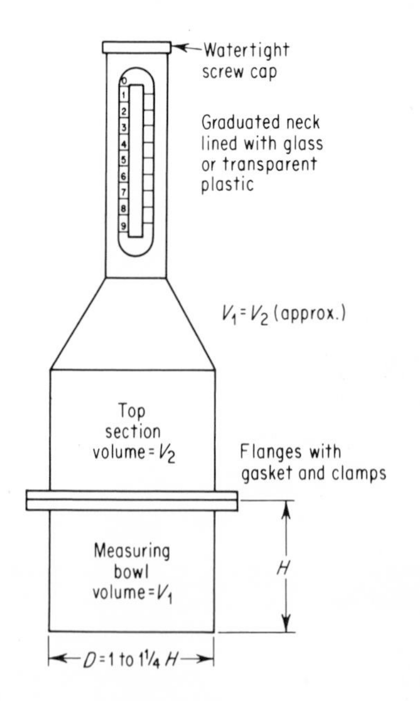

Fig. 5.7. Apparatus for measuring air content of fresh concrete by volumetric method. (*From American Society for Testing and Materials* [C173].)

Gravimetric method [C138]. In the gravimetric method, the volume of air (entrained and entrapped) is determined as the difference between the actual volume of a standard sample and the total air-free volume computed from the proportions and specific gravities of the ingredients including water and admixtures. It requires accurate knowledge of the amount and specific gravity of each ingredient and on that account is seldom used in the field. It is impractical for concretes containing lightweight or vesicular aggregates, for which determinations of specific gravity are indefinite.

Other methods. Other methods for measuring air content of fresh concrete have been developed, differing in detail but not in principle from the ASTM pressure and volumetric methods [525]. A pocket-size indicator measures volumetric displacement in alcohol of a sample of mortar from the concrete; the test is less accurate than those previously mentioned, but it is rapid and is useful for estimates and for detection of changes in air content in the field. The standard test for unit weight of concrete [C138] is often used to check possible changes in air content or in mix proportions from batch to batch or from time to time.

COMPOSITION: PROPERTIES

5.13 Unit weight; yield; cement factor. The density, in terms of weight per cubic foot, of freshly mixed concrete is necessary for a number of computations pertaining to mix design and control. The procedure is simple, involving the use of a calibrated container and the observance of such precautions as securing a representative sample and properly consolidating and screeding it. The procedure [C138] is used in several of the tests described in Part II herein.

From the unit weight of fresh concrete, the volume of concrete produced per batch can be computed as follows:

$$V = \frac{(N \times 94) + W_f + W_c + W_w}{W} \tag{5.1}$$

in which V = volume of concrete produced per batch, cu ft
N = number of sacks of cement in batch
94 = net weight of sack of cement, lb
W_f = total weight of fine aggregate in batch in condition used, lb
W_c = total weight of coarse aggregate in batch in condition used, lb
W_w = total weight of mixing water added to batch, lb
W = unit weight of concrete, pcf

The batch volumes thus calculated will not necessarily check with volumes measured in the forms because of absorption and losses in conveying and placing.

The volume of concrete can also be computed from the summation of the solid volumes (see Art. 3.5) of the cement and aggregates and the volume of water in the mix, and adding 1 or 2 percent of that summation to allow for some entrapped air, as shown in the following example:

Given is a concrete mix in which the proportions of cement to fine aggregate to coarse aggregate are 1:2.5:3.5 by weight and for which the water-cement ratio W/C is 6 gal per sack as determined by methods explained in Chap. 6. Since 1 cu ft equals 7.48 gal and since 1 sack of cement is assumed to be 1 cu ft, the cubic feet of water per sack of cement (W/C by volume) is $6.00/7.48 = 0.80$. Also, since there are 8.33 lb of water per gal and 94 lb of cement per sack, the W/C by weight is $6.00 \times 8.33/94 = 0.53$. The volume of each ingredient in a 1-sack batch is computed as shown in the following tabulation.

	Water	Cement	Sand	Gravel
Proportions, by wt	0.53	1.00	2.50	3.50
Wt for 1-sack batch, lb	$0.53 \times 94 = 50$	94	$2.50 \times 94 = 235$	$3.50 \times 94 = 329$
Sp gr (assumed)	1.00	3.10	2.65	2.65
Volume for 1-sack batch, cu ft $= \dfrac{\text{wt}}{62.4 \times \text{sp gr}}$	0.80	0.49	1.42	1.99

$$\text{Total volume} = 4.70 \text{ cu ft}$$

Assuming 1 percent air voids,

$$\text{Total volume} = 4.70 \times 1.01 = 4.75 \text{ cu ft}$$

$$\text{Total weight of batch} = 708 \text{ lb}$$

$$\text{Unit weight} = \frac{708}{4.75} = 149.0 \text{ pcf}$$

The "yield" of concrete is the amount of fresh concrete produced per sack of cement. The yield Y in cubic feet per sack is

$$Y = \frac{V}{N} \tag{5.2}$$

In the foregoing example

$$Y = \frac{4.75}{1} = 4.75 \text{ cu ft/sack}$$

The "cement factor" for a concrete mix is the cement content expressed in terms of sacks of cement per cubic yard of concrete. The cement factor

CF is

$$\mathrm{CF} = \frac{27}{Y} \quad \text{or} \quad \frac{27N}{V} \tag{5.3}$$

In the foregoing example

$$\mathrm{CF} = \frac{27}{4.75} = 5.68 \text{ sacks/cu yd}$$

The effective cement factor for concrete in place, computed from yardage in the forms and number of sacks delivered to the mixer, will differ somewhat from the cement factor as computed above, owing to settlement of concrete and bulging of forms.

Because of the trend to the metric system, because a "sack" means different amounts in different countries, and because cement is now batched from sacks only on small jobs, the practice is increasing of stating proportions and amounts of concrete and its components in terms of weight.

5.14 Proportions of components. On occasion it becomes desirable to check the actual proportions of one or more components of a field mix in order to evaluate uniformity, check mixer efficiency, or detect segregation. In effect, the mix is "unscrambled," and the amounts of the various components are measured.

The most widely used method, in which all components (except air and admixtures) are determined, was developed by Dunagan [536]; it is based on the specific gravities of the components. In brief, the procedure is as follows:

1. A sample of fresh concrete is weighed in air.
2. The sample is placed in a container with some water, stirred to drive out air, and (still in the container) weighed under water.
3. The sample is washed onto nested No. 4 and No. 100 sieves. The sample is thus separated into coarse aggregate, fine aggregate, and cement (with some fine sand) which with the water is washed away.
4. The coarse aggregate and fine aggregate are weighed under water.
5. The proportions are computed from the weights, making use of the relations (*a*) that the apparent absolute volume of a material is equal to the weight (in metric units) of water displaced and (*b*) that the specific gravity is equal to the weight of the material in air divided by the weight of displaced water.

The immersed weight of the cement is determined as the difference between the immersed weight of the total sample and that of the coarse and fine aggregates.

If a sieve analysis of the sand is available, the amounts of cement and sand can be corrected by the amount of sand finer than No. 100, which was washed out with the cement. Usually this amount can be assumed and in some cases even neglected.

The apparatus (containers, scales, sieves, etc.) for the Dunagan test is relatively simple, and the test can be made in a short time, perhaps 15 minutes. However, the method is subject to several errors of testing.

Several other methods of unscrambling fresh concrete by means of sieves have been devised. However, in general they are cumbersome and are subject to a number of errors; they are little used.

In a test used by the U.S. Bureau of Reclamation [106], a sample of concrete is washed on a No. 4 sieve and the air-free unit weight of the mortar is computed; samples taken from different parts of a batch indicate the uniformity of the concrete. The amount of the portion retained on the No. 4 sieve may also be determined.

The cement content of fresh concrete can be determined by a method of separation developed by Hime and Willis [538]. The cement and some fine sand are washed from a concrete sample through a No. 30 sieve, dried, and then separated by being centrifuged in a liquid of density greater than that of the sand but less than that of the cement. The test is rather involved and is subject to a number of errors of sampling and testing; it is not widely used.

The water-cement ratio of freshly mixed concrete can be compared with that of corresponding concrete of known water-cement ratio by measuring the electrical resistance of the two concretes under fixed conditions [539]. This rapid method can also be used to detect fluctuations from batch to batch without necessarily knowing the actual water-cement ratio. Other methods of determining water-cement ratio, which have been proposed but which so far have not been developed for use, involve neutron attenuation for measurement of water content and x-ray spectroscopy for measurement of cement content.

5.15 Uniformity. During production, fresh concrete may vary in composition between batches at various times or from various mixers because of such factors as uncontrolled differences in materials, proportions, mixing, and temperature. Further, concrete within a given batch may vary because of faulty mixing equipment, improper loading, or insufficient mixing. In order to keep such batch-to-batch variations and within-batch variations within reasonable limits, methods of field testing have been used to some extent [540]. Usually they involve determinations of proportions, unit weight, air content, and/or consistency, all of which are described briefly in foregoing articles of this chapter.

Batch-to-batch variations are determined by conventional methods and are checked against specified allowable limits. For example, the allowable variation in slump is often set at 1 in. and in air content at 1 percent.

Within-batch variations are of importance in checking efficiency and performance of mixers, especially truck mixers. Based on a comprehensive investigation of truck mixers, Bloem, Gaynor, and Wilson [535] recommend that samples be taken at about 1/6 and 5/6 the length of the mixer and that, as a minimum, tests be made to determine slump, unit weight, air content, and the percentage of coarse aggregate by washout test. They consider that the approximate maximum differences likely to occur in reasonably uniform batches of ready mixed concrete are: slump, 2 in.; air-free unit weight of mortar, 2 pcf; air content, 1 percent, and coarse-aggregate content, 5 percent.

The U.S. Bureau of Reclamation, which is primarily concerned with large stationary mixers, limits within-batch variations in unit weight of air-free mortar to 0.8 percent, which is approximately equal to the maximum variation in truck mixers given in the preceding paragraph. Few organizations are concerned with within-batch variations in stationary or paving mixers, but probably an allowable maximum variation in slump would be ½ or 1 in. and in air content would be 1 percent.

5.16 Temperature of components and mix. In cold weather it may be desirable to produce concrete at a temperature above the ambient temperature by heating the mixing water, the aggregates, or both. In hot weather, sometimes the fresh concrete is made with cooled mixing water and, in the case of dams, with cooled aggregate. An estimate [106] of the temperature of the concrete as produced at the mixer can be made as follows:

$$T = \frac{S(T_aW_a + T_cW_c) + T_fW_f + T_mW_m}{S(W_a + W_c) + W_f + W_m} \tag{5.4}$$

in which the symbols are given in the following tabulation:

	Weight	Temperature
Aggregates (surface-dry basis)	W_a	T_a
Cement	W_c	T_c
Free moisture on aggregates	W_f	T_f
Mixing water (added)	W_m	T_m

S = specific heat of solid materials (may be assumed to be 0.20 to 0.22 for practical purposes)

The foregoing equation may be used with any system of units. However, it is applicable only as long as temperatures of all water, including

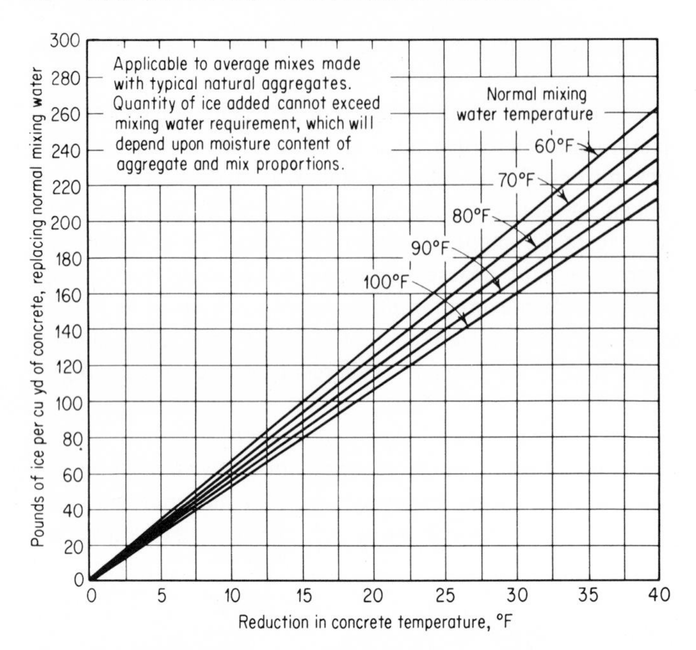

Fig. 5.8. Effect of ice in mixing water on concrete temperature. (*From National Ready Mixed Concrete Association.*)

moisture in aggregates, are above 32°F. If the water alone is heated, for each degree that the water temperature is raised, the temperature of ordinary fresh concrete is raised about ¼°F. Depending on heat losses during batching and mixing and on heat generated by mixing, wetting, and early chemical activity of the cement, the actual temperature of the concrete will differ from that computed by the equation. For given working conditions a correction factor can be determined.

If ice is used as part of the mixing water, the heat of fusion of the ice should be subtracted from the numerator. In British units the heat of fusion of ice is 144 Btu per lb. Figure 5.8 shows the effect of ice in mixing water on concrete temperature, within stated limits.

Cold-weather concreting is discussed in Art. 8.14 and hot-weather concreting in Art. 8.16.

QUESTIONS

1. Discuss four principal factors governing the plasticity of a concrete mixture.

2. Distinguish between workability and consistency. Why are the tests for these properties relative, not absolute?

3. Describe the changes which take place in a concrete mix as increments of water are added.

4. In qualitative terms, what consistencies are considered optimum for the principal types of construction?

5. Describe briefly the slump, flow, and ball tests, giving comments on the advantages and disadvantages of each.

6. For a concrete of medium consistency, say a 3 to 4-in. slump, what are the roughly corresponding values of flow? Of ball penetration?

7. Describe the Powers remolding test. What is its significance?

8. State the principles on which two or more of the other attempts to measure workability of concrete are based.

9. What are the undesirable effects of bleeding? What effects are desirable, if uniform or controlled?

10. How can bleeding be minimized?

11. What are the causes of plastic settlement and shrinkage? The effects?

12. Trace the gradual hardening of concrete up to the age of, say, 1 day. What is the basis of the specified time of initial set? Final set?

13. Describe the standard penetration-resistance method of determining times of set of concrete.

14. What is the effect of temperature on rate of setting?

15. Discuss the variation in lateral pressure at a given elevation in a deep form as concrete continues to be placed above that elevation. What principal factors affect the pressure?

16. Discuss the beneficial and adverse effects of entrained air on the properties of fresh concrete. Distinguish between entrained and entrapped air.

17. What factors affect the amount of entrained air? In what direction?

18. Briefly describe the three standard methods of measuring entrained air in fresh concrete, and comment on where and why each would be used.

19. How may the volume of fresh concrete produced be computed from the weights of the component materials and the unit weight of the concrete? From the weights and specific gravities of the materials?

20. Define "yield" and "cement factor," and state the equation which relates them.

21. What is the general process of determining the proportions of components of fresh concrete? Discuss the basis and general procedure of two of the methods.

22. Discuss batch-to-batch and within-batch variations in fresh concrete. What tests are sometimes made to investigate each type? What are reasonable values of allowable variation in properties for each type?

23. What three values must be known for each component in order to compute the temperature of fresh concrete? State in your own words the relationship in the equation.

SIX
PROPORTIONING OF CONCRETE MIXES

6.1 General. Proportioning of a concrete mix consists in determining the relative amounts of materials to be used in batches of concrete for a particular purpose. The process is sometimes called "design" of mix, although this term tends to be confused with structural design. Proportioning may be entirely empirical, that is, it may depend on experience or observation alone, or it may have a technical basis of tests and calculations. It involves consideration of available materials and their costs, requirements of placing and finishing the concrete, and properties of the hardened concrete such as durability, strength, and volume constancy. The aim is to produce economical concrete of satisfactory workability while still fresh, and of satisfactory quality after hardening.

In this chapter are discussed the conventional methods of expressing proportions, fundamental relationships between aggregate and cement-water paste, effects of changes in one variable on other variables, a procedure for proportioning by direct trial and observation, the current American Concrete Institute (ACI) methods of selecting proportions, and (briefly) several other methods which are chiefly of historical interest. It is assumed that reasonably well-graded aggregates are available conforming to the requirements discussed in Chap. 3, that all materials are acceptable, and that the properties of fresh concrete (Chap. 5) are understood.

The proportioning of the ingredients of concrete mixtures is a highly important phase of concrete technology, in that it provides the means of meeting the fundamental requirements of quality and economy. Concretes varying widely in quality and cost can be made with a given set of materials; also materials of widely varying make-up can be combined to produce concretes of acceptable quality. Hence it is important that the interrelationships be understood and controlled to produce the optimum combination of properties at reasonable cost.

It is not essential that all concretes be of the highest possible quality as judged by some fixed standard. For example, some parts of a structure can well be stronger than others, and interior concrete must primarily meet strength requirements, whereas exterior (exposed) concrete may need primarily to be weather-resistant. Likewise, a satisfactorily workable concrete for placement in large unobstructed forms may be unsuitable for placement in small heavily reinforced members.

It is customary to evaluate the quality of concrete in terms of its strength, even when strength is considered merely as an indicator of other properties such as resistance to weathering or resistance to cracking. Appropriate strengths for various types of structures are fairly well established by experience (see Art. 10.5). The average strength for which concrete is proportioned should be somewhat higher than that assumed in the structural design (Art. 17.10).

The relative economy of concrete mixes in construction depends primarily on cost of materials rather than on cost of labor and other related factors. Although the labor and other costs of forming, transporting, and placing concrete are a large part of the total, *differences* as between different concretes are small, particularly with modern mechanical methods of handling. Therefore, in proportioning, attention is directed principally (1) to the costs of available materials as related to the amounts needed in the desired mix and (2) to producing a mixture of appropriate workability so as to keep placement and finishing costs at a minimum consistent with securing quality in the finished structure.

An outstanding relationship affecting cost as governed by ingredients is that the cost of portland cement, *per pound,* is of the order of seven or eight times that of aggregates; hence, with regard to economy, effort is made to use as little cement as possible consistent with strength and other requirements. Differences in cost as between aggregates are usually secondary, although in some localities or with some special aggregates they are sufficient to influence selection and proportioning. The cost of mixing water usually has negligible influence. The cost of admixtures may be important because of their potential effect on proportions of cement and aggregates.

The following generalization is useful: Other conditions (such as materials and consistency) being equal, stronger concretes require more cement and are therefore more expensive per unit of volume.

In structural design, the higher unit cost of stronger concrete may be more than offset by reduction in size of members—columns, for example—due to use of proportionately higher permissible working stresses.

6.2 Methods of expressing proportions. The ingredients of concrete are cement, water, aggregate, and in some cases admixtures or air which is entrained by means of admixtures. Fine and coarse aggregates are handled separately, in part to minimize segregation but principally as a means of regulating the gradation of the mixed (total) aggregate; also usually sand (consisting of rounded particles) is used as fine aggregate even when the coarse aggregate consists of crushed (angular) material. In some cases the fine aggregate may consist of more than one sand. Likewise, the coarse aggregate may consist of two or more materials or size fractions, particularly when the maximum size is greater than, say, 1½ in.; for mass concrete the coarse aggregate may contain as many as four fractions ranging up to a maximum size of 6 in. In this chapter, however, proportioning is discussed for the simplest case of one fine and one coarse aggregate, under the assumption that each will be reasonably well graded.

One common method of expressing the proportions of the granular materials in concrete mixes is in the form of parts, or ratios, of cement, fine aggregate, and coarse aggregate, with cement taken as unity; for example, a 1:2:4 mix contains 1 part of cement, 2 parts of fine aggregate, and 4 parts of coarse aggregate. This practice was established before the important influence of amount of water on quality was understood, and concrete was considered to be a mixture of granular materials with water merely added for fluidity. Although this concept has been replaced by one that basically visualizes a two-component mix—paste and aggregate—as explained in Chap. 1, the original form of expression is still convenient for a number of purposes, and it continues to be used. Carried to its logical conclusion, the expression should contain four terms, with water included; thus a 0.7:1:2:4 mix would contain the proportions of granular materials previously explained but with the relative amount of water (first) stated also. However, current common practice is to state only the amounts of cement, fine aggregate, and coarse aggregate by parts, with the water content expressed separately (if at all) in terms of water-cement ratio. Likewise, the content of entrained air or an admixture is expressed separately.

In order to be definite, the *basis* of proportions should be stated; otherwise it is not always clear whether they are by weight, by solid (sometimes called "absolute") volume, or by overall (bulk) volume. Further, if by overall volume, it should be understood whether the aggregates are loose or compacted, and—particularly for sand—whether they are damp or dry. When proportions are stated by overall volume, it is customary in the United States to consider that one sack (94 lb) of cement has an overall volume of 1 cu ft.

Water-cement ratio may be expressed by weight or by volume, but is most often expressed in terms of gallons of water per sack of cement.

The amount of air accidentally entrapped or purposely entrained in concrete is expressed as a percentage of the volume of concrete. Amounts of admixtures are expressed relative to the weight of cement.

Alternative forms of expressing proportions are (1) by cement factor, or sacks of cement per cubic yard of concrete (Art. 5.13), and (2) by ratio of cement to the sum of fine and coarse aggregates, as 1:6.34.

For purposes of proportioning, computations are customarily made on the basis of corresponding amounts either per sack of cement (a "one-sack batch") or per cubic yard of concrete, the traditional unit.

The fundamental basis for proportioning is by solid volumes, since essentially the effort is to produce an arrangement of graded particles in space and since the actual space occupied by the materials is of significance in computations of yield. When proportions are expressed by weight, the specific gravities of the aggregates and cement are involved. Proportioning by overall volume is indefinite because of differences and variations in compaction and bulking, and it should be avoided except perhaps on small unimportant jobs.

6.3 Aggregate-paste relationships. It has been shown in Chap. 1 that concrete consists essentially of cement-water paste which is the "active" ingredient, and "inert" mineral aggregate. (In this chapter it is assumed that the aggregates are not reactive; see Arts. 3.15 and 11.16.) The internal structure may be further visualized by reference to Fig. 6.1*a*, representing a cross section of concrete in which A designates pieces of aggregate (large and small) and P designates paste. Workability of the mass is provided by the lubricating effect of the paste, and is influenced by the amount and dilution of paste. Strength of concrete is limited by strength of paste, since mineral aggregates, with rare exceptions, are far stronger than the paste component. Essentially the permeability of concrete is governed by the quality and continuity of the paste, since but little water flows through aggregate either under pressure or by capillarity. Further, the predominant contribution to drying shrinkage of concrete is that of the paste.

Dilution of paste. Since the properties of concrete are governed to a considerable extent by the paste, it is helpful to consider more closely the structure of paste. The fresh paste is a suspension, not a solution, of cement in water. In Fig. 6.1*b*, which is an enlargement of the small square in Fig. 6.1*a*, the cement particles (large and small) are indicated by C and the water-filled space between by W. The more dilute the paste, the greater the spacing between cement particles, and thus the weaker will be the ultimate paste structure at any stage of hydration. This qualitative statement is borne out by the well-known relationship that, other conditions being equal, for plastic mixes the strength of concrete varies as an inverse function of the water-cement ratio, which is essentially a way of expressing the degree of dilution of paste. Figure 6.2 gives the general form of the relationship; the numerical relationships expressed by the curve depend on such factors as the particular materials (principally cement),

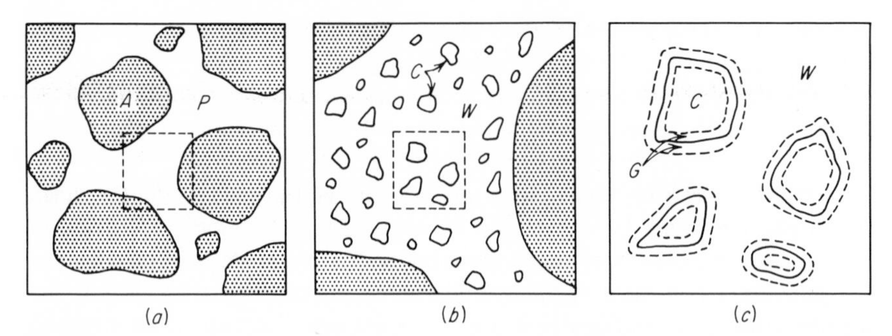

Fig. 6.1. Interrelationship of aggregate (A), paste (P), cement (C), water (W), and gel (G).

length of curing period, and moisture and temperature conditions of curing and testing. Typical relations between water-cement ratio and strength of standard-cured concrete containing type I portland cement are shown in Fig. 10.9.

The so-called "water-cement ratio law" is essentially that for given materials and given conditions of placing and curing of originally plastic concrete, the strength is governed principally by the ratio of water to cement. Strength is also affected by the characteristics of the cement, by the extent of hydration, and (to a lesser degree) by the shape, surface texture, maximum size, grading, stiffness, and amount of aggregate. In spite of these other influencing factors and the limitation to plastic mixes, however, the primary relationship between water-cement ratio and strength is most useful in proportioning, and it is widely used.

Likewise, it can be visualized in Fig. 6.1*b* that the more dilute the paste, the more readily can water be forced through the concrete under pressure. Permeability tests confirm this relationship. Further, the lower

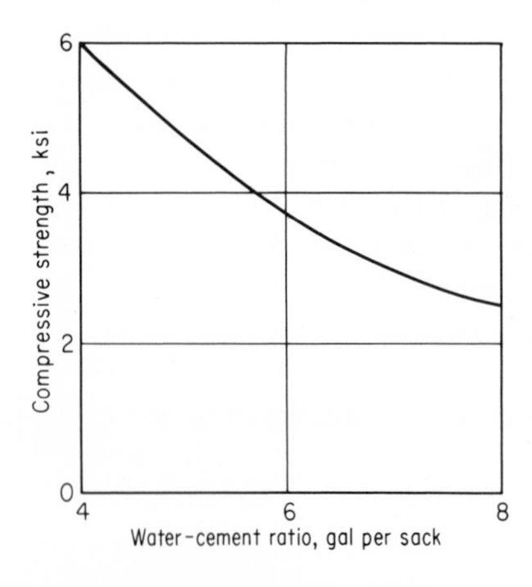

Fig. 6.2. Typical form of relationship between water-cement ratio and compressive strength. (For illustration only, to show trends; values will differ with materials and test conditions. See Fig. 10.9.)

the water-cement ratio, the greater the resistance of concrete to the effects of freezing and thawing.

Hydration. A still finer representation of the structure of paste is illustrated by Fig. 6.1*c*, which is an enlargement of the small square in Fig. 6.1*b*. As hydration progresses, the water and material on the surface of the cement particles form a gel, indicated as G. This cement gel may have more than twice the volume of cement from which it is produced. As hydration takes place, the gel extends outward from each particle to join with gel from other particles, and a connected skeletal structure is formed as setting takes place. So long as sufficient water remains in the mass to permit hydration to proceed, the amount of gel increases and the paste (and therefore the concrete) becomes stronger, although the rate diminishes as the smaller particles of cement are used up and the unhydrated cores of the larger particles become smaller. A generalized relationship between strength and period of hydration is shown in Fig. 6.3. The important point to be observed here is that hydration is far more rapid at early ages than at later ages; more detailed discussion is given in Chap. 10.

The influence of fineness of grinding of cement on rate of strength development is readily understood from the foregoing discussion. For a given dilution of paste, the finer the particles, the closer together they are initially and—because of their greater area of surface exposed to the water of hydration—the more rapid the formation of gel. Often a so-called high-early-strength cement is merely one which is ground finer than the normal cement from a given supply of clinker.

Roughly, the amount of water required for complete hydration of cement is half of that used in making an average concrete. The excess water over that sufficient for hydration is required for purposes of workability; in some special processes such as the application of absorbent coverings or the vacuum process, a part of or all the excess water may be extracted after the concrete has been placed. In any case, so far as moisture is

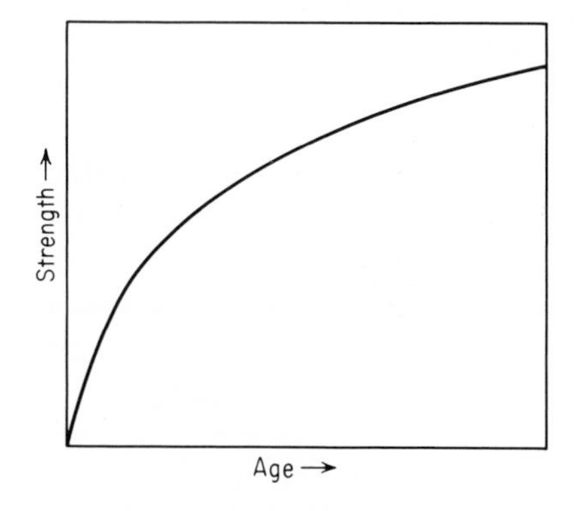

Fig. 6.3. Typical development of strength of moist-cured concrete with age. (For illustration only, to show trends; values will differ with materials and test conditions. See Figs. 10.8 to 10.11.)

concerned, "curing" consists essentially in retaining water which is already present, and curing may usually be accomplished by effective sealing of the surface. In exceptional cases of originally low water content, additional water may be needed to ensure hydration.

6.4 Variables in proportioning. With given materials, the four variable factors to be considered in specifying a concrete mix are (1) water-cement ratio, (2) cement content or cement-aggregate ratio, (3) gradation of the aggregate, and (4) consistency of the fresh concrete. In general, all four of these interrelated variables cannot be chosen or manipulated arbitrarily; usually two or three of the factors are specified, and the others are adjusted to give maximum workability and economy. The principal requirements are usually dictated by general experience with regard to structural design conditions, durability, and conditions of placing.

As previously stated, water-cement ratio expresses the dilution of paste. The cement content varies directly with the amount of paste of given dilution; it may be varied with concomitant changes in one or more of the other factors. So far as the practical proportioning of a given set of materials is concerned, the gradation of aggregate is controlled by varying the amounts of the given fine and coarse aggregate (or aggregates); the common means of expression is the percentage of sand (fine aggregate) in the total aggregate (not in the concrete). The consistency is established by the practical requirements of placing and hence is not subject to manipulation in proportioning as are the other three variables; it is usually expressed in terms of slump or of ball penetration (Art. 5.4). The stiffest consistency that can be placed effectively should be employed.

Although the variables in concrete proportioning could be expressed in other terms, those given above are of primary significance, and they are commonly used.

Further consideration of the amount of paste in concrete is desirable at this point. In brief, the effort in proportioning is to use a minimum amount of paste (and therefore cement) that will lubricate the mass while fresh and that after hardening will bind the aggregate particles together and fill the spaces between them. Any excess of paste involves greater cost, greater drying shrinkage, and greater susceptibility to percolation of water and therefore attack by aggressive waters or weathering action.

In order to minimize paste content, it is desirable that the gradation of aggregate particles be such that the amount of "voids," or spaces between particles, be relatively small. Ideal gradations are not economically feasible, and in practice the available fine and coarse aggregates are combined in proportions which, so far as practicable, require minimum paste. In the common case of a combination of one fine and one coarse aggregate, the percentage of sand (in the total aggregate) which requires minimum paste is called the "optimum." The optimum value depends on the particular materials and concrete; for an average concrete (No. 4 to 1½-in. smooth well-rounded gravel, 0 to No. 4 natural sand, a medium consistency, and a water-cement ratio of 6½ gal per sack) it is approximately 40 percent. A lower percentage of sand can be used when the maximum size of graded coarse aggregate is increased (thus reducing voids), when a finer sand is used (thus distending the coarse aggregate less), when a richer mix (of lower water-cement ratio) is employed, or when the concrete is air-entrained. For flat, rough, or angular (crushed) aggregates, the optimum percentage of sand is higher.

Figure 6.4 illustrates the influence of proportion of sand on workability, economy, and quality of concrete. In mixture A (at top) there is not enough cement-sand mortar to fill all the spaces between the pieces of coarse aggregate. Such a mixture will be difficult to handle and place and will result in rough, honeycombed surfaces and porous concrete.

Mixture B contains a correct amount of mortar. With light troweling all spaces between pieces of coarse aggregate are filled with mortar. Note the appearance of some coarse aggregate at the edges of the pile. This is a good workable mixture and will give maximum yield of concrete with a given amount of cement.

In mixture C there is an excess of mortar. While such a mixture is plastic and workable and will produce smooth surfaces, the yield of concrete will be low and consequently uneconomical. Such concrete is also likely to be porous.

6.5 Trial method of proportioning. With a basic understanding of the relationships and factors involved, concrete may be proportioned effectively by direct trial based on control of water-cement ratio, and this method is widely used. Herein one simple approach will be described; it may be expanded to include comparative tests of different materials and mixes, and for large jobs to include tests of significant properties in addition to strength of the hardened concrete.

The method requires that samples of the cement, fine aggregate, and coarse aggregate be available and that the relation between water-cement ratio and strength be known or assumed.

Briefly, the process of proportioning given materials for a given purpose by the trial method in advance of actual construction comprises the following steps:

1. Selection of the water-cement ratio appropriate to the requirements of strength and other properties such as durability
2. Making one or more small trial batches of concrete having this water-cement ratio and the required consistency, in order to determine the

Fig. 6.4. Appearance of (A) undersanded, (B) properly sanded, and (C) oversanded concretes. *(Portland Cement Association.)*

optimum proportions and amounts of the aggregates that will produce a workable mix with a minimum of paste (and therefore cement)

Example. Let us design a concrete mix to be used in some part of a building, say, a column. The structural designer has based his allowable working stress on an assumed compressive strength, such that, with an appropriate margin of safety, the average compressive strength for which the design is to be made is 4,000 psi. From experience or by referring to values published by technical societies (e.g., Ref. 601 and Table 6.6 herein) the corresponding water-cement ratio is taken to be 6 gal per sack. It is decided that a hand-mixed batch containing $\frac{1}{16}$ sack of cement will be large enough for trial purposes. This amount of cement (5.87 lb) and $\frac{1}{16}$ of 6 gal of water (3.12 lb) are measured out and mixed by hand in a pan to form a paste.

Quantities of the two aggregates have been brought to the saturated surface-dry condition so that they will not contribute water to, nor absorb water from, the paste. From supplies of known weight the aggregates are added in amounts dictated by judgment, with intermittent mixing, until the batch is brought to the desired consistency. As the limiting consistency is approached, the batch is carefully examined in order to judge which aggregate to add. The correct amount of sand is the minimum which will produce enough mortar to fill the spaces between pieces of coarse aggregate and provide a slight excess for workability; guides to judgment in this respect are afforded by Fig. 6.4. In general, undersanding creates harshness whereas oversanding decreases yield; hence a "balanced" mix is desirable. The optimum percentage of sand is not sharply defined; a variation of 3 above or below this percentage makes a difference of only about 1 percent in paste content. The operator can soon develop a sense of judgment regarding the proper proportions.

When the batch is judged to be satisfactory, the remaining supplies of aggregate are weighed, and by difference the amounts used in the batch are computed. In this example, suppose that it is found that 11.94 lb of sand and 18.70 lb of coarse aggregate have been used. The proportions of cement and aggregates are thus fixed and can be expressed as desired in terms of parts by weight (1:2.03:3.18) or pounds per one-sack batch (94:191:299). From any of the three foregoing sets of values, the percentage of sand in the aggregate is computed to be 39.

The quantities of materials to be used in field batches may now be computed, provided the moisture content of the aggregates is known. Suppose that one-sack batches are to be used, that the job sand contains 4 percent of surface moisture by weight, and that the job coarse aggregate is air-dry and will absorb 1 percent of water by weight. The surface moisture in the job sand then amounts to $0.04 \times 191 = 8$ lb, and the batch weight of damp sand must be $191 + 8 = 199$ lb. The coarse aggregate will absorb $0.01 \times 299 = 3$ lb of water, and the corresponding batch weight must be $299 - 3 = 296$ lb. The amount of water to be added directly at the mixer to obtain 6 gal (50 lb) net is

$$50 - 8 + 3 = 45 \text{ lb}$$

To summarize, the batch quantities, in pounds, are as follows:

	Cement	Water	Sand	Coarse aggregate	Total
Saturated surface-dry aggregates	94	50	191	299	634
Job aggregates	94	45	199	296	634

Further, the yield and cement factor of the concrete may now be computed as described in Art. 5.13. Suppose that in this example the unit weight of the fresh concrete is determined by test to be 151.2 lb per cu ft. Then a one-sack batch contains its total weight divided by its unit weight, or $634/151.2 = 4.19$ cu ft of concrete, and the corresponding cement factor is $27/4.19 = 6.44$ sacks per cu yd. Slide-rule computations are sufficiently precise, to three (or the lower range of four) significant figures.

There are many variations of the simple procedure just described. The order of adding the ingredients is immaterial; the order just described was used as an aid in visualizing the process. Damp and/or absorbent aggregates may be used instead of saturated surface-dry aggregates provided the moisture contents are known and the final batch has the correct water-cement ratio; this procedure, however, usually would require more than one trial batch—the first trial being used as a guide to the second. If the operator is undecided as to the optimum percentage of sand, other batches may be made using somewhat more and somewhat less sand, and the workability and paste content of these batches compared. Likewise, various available aggregates may be compared by means of trial batches.

It should be borne in mind that the workability of small handmade batches may not be reflected precisely in large field-mixed batches owing to differences in mixing, amount of paste adhering to the apparatus, appearance of edges of batch, etc.

An important variation of the trial-batch method is its use on the job, in the regular construction mixer, with the aggregates in the field-moisture condition, and without stopping the work. The moisture content (positive or negative) of the aggregate is either estimated or determined by test, depending on the precision desired. With the measured amounts of cement, aggregates, and added mixing water known, the net water-cement ratio can be computed. If this is greater or less than the desired amount, the water added per batch is decreased or increased accordingly, while at the same time the amount of aggregate is increased or decreased respectively to maintain the desired consistency. If inspection of the concrete being transported and placed indicates that it is undersanded or oversanded (see Fig. 6.4), the proportions of aggregates in succeeding batches are adjusted to correspond. All these common-sense adjustments are interrelated, but the process of continual adjustment by successive approximations is simple and practical. Eventually the mix is brought to the correct water-cement ratio, consistency, and percentage of sand; and it will remain so until variations in gradation or moisture content of aggregate in succeeding batches disturb the balance and require compensating corrections.

In practice, it has been found that a fixed quantity of water in successive batches is not always practical, as unavoidable variations in moisture content or grading of aggregates result in variations in consistency before the difficulty can be detected and corrected. Hence, once the proportions of a mix having the desired water-cement ratio are established, it is customary to bring succeeding batches to the same consistency under the assumption that the water-cement ratio will thus be maintained within a permissible tolerance of approximately ¼ gal per sack.

When the concrete is to contain purposely entrained air, the procedure is identical, except that machine mixing is required for proper air entrainment. There is likely to be some discrepancy between the air content of small laboratory batches and full-size field batches, for which adjustment must be made as construction progresses.

6.6 Mix adjustments. The effects of accidental or intentional changes in the consistency and/or proportions of a concrete mix may be compensated or controlled in various alternative ways. For example, if the *water-cement ratio* is to be held constant, an increase in slump is accomplished by increasing the amount of paste (or more convenient by decreasing the amount of aggregate), resulting in richer (and therefore more expensive) mix. If the *cement content* is to be held constant, however, an increase in slump is accomplished by adding water, thus increasing the water-cement ratio and resulting in a weaker concrete.

Over a reasonable range, cement and sand may be interchanged by equal solid volumes, resulting in changes in both cement content and water-cement ratio without affecting total water content or slump.

Approximate quantitative adjustment for changes in mix conditions may be made in accordance with the values in Table 6.1. The variations are from a base mix containing rounded aggregates of average grading and having a water-cement ratio of 5.2 gal per sack and a slump of 3 in.

Table 6.1 Effect of changes in mix conditions on optimum percentage of sand and water content of concrete*

Change in conditions	Effect on optimum percentage of sand	Effect on unit water content
Each 0.05 (½ gal/sack) increase in water-cement ratio (water constant, cement variable)	+1	0
Each 0.1 increase in fineness modulus of sand	+0.5	0
Each 1-in. increase in slump	....	+3%†
Each 1% increase in air content	−0.5 to 1.0	−3%
Each 1% increase in sand content	....	+2.5 lb/cu yd
Angular coarse aggregate	+3 to 5	+15 to 25 lb/cu yd
Manufactured sand (sharp and angular)	+2 or 3	+10 to 15 lb/cu yd
For less workable concrete, as in pavements	−3	−8 lb/cu yd

* From earlier editions of Ref. 617 (U.S. Bureau of Reclamation).

† Average; varies from 4½ percent at 2½-in. slump to 2 percent at 5½-in. slump.

6.7 ACI method of proportioning. For the purpose of predetermining, on a more quantitative basis, the proportions of materials to produce mixes of desired properties, a number of semianalytical procedures have been developed. While these methods are also empirical in the sense that they depend on data tabulated from observation of a large number of trial mixtures, they do permit fairly close calculation of mix quantities for a variety of conditions.

In recent years, the procedure described in this article, or some modification thereof, has come into fairly common use in the United States. The procedure is that adopted by the American Concrete Institute and published as a "Recommended Practice for Selecting Proportions for Concrete" [601]. Six tables are given which, together with laboratory tests to determine fineness modulus of fine aggregate, maximum size of coarse aggregate, unit weight of dry-rodded coarse aggregate, and specific gravity of aggregates yield information for selecting proportions of materials (dry-aggregate basis). The specific gravity of cement is assumed to be 3.15. It must be known whether air is to be entrained in the concrete. As with other methods, the proportions are considered as approximate only, should preferably be checked by trial batches, and should be modified during construction if necessary to maintain the correct water-cement ratio and optimum yield at the established consistency. The method has proved practical and is in wide use. It has the advantage of simplicity in that it applies equally well, and with identical procedure, to rounded or angular coarse aggregates, to regular or lightweight aggregates (see also Art. 6.9), and to air-entrained or non-air-entrained concretes. Recommended practices for proportioning

Table 6.2 Recommended slumps for various types of construction*

Types of construction	Slump, in.†	
	Maximum	Minimum
Reinforced foundation walls and footings	5	2
Plain footings, caissons, and substructure walls	4	1
Slabs, beams, and reinforced walls	6	3
Building columns	6	3
Pavements	3	2
Heavy mass construction	3	1

* From Ref. 601.

† When high-frequency vibrators are used, the values given should be reduced about one-third.

structural lightweight concrete (Art. 6.9) and no-slump concrete (Art. 6.10) are also published by the ACI.

As one distinctive feature, the regular ACI method makes use of the established fact that, over a considerable range of practical proportions, fresh concrete of a given slump and containing a reasonably well-graded aggregate of given maximum size will have practically a constant total water content regardless of variations in water-cement ratio and cement content, which variations are necessarily interrelated. It is desirable to use the lowest water content consistent with the requirements of placing. Table 6.4 shows the total free water content corresponding to each of a range of slumps and maximum sizes.

Another distinctive feature makes use of the relation that the optimum dry-rodded volume of coarse aggregate per unit volume of concrete depends on its maximum size and the fineness modulus of the fine aggregate as indicated in Table 6.7, regardless of shape of particle [609]. The effect of angularity is reflected in the void content; thus angular coarse aggregates require more mortar than rounded coarse aggregates.

Briefly, the customary procedure consists of the following steps:

1. The consistency (in terms of slump) appropriate to the conditions of placing is selected from Table 6.2. Concrete of the stiffest consistency that can be placed efficiently should be used.

2. The maximum size of aggregate recommended for the type of construction is selected from Table 6.3. Within the limits shown, the largest permissible maximum size should be employed, except as dictated by availability and economy. The fineness modulus of the selected sand, the unit weight of dry-rodded coarse aggregate, and the specific gravity and absorption of both aggregates are determined by test.

3. The total free water (gallons) per cubic yard of concrete is read from Table 6.4, entering the table with the selected slump and selected maximum size of aggregate. Table 6.4 gives also the approximate amount of accidentally entrapped air in non-air-entrained concrete as well as the recommended average total air content of air-entrained concrete.

4. The water-cement ratio is selected from a consideration of the limitations imposed by requirements of durability and strength. Table 6.5 gives recommended maximum water-cement ratios permissible from considerations of durability, for the particular type of structure and degree of exposure. Also, from Table 6.6 is taken the maximum permissible water-cement ratio corresponding to the desired compressive strength. The lower of these two values of water-cement ratio must be used.

5. The cement content is computed by dividing the total free-water content by the water-cement ratio.

6. From Table 6.7 is taken the bulk (overall) volume of dry-rodded coarse aggregate per unit volume of concrete, for the particular maximum size of aggregate and fineness modulus of sand. The solid volume of coarse aggregate per cubic yard of concrete is then computed from this volume and the specific gravity and weight of the aggregate (Art. 3.5).

7. The solid volume of sand is computed by subtracting from the total volume of concrete the solid volumes of cement, coarse aggregate, water, and entrapped or intentionally entrained air.

8. Batch quantities are computed on the basis of dry aggregates and are converted into terms of the quantities of the various materials as batched at the job.

9. The corresponding field proportions are computed to take account of absorption or free-moisture content of aggregates.

10. In the field, adjustments are made as necessary to meet the requirements for water-cement ratio and consistency.

Example. It is desired to establish proportions for non-air-entrained concrete for the interior of a building. The concrete will not be exposed to weathering, and the design requirements call for a compressive strength of 3,500 psi. The following steps in the procedure are numbered corresponding to steps in the foregoing explanatory procedure:

1. The conditions of placing are such that, from Table 6.2, it is decided that a slump of 3 to 4 in. should be used.

2. The dimensions of the section and the amount of reinforcement are such that, from Table 6.3, it is decided that the appropriate maximum size of aggregate is 1½ in. Suitable graded aggregates are available locally. The selected coarse aggregate has a specific gravity, bulk dry, of 2.68, an absorption of 0.5 percent, and a dry-rodded unit weight of 100 pcf.

Table 6.3 Maximum sizes of aggregate recommended for various types of construction*

Minimum dimension of section, in.	Maximum size of aggregate, in.† Reinforced walls, beams, and columns	Unreinforced walls	Heavily reinforced slabs	Lightly reinforced or unreinforced slabs
2½–5	½–¾	¾	¾–1	¾–1½
6–11	¾–1½	1½	1½	1½–3
12–29	1½–3	3	1½–3	3
30 or more	1½–3	6	1½–3	3–6

* From Ref. 601.

† Based on square openings. Maximum size should not be larger than one-fifth of the narrowest dimension between sides of forms nor larger than three-quarters of the minimum clear spacing between reinforcing bars or between bars and forms.

Table 6.4 Approximate free mixing-water requirements for different slumps and maximum sizes of aggregates*

Slump, in.	Water, gal/cu yd of concrete for indicated maximum sizes of aggregate							
	⅜ in.	½ in.	¾ in.	1 in.	1½ in.	2 in.	3 in.	6 in.
Non-air-entrained concrete								
1 to 2	42	40	37	36	33	31	29	25
3 to 4	46	44	41	39	36	34	32	28
6 to 7	49	46	43	41	38	36	34	30
Approximate amount of entrapped air in non-air-entrained concrete, %	3	2.5	2	1.5	1	0.5	0.3	0.2
Air-entrained concrete								
1 to 2	37	36	33	31	29	27	25	22
3 to 4	41	39	36	34	32	30	28	24
6 to 7	43	41	38	36	34	32	30	26
Recommended average total air content, %	8	7	6	5	4.5	4	3.5	3

* From Ref. 601. These quantities of mixing water are for use in computing cement factors for trial batches. They are maxima for reasonably well-shaped angular coarse aggregates graded within limits of accepted specifications. If *more* water is required than shown, the cement factor, estimated from these quantities, *should* be increased to maintain desired water-cement ratio, except as otherwise indicated by laboratory tests for strength. If *less* water is required than shown, the cement factor, estimated from these quantities, *should not* be decreased except as indicated by laboratory tests for strength.

Table 6.5 Maximum permissible free water-cement ratios (gallons per sack) for different types of structures and degrees of exposure*

	Exposure conditions†					
	Severe wide range in temperature, or frequent alternations of freezing and thawing (air-entrained concrete only)			Mild temperature rarely below freezing		
		At the water line or within the range of fluctuating water level or spray			At the water line or within the range of fluctuating water level or spray	
Type of structure	In air	In fresh water	In sea water or in contact with sulfates‡	In air	In fresh water	In sea water or in contact with sulfates‡
Thin sections, such as railings, curbs, sills, ledges, ornamental or architectural concrete, reinforced piles, pipe, and all sections with less than 1-in. concrete cover over reinforcing	5.5	5.0	4.5¶	6	5.5	4.5¶
Moderate sections, such as retaining walls, abutments, piers, girders, beams	6.0	5.5	5.0¶	§	6.0	5.0¶
Exterior portions of heavy (mass) sections	6.5	5.5	5.0¶	§	6.0	5.0¶
Concrete deposited by tremie under water		5.0	5.0	...	5.0	5.0
Concrete slabs laid on the ground	6.0	...	...	§		
Concrete protected from the weather, interiors of buildings, concrete below ground	§	...	...	§		
Concrete which will later be protected by enclosure or backfill but which may be exposed to freezing and thawing for several years before such protection is offered	6.0	...	...	§		

* From Ref. 601.

† Air-entrained concrete should be used under all conditions involving severe exposure and may be used under mild exposure conditions to improve workability of the mixture.

‡ Soil or ground water containing sulfate concentrations of more than 0.2 percent.

¶ When sulfate-resisting cement is used, maximum water-cement ratio may be increased by 0.5 gal per sack.

§ Water-cement ratio should be selected on basis of strength and workability requirements.

Table 6.6 Compressive strength of concrete for various water-cement ratios*

Water-cement ratio		Probable average compressive strength at 28 days, psi	
Gal per sack of cement	By wt	Non-air-entrained concrete	Air-entrained concrete
4	0.35	6,000	4,800
5	0.44	5,000	4,000
6	0.53	4,000	3,200
7	0.62	3,200	2,600
8	0.71	2,500	2,000
9	0.80	2,000	1,600

* From Ref. 601. These average strengths are for concretes containing not more than the percentages of entrained and/or entrapped air shown in Table 6.4. For a constant water-cement ratio, the strength of the concrete is reduced as the air content is increased. For air contents higher than those listed in Table 6.4, the strengths will be proportionally less than those listed in this table. Strengths are based on 6 by 12-in. cylinders moist-cured under standard conditions for 28 days.

The selected fine aggregate has a specific gravity, bulk dry, of 2.64, an absorption of 0.7 percent, and a fineness modulus of 2.80.

3. From Table 6.4, the required total free water content is 36 gal per cu yd, and the approximate amount of entrapped air is 1 percent.

4. Since the concrete is not to be exposed, Table 6.5 is not applicable (or, see last footnote of the table). By interpolation from Table 6.6, the water-cement ratio corresponding to a compressive strength of 3,500 psi is approximately 6.6 gal per sack, and this value will be used.

5. From the information in the two preceding paragraphs, the required cement content is computed to be $36/6.6 = 5.5$ sacks per cu yd.

6. From Table 6.7, the bulk volume of dry-rodded coarse aggregate per unit volume of concrete is found to be 0.72. Since the coarse aggregate weighs 100 pcf, the weight of coarse aggregate in 1 cu yd (27 cu ft) of concrete will be $0.72 \times 27 \times 100 = 1{,}940$ lb.

7. The total space occupied by each component except sand in 1 cu yd of concrete is computed as follows:

$$
\begin{aligned}
&\text{Solid volume of cement} && = \frac{5.5 \times 94}{3.15 \times 62.4} && = 2.63 \text{ cu ft} \\
&\text{Volume of water} && = 36/7.5 && = 4.80 \text{ cu ft} \\
&\text{Solid volume of coarse aggregate} && = \frac{1940}{2.68 \times 62.4} && = 11.60 \text{ cu ft} \\
&\text{Volume of entrapped air} && = 0.01 \times 27 && = \underline{0.27 \text{ cu ft}} \\
& && \text{Total, except sand} && = 19.30 \text{ cu ft}
\end{aligned}
$$

Table 6.7 Bulk volume of dry-rodded coarse aggregate per unit of volume of concrete*

Maximum size of aggregate, in.	Bulk (overall) volume of dry-rodded coarse aggregate per unit volume of concrete for different fineness moduli of sand			
	2.40	2.60	2.80	3.00
3/8	0.46	0.44	0.42	0.40
1/2	0.55	0.53	0.51	0.49
3/4	0.65	0.63	0.61	0.59
1	0.70	0.68	0.66	0.64
1 1/2	0.76	0.74	0.72	0.70
2	0.79	0.77	0.75	0.73
3	0.84	0.82	0.80	0.78
6	0.90	0.88	0.86	0.84

* From Ref. 601. These volumes are selected from empirical relationships to produce concrete with a degree of workability suitable for usual reinforced construction. For less workable concrete such as required for concrete pavement construction they may be increased about 10 percent.

The solid volume of sand required is then $27 - 19.30 = 7.70$ cu ft; and the corresponding weight of dry sand is $7.70 \times 2.64 \times 62.4 = 1{,}270$ lb.

8. The weights of materials in 1 cu yd of concrete are as follows:

	Lb
Cement (5.5 sacks)	517
Water (36 gal)	300
Sand (dry basis)	1,270
Coarse aggregate (dry basis)	1,940

9. *Field proportions.* The corresponding field proportions may now be computed to take account of free-moisture content or absorption of the aggregates. The weights of sand and coarse aggregate are to be increased in proportion to the "damp" weight, including the *total* moisture in the aggregates. The weight of water is reduced, however, only by the amount

Ingredient	Quantity per cu yd of concrete	
	Computed (dry-aggregate basis)	Used in field (moist-aggregate basis)
Cement	517 lb (5.5 sacks)	517 lb (5.5 sacks)
Water	300 lb (36.0 gal)	235 lb (28.2 gal)
Sand	1,270 lb (dry)	1,334 lb (moist)
Coarse aggregate	1,940 lb (dry)	1,959 lb (moist)
Total	4,027 lb	4,045 lb

of *surface* moisture in the aggregates. (Or, throughout the tests and calculations the specific gravity and weight of aggregates could have been taken on a saturated surface-dry basis, in which case only surface moisture would need to be considered in computing the field batch weights.)

Continuing the example, assume that field tests show the sand to contain 5.0 percent and the coarse aggregate 1.0 percent total moisture. Since the required quantity of dry sand is 1,270 lb, the amount of moist sand to be weighed out must be $1{,}270 \times 1.05 = 1{,}334$ lb. Similarly, the weight of moist coarse aggregate must be $1{,}940 \times 1.01 = 1{,}959$ lb.

The free water on aggregates in excess of their absorption must be considered as part of the mixing water. The free water on the sand is $5.0 - 0.7 = 4.3$ percent, and the free water on the coarse aggregate is $1.0 - 0.5 = 0.5$ percent. Therefore the mixing water contributed by the sand is $0.043 \times 1{,}270 = 55$ lb and that contributed by the coarse aggregate is $0.005 \times 1{,}940 = 10$ lb. Then the quantity of mixing water to be added is $300 - (55 + 10) = 235$ lb, or 28.2 gal. The accompanying tabulation affords a comparison of the batch quantities; the excess of total field weight over total computed weight represents the 18 lb of water absorbed within the aggregates.

10. *Adjustments.* If in the field it is found that *more* water than the total of 36 gal, say 38 gal per cu yd, is required to produce the desired slump, then in order to maintain the water-cement ratio of 6.6 gal per sack, the cement factor must be increased from 5.5 to $38/6.6 = 5.8$ sacks per cu yd and the batch quantities revised acordingly. If *less* water is required, say, 34 gal, it is recommended that the cement factor be not reduced (although theoretically this would be possible) but that the solid volume of sand be increased by the amount of the reduction in water (in this case, $2/7.5 = 0.27$ cu ft). The amount of entrained air should be checked by test or computation; a variation of ± 1 percent is considered acceptable. A

variation in water-cement ratio of ± 0.25 gal per sack is also considered acceptable.

11. *Variations.* When the conditions of exposure are such that the maximum permissible water-cement ratio is lower than that required for strength, the lower value governs; otherwise the procedure of selecting proportions is the same as that just described and illustrated. When air-entrained concrete is to be used, the procedure is likewise similar, the recommended air content for the particular maximum size of aggregate being taken from Table 6.4. As previously stated, tests and computations can be made on a basis of saturated surface-dry aggregates if desired, even though Table 6.7 is based on dry coarse aggregate. The bulk volume of coarse aggregate is affected little by moisture content.

If the specifications call for a fixed minimum cement content, the tables may be used with appropriate modifications in procedure. For example, under conditions as in the foregoing example but with 6.0 sacks of cement per cu yd of concrete specified, the water-cement ratio would be $36/6.0 = 6.0$ gal per sack and the corresponding probable compressive strength would be 4,000 psi. The remaining computations would be identical, and the field adjustments would necessarily maintain the required cement content with corresponding variations in water-cement ratio.

6.8 ACI method for small jobs. For small jobs where time and personnel are not available to determine proportions in accordance with the foregoing procedure, the ACI has published values determined in conformity with it, for aggregate of average specific gravity [601]. The mixes shown in Table 6.8 will provide concrete that is amply strong and durable if the amount of water added at the mixer is never so much as to make the concrete overwet.

For each given maximum size of aggregate (for selection, see Table 6.3), three mixes are listed in Table 6.8. At the beginning of the job, mix B is used. If it is found that this mix is oversanded, mix C is used instead; or, if undersanded, mix A is used instead. Amounts of damp sand corresponding to the given weights of dry sand are to be employed.

The amount of water added at the mixer should be the minimum that will permit ready working of the concrete into the forms without objectionable segregation; concrete should slide, not run, off a shovel.

The approximate cement contents listed in Table 6.8 are helpful in estimating cement requirements for the job.

Procedure. Select the proper maximum size of aggregate and then, using mix B, add just enough water to produce a sufficiently workable consistency. If the concrete appears to be undersanded, change to mix A, and if it appears to be oversanded, change to mix C.

Table 6.8 Concrete mixes for small jobs*

Maximum size of aggregate, in.	Mix designation	Approximate sacks of cement per cu yd of concrete	Aggregate, lb per 1-sack batch		
			Dry sand† (air-entrained concrete‡)	Dry sand† (concrete without air)	Gravel or crushed stone
½	A	7.0	235	245	170
	B	**6.9**	**225**	**235**	**190**
	C	6.8	225	235	205
¾	A	6.6	225	235	225
	B	**6.4**	**225**	**235**	**245**
	C	6.3	215	225	265
1	A	6.4	225	235	245
	B	**6.2**	**215**	**225**	**275**
	C	6.1	205	215	290
1½	A	6.0	225	235	290
	B	**5.8**	**215**	**225**	**320**
	C	5.7	205	215	345
2	A	5.7	225	235	330
	B	**5.6**	**215**	**225**	**360**
	C	5.4	205	215	380

* From Ref. 601. May be used without adjustment.

† Weights are for dry sand. If damp sand is used, increase weight of sand 10 lb for 1-sack batch; if very wet sand is used, add 20 lb for 1-sack batch.

‡ Air-entrained concrete should be used in all structures which will be exposed to alternate cycles of freezing and thawing. Air-entrained concrete can be obtained either by the use of an air-entraining cement or by adding an air-entraining agent. If an agent is used, the amount recommended by the manufacturer will, in most cases, produce the desired air content.

6.9 ACI method for structural lightweight concrete. The conventional ACI method of proportioning (Art. 6.7) can be used without difficulty for concretes containing lightweight aggregates which are relatively rounded, smooth, and low in absorption. However, aggregates having angular shapes, vesicular surfaces, high absorption, and/or differences in specific gravity of various fractions present certain problems. It is difficult to determine the surface-water content of vesicular aggregates, and the absorption may proceed so rapidly that the consistency of the concrete is stiffened during the period of mixing and placing. When the specific gravity of the finer fractions of fine aggregate is greater than that of the coarser fractions, as often happens, the fineness modulus computed by weight in the usual manner is considerably less than the fineness modulus by volume, which is the proper basis for use in proportioning inasmuch as void content and paste content are functions of volume.

In order to take account of such special conditions the ACI has developed a "Recommended Practice for Selecting Proportions for Structural Lightweight Concrete (ACI 613A)" [602]. Structural lightweight concrete is defined as that having a 28-day compressive strength greater than 2,000 psi and an air-dry unit weight less than 115 pcf. The procedure does not require the use of values of specific gravity or absorption of the aggregates but makes use of a "specific gravity factor" determined by trial. This factor relates the dry weight of an aggregate to the space it occupies in the concrete, *assuming that no water is absorbed during mixing.* The proportions of fine and coarse aggregate to produce minimum void content are determined by a modification of the maximum-density method (see Art. 6.12) even when the specific gravity of the fine aggregate differs from that of the coarse aggregate. It is roughly estimated that the net (free) water required to produce a 2-in. slump in non-air-entrained concrete averages about 390 lb per cu yd of concrete. Slumps less than 4 in. are recommended. An approximate relationship between the cement content and the compressive strength of structural lightweight concrete is taken at the values shown in the following tabulation.

Compressive strength, psi	Cement content, sacks per cu yd
2,000	4–7
3,000	5–8
4,000	6–9
5,000	7–10

In preparation for trial mixes, the dry loose unit weight and the moisture content of each aggregate, as it is to be used, are determined. A trial mix is then made as illustrated by the following example.

Example. A mix is to contain 6 sacks of cement per cu yd. The dry loose unit weight of the fine aggregate is 56 and that of the coarse aggregate, 45 lb per cu ft. Experience has shown that the sum of separate dry loose fine and coarse aggregates required to produce 1 cu yd of concrete is about 32 cu ft. The fine and coarse aggregates will be used in equal volumes, as indicated by experience.

The amounts of cement and aggregates required for a trial batch of approximately 1 cu ft of concrete will be $\frac{1}{27}$ of the amounts assumed per cu yd, as follows:

Cement	$\dfrac{6 \times 94}{27} = 20.9$ lb
Fine aggregate (dry)	$\dfrac{16 \times 56}{27} = 33.2$ lb
Coarse aggregate (dry)	$\dfrac{16 \times 45}{27} = 26.7$ lb

These quantities are mixed together while water is added to produce the required slump. The amount of water, including that which is absorbed during mixing, is found to be, say, 17.8 lb. Then the total weight of the batch is 98.6 lb.

The unit weight of the fresh concrete is found by test to be, say, 97.0 lb per cu ft; and the air content (entrapped air in this case) is found by test to be, say, 2.5 percent.

The yield of the batch is 98.6/97.0 = 1.016 cu ft. Quantities per cubic yard of concrete are computed in proportion by multiplying the batch quantities by 27/1.016, with results as follows:

	Lb per cu yd of concrete
Cement	556 (5.91 sacks)
Fine aggregate (dry)	882
Coarse aggregate (dry)	710
Water	473

The "specific gravity factor" is then computed as follows:

	Cu ft per cu yd of concrete
Cement	$\dfrac{556}{62.4 \times 3.15} = 2.83$
Water	$\dfrac{473}{62.4} = 7.58$
Entrapped air	2.5 percent = 0.67
	Total 11.08

The aggregates occupy the remaining space, or 27 − 11.08 = 15.92 cu ft; and the fine and the coarse aggregate each occupy 7.96 cu ft. (This is not precisely correct, but the error is unimportant for the purposes of this method.) The specific gravity factor is computed in the same manner that specific gravity would be computed, as follows:

	Specific gravity factor
Fine aggregate	$\frac{882}{62.4 \times 7.96} = 1.78$
Coarse aggregate	$\frac{710}{62.4 \times 7.96} = 1.43$

The specfic gravity factor is *not* equal to the specific gravity because no account is taken of any water absorbed during mixing. However, in subsequent mixes with this aggregate in the same moisture condition, the volume of water absorbed during mixing is nearly constant, and the factor can be used as if it were the apparent specific gravity for proportioning by the conventional ACI absolute-volume method (Art. 6.7).

Additional trial batches are then made with different cement contents, and are adjusted as necessary to secure the desired workability and cement factor.

Air-entrained concrete mixes are proportioned in the same manner but with the water content reduced 2 to 3 percent for each 1 percent of entrained air.

Job control is maintained by keeping constant the cement content, slump, and volume of dry aggregate per cubic yard of concrete, regardless of variations in absorbed or surface moisture. Frequent determinations are made of the moisture content and gradation of the aggregates and of the unit weight and air content of the concrete.

6.10. ACI method for no-slump concrete. The conventional ACI method of proportioning applies to concretes having a slump of 1 in. or greater. For drier consistencies, for which the slump test is impractical, ACI has developed a "Proposed Recommended Practice for Selecting Proportions for No-slump Concrete" [603]. Three types of equipment for measuring such consistencies are described, namely, the Vebe apparatus, the compacting-factor apparatus, and the Thaulow drop table (Art. 5.6). Tables similar to, and in conjunction with, those in the conventional ACI method are provided; together with laboratory tests for physical properties of fine and coarse aggregate these tables yield information for selecting proportions of materials for a trial mixture. The method is too extensive and detailed to be covered herein. Examples are given in the recommended practice.

6.11 Arbitrary proportions. The original method of proportioning materials for concrete was that of arbitrary proportions such as 1:2:4 (by overall volume of the cement, sand, and coarse aggregate). Strength was roughly controlled by varying the cement content; for example, a rich mix was 1:1:2 and a lean mix 1:3:6. This method, although long used with varying success, was in general unsatisfactory because it took no account of water content or of differences in shape, gradation, maximum size, or bulking of aggregates. The customary 1:2 ratio of sand to coarse aggregate (33 percent sand in aggregate) was usually too low for workability, especially in view of the bulking of sand, and much honeycombing resulted. The condition could be corrected somewhat by increasing the arbitrary percentage of sand, but obviously no single fixed proportion can be applicable to the wide variety of aggregates commercially available. Differences in gradation, shape, or surface texture of fine and coarse aggregate, in richness of mix, and in consistency of concrete all affect the optimum percentage of sand.

At present, arbitrary proportioning is limited to small jobs where suitable proportions have been established by experience and observation. Various organizations publish tables of recommended mixes for small jobs, the proportions for which vary with shape of aggregate (rounded or angular), maximum size of aggregate, and fineness of sand [601, 614]. Such proportions serve as a reasonable guide but should be varied during construction if observation indicates possibility of improvement.

6.12 Proportioning by maximum density of aggregate. Since it is desirable to use a minimum amount of cement-water paste in a concrete mix, one logical basis for proportioning the aggregates is that of using the combination which has minimum voids, or maximum density. In the early days of concrete construction in the United States, much use was made of Fuller and Thompson's maximum-density method [607], in which sieve analyses were made to obtain proportions of aggregate (and cement) giving maximum density of concrete. An "ideal curve" of gradation was published, and the available aggregates were made to conform as closely as practicable to the curve. This method is not in use today, although some organizations have established "ideal" gradations of aggregate which are approximated with the available aggregates.

It has been observed that the optimum percentage of sand, which requires minimum paste, is practically always somewhat below the percentage for maximum density (minimum voids) of mixed aggregate; the difference between the two percentages results from the separation of aggregate by

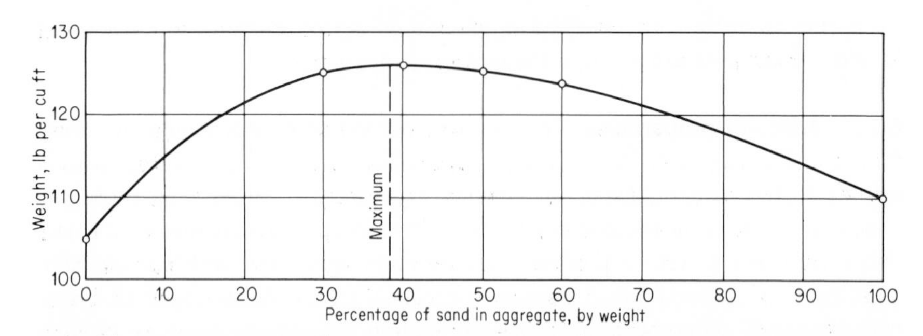

Fig. 6.5. Unit weight of dry-rodded fine and coarse aggregates mixed in various proportions. (For illustration only, to show trends; values will differ with aggregates.)

the paste. The amount to be subtracted from the maximum-density percentage of sand to obtain the optimum percentage is rarely as much as 8, and usually it will be of the order of 2 to 5; hence 3 can be taken for working purposes without error of consequence, particularly since the variation in paste content near the optimum is relatively small.

A test for maximum density of combined aggregates is easily made by weighing several proportions of dry-rodded mixed aggregate (containing say 0, 30, 40, 50, and 100 percent of sand), plotting a curve, and observing the percentage of sand at which the unit weight is a maximum. For example, in Fig. 6.5 the maximum weight of mixed aggregate occurs when 38 percent of sand is used; the optimum percentage of sand is then likely to be $38 - 3 = 35$.

If the fine and coarse aggregates differ in specific gravity, the proportion at which the weight is maximum is not exactly that at which the voids are minimum; however, the difference is small, and for the purposes of this approximate method it may be neglected when regular (not lightweight) aggregates are used. The *difference* between the peak value and the optimum value is affected but little as between maximum weight and minimum voids.

6.13 Proportioning by surface area of aggregate. Prior to 1918, L. N. Edwards [606] deduced from tests that the quantity of cement to produce a given strength of concrete is proportional to the surface area of the aggregate particles. He published the surface areas for various sieve-size fractions of sand, rounded coarse aggregate, and crushed coarse aggregate; thus from the sieve analysis the total surface area could be computed. The corresponding amounts of cement to produce concrete of desired strength were determined empirically.

The surface-area concept has some logical basis in that the paste requirement is undoubtedly affected by the amount of surface which must be coated; however, the paste requirement is also affected by the amount of voids. The method has been used but little.

6.14 Proportioning by fineness modulus of aggregate. In connection with the water-cement ratio basis for proportioning, Duff A. Abrams developed in 1918 a method of establishing proportions of fine and coarse aggregate based on the fineness modulus of aggregates as determined by sieve analysis [600]. Later, charts based on extensive laboratory tests were published by the Portland Cement Association, showing for each of a range of slumps the interrelation between fineness modulus of mixed aggregate, volume of dry-rodded mixed aggregate per volume of cement, strength of concrete, and maximum size of aggregate [614].

To proportion a mix by this method, the chart for the established slump was entered with the known maximum size of aggregate and the desired compressive strength of concrete, and the fineness modulus and volume of dry-rodded mixed aggregate were read. The percentage of sand was computed by means of the relations explained in Art. 3.10, and the quantities of the separate fine and coarse dry-rodded aggregate were computed and were converted into terms of quantities of aggregate as batched at the job. The charts were prepared for rounded coarse aggregates, with a correction for flat or angular aggregates which resulted in somewhat lower fineness modulus and correspondingly richer mix or lower yield. Others [615, 619] have extended the application of the method by limiting the maximum permissible fineness modulus of the portion finer than any intermediate size as well as that of the total aggregate, and/or by including the cement in the computation of fineness modulus.

The water-cement-ratio and fineness-modulus concepts were developed at about the same time, and were often confused. Actually water-cement ratio is basic, and its use has continued, whereas fineness modulus is merely one means of expressing gradation by a single number and of arriving at the optimum percentage of sand; although widely used for years, fineness modulus as a means of proportioning has largely been superseded by other methods.

The formerly published charts and tables for designing mixes by the fineness-modulus method are still useful, especially for conveniently comparing different possible mixes in advance of a job, provided the strength values are replaced by the corresponding water-cement ratios as of the period of publication. For present-day use, also, the ordinates are preferably expressed in terms of yield rather than volumes of mixed aggregate. These conversions have been made in Fig. 6.6 for two ranges of slump. To illustrate the use of the charts, suppose it is desired to select proportions for a mix containing aggregate of maximum size 1½ in. and having a water-cement ratio of 6 gal per sack and a slump of 3 to 4 in. From the chart,

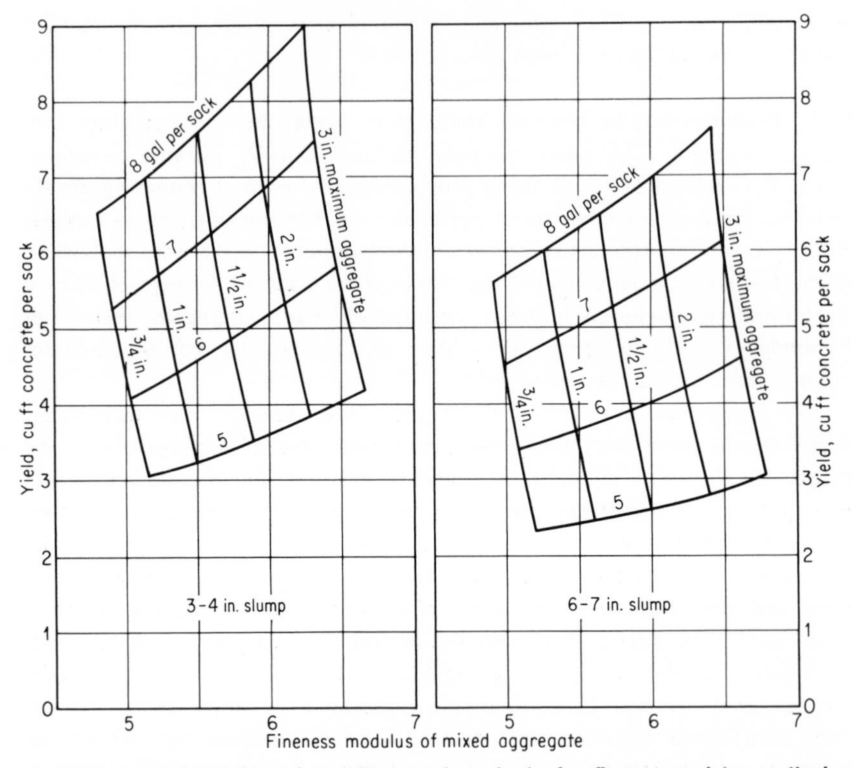

Fig. 6.6. Interrelationship of variables used as basis for fineness-modulus method of proportioning. (For rounded aggregates only; if coarse aggregate is flat or angular, proceed as if limiting curves of maximum size were 0.25 in fineness modulus to the left.) *(Adapted from an early edition of Ref. 614.)*

the maximum permissible fineness modulus is 5.75; the corresponding proportions of fine and coarse aggregates can now be computed from their separate fineness moduli, as described in Art. 3.10. Also from the chart the yield is 4.9 cu ft of concrete per sack; the corresponding cement factor is 27/4.9 = 5.5 sacks per cu yd.

To illustrate the use of the charts for comparative purposes, suppose that the conditions of placing the concrete just discussed require a slump of 6 to 7 in. The corresponding maximum permissible fineness modulus of mixed aggregate is 5.85, thus requiring a lower percentage of sand. The yield is only 3.9 cu ft of concrete per sack, and the corresponding cement factor is 27/3.9 = 6.9 sacks per cu yd; thus the use of the wetter slump in this case involves a "penalty" of 1.4 sacks per cu yd, or approximately 25 percent more cement. Other comparisons can be made to estimate the effect of using a smaller aggregate, a higher water-cement ratio, etc.

6.15 Proportioning by voids-cement ratio and mortar voids. During the early 1920s, A. N. Talbot and F. E. Richart developed the "voids-cement ratio" basis for proportioning and the "mortar-voids" method of applying it [616, 613]. They considered as "voids" in concrete or mortar all space not occupied by aggregate or cement; thus the volume of voids is the sum of the volumes of water and entrapped air. They also assumed that the strength of concrete is equal to the strength of the mortar in it.

The voids-cement ratio is a basic relationship analogous to the water-cement ratio. Curves showing the relationship between strength and voids-cement ratio are similar and close to those for water-cement ratio, as would be expected in view of the fact that the amount of entrapped air is relatively small. The voids-cement ratio is considered more basic, especially since it applies consistently even to rather harsh mixes. The customary form of expression for voids-cement ratio is cubic feet of voids per sack of cement.

The mortar-voids method is one means of proportioning aggregates to secure the optimum mix having the desired voids-cement ratio. It is rather complicated, and it has not been widely used. In order to apply the method, it is necessary to determine by means of trial mortars the voids-cement ratio and strength over a range of mixes each with several water contents. All computations are based on "absolute" (displacement or solid) volumes of the ingredients and mortar.

One important concept of the mortar-voids method is that of "basic water content." If small increments of water are added to a mixture of cement and dry sand and the volume measured (after mixing), at first the volume increases because of surface films on the particles of cement and sand. Additional water causes the volume to decrease, as the surface films are broken, until the voids are filled with water. Still further additions of water cause the volume of the mixture to increase by the amount of the added water. The water content at which the voids are just filled with water and the volume is a minimum is called the basic water content, to which other "relative" water contents are referred. The amount of water required for workability is always greater than the basic water content.

The first step of the mortar-voids procedure is to establish the general relationship between the strength of mortar and the voids-cement ratio over a range, for the particular materials and conditions. Trial mortars having several ratios of sand to cement and several water contents are mixed; the volume of voids in a given volume of each is determined by measurement; and specimens of each are prepared and tested to determine strength. Normally the sand-cement ratios by absolute volume are 1 , 2, 3.5, and 5; with each of these mortars the basic water content and relative water contents of 1.2 and 1.4 are employed. For convenience in proportioning, the relationships are plotted in the form of the following diagrams:

1. Voids-cement ratio vs. strength (one curve for each water content)
2. Sand-cement ratio vs. voids-cement ratio (one curve for each water content)

3. Sand-cement ratio vs. voids per unit volume of mortar
4. Sand-cement ratio vs. water per unit volume of mortar

The next step is to assume from previous experience the relative water content that will produce the desired consistency of concrete. With this and the desired strength, the required voids-cement ratio and sand-cement ratio, respectively, are taken from the first two diagrams listed above. Then the volume of voids and the volume of water per unit volume of mortar are taken from the third and fourth diagrams, respectively. The corresponding volumes of cement and sand per unit volume of mortar are then computed by difference, with the sand-cement ratio known.

The next step is to determine the solid volume b of coarse aggregate to be used in a unit volume of concrete. The exact amount is determined by trial or experience. For crushed aggregate it usually lies between 38 and 43 with a recommended upper limit of 40 percent and for rounded aggregate between 40 and 46 with a recommended upper limit of 45 percent. The solid volume b_o of coarse aggregate in a unit volume of coarse aggregate is determined by test and computation.

Finally, the solid (absolute) volume of each of the other materials in a unit volume of concrete is computed, and the quantities of all materials are converted into terms of those used in batching at the job.

6.16 Proportioning by void content of coarse aggregate. In 1942, A. T. Goldbeck and J. E. Gray published a method of proportioning [609] which included an extension of the Talbot-Richart method just described. A distinctive feature was a table of recommended bulk volumes b/b_o of coarse aggregate per unit volume of concrete for a range of maximum sizes of coarse aggregate and a range of fineness moduli of sand. The table was applicable to all cement contents and equally well to crushed or rounded aggregates. With modification it was incorporated into the ACI method of proportioning and appears as Table 6.7 herein. The Goldbeck-Gray method also made use of the constancy of water content per cubic yard of concrete for given aggregates and consistency.

QUESTIONS

1. What are the principal requirements of fresh and hardened concrete toward the fulfillment of which proportioning is directed?

2. What is the principal relationship in proportioning with regard to cost of materials for concrete?

3. Why do stronger concretes cost more than weaker concretes? How can this be offset in some structures?

4. Why is the use of a fixed proportion of sand to coarse aggregate for all concretes unsuitable? What factors affect the optimum proportion?

5. Which would be the ideal basis for expressing proportions—overall volume, solid volume, or weight? Why?

6. List the functions of paste and aggregate in concrete. How does dilution of the paste affect these functions? Why is an excess of paste undesirable?

7. In your own words, formulate a statement of the relationship between water-cement ratio and strength, including its limitations.

8. What are the four principal variables in proportioning of a concrete mix with given materials?

9. How is gradation of available aggregates controlled in practice? Toward what end is control of gradation directed?

10. Why is the trial method of proportioning so widely used when several methods based on experience and/or laboratory tests are available?

11. Describe breifly the trial method of proportioning (*a*) in advance of construction and (*b*) during construction.

12. Discuss two distinctive features of the current ACI method of proportioning.

13. To what extent is the total water content of concrete affected by variations in water-cement ratio, provided the same aggregates and same consistency are employed?

14. What information obtained by test or assumption is needed for the current ACI method of proportioning?

15. In the ACI method, how is the solid volume of sand per cubic yard of concrete determined?

16. By the ACI method, proportion a mix for a heavy bridge pier which will be exposed to fresh water in a severe climate. Air-entrained concrete is to be used. A 28-day compressive strength of 4,000 psi is specified, and placement conditions will permit a slump of 1 to 2 in. and the use of aggregate of maximum size 3 in. The coarse aggregate economically available has a dry-rodded unit weight of 110 lb per cu ft, a specific gravity of 2.68, and an absorption capacity of 0.4 percent. The available sand has a fineness modulus of 2.8, a specific gravity of 2.64, and an absorption capacity of 0.6 percent. Assume that in the field the total moisture content of the sand will average 4.5 percent and that of the coarse aggregate will average 1.1 percent.

17. Analyze Table 6.8 with regard to (*a*) effect of maximum size of aggregate on cement content, (*b*) effect of cement content on percentage of sand in aggregate, and (*c*) use of air entrainment on percentage of sand.

18. What is the significant difference between the "specific gravity factor" of an aggregate and its apparent specific gravity?

19. If a concrete mix contains fine and coarse aggregates proportioned to give maximum density (minimum voids) of mixed aggregate, is the mix oversanded or undersanded? Why? To approximately what extent?

20. Surface area of aggregate is one factor affecting paste requirement. What is the other principal factor so far as aggregates are concerned?

21. If the optimum fineness modulus of mixed aggregate for a given concrete is known or assumed, how can the corresponding percentage of sand be computed? (Hint: see Art. 3.10.)

22. Distinguish between water-cement ratio, fineness modulus, and voids-cement ratio.

23. In what way are voids-cement ratio and the mortar-voids method of proportioning related?

24. What important assumption was made by Talbot and Richart regarding the relation of concrete strength to mortar strength?

25. What is b/b_o and what is its value in proportioning?

SEVEN
MANUFACTURE OF CONCRETE

BATCHING

7.1 Batching. On all first-class work, batching is done by weight, rather than by volume, as the amount of solid granular material in a cubic yard is an indefinite quantity. A volume of moist sand in a loose condition weighs considerably less than an equal volume of dry compact sand. However, a ton of aggregate is a fairly definite quantity. The net water-cement ratio is computed upon the basis of saturated surface-dry aggregates, so that free moisture on, or absorption by, aggregates must be allowed for at the mixer if a water-cement ratio specification is being followed. Some small jobs still employ volumetric measurements of quantities; here, moisture in the sand produces bulking, which must be allowed for. Proportions by volume are usually specified in terms of a dry-rodded condition, but the batch quantities must be given in the damp, loose condition. If the cement is supplied in sacks, the batches should be such as to use integral multiples of one sack. One sack of cement weighs 94 lb net and is assumed to hold 1 cu ft. The cement is weighed only on jobs where it is supplied in bulk. Water is usually measured by volume in gallons, although weighing is sometimes used.

Proper batching requires that the fixed proportions of materials determined as outlined in Chap. 6 be measured uniformly from batch to batch. Fine and coarse aggregate must be uniform in grading and moisture content on arrival at the batching equipment if uniform concrete is to be attained. Unless such aggregate is assured through appropriate specifications and effective inspection of selection, preparation, and handling of aggregates, production of uniform concrete is extremely difficult and to a large degree unlikely, despite a high order of accuracy in measuring and superior performance in mixing and placing. At intervals, tests for moisture content and grading of the aggregate should be made, and the batch proportions adjusted to maintain a uniform mix.

The first batch of concrete should be somewhat richer and wetter than normal, because mortar sticks to the mixer and the conveying equipment and the first concrete deposited in the forms has no previous bed of mortar with which to make contact. Ordinarily these conditions can be remedied by simply leaving out part of the coarse aggregate in the first batch.

7.2 Weight-batching equipment. As stated previously, the batching of materials for most large and important jobs is done by weighing. Specifications usually permit the use of a separate batcher for each material or the cumulative weighing of each material, except water and cement, in a single batcher, as shown in Fig. 7.1. The use of separate hoppers as shown in the "good arrangement" of Fig. 7.1 is preferred, as it accomplishes some mixing of the materials before they enter the mixer. For a high degree of accuracy, the gate-charging automatic batching equipment is sometimes arranged to

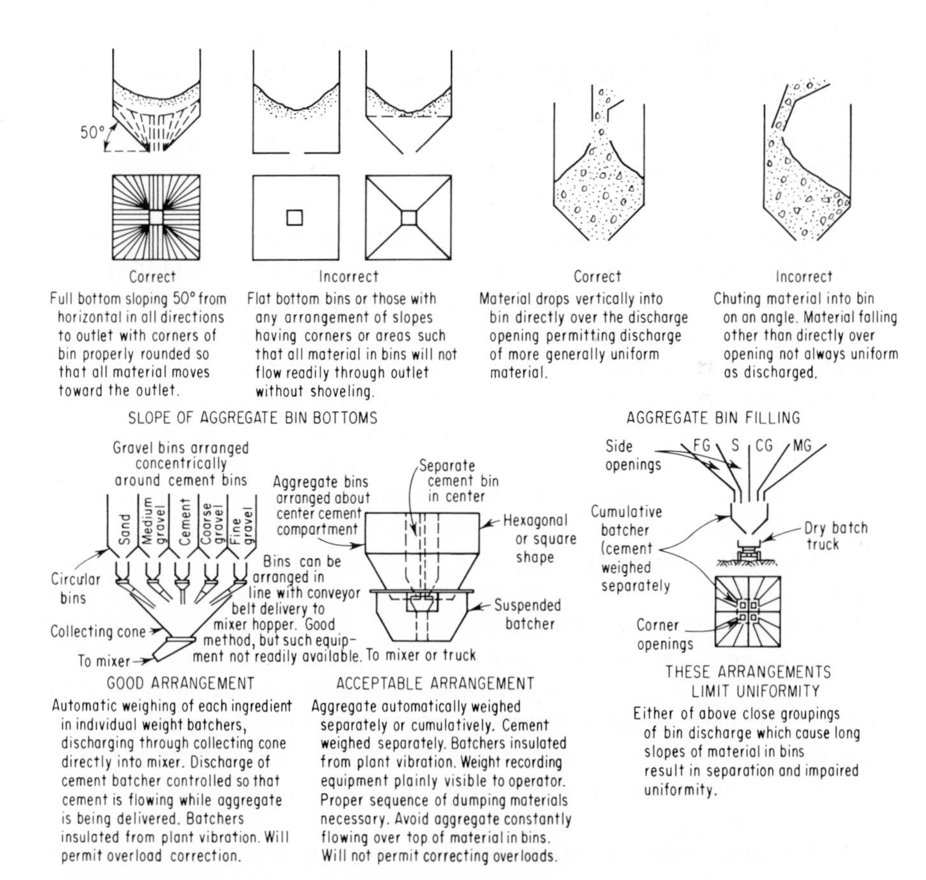

Fig. 7.1. Uniform batching of concrete requires adequate arrangement of supply bins and weighing hoppers. (*From ACI Committee* 614 [700].)

operate with a suitable "dribble" after there is nearly the desired quantity of material in the batcher.

Many batching units are operated manually. Such units usually have a dial scale which cumulatively weighs up to three sizes of aggregates in the main bin and the cement in a side bin. The springless full-reading dial-scale type has certain advantages over the beam type of scale, as the former gives indications of what is occurring during the entire filling and emptying operation. It shows not only whether hoppers are properly charged, but also whether they are completely discharged, an equally important function. Also, it shows any irregularities in flow of the materials which may be caused by arching, jamming, leaky gates, formations of encrustations in the hopper, or other conditions. All weighing hoppers should be constructed in such manner that they may be inspected readily and so that samples or overweight material in excess of prescribed tolerances may be easily removed.

To obtain good results with weight-batching equipment, it is essential that all working parts, particularly the knife edges, be kept clean and in good condition. The design should provide for ready accessibility to all critical parts for inspection and cleaning, and should provide for the protection of such parts from falling material. All nuts that might work loose in operation should be protected by locking devices. It is essential that the weighing devices be equipped to show when the correct amount of material is in the hopper and that they be in full view of the operator [1504].

The batching units for most modern plants are automatic in action. One type of electronic control unit for automatic batching, premixing if desired, and truck loading is shown in Fig. 7.2. Various mix proportions, including admixtures, are preset in advance of daily operations. To activate the plant, the operator channels in the proper batch-selector switch, dials in the number of yards on the batch proportioner, and pushes the batch start control. The console electronically adjusts the quantity of sand and water to compensate for variations in moisture in the sand.

A typical automatic batching plant for charging batcher trucks or transit mixers is shown in Fig. 7.3. The aggregate storage bins for this plant are filled by means of a bucket elevator.

Some small jobs use ordinary platform scales for weighing aggregates in wheelbarrows or carts, the cement being measured by the sack. The aggregate scales are placed at the stockpile or on runways between stockpiles and the mixer. To avoid continual changing of the scale setting, it is desirable to use a scale having a separate scale beam for each aggregate.

7.3 Checking weighing equipment. The weighing equipment should be maintained accurate within a maximum tolerance of about 0.4 percent of the net load being weighed. For checking the scales before operations

Fig. 7.2. Electronic console of fully automatic batching plant. *(Noble Company.)*

are started, and also periodically during operations, each batcher plant should be provided with at least 10 standard 50-lb weights. With these weights the scales may be checked over their full range. The scale should first be balanced at zero load. After the scale has been checked to the limit of the weights, the weights are removed and enough material run into the hopper to give the same scale reading. The weights are then reused to check the scale at higher loads. From a record of applied loads and indicated weights, scale errors may be determined for use in batching operations. The scale should be inspected periodically for sluggishness or inaccuracy; and occasionally each day the scale should be balanced at zero load.

For a batcher equipped with automatic feeding and cutoff devices, two checks are customarily made, the first of which is conducted as described previously to determine the weighing error (difference between actual load and scale reading). The cutoff equipment is tested during regular batching operations to determine the feeding error (difference between cutoff setting and corresponding scale reading). The weighing error and the feeding error are considered jointly; they should be added regardless of sign and should not be regarded as compensating errors [106]. Ordinarily, operation should

Fig. 7.3. Automatic batching plant for loading batcher trucks or transit mixers. (*Noble Company.*)

be required within a degree of accuracy of 1 percent for cement and water, 2 percent for individual aggregates and 1 percent for total aggregate [700, C947].

7.4 Volumetric batching equipment. Volumetric batching equipment for the aggregates is used much less than formerly. When used it should conform with requirements similar to those for weight batchers. It should be accurate within the prescribed tolerance, in good operating condition, readily adjustable, accessible for inspection and observation of its performance, and have complete discharge. Its accuracy should be checked by computations of the volume for various settings, or by drawing off sample batches and measuring in containers of known volume. As some hoppers tend to bulge when filled, measurements of an empty hopper may not indicate the effective volume.

Carts or wheelbarrows are often used on small jobs for measuring aggregates by volume, but they should be calibrated with measured volumes of material screeded to a level plane. If the aggregate is not leveled off, the heaping of the material will vary and introduce errors.

7.5 Batching cement. Sacked cement for jobs batched manually should be measured in units of not less than one sack unless the fractional bags are weighed. Cement for larger jobs should be handled in bulk and weighed separately with automatic batching equipment fed by controlled screw conveyors or other effective devices that will permit a precise cutoff. A vibrator should be provided to aid in securing complete discharge of the batch. Bulk cement is often lost or scattered indiscriminately in dropping from the batcher to batch trucks below. This can be prevented by enclosing the discharging cement in a canvas boot of such length that its outlet may be quickly buried in the cement when loading portable batch compartments. Such compartments should contain a separate section for cement, attached to and operating with the individual batch-release gate as shown in Fig. 7.4. If a separate section is not provided for the cement in each compartment, or the cement is not enclosed in the aggregate, tarpaulin covers should be provided during transportation [700].

7.6 Irregularities in batching. To maintain uniformity of the materials delivered to the mixer there is need for attention to various incidental operations. For example, to ensure that batches accurately assembled arrive as uniform batches in the mixer, it is necessary to eliminate (1) overlap of batches in loading and discharging multiple batch trucks, (2) loss of material in transferring batches to the mixer, and (3) loss or "hang-up" of a part of one batch, or its inclusion with another, when dry batches are transferred by means of belts and hoppers.

7.7 Water- and admixture-measuring equipment. The mixing water is commonly batched by tanks or by water meters. If a measuring tank is used, it should be a vertical unit with central overflow for regulating the filling of the tank and with a central siphon discharge. This type of tank is unaffected by ordinary changes in slope such as are likely to occur with portable mixers. Also, with such an installation a leaky valve cannot dribble excess

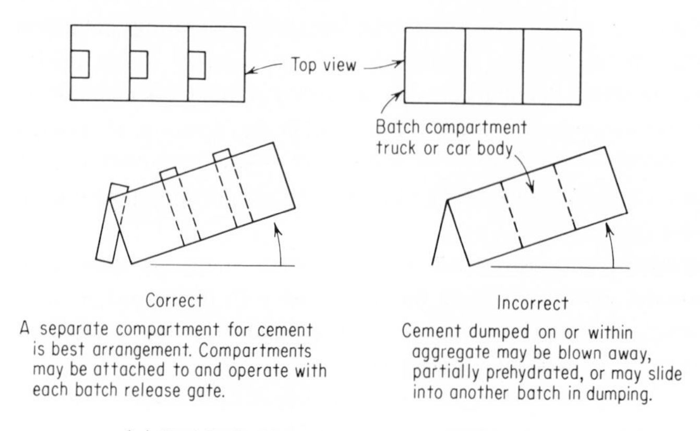

(*a*) Provision for cement in dry-batch compartments

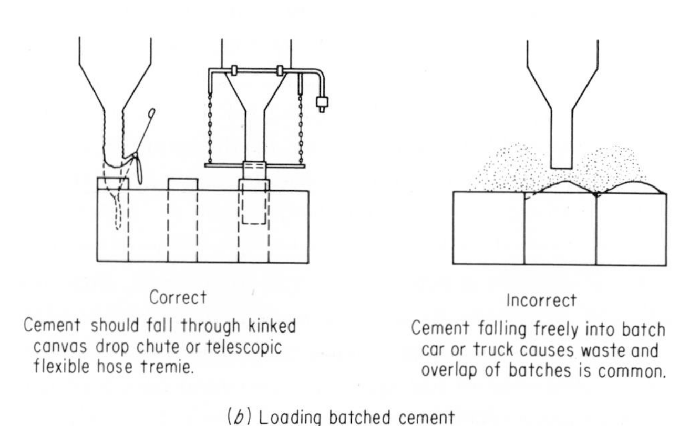

(*b*) Loading batched cement

Fig. 7.4. Proper care in handling batched cement prevents waste and produces more uniform concrete. (*From ACI Committee* 614 [700].)

water into the mixer. Water tanks should be equipped with a gage glass and a graduated scale to permit a direct reading of the mixing water used.

Water meters of suitable construction are available which permit the water flow to be cut off automatically at some predetermined discharge, or stopped normally at any time. They register the water per batch and also have a totalizer which is particularly useful in determining the average water requirement and water-cement ratio for a group of concrete batches. To time the entry of the water into the mixer most satisfactorily, some meter installations provide an auxiliary tank into which the meter flow is discharged. Some water meters have given unsatisfactory service because of inaccuracies at low or variable pressures, the effects of scale or sediment often found in construction water-supply systems, and the high water temperatures which may be necessary during winter construction.

All water measuring devices should be calibrated by measuring or weighing sample batches of water drawn off for various settings, and should be capable of routine measurement within an accuracy of 1 percent under all operating conditions [700]. They should be checked for any leakage from the pipe entering the mixer. No system of valves should be used which allows unmeasured water to flow into the mixer from a tank which is being charged or discharged.

As measurement of the correct amount of water is so intimately related to variations in sand moisture, automatic equipment [732] now available for indicating the amount of moisture in sand as it is batched should be used on projects large enough to warrant its use.

Calcium chloride (if used) should first be made into a solution which then can be batched into the mixing water. On small jobs the correct amount of solution may be poured directly into the mixer after charging the other materials.

Air-entraining agents in either liquid or powdered form are commonly introduced by special dispensers obtainable from suppliers of air-entraining agents. Care should be taken to avoid mixing solutions of air-entraining agents and calcium chloride while batching or storing, as a gummy precipitate will form and cause the plugging of valves and meters.

MIXING

7.8 Mixing. The principal requirement with regard to mixing is that it be thorough so as to produce uniformity of consistency, cement and water content, and aggregate grading from beginning to end of each batch *as discharged.* "Mixing action" involves a working or vigorous rubbing of the cement paste onto the surface of the aggregate particles as well as a general blending of all the ingredients. To accomplish this, the mixer must be properly designed (or with hand mixing the procedure must be appropriate), and the process must be carried on for a sufficient length of time. To do the work within a reasonable time the mixer should be clean and in good condition, operated at the optimum speed as recommended by the manufacturer, fed efficiently, and not overloaded.

7.9 Types of mixers. In the field, machine mixing is done in batch mixers of the revolving drum type. Mixers of both stationary and portable, nontilting and tilting types are in common use. The tilting mixers usually have a conical or bowl-shaped drum, whereas the nontilting type usually has a cylindrical drum and a manually operated swinging discharge chute. Either type may be equipped with a loading skip or may be charged by means of a feed hopper, the former being preferable for portable mixers.

Mixers are classified as construction mixers or as paving mixers. In each case the size is designated by a number equal to its rated capacity in cubic feet of mixed concrete, but its guaranteed capacity shall be 10 percent greater. For construction mixers a letter "S" is placed after the number. Nine sizes of construction mixers are standard; they are designated 3½-S, 5-S, 7-S, 10-S, 28-S, 56-S, 84-S, and 112-S. For paving mixers a letter "E" is placed after the number. The three sizes of single-compartment paving mixers are the 13-E, 27-E and 40-E. The two sizes of two-compartment paving mixers are the 27-E dual and the 34-E dual. Many organizations contend that tilting mixers are more efficient for large jobs than the nontilting type, principally because they can be discharged in a short time and because the concrete is discharged in a bulkier mass which has less opportunity to segregate. Furthermore, tilting mixers can be cleaned more readily and thus kept in better operating condition [106].

Drum mixers should have a combination of blade arrangement and drum shape to ensure an end-to-end exchange of material parallel to the axis of rotation, as well as a rolling, folding, or spreading movement of the mix over on itself as the batch is mixed.

The newer types of portable mixers have introduced various improvements in design to increase their usefulness. These features include vertical water tanks, more efficient arrangement of blades, automatic skip shakers to aid in emptying the skip, batch meters, dustproof antifriction bearings, and pneumatic tires for greater portability.

Large stationary mixers equipped with separate batchers for each ingredient can be operated most efficiently when the solid materials (particularly the cement and sand) are fed to the mixer simultaneously and over the same period of time, as this accomplishes some premixing of the materials. Also, for best results about 5 to 10 percent of the water should precede, and a similar percentage should follow, the introduction of the solid materials, the remainder of the water being added uniformly and simultaneously with the solid materials [106]. The flow of water after the last of the dry batch has entered the drum should not continue for more than the first 25 percent of the mixing time [700].

All mixers should be so arranged that the operator may see into the mixer drum in order that he may observe the consistency of the concrete. Otherwise changes in the moisture content of the aggregates may go unnoticed.

Small drum mixers are not well adapted to laboratory use, because proper mixing action is difficult to obtain and because too large a percentage of the mortar in a small batch sticks to the drum and to the blades; for these reasons, in the past hand mixing has usually been resorted to in the laboratory. Recently, a mixer has been developed which has a flat-bottom pan rotating clockwise on rollers with mixing blades mounted on a vertical shaft rotating counterclockwise. The smaller units for laboratory use have no discharge opening, the entire pan being readily removable. Larger commercial units have a gate opening in the bottom of the pan. An advantage of this type is that the entire batch is readily discharged.

7.10 Time of mixing. With machine mixing there is an increase in strength of concrete with time of mixing up to perhaps 5 or 10 min. With moderate size mixers and for mixing times up to 1 min, the increase in strength is large; with times in excess of 2 min there seems to be insufficient gain in strength to justify the cost of longer mixing. The necessary mixing time varies with the size and depends somewhat upon the type of mixer. Based on test results, current practice usually requires a minimum mixing time of not less than 1 min for common mixers having a capacity of 1 cu yd or less, and an additional 15 sec for each additional cu yd or fraction thereof [106 ,700].

Although this is normally satisfactory for gravel concrete of medium consistency containing 1.5 bbl or more of cement per cu yd, for leaner mixes, drier consistencies, or mixes containing harsh aggregates, the mixing time should be increased up to as much as 50 percent according to the conditions and based on the results of the mixer efficiency test (see Art. 7.11). The mixing time should be measured from the time when all the solid materials are in the mixer, provided that all the mixing water has been introduced before 25 percent of the mixing time has elapsed. No portion of the time required for discharging any mixer should be considered a part of the required mixing time.

To control the mixing operation it is desirable to have a batch timer and counter, including an automatic lock which will release the discharge lever only at the end of the proper mixing period, in operation on each concrete mixer. The batch timer should be checked to see that it functions properly. If a timing device is not used, the time of mixing should be checked frequently, as often the mixer is the bottleneck of the job.

Whenever delays occur in the conveyance or placement of the concrete it may be necessary to continue mixing a batch much longer than normal. Excessive overmixing is harmful because some grinding of the aggregates occurs during mixing which causes a higher water content to maintain the desired consistency. Also, a reduction in any entrained air may result. Therefore, a maximum allowable mixing time of three times the normal should be specified, and if discharge of the batch is to be delayed longer, the mixer should be operated intermittently. For this reason all mixing equipment should be so arranged that it can be started under full load.

7.11 Mixer efficiency. The mixer efficiency as measured by the thoroughness of mixing of the constituents in a batch of concrete can be determined by a procedure described in Ref. 106. Prior to discharge at the end of the prescribed mixing period, two samples, one taken at the front and one taken at the rear, should not exceed the following limits of uniformity:

1. Unit weights of air-free mortar from the two samples should not vary more than 0.8 percent from the average of the two mortar weights.
2. Weights of coarse aggregate retained on a No. 4 screen from the two samples should not vary more than 5 percent from the average of the two weights of coarse aggregate.

The results of adequate mixing may be nullified unless care is taken in discharging the mixer to prevent the concrete from becoming segregated as it drops into hoppers, buckets, or other receptacles. The relatively small discharge stream from nontilting mixers is particularly susceptible to this condition. To prevent segregation from this cause, a vertical section of pipe should be provided at the end of the discharge chute, so that the concrete will fall vertically into the center of the receptacle as shown in Fig. 7.5. Should the last fraction of the batch contain an excessive amount of coarse aggregate, this portion should be retained and mixed with the succeeding batch. Batch size in this case should be reduced by an amount corresponding to the quantity withheld.

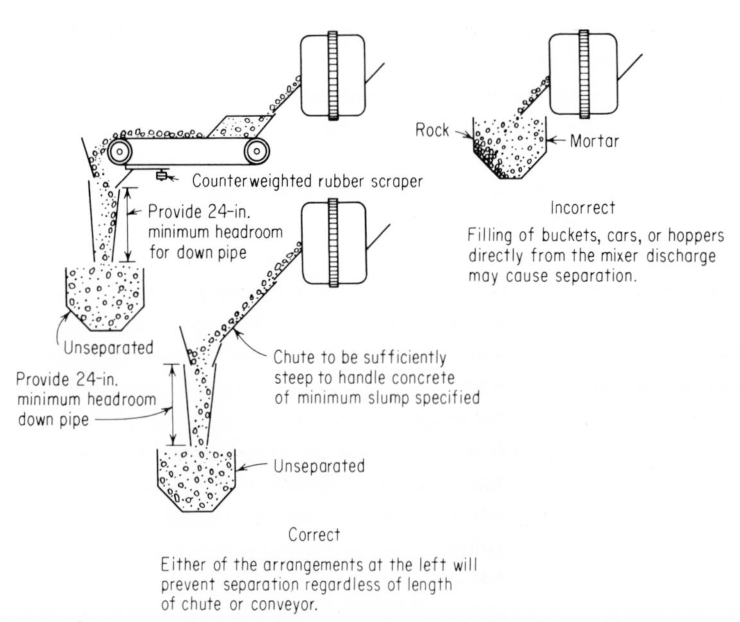

Fig. 7.5. Coarse aggregate will separate from mortar unless discharge from mixer is handled properly. (*From ACI Committee* 614 [700].)

7.12 Hand mixing; retempering. If concrete is to be mixed by hand, the procedure should be as follows: Level off the pile of aggregate, and spread the cement evenly over the top. Turn the dry materials at least twice, always shoveling from the toe of the pile and distributing in a thin layer on the new pile. Add the water while turning again, and then turn further until a homogeneous mixture of the required consistency is obtained.

Retempering or the indiscriminate addition of water to batches which have become stiffer than normal for any reason—such as delays in placing—should be prohibited, whether or not additional cement and aggregates are added. With adequate supervision and followed by additional mixing equal to half of minimum required mixing time, water may sometimes be added to improve the consistency of the concrete provided the maximum allowable water-cement ratio is not exceeded. The results of a study of retempering are given in Ref. 740.

7.13 Ready-mixed concrete. On most jobs, particularly in large cities and on certain highway and canal projects, it has been found practical to transport concrete to the job "ready-mixed" [C94]. Installations for this type of operation (see Fig. 7.3) are usually adequately equipped for handling and batching materials, and are usually well supervised. Thus, small jobs supplied with ready-mixed concrete may use control facilities normally available only to large jobs.

Under various systems the concrete is (1) "central-mixed," i.e., mixed completely in a stationary mixer and then hauled in a truck agitator, (2) "transit-mixed" (mixed completely in a truck mixer), or (3) "shrink-mixed," in which case it is mixed partially in a stationary mixer, and the mixing completed in a truck mixer. In most cases the concrete is mixed or at least agitated sufficiently during transportation to prevent segregation. For this work special truck mixers (also called transit mixers), or agitator trucks which produce a mild mixing action, are used. On some jobs where short hauls were involved central-mixed concrete has been transported in nonagitating equipment, but such practice has not received general approval.

Central-mixed concrete presents about the same problems as job-mixed concrete which must be transported for considerable distances from the mixer to the forms. The principal condition requiring consideration is the possible loss in consistency between the time of mixing and placing. Although this factor is generally negligible for normal conditions, care should be taken to minimize loss of slump by expediting delivery and, in warm weather, keeping the concrete as cool as practicable by using cold mixing water, avoiding hot cement, and by shading and sprinkling the aggregates.

It is more difficult to control the consistency in truck-mixing operations than in central plants, because the consistency cannot be observed until the concrete is delivered and discharge is started. Nevertheless, variations in slump may be minimized by:

1. Reducing variations in grading and moisture content of the aggregates to the fullest extent by means of suitable handling
2. Accurate control of the mixing water used by the truck mixer
3. Withholding some of the mixing water until arrival at the job, then adding the final water, within specification limits, after checking on the water requirement of the previous load, and doing additional mixing as necessary to produce a homogeneous mix
4. Telephone or radio connection between the job and the batching plant to ensure prompt response to any change orders
5. Careful technical control of all operations

Transit mixers should be operated at the mixing speed and number of revolutions found by performance tests (Art. 7.11) to accomplish thorough mixing and be capable of operating at a suitable slower speed for agitation. In transit mixing and shrink mixing, no more revolutions of the drum should be made at mixing speed, (usually within 50 to 100 revolutions for transit mixing) than are necessary to meet the mixer performance requirements. Mixing for agitation only should be at agitator speed. The batch volume should not exceed 63 percent of the gross for transit mixing, 70 percent for shrink mixing, and 80 percent for agitation. Agitation of mixed concrete should be kept to the minimum required for homogeneity at delivery; usually 10 revolutions at mixing speed or agitation during the last ½ mile before delivery, is sufficient.

Up to about 6 hr, the length of time over which the fresh concrete is agitated apparently has no effect upon the strength, as long as the mix remains workable and provided no water other than the original mixing water is added. However, even when agitated, concrete tends to stiffen considerably over a period of a few hours and becomes unworkable unless water is added. With such additions the strength of remixed concretes is governed by the effective total water-cement ratio [740].

Current specifications for ready-mixed concrete require that the concrete be discharged from the truck within 1½ hr or before the drum has had 300 revolutions (whichever comes first) after the water is added to the batch, or the cement to the moist aggregate. Under specially favorable conditions, periods up to 2 and 3 hr may be allowed. Conversely, under unfavorable conditions where air temperatures are unusually high, or the ingredients of the concrete are such that an unusually quick time of set or loss of plasticity may occur, it may be necessary to substitute a shorter period [C94].

CONVEYING

7.14 Conveying. Dependent upon the type of work and equipment available, various methods are employed to transport the concrete from the mixer to the forms; these include the use of wheelbarrows, buggies, chutes, dump buckets, trucks, endless-belt conveyors, monorail systems, pneumatic-pressure equipment, and displacement pumps. The method selected must be suitable for use with the consistency of concrete employed.

It is important that the concrete be handled without segregation of the constituent materials, a condition which may easily occur unless special attention is given to its prevention. When concrete is transferred from one conveyance to another, the use of hoppers, baffles, and short vertical drops through a pipe to the center of the receiving container as shown in Figs. 7.6 and 7.7, rather than long unconfined drops, is recommended as an aid in preventing segregation in concretes of the rather stiff consistency now frequently employed with vibratory tamping [700]. No consideration should be given to the common erroneous belief that separation occurring in handling will be eliminated in the course of other operations. Separation should be prevented—not corrected after its occurrence.

Segregation of concrete occurs because it is not a homogeneous combination but a mixture of materials differing in particle size and specific gravity. Consequently, as soon as the concrete is discharged from the mixer, there are internal and external forces acting to separate the dissimilar constituents. Any lateral movement such as occurs when concrete is deposited at one point and allowed to flow within the forms, or when the concrete is projected forward by the conveying equipment, causes the coarse aggregate and mortar to separate. If overly wet concrete is confined in some container or in restricting forms, the coarser and heavier particles tend to settle and the finer and lighter materials—particularly the water—tend to rise. These movements continue until they are finally arrested by solidification of the mass through chemical reaction between the cement and water [106].

The equipment for conveying concrete from mixer to forms should be such as to result in uniform concrete of the desired consistency at the forms. The method of conveying should not require a wetter concrete than is required for placing, nor should it produce segregation or excessive drying or stiffening.

7.15 Batch containers. Buckets, cars, and trucks are used as batch containers. Large buckets containing up to 12 cu yd are commonly used for the construction of dams and other massive structures. They may be designed to give a bottom opening the full size of the bucket to permit rapid discharge or may be arranged for partial opening and closing again to give

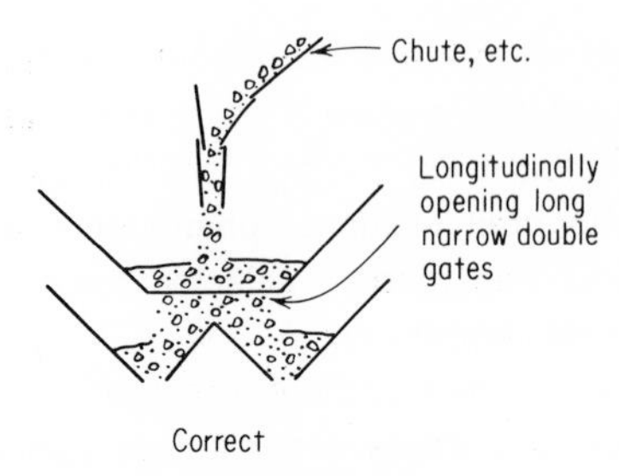

Correct

Single discharge hoppers should be used whenever possible, but the above arrangement shows a feasible method if a divided hopper must be used.

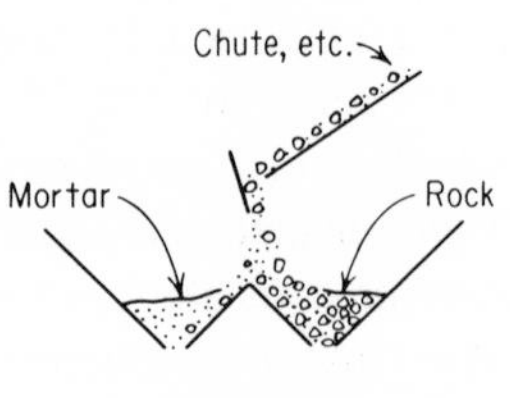

Incorrect

Filling divided hopper as above invariably results in separation and lack of uniformity in concrete delivered from either gate.

(*a*) Divided hoppers

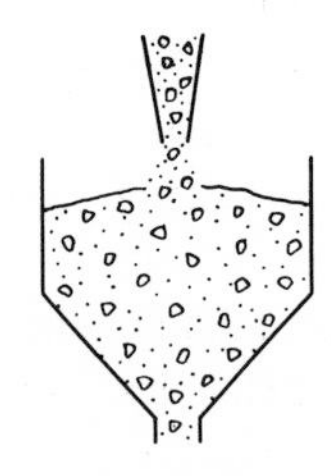

Correct

Concrete should be dropped vertically directly over gate opening.

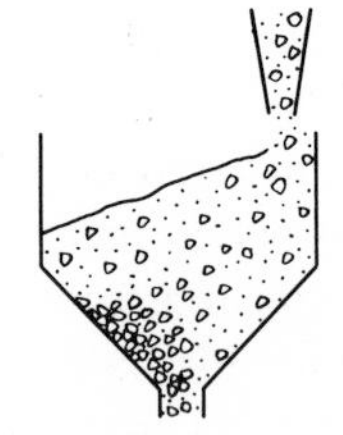

Incorrect

Dropping concrete on sloping sides of hopper should be avoided.

(*b*) Filling hoppers or buckets

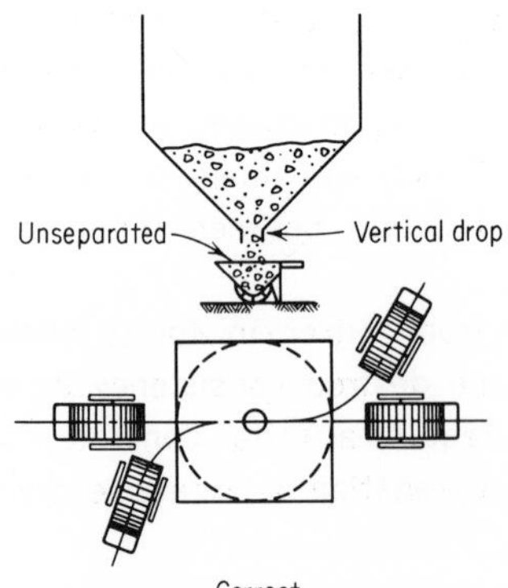

Correct

Discharge from hopper should be from center opening with vertical drop into center of buggy. Alternate approach from opposite sides permits as rapid loading as may be obtained with the objectionable divided hoppers having two discharge gates.

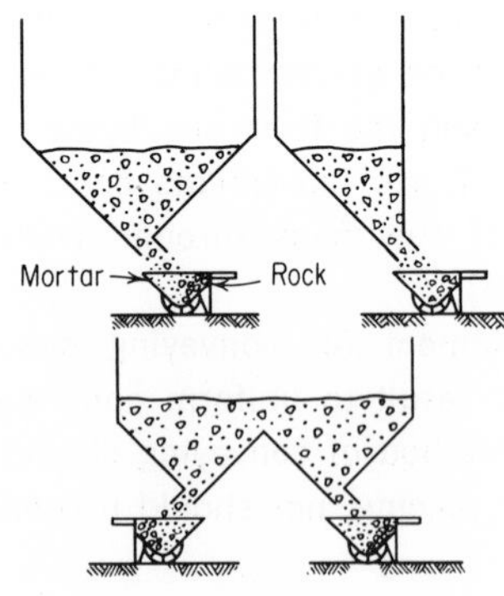

Incorrect

Sloping hopper gates which are, in effect, chutes without end control cause objectionable separation in filling the buggies.

(*c*) Discharge of hoppers for loading buggies

Fig. 7.6. Proper methods of loading hoppers and buggies prevent segregation of concrete. (*From ACI Committee* 614 [700].)

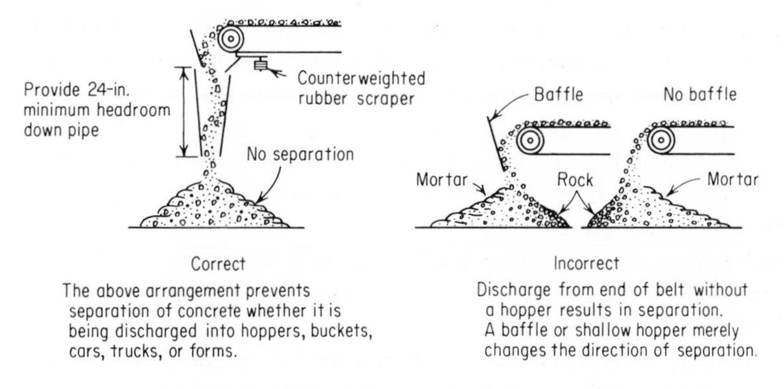

(*a*) Control of separation of concrete at end of conveyor belts

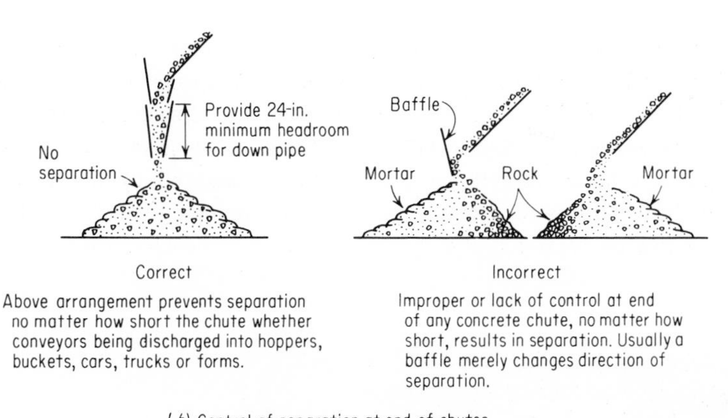

(*b*) Control of separation at end of chutes

Applies to sloping discharges from mixers, truck mixers, and to longer chutes, but not when concrete is discharged into another chute or onto a conveyor belt.

Fig. 7.7. Segregation of concrete at ends of chutes and conveyors can be controlled by correct use of downpipes. (*From ACI Committee* 614 [700].)

better control of the placing operations. Buckets may be handled by cranes or by overhead cableways. Cableway operation requires considerable care to avoid too rapid rebound when dumping and consequent improper control of the shape, character, and position of the dumped concrete. In general, well-designed buckets properly operated offer a very satisfactory means of conveying and placing concrete, particularly mass concrete containing large aggregate, as such concrete requires special care and confinement in handling to prevent segregation.

The use of buckets for this purpose is satisfactory, provided (1) only complete mixer batches are placed in the bucket; (2) segregation is avoided in filling the buckets; (3) they are of a size, and may be discharged in a manner and with such frequency, that the concrete may be placed in small amounts or in approximately horizontal layers while the previous layer is still soft; (4) they are capable of discharging concrete of the stiffest consistency specified; (5) successive batches are so placed at to afford opportunity for thoroughly working the concrete by means of internal vibrators; and (6) discharge is controllable so that it will cause no damage to, or misalignment of, the forms [700].

Special railway cars or motor trucks are used satisfactorily on some jobs for transporting the concrete from the mixer to a point directly above or near the forms. Such cars and trucks are specially designed to facilitate unloading and may be equipped with agitators for use on long hauls. The use of drier mixes and air entrainment are very helpful in reducing segregation.

7.16 Pump and pipeline. Very successful transportation of concrete has resulted from pumping through steel pipelines. A special type of pump for this work is shown in Fig. 7.8, and a schematic drawing showing the action of its valves is presented in Fig. 7.9. Concrete feeds from the hopper into the pump cylinder largely be gravity, but is assisted by the vacuum created on the suction stroke of the piston. The pumping rate is in direct ratio to the degree to which the cylinder is filled on each stroke, so the

Fig. 7.8. Rex concrete pump. *(Chain Belt Co.)*

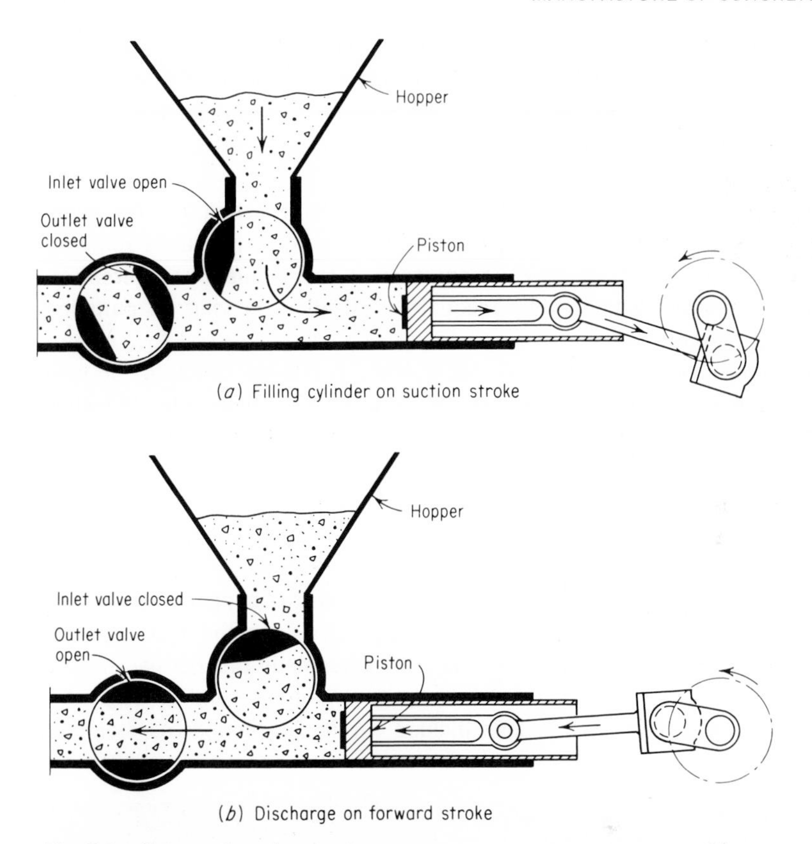

Fig. 7.9. Valve action of concrete pump.

flowability of any concrete affects the hourly capacity of the pump. Concrete of average cement content having a well-graded aggregate and mixed to a workable consistency will feed more readily than a harsh, dry, or sticky mix.

The valves do not make full closure when pumping concrete but are set to the size of the largest aggregate in the mix. They can be adjusted to full closure for pumping grout, but the piston starts forward before the inlet valve completes its travel and is on the return stroke before the outlet valve closes completely. This accounts for the fact that the concrete pump will not handle liquids and explains why production falls off with overly wet concrete and small aggregates, particularly when the valves and liners are worn excessively.

The concrete pump will work satisfactorily on most concrete mixes which are not undersanded or poorly graded, and not too dry or wet. in general, concrete of the most pumpable consistency will have a slump of 3 to 4 in. Wetter and drier consistencies can be handled satisfactorily, but the rate of pumping may be somewhat slower. With wet, dry, or harsh mixes there is danger of plugging the pipeline with attendant delays in resuming operations. In general, pumpability is a relative term. The shorter the distance concrete is to be pumped, the less its pumpability will be affected by harshness of the mix.

In starting pumping operations it is common practice to use some mortar ahead of the regular concrete. About 1 cu yd of mortar is required to lubricate 1,000 lin ft of pipe irrespective of the diameter. Thereafter, the regular concrete will provide adequate lubrication as long as pumping of a suitable mix continues.

Gravel aggregate usually produces a more pumpable concrete for long lines than crushed stone, particularly for slumps less than 3 in., because of its tendency to produce more workable concrete with less likelihood of the formation of segregated stone pockets. However, well-graded crushed stone having only a few flat, elongated particles is about equally satisfactory. For both kinds of concrete, sand with an adequate percentage of fines that will hold water and prevent "bleeding" is desirable.

Segregation of mortar and coarse aggregates in the hopper of the pump is likely to cause clogging of even short pipelines. As the last concrete out of the mixer is especially likely to be segregated, the hopper should be kept partially full at all times. To eliminate segregation and its attendant ill effects, all new large installations include a remixing hopper as standard

Fig. 7.10. Delivery of concrete by pipeline. (*Chain Belt Co.*)

equipment. In addition to eliminating segregation, remixing often increases the pumping rate, as it tends to make the concrete flow more readily into the cylinder.

Air-entrained concrete may require somewhat more sand and perhaps an inch greater slump than would likely be necessary without air entrainment for successful delivery through long lines. These are required to offset reduction in workability as a result of compression of the entrained air while the concrete is moving in the pipeline. Lightweight aggregate is so highly absorptive when under pressure that lightweight concrete may cause plugging of the pipe line. A procedure for comparing the pumpability of concrete mixtures is given in Ref. 751.

Increases in slump of the concrete sometimes observed between the pump hopper and the end of the pipeline have been traced to leakage of water into the concrete from the water supply which serves to keep the cylinder and piston clean. Wear of the rubber piston head (after pumping about 1,000 cu yd of concrete) permits this leakage to occur. Replacement of the piston head or other worn parts will correct this condition.

Tests of operations at zero temperatures indicate a 1 to 2°F loss per 100 ft of exposed pipe. To reduce the chilling effect of a cold pipe, steam may be used to heat the pipe at the beginning of a run. In some instances, overheating the mixing water or the aggregates has caused the concrete to stiffen so rapidly as to cause plugging of the pipe. Operations during hot weather have also been interfered with by rapid stiffening of the concrete. This can be corrected by covering the pipe exposed to the sun with burlap and keeping it sprinkled with water.

7.17 Pump sizes. Concrete pumps using 6 or 8-in.-diameter pipes are made in various sizes capable of delivering from 15 to 65 cu yd per hr. The maximum size of aggregate which can be handled is about 3 in., but the use of aggregate larger than 1½-in. maximum size is usually not advisable. The pumps can be used for horizontal runs up to 1,000 ft, for vertical lifts of 120 ft, or for combinations of the two considering each foot of rise to be the equivalent of 8 ft of horizontal run. Large radius curves can be readily used, but add considerably to the pipe friction losses. As an approximation, a 10-ft piece of pipe bent to form a 5-ft-radius 90° elbow has about the same resistance as 40 ft of straight pipe. Hence, the pipeline should be laid with as few bends as possible, and changes in direction should be accomplished preferably with bends of 45° or less. Standard pipe sizes are 6, 7, and 8 in.; pipe lengths are 1, 2, 3, 5, and 10 ft. Also 90, 45, 22½, and 11¼° bends are available. With this assortment, almost any desired point can be reached. The abrasiveness of the concrete causes wear in the pipeline, but most of the pipe, including the bends, will be good for more than 50,000 cu yd.

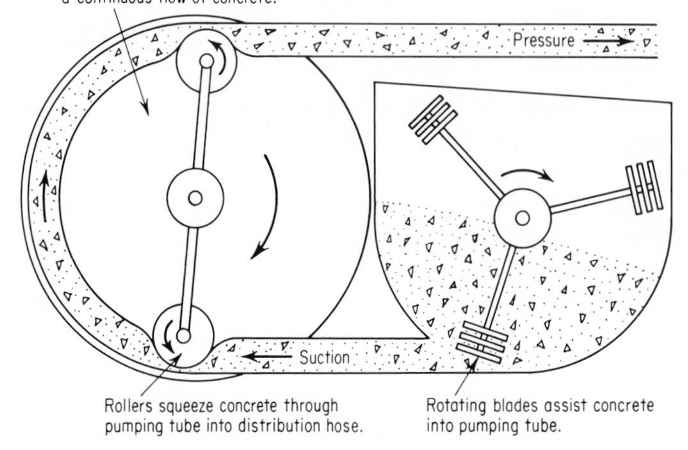

Fig. 7.11. Schematic drawing showing how squeeze-type, small-line pumping system operates.

Various new types of small pumping units are now available which pump concrete, having a 1-in. maximum aggregate, through 3-in. rubber hoses at rates of 10 to 30 cu yd per hr. Some use a piston-type pump while others use a squeeze-type unit shown in Fig. 7.11. In the latter a pair of rubber rollers, operating on a planetary drive principle, roll over a specially designed 3-in.-diameter "pumping tube."

7.18 Cleaning the concrete pump. For the larger pumps emptying and cleaning the pipeline at the completion of a run is accomplished by the use of a "go-devil," which is propelled through the line by water or air pressure. The go-devil is a dumbbell-shaped device with a rubber cup at each end to produce a seal. To insert the go-devil, a short length of pipe is disconnected next to the pump. Water pressure may be supplied by a separate pump or by the concrete pump after it is converted into a high-pressure water pump by the addition of water valves above the inlet concrete valve and beyond the outlet concrete valve. This change requires only 5 to 15 min.

As the go-devil moves through the pipeline, its location can be determined by tapping the pipe. Just before it reaches the end of the line, the pump is stopped and the last length or two of pipe dumped by hand. Care must be taken that no flushing water mixes with the concrete. It should be wasted outside the forms or drained back to the pump. On some jobs where it is difficult to dispose of the flushing water, air can be used to propel the go-devil. In that case, the expansion of the air when the go-devil is near the end of the line may eject the go-devil with such velocity as to cause damage.

On the squeeze-type pump, the emptied hopper is filled with water and a large sponge is inserted in the hopper outlet. When the pump is started again, the sponge and the water which follows clear the hose of concrete and clean it at the same time.

7.19 Pneumatic method. The transportation and placement of concrete by means of pipeline and compressed air has been used quite commonly for tunnel linings and similar structures, but this practice is followed much less since the advent of the concrete pump. In general, to obtain good results, the volume of concrete at each discharge of the gun should not exceed about 7 cu ft, there should be no segregation of the mix delivered to the gun, the length of pipe should be as short as possible and never over 1,000 ft, and the discharge line should never be inclined downward. The nozzle should be directed so that the concrete does not impinge directly upon the reinforcement or produce excessive rebound. As excessive segregation will occur at the point of discharge when the concrete is violently shot from the open end of the pipe, the discharge end should either be buried in the concrete or be suitably baffled until it can be buried. The use of a smaller proportion of coarse aggregate at the beginning of a run will assist in preventing segregation.

A general objection to the use of the pneumatic method is that an appreciable loss of slump, commonly amounting to 2 or 3 in., occurs during the shooting operation. Hence a wetter consistency than that required in the forms is necessary at the charging gun. This results in a weaker concrete less resistant to freezing and thawing or requires a richer concrete subject to greater shrinkage than if placed by other approved methods.

For a discussion of the cement gun for spraying mortar see Arts. 14.8 to 14.15.

7.20 Chutes and belts. The use of long chutes has met with such disfavor by engineers that few jobs now permit their use. This situation has developed because many mixes are too stiff to be handled by chutes and because marked segregation tends to occur in mixes which are wet enough to be handled by chutes. This segregation occurs not only along the chute but at any transfer point, as shown in Fig. 7.7. In no case should the consistency of concrete be increased above that required for proper placement, merely to make it wet enough to flow along a given chute. When chutes are used, they should be amply steep (about 1 vertical to 2 horizontal), metal or metal-lined, round-bottomed, of large size, and protected from overflow.

Conveyor belts have been used for handling relatively stiff concretes, but there is a tendency for segregation on steep inclines, as the belt passes over the rollers, and at transfer points (see Fig. 7.7).

Both chutes and belts tend to cause drying and stiffening of the concrete, particularly when they are long, as a result of exposure of a thin stream to sun and wind. In using either type of equipment, the concrete should be passed through a vertical pipe placed at the end of the line to reduce segregation of the mix. For both systems the flow should be as nearly continuous as possible to minimize the effects of segregation. A receiving hopper placed near the forms may be necessary when the flow is intermittent. Care must be taken that any water used in flushing the chutes or belts does no drain inside the forms.

QUESTIONS

1. Describe various methods of measuring cement, sand, coarse aggregate, and mixing water for a concrete mix.
2. What are the advantages of batching by weight as opposed to batching by volume?
3. How does the first batch of concrete for any day's run differ from subsequent batches, and why?
4. Are there any advantages of dial-scale type batching units in comparison with beam-scale units? Explain.
5. What types of concrete mixers are in common use?
6. What are the relative merits of tilting and nontilting mixers?
7. How does the time of mixing affect the strength of concrete?
8. What time of mixing is recommended?
9. What are the three types of ready-mixed concrete, and how are they produced?
10. What is the effect of long agitation of concrete in a truck mixer?
11. What is "segregation" in wet concrete? Why is it undesirable?
12. Why are long unconfined drops of fresh concrete undesirable?
13. Discuss the relative tendency of dry and wet concretes to segregate while being transported.
14. Describe the operation of the concrete pump.
15. Can the concrete pump be used to pump water? Explain.
16. Why are chutes ordinarily undesirable for transporting concrete?

EIGHT
PLACING AND CURING CONCRETE

8.1 Preparations for placing. Before placement of concrete the foundation or base must be suitably prepared. Earth foundations should be compacted by rolling or tamping and should be thoroughly moistened to a depth of about 6 in., by intermittent sprinkling, to prevent excessive loss of moisture from the concrete. In no case should a muddy surface be created, as contamination of the concrete may result. Rock foundations, to which the concrete is to be bonded, should be roughened where necessary and thoroughly cleaned of all weak, loose, or disintegrated material. The rock surface should be completely surface-dried by air jets as any surface water will prevent proper bonding to the surface.

Where concrete is to be bonded to a previous lift of concrete, special precautions must be taken to clean the surface of all foreign matter. The actual procedure to follow depends largely upon the type of structure and the quality of concrete at the top of the earlier lift. If the concrete in the previous lift was of a wet consistency and was worked excessively during or after placement, then the top of that lift will be relatively porous and weak. The laitance, or scum, on the surface and the fine inert materials carried near to the surface make it difficult to secure good bond. On the other hand, the placement of stiffer concrete in the lower lift and the minimizing of undesirable working of the joint surface or walking over it will result in a concrete of higher quality, so that better bond may be developed.

In mass-concrete construction, as in dams, two principal methods are employed in preparing the joint surface. For surfaces having excessive laitance, it has been common practice to remove all laitance and inferior surface concrete and to wash the mortar from protruding aggregate by means of a high-velocity jet of air and water as soon as the concrete has hardened sufficiently to prevent the jet from raveling the concrete below the desired depth. Ordinarily, the surface is cut about ⅛ in. The time interval between placing and cleanup operations may range from 4 to 12 hr dependent upon the temperature, humidity, and the setting characteristics of the concrete. Because of leaving cloudy pools of water which upon drying leave a surface film, contamination from traffic, and carbonation, the initially clean surface must be restored just previous to placement of the new lift. The final cleanup is most effectively accomplished by wet sandblasting and washing. On surfaces of concrete which have been properly placed at the right consistency, modern practice favors the omission of the initial cleanup, using only the wet sandblasting and washing and drying of the surface immediately prior to placement of fresh concrete. The advantages of this method are its relative simplicity, since only one cutting and washing operation is used; dependability, in that good results can always be obtained; and relative economy, as a result of the development of efficient sandblasting equipment and methods.

The cleanup practice for joints in reinforced-concrete construction must be modified from the above as embedded fixtures, the erection of forms, placement of steel, and the usual narrowness of the formed space add to the difficulties of the cleanup operation and lengthen the time interval between lifts, so that in many instances the curing period is completed and the joint surface hard and dry before placement operations are resumed. The problem is further complicated, as owing to greater difficulties of placement, structural concrete is of a wetter consistency and is usually worked more than mass concrete, so that it may be consolidated properly. This results in an inferior concrete near the joint, and a surface coated with considerable laitance. But most construction joints in reinforced-concrete structures are not subjected to hydraulic pressure and therefore not subject to pronounced leakage. In such cases a thorough cleaning by scraping and brushing of the surface before erecting forms and reinforcement, followed by a thorough washing or air blasting to remove sawdust, chips, etc., just before concreting is resumed, will usually suffice to clean the surface. For joints which are to resist water pressure the preferable treatment involves curing as long as possible and sandblasting just before the erection of the enclosing form. Thereafter, it is desirable that the surface be covered with damp sand, which should be washed or blown out just prior to beginning placement of the new lift.

Concrete surfaces which have dried out after curing for the specified period should be wet before placing operations are resumed, but there should be no pools or film of water on the surface when the mortar layer is applied. At dried joints requiring a high bond strength, the surface should be kept wet for a few hours and then surface dried just before resuming concreting operations.

Immediately prior to the placement of new concrete, the cleaned surface of the previous lift should have scrubbed into it a mortar layer about ½ in. thick having the same water-cement ratio as for the regular concrete. The cement-sand proportions should result in a slump of 6 to 8 in. Mortars which are wetter than this tend to segregate, and drier mortars cannot be broomed or otherwise worked effectively into the irregularities of the surface.

The purpose of applying the mortar layer is to ensure good bond and watertightness at all joints, as, in hydraulic structures, these locations are most likely to develop leakage. Also, the mortar layer facilitates the placement of the regular drier concrete, serving to bed the coarse aggregate and to minimize the development of air voids and rock pockets at the bottom of a lift.

The reinforcement should be freed of all loose rust or scale by wire brushing, sandblasting or other effective means. Other objectionable coatings likely to be found on parts of the reinforcement are paint, oil, grease, dried mud, and dried mortar splashed on the bars ahead of the concrete being placed. All such coatings should be removed. The reinforcement should be checked for size, bending, spacing, and location. It should be properly spliced, anchored firmly in position, and embedded a given minimum distance from the surface.

Anchor bolts, inserts, pipe sleeves, pipes, conduits, wiring, manhole-cover frames, and other embedded fixtures should in general be firmly fixed in position before the concrete is placed. Such fixtures should not be permitted to affect the position of metal reinforcement except as shown on the plans, nor to be so located as to reduce appreciably the strength of the construction.

8.2 Placing. The placing and compacting operations are, perhaps, the most critical field activities performed by the concrete organization. Unless care is exercised in these operations, a very poor job may result, even though the concrete leaving the mixer is perfect. First of all, an effort should be made to avoid segregation of the concrete, but if it occurs, remedial measures should be taken to eliminate all traces of segregation, so that the concrete in place will be homogeneous and free from honeycomb. Proper methods of placement will avoid displacement of forms and reinforcement, develop a good bond between successive layers, minimize shrinkage cracking, and ensure a neat appearance of the structure. No concrete should be placed where rain falls upon it as this will cause an overly wet, weak concrete, especially at the surface.

When concrete is placed in deep forms, it is too frequently the practice just to let it drop, regardless of height. This may result in segregation, and damage to forms and embedded fixtures. Also, the forms and reinforcement above the level of placement become coated with mortar which may dry before the concrete builds up to its level. This condition may be avoided by dropping the concrete into an outside pocket and allowing it to flow over into the form without segregation as shown in Fig. 8.1. If the height of drop is not great, a hopper feeding into a vertical pipe will avoid separation and will keep the reinforcement clean. In general, dropping concrete a considerable distance through a vertical pipe is less harmful than letting it fall a much shorter distance but strike against the form, causing it to separate as shown in Fig. 8.1. A good practice in filling deep narrow forms is the gradual use of drier concrete as the upper lifts are reached; otherwise, water gain tends to make the upper layers excessively wet and markedly to reduce the quality of the concrete.

Concrete should be deposited in horizontal layers close to its final location; it should not be dumped in large quantities at any point and allowed to flow within the forms, as this causes segregation, poor consolidation, and sloping work planes. Each layer should be compacted thoroughly before the succeeding layer is placed. The depth of layers will depend upon the character of the work. In reinforced structures the depth may vary from only 6 to 12 in. to give ample opportunity to consolidate the concrete around the steel, but in more open structures the depth may be as much as 2 ft. In all cases the placing operations should be so timed that a new lift is placed while the earlier concrete is either still plastic or after it has hardened thoroughly; if it is in a semihardened condition, there is danger of its being damaged by subsequent placing operations.

When concrete is placed on an appreciable slope by use of a chute, there is a tendency for segregation and for the concrete to be carried downslope. This can be avoided as shown in Fig. 8.2 by using a baffle at the end of the chute. Also, a weighted, unvibrated steel-faced slip-form screed working up the slope, as shown in Fig. 8.2, is useful.

Where concrete must be placed on a slight slope, placing should begin at the lower end of the slope, thereby permitting consolidation of the layer by vibration or other suitable means. When placement is begun at the top of the slope, vibration tends to shift it downhill, as shown in Fig. 8.3. Rock pockets can be eliminated by shoveling the segregated coarse aggregate onto softer concrete and blending in place by vibration or other means; as shown in Fig. 8.3 shoveling mortar and soft concrete over the top of the segregated rock will not correct this type of defect. It is a common fallacy that separation occurring in handling will be eliminated in the course of other operations. Separation must be prevented—not corrected after its occurrence.

The rate of placement should not be so rapid that the placing crew cannot compact it properly; however, the faster it can be placed without damage to the forms and with ample vibration, the better the results generally obtained. If thin walls or columns are rapidly filled with concrete of a wet consistency, there is danger of movement or failure of the forms due to excessive lateral pressure (see Art. 9.3). But where concrete-production

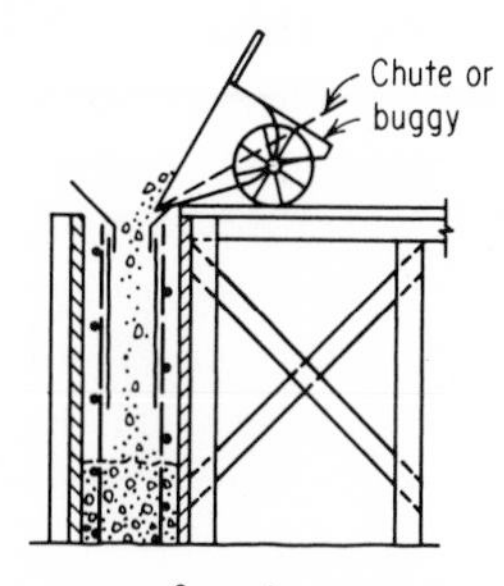

Correct

Separation is avoided by discharging concrete into hopper, feeding into drop chute. This arrangement also keeps forms and steel clean until concrete covers them.

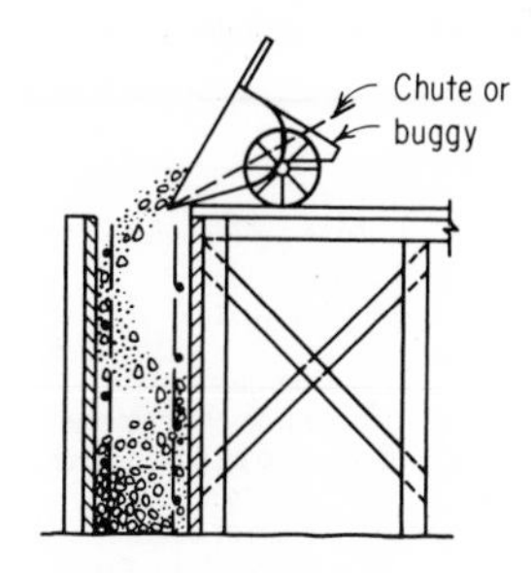

Incorrect

Permitting concrete from chute or buggy to strike against form and ricochet on bars and form faces causes separation and honeycomb at the bottom.

(*a*) Placing in top of narrow form

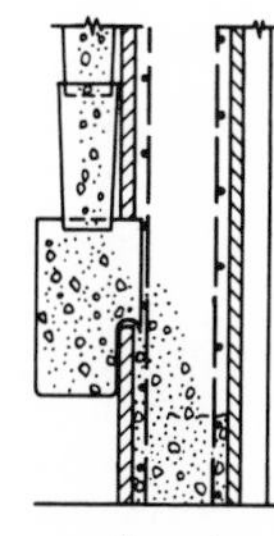

Correct

Drop concrete vertically into outside pocket under each form opening to let concrete stop and flow easily over into form without separation.

Drop chute to movable pocket or opening in form

Incorrect

Permitting rapidly flowing concrete to enter forms on an angle invariably results in separation.

(*b*) Placing in deep, narrow wall through port in form

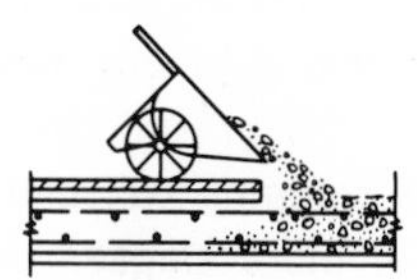

Correct

Concrete should be dumped into face of previously placed concrete.

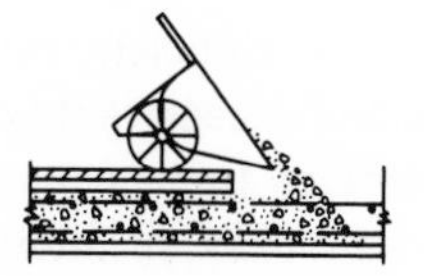

Incorrect

Dumping concrete away from previously placed concrete causes separation.

(*c*) Placing slab concrete from buggies

Fig. 8.1. Proper introduction of concrete into forms will prevent separation of constituents. (*From ACI Committee* 614 [700].)

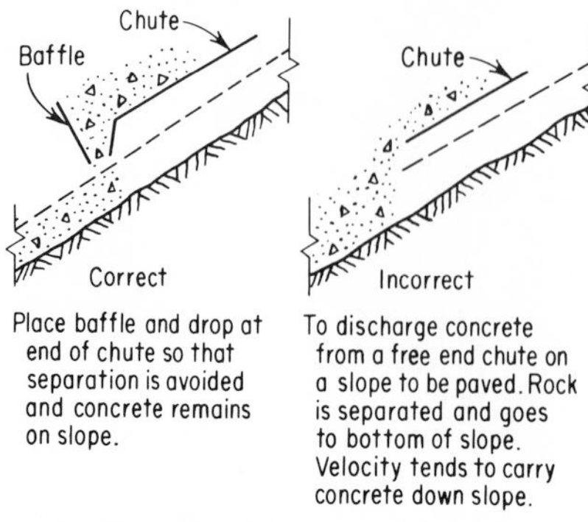

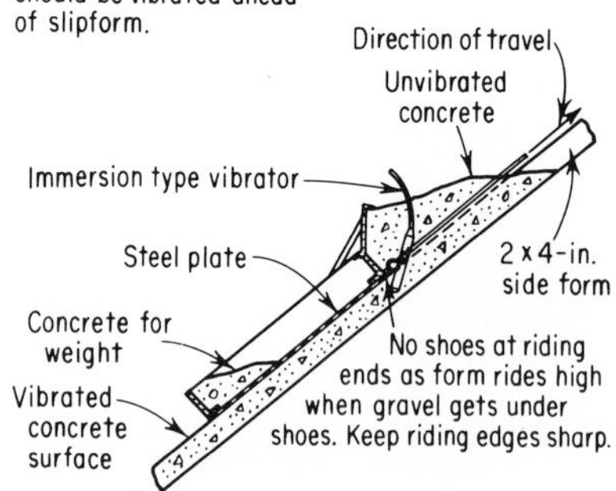

Fig. 8.2. Correct and incorrect methods of placing concrete on a sloping surface. (*From ACI Committee* 614 [700].)

facilities are adequate and it is otherwise practicable, it is desirable to place concrete to full height in one lift and thus avoid construction joints and cleanup problems. As long as the rise of concrete in the forms does not exceed 5 ft per hr in warm weather or 3 ft per hr in cold weather, concrete stiffens at a sufficient rate to permit placement to any height in forms without causing excessive fluid pressures.

The settlement of fresh concrete in walls and columns will cause the formation of cracks unless an interval of 1 to 3 hr or more elapses between the completion of walls and columns and the placing of beams or slabs supported by them. If the beams or slabs are haunched, the stop should be made at the bottom of the haunch. Wherever possible, built-in window frames should be allowed to settle slightly with the concrete to avoid cracking the concrete.

The placement of concrete under water is discussed in Art. 14.19.

The method of placing the concrete should avoid shock to the form. Impact loading due to suddenly dropping a large mass of concrete through any appreciable distance may cause serious displacement or even collapse of the forms.

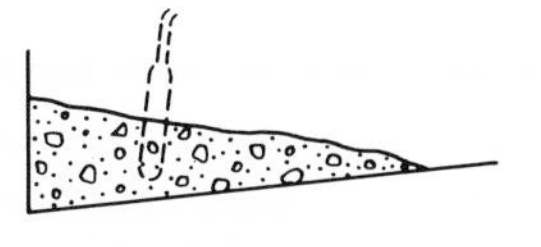

Correct

Start placing at bottom of slope so that compaction is increased by weight of newly added concrete. Vibration consolidates.

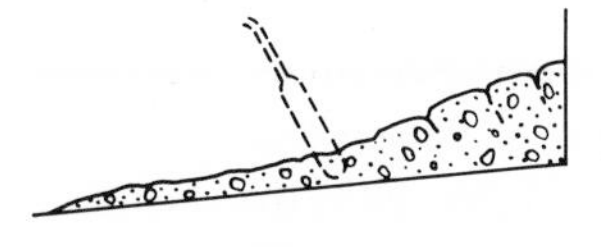

Incorrect

To begin placing at top of slope. Upper concrete tends to pull apart especially when vibrated below, as vibration starts flow and removes support from concrete above.

(*a*) When concrete must be placed in a sloping lift

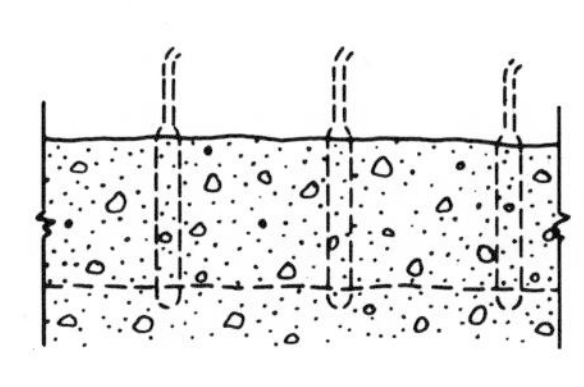

Correct

Vertical penetration of vibrator a few inches into previous lift (which should not yet be rigid) at systematic regular intervals found to give adequate consolidation.

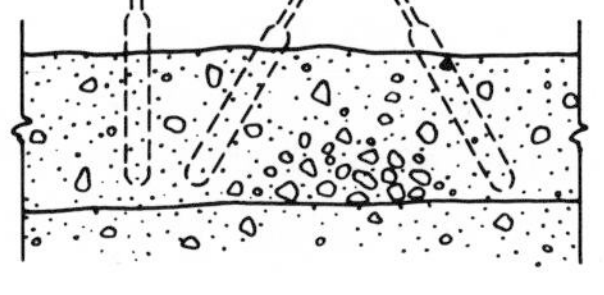

Incorrect

Haphazard random penetration of the vibrator at all angles and spacings without sufficient depth to assure monolithic combination of the two layers.

(*b*) Systematic vibration of each new lift

Correct

Shovel rocks from rock pocket onto softer, amply sanded area and tramp or vibrate.

Incorrect

Attempting to correct rock pocket by shoveling mortar and soft concrete on it.

(*c*) Treatment of rock pocket when placing concrete

Fig. 8.3. Proper placement and vibration of the concrete will ensure uniform compaction. *(From ACI Committee* 614 [700].)

COMPACTION

8.3 Compaction. Immediately upon placement of concrete in the forms it should be compacted to assure close contact of the constituent materials with themselves, as well as with the forms, reinforcement bars, and any embedded fixtures. This consolidation may be accomplished with the use of hand tools, but vibrators are much preferred.

8.4 Hand tamping. Dry concrete to be consolidated by hand operations should have the surfaces well rammed with heavy flat-faced or grid-type tools until a thin film of mortar is brought to the surface, indicating that consolidation is complete.

For plastic mixes, the hand tool should penetrate the full depth of the last layer placed, and if the underlying layer is still plastic, even into such material to bond the layers together. Slicing and spading near the vertical form faces serves to bring a mortar film to the forms and produce a smoother surface, but overworking these areas often develops a thin layer of cement paste next to the forms, and crazing may result upon drying.

8.5 Vibrators. Hand tamping was formerly the only means of compacting concrete in place on the ordinary job. On many classes of work, however, high-frequency power-driven vibrators have come into more general use. There are two general classes of vibrators, the internal, or immersion, type and the external type. In most cases the vibratory action is caused by the rotation of an eccentrically weighted shaft at high speed, usually over 7,000 rpm when immersed in the concrete. The internal vibrator consists essentially of a steel tube enclosing the vibrating element, which is inserted into the fresh concrete. For use in deep forms, some internal vibrators are attached to the end of a long flexible shaft inside a rubber tube. External vibration may be applied to the top surface of a concrete mass by means of a vibrated platform or screed, it may be applied through the forms by vibrators or pneumatic hammers, or, in the case of precast products, it may be applied by vibrating the molds.

Internal-type vibrators are more efficient than those attached to the forms, as in the latter type much of the energy is absorbed by the forms. Also, internal vibrators are more effective than those applied to a vibratory platform. As surface vibrators may bring an excess of fines to the top, resulting in a layer of low-quality material, they should be used only where internal vibration is impracticable. Internal vibrators should be used wherever the sections are sufficiently large for insertion and manipulation of this type of device. This includes practically all concrete work except some thin walls, pipe, and other precast products. Surface vibrators may be employed to supplement internal vibrators when experience shows their use to be beneficial. Form vibrators should be used only where it is not practicable to use either the internal or surface type, as the film of cement paste which they tend to bring to the formed surface causes crazing. Also, they are more difficult to move, and they develop greater stresses in the forms.

Gasoline-engine-driven vibrators generally are less convenient to use and require more servicing than pneumatic or electric units. Air motors are much lighter than electric motors for a given power input. This feature makes them adaptable over a wide range of service, but it is essential that the air supply be adequate. Electric vibrators have proven to be very satisfactory, especially in the medium and smaller sizes.

8.6 Vibrator efficiency. The efficiency of any vibrator depends very much upon its frequency or speed of operation. Tests have shown that for complete compaction a concrete of ½-in. slump required 90 sec vibration at 4,000, 45 sec at 5,000, and 25 sec at 6,000 rpm. These frequencies were developed when the vibrator was submerged in the concrete and operating under full load. Frequencies of underpowered vibrators may be much less under load than when operating free of load. Vibrator speeds should be regularly checked by the inspectors. Pencil-size vibrating reeds to determine vibration frequency are available commercially for this purpose.

The amplitude of vibration also effects the results obtained although specific data are lacking. Too great an amplitude of form vibrators may cause so much movement of forms as to pump air into the concrete [840]. The selection of type, number, and size of vibrators for a particular job depends upon the structure, size of batch, number of batches per hour, size of aggregate, amount of reinforcement, the mix proportions (particularly the percentage of sand), and consistency of the concrete. On the average, high-speed small vibrators can compact up to 20 cu yd per hr, and the large heavy-duty, two-man type can handle up to 50 cu yd per hr in open forms [106, 840].

Sufficient compaction equipment and operations should be provided to handle the entire mixer output without delay; otherwise the concrete may stiffen excessively before it is compacted, with the result that it may remain segregated and porous.

In studies of the effect of internal vibration in comparison with hand tamping upon the final homogeneity of a mass of artificially segregated concrete, it was found, in general, that with proper manipulation of the vibratory tamper, rock pockets could be entirely eliminated. With hand tamping, this elimination of rock pockets is often impossible once segregation has occurred.

8.7 Concrete mix for vibratory compaction. The use of vibration in the compaction of concrete has been characterized by one authority as the greatest advance in the art of concrete making since the invention of the concrete mixer. It must be intelligently used, however. It is not a cure-all. The mix should be designed for use with vibratory compaction, and proper technique in the use of vibratory equipment must be observed [840].

In general, drier mixes can be placed by the use of vibration than by hand tamping. The relation of water-cement ratio to strength holds for vibrated concrete the same as it does for hand-placed concrete. Vibration merely extends this relationship to the range of stiff mixes which cannot be placed by hand. Hence, through the use of drier consistencies vibration permits the attainment of higher strengths with the same mix proportions; with the same water-cement ratio, a leaner mix may be used where vibration is employed. Thus, certain definite advantages in the way of economy of materials and quality of concrete may be obtained through the use of vibration, even though it imparts no new properties to concrete.

The consistency of the concrete suitable for compaction by vibration depends upon such factors as the character of the mix, placing conditions, and the effectiveness of the vibrators, but some authorities require that vibration should not be used when the slump of concrete exceeds 3 in., except where placing conditions are difficult. In no case should the concrete be so dry as to make placement difficult, nor so wet as to cause water gain on the surface.

Although experience has shown that, when vibration is used, the proportion of sand in the concrete mix can be reduced about 4 to 8 percent in terms of the weight of total aggregate, it is necessary to avoid undersanding. The small economies or increases in quality which are possible may be offset by increased difficulties in handling and placing if undersanding is approached [840].

Water gain may occur more readily with vibration than with hand compaction. This water gain is due to a lack of fine particles in the cement and sand or to an excess of water. It should be prevented by using a nonbleeding cement, more fine sand particles passing the No. 50 and No. 100 sieves, or less water. While less sand can be used than is required for hand-rodded concrete, still there must be sufficient fines in the mix to prevent excessive water gain.

Sand streaks on vertical surfaces of concrete are defects closely related to the bleeding tendencies of the concrete but are also influenced by the tightness of the forms, and the methods of working the concrete. The use of wet, lean, and undersanded mixes, sands lacking sufficient fines, coarsely ground cement, and excessive working of the concrete all contribute to the formation of sand streaks [106].

8.8 Proper use of vibration. Vibrators should not be used to transport concrete in the forms, as segregation is likely to result. Internal-type vibrators should be inserted vertically at fairly close intervals, about 18 to 30 in.

apart, for short periods of about 5 to 15 sec, rather than for longer periods at greater distances apart [700]. They should be withdrawn slowly at each location and should be operated continuously while being withdrawn, in order that no hole will remain in the concrete. Vibrators should be inserted to the full depth of each layer and even into the previous layer as shown in Fig. 8.3, provided that layer will become plastic under the vibratory action, to ensure good bonding of the layers. Bonding of new concrete to concrete that has hardened and has been properly cleaned is essentially a matter of thoroughly vibrating the new concrete close to the joint surface.

Usually, vibratory compaction is complete when mortar just begins to flush to the surface adjacent to the vibrator, at the forms, or at the surface of the steel reinforcement. Another indication is the development of a level top surface with sufficient mortar for finishing [840]. Overvibration of concrete, particularly if the slump exceeds 4 in., should be avoided, as it causes settling of the coarse aggregate to the bottom and the accumulation of water or mortar at the top. Excessive working of the concrete also is undesirable, as such action causes the water and fines to accumulate at the lower surfaces of large aggregates or reinforcement bars and results in a lower quality of concrete at such locations. Efforts to avoid overvibration often result in inadequate vibration. Experience indicates that objectionable results are much more likely to be obtained from undervibration than from overvibration. Judgment is required in determining when compaction is complete.

Revibration of concrete, or steel embedded in it, is not harmful provided the concrete has not set so much that it will not again become plastic by vibration. There is some evidence that revibration may actually increase the strength of the concrete as well as its bond with the reinforcement bars [700]. Revibration could well be more widely practiced to eliminate settlement cracks and the effects of internal bleeding, and also as an aid in making tight concrete repairs in structures.

Where air bubbles are objectionable on a vertical surface, experience has shown they will be largely eliminated by using up to 100 percent more vibration than is necessary merely to insure solid filling. If this extra vibration appears to overvibrate the concrete, less water should be used in the concrete mix. With plastic mixtures, better results are obtained by placing layers less than 12 in. thick which are given a light vibration followed by spading along the forms [840]. Care is required to avoid overworking, as mortar drawn to the surface may later craze, crack, or scale.

CURING

8.9 The curing period. The last step, and an exceedingly important one in the manufacture of concrete, is the curing. As hydration of cement takes place only in the presence of moisture and at favorable temperatures, these conditions must be maintained for a suitable time interval called the curing period.

At the time concrete is mixed, sufficient water is added to give workability. The amount of mixing water actually used is ordinarily in excess of 50 percent of the weight of the cement, while the amount of water required for reasonably complete hydration of the cement is considerably less than 50 percent. Therefore, if the original water can be retained, there is more than sufficient for curing purposes. Curing may be said to consist of preventing the evaporation of the mixing water.

Concrete gains strength most rapidly at early ages, as shown in Figs. 10.10 and 10.11, so that the greatest benefit from curing is secured during this period, and each additional day is of lesser importance than the preceding one. The desired strength of concrete is usually not developed within the curing period specified for most concrete jobs, but it is not generally considered worth the cost to keep the concrete wet for longer periods. Furthermore, since it requires many days for partially hardened concrete of ordinary thickness to lose its water by evaporation, considerable hydration will occur after the stated curing period.

Specifications usually require that the surfaces of concrete be protected to prevent loss of moisture for at least 7 days where normal cement is used, and some specifications require curing for 14 days or more. Where high-early-strength cements are used, the curing period may be reduced about half, while for slow-hardening cements the time should be longer than for normal cements.

8.10 Curing methods. A common method of preventing loss of moisture from exposed surfaces of concrete is to keep the surfaces continually damp by frequent sprinkling, ponding with water, or covering with continuously wetted burlap or its equivalent. Other methods for preventing loss of moisture involve the use of liquid seal coats, or tight covers such as light-colored waterproof paper or an impervious plastic film [850]. A tentative method for testing the effectiveness of curing materials and procedures is given in ASTM C156.

Too rapid drying of exposed surfaces before they have hardened sufficiently to stand sprinkling with water or covering with damp burlap may result in serious checking and crazing of the concrete. To prevent such rapid drying some specifications require that concrete be protected from drying winds and direct rays of the sun for the first day after placement, until adequate curing is begun. Plastic-shrinkage cracks may occur even before a surface is finished. In this case covering of surfaces with plastic films during the interval between placing and finishing is helpful. In extreme situations, fog sprays may be used to maintain high humidity over the concrete. Alternate drying with rewetting of slabs during curing must be avoided as it results in hairline cracking of the surface.

Forms, if used on structures in which the concrete will not be cured after removal of the forms, should be left in place as long as possible to protect the surface and aid in delaying the loss of moisture in the concrete. Such protection is desirable, as other types of curing are not always applicable to structures using forms. Wetting wooden forms periodically serves to prevent their shrinkage and opening of cracks between boards, and thus further aids the retention of moisture in the concrete. Ordinarily, buildings are not kept wet after removal of the forms because of the difficulty involved and the inconvenience to workers, although the quality of concrete is unquestionably lower than it might otherwise be, because of the resulting lack of moisture.

If the concrete surface can be cured properly after removal of the forms, then it is desirable to remove them as soon as possible, as forms made of narrow boards are not fully effective in preventing loss of moisture from the concrete. Furthermore, early form removal permits better repairs, if any patching or other repairs are necessary, as then the concrete is still green, i.e., in the early stage of hydration, so that repairs bond to it more readily.

8.11 Curing of pavements and other structures. Concrete pavements present a difficult curing problem, as unless protected they are subject to direct exposure to sun and wind. The top surface, which would be exposed to the severest drying, is the portion of the concrete requiring the greatest strength and resistance to wear. Various methods which have been employed for the curing of concrete pavements are mentioned below. Some of these, with certain limitations, are used for other types of structures as well. The three methods of water curing are generally accepted as satisfactory. Some difference of opinion exists as to the effectiveness of methods 4 and 5, which require no additional water.

1. Damp earth or straw covering. A 2-in. layer of earth or sand, or a 3-in. layer of straw, is spread over the surface as soon as it is sufficiently hard. This is kept damp by as frequent wetting as is necessary. These covers are effective for curing but in recent years have been largely discontinued as a result of their high cost.

2. Ponding. Earth dikes are placed across and along the pavement and the pavement flooded between them. Level slabs only can be cured in this manner. Tunnel linings and cast-in-place pipe on a flat slope can be effectively cured, and cracking can be materially reduced by ponding combined with closure of any openings.

3. Sprinkling. Continuous sprinkling is used to keep the surface wet. The water should be applied to unformed surfaces as soon as the surface can be wet without damage, and to formed surfaces immediately after removal of the forms. With intermittent sprinkling the use of wet burlap

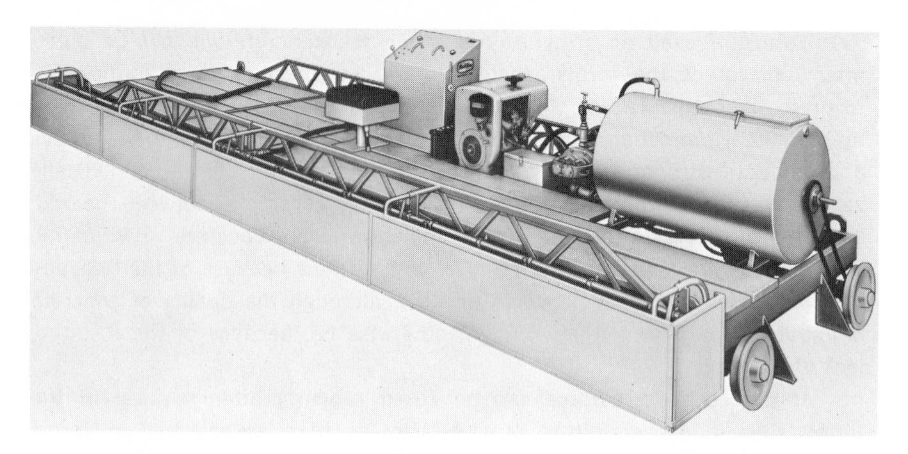

Fig. 8.4. Machine for spraying sealing compounds over full width of highway slab.

or cotton mats will increase the time interval between wettings, but even then periodic inspections should be made to ensure against drying of any surfaces. This method, as well as the following, is also of use for vertical or inclined surfaces.

4. Sealing compounds. Asphaltic and coal-tar cutbacks can be sprayed on fresh concrete to seal the surface. Bituminous emulsions also are used as coatings. After exposure to the air the emulsion breaks, leaving a bituminous film which, in general, is fairly impervious to moisture. Figure 8.4 shows a modern machine for applying such materials to the full width of an 11 to 25-ft-wide highway as it moves forward at a speed of up to 60 ft per min. The use of plain black coatings is often restricted because the resulting appearance of the structure is undesirable, and in hot weather the black-coated surfaces permit the absorption of so much heat that the temperature of the concrete may be excessive, resulting in some loss of moisture and increasing the tendency toward cracking and crazing. The excessive heat absorption may be overcome temporarily by applying a coat of whitewash or other heat-reflecting material, but this is soon removed by weathering. Even so, this method is satisfactory for coating exterior surfaces which will later be backfilled. Black compounds without light-colored coatings are used successfully on interiors of culverts, tunnels, and other chambers which will not be illuminated.

Clear sealing compounds are in common use on certain classes of structures where water curing is not practicable and where the black coatings are not desirable. Although marked differences exist in the effectiveness of the clear compounds, some of them are satisfactory. These compounds cause only minor staining of the surface, but even this is enough to make their use undesirable on some structures. For best results with clear compounds the surfaces should be shaded for the first 3 days to further aid in preventing evaporation of water from the concrete. As shading may be difficult of accomplishment, particularly during strong winds, various white-pigmented compounds have been developed which are sufficiently opaque that they are the equivalent of clear compounds with shading and are accepted as a substitute for water curing on certain classes of work.

When a sealing compound is to be applied to a surface on which forms have been in place for a day or more, the surface should be water soaked for a few hours before the seal coat is applied, as forms permit some drying to occur.

Under job conditions it is impossible to apply a uniform coverage to all areas. Furthermore, when only one coat is applied, pinholes are quite common. To seal the surface effectively two coats should be applied. One coat is applied by moving the spray gun back and forth in one direction, and the other coat by spraying at right angles to this direction. Some of the clear compounds contain quick-fading dyes to assist in securing uniform coverage and in distinguishing between successive coats. Ordinarily a total coverage of about 150 sq ft per gal is specified. Even this coverage is no assurance of an effective seal, as there are marked differences in the available compounds. Tests of their sealing qualities under service conditions should be determined before they are used.

5. Waterproof covers. Moistureproof papers and pigmented plastic films have been used with success for curing both highway and building-floor slabs. Plastic film has two advantages: it is light in weight and can easily be draped over complex shapes. The cover is put in place as soon as possible without damage to the concrete at the surface, and the lap joints between adjoining strips tightly sealed. The cover serves a dual purpose of preventing loss of water from the concrete and protecting the surface from damage. Care must be taken that the cover is not torn. Occasional inspections should be made to be certain that the slab is damp. If dry, it should be rewetted and then resealed. The method is effective and relatively economical.

8.12 Curing temperatures. The favorable range in curing temperature for most concrete appears to be from about 60 to 90°F, which includes the average temperature existing on many concrete jobs; for mass concrete, fairly low initial temperatures are usually the most favorable. In general, for curing tempertures below 90°F, the lower the temperature at which concrete is continuously cured, the lower the strength at any given age, although temperatures as low as 40°F will ultimately produce satisfactory strengths provided the concrete is kept moist for sufficiently long periods (Art. 10.13 and Fig. 10.11a).

Concretes stored at low temperatures (above freezing) will show appreciable increases in strength upon later storage at normal temperature provided moisture is continuously available.

Temperatures below freezing are decidedly harmful to fresh concrete, as the expansion of the water when transformed into ice causes separation of the solid particles, thus reducing their bond at all ages. The results of a comprehensive investigation of the effect of freezing on concrete [874] show the effects of immediate freezing to be as follows:

1. Concrete frozen immediately after placing but later stored at favorable temperatures attains on the average about 50 percent of the strength of the normal unfrozen concrete at the same age. There is little difference in this respect between a 1-day and a 7-day freezing period.

2. There is some indication that concretes tested at later ages have less proportionate loss of strength due to fresh concrete freezing immediately after placement than those tested at earlier periods.

3. There is some indication that dry concretes suffer less injury than wet concretes, especially at early ages.

4. There is little difference between a rich mix and a lean mix in their resistance to immediate freezing.

5. The rate of leakage of water in a permeability test may be taken as a measure of the relative porosity of the concretes. Concretes which had little porosity unfrozen showed relatively great porosity after being frozen before hardening.

6. There seems to be a great deal more leakage for concretes frozen for 7 days than for 1 day.

7. The leakage for frozen specimens tested at 1 year was much less than when tested at 7 days.

From this same investigation the following effects of delayed freezing were shown:

1. Concretes of relatively dry consistency (2-in. slump with 1½-in. maximum aggregate) may be frozen solid without practical injury if cured for 24 to 48 hr in warm, dry air before freezing.

2. Concretes of very wet consistency (9-in. slump) require a considerably longer time of curing before freezing in order to be uninjured by freezing.

3. In order that the concrete shall be uninjured by freezing, moist-cured specimens require a longer period of curing before freezing than dry-cured specimens, as the moist-cured specimens contain a larger percentage of free water available for conversion into ice.

8.13 Steam curing. In the mass production of precast units such as blocks, pipe, prestressed units, etc., it is often desirable to hasten the process of hydration and hardening so that units may be installed within a few days after manufacture and thus avoid storage problems and delays inherent in normal curing. For this purpose steam curing is very effective. Steam-cured precast units attain strength so rapidly that the forms can be removed and reused very soon after placing the concrete. In use are high-pressure, low-pressure continuous, and low-pressure intermittent processes.

The high-pressure process requires a specially constructed steel chamber, or autoclave. Excess water is provided to ensure a saturated atmosphere. The usual temperatures are about 250 to 350°F, and the steaming period is relatively short, usually about 8 to 12 hr inclusive of a 2-hr period at room temperature, a gradual warming-up period of about 2 hr, and a fall to atmospheric conditions in not less than 15 min. Unless pozzolans are used as a component of the mix, the later-age strengths will not be as high as for normally cured concrete; therefore it is common practice to use a pozzolan to cement weight ratio of about 0.40 to 0.60. The following effects of the high-pressure process have been determined [864]:

1. The resulting products are drier and lighter in color than moist-cured units.
2. The compressive strength at 1 day is at least equal to that of moist-cured products at 28 days.
3. There is a tendency to stabilize unsound materials that might otherwise cause popping or spalling in service.
4. Increased resistance to sulfate action is developed.
5. Leaching and efflorescence are practically eliminated when a pozzolanic admixture is used.
6. Shrinkage is about half that for moist-cured units.
7. The bond strength between steel and concrete is reduced.

The low-pressure continuous process is well suited for mass production. The concrete units are placed on a conveyor which transports them through a long steam-heated tunnel. The low-pressure intermittent process is carried out in a series of curing chambers. Each chamber is usually designed to take a day's output, and steaming is begun at the end of the day. The results of many tests of steam-curing conditions are given in Refs. 860 and 865.

The effects on early-strength development of concrete steam-cured at various temperatures from 100 to 160°F and for delay periods of 1 and 3 hr are shown in Fig. 8.5. The higher temperatures produce higher strengths at early ages, but after about 14 days the strength is less than

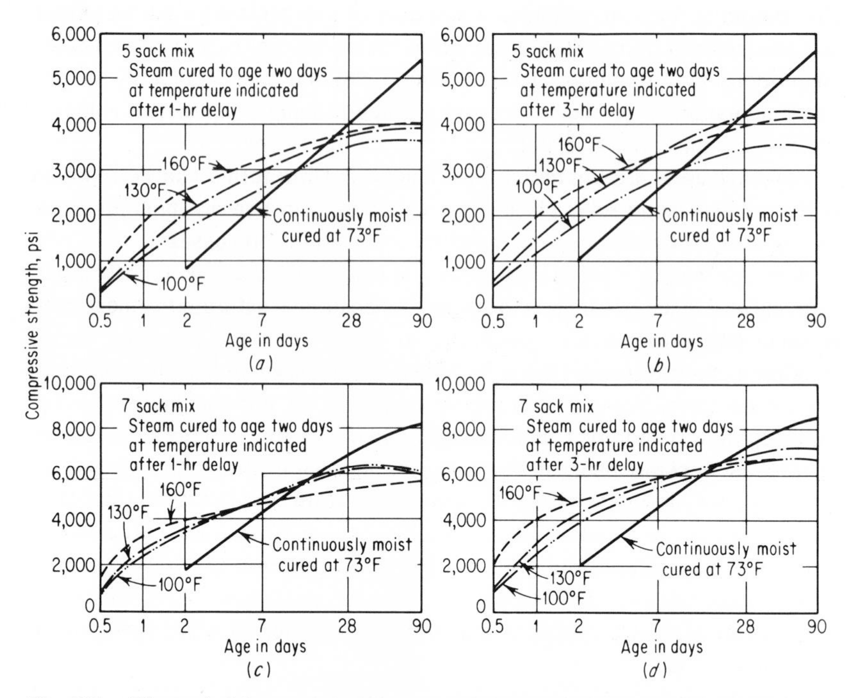

Fig. 8.5. Effect of delay period and temperature on the compressive strength of steam-cured concrete. (*From ACI Committee* 517 [860].)

for normal moist curing. In general, higher strengths at 24 hr will be developed if steam curing is delayed 2 to 6 hr [860]. Steam curing is usually not continued beyond 18 to 24 hr.

In all steam-curing operations the early temperature rise should not be too rapid, and both faces of any section should be simultaneously exposed to the steam curing in order to avoid stress-producing temperature differences in the concrete which may adversely affect its strength.

8.14 Concrete work during cold weather. To produce a suitable temperature of fresh concrete during cold-weather construction, it is common practice to heat the mixing water, and even the aggregates if necessary. The water can be heated more readily than the aggregates, and each pound of water at a given temperature has about five times as many heat units as a pound of aggregate or cement at the same temperature. The temperature of the water should not be above 165°F, as a flash set of the cement may occur. The approximate temperature of concrete can be calculated from the temperatures of the ingredients, as shown in Art. 5.16.

On some jobs it is necessary to heat the aggregates to remove ice or to increase the temperature of the concrete. The use of steam coils, hot-water coils, or steam jets for this purpose is preferable to the use of fire flues, as the latter tend to cause nonuniform heating. Excessive heating of aggregates may result in some cracking of the particles. Excessive drying of the aggregates, so that absorption causes rapid stiffening, should be overcome by uniform prewetting.

The concrete should not be overheated, because this would cause excessive loss of slump or increase the water requirement for a given slump (see Fig. 5.2) and thus cause a reduction in strength as well as an increase in drying shrinkage. Furthermore, the higher the temperature of the concrete as placed, the greater the drop to its final temperature, and the greater the thermal shrinkage.

For ready-mixed concrete to be delivered in cold weather, the minimum temperature of the concrete specified [ASTM C94] is shown in Table 8.1. Its maximum temperature should not exceed 90°F.

The minimum temperature of mass concrete may be permitted to be as low as 40°F when placed, if adequate protection against freezing of the surfaces is provided, as the heat of hydration is not dissipated so readily with this type of concrete. However, as not much of this chemical heat is available during the first day, surface protection is as necessary for mass concrete as for regular concrete during this period, but less protection is required later.

Before placing concrete at low temperatures, all subgrade, form, or reinforcement surfaces with which the concrete may make contact should be heated to remove any ice or snow and to prevent freezing of the concrete. This heating is best accomplished with steam.

When the temperature of the fresh concrete in place is about 70°F, setting soon begins, and the heat of hydration will maintain suitable temperatures to cause the hardening to progress at a normal rate, provided the concrete is insulated for several days to prevent dissipation of heat more rapidly than it is generated. This procedure is superior to that of placing the concrete at lower temperatures and then attempting to heat it in place with salamanders (stoves) or other means; however, when air temperatures

Table 8.1 Temperature of concrete for cold-weather construction

	Minimum concrete temperature, °F	
Air temperature, °F	Thin sections and unformed slabs	Heavy formed sections and mass concrete
30 to 45	60	50
0 to 30	65	55
Below 0	70	60

are very low, it may be necessary to use external heat as well as to provide insulation because concrete at low temperatures, in comparison with normal conditions, gains strength slowly, as shown in Fig. 10.11. Steam released under a tarpaulin or other enclosure provides excellent protection since it not only aids in maintaining proper temperatures but also affords a moist atmosphere favorable to curing. In some winter construction work, the concrete has been successfully heated internally by electrical means [870] instead of steam. Salamanders are not very satisfactory, as they are a fire hazard and, in addition, cause nonuniform temperatures within the enclosure; some areas may even be overheated. All heating devices other than steam jets are subject to the criticism that they lower the relative humidity of the air to such a degree that drying of the concrete may occur, unless all formed and exposed concrete surfaces are kept continuously moist by sprinkling. For uniform and efficient heating all enclosures should be tight, and ample space provided between concrete and enclosure to permit free circulation of the warmed air.

Thin concrete members which have a large surface area per unit volume need much more protection than do massive structures such as piers or dams. Corners and edges are most vulnerable; at such locations it is necessary that the initial as well as later temperatures of the forms be above 32°F; otherwise they may absorb heat from the concrete, and the mean temperature may fall below freezing. This is especially likely if metal forms are used.

When sealing compounds are used, the concrete should be protected against low temperatures in the same way as for water-cured concrete. When using dry heat, the resulting low humidity may permit so much of the mixing water in the concrete to pass through the coating that it may be necessary to use thicker coatings, or increase the humidity of the air, or use water curing for a few days before applying the compound.

Air temperatures should be maintained well above freezing because the surface temperature of damp concrete may be several degrees below the temperature of the adjacent air, owing to evaporation. The curing period at temperatures above freezing should be of sufficient length so that the concrete will not be injured when exposed to temperatures below freezing at the end of the curing period. At the end of the curing period the concrete should be cooled gradually within the enclosure, as surface cracking will result if heated concrete is suddenly exposed to low temperatures. Determination of the age at which supporting forms may be stripped safely from concrete cured at low temperatures requires experience and good judgment, as the strength is uncertain. Usually the engineer must give approval before forms may be removed.

The principal recommendations of ACI Committee 306 for cold weather concreting [872] are:

1. Entrained air should be used.

2. The minimum temperature of fresh concrete *as placed* in thin sections should be about 50°F, and the maximum temperature should not exceed this by more than about 10°F.

3. All concrete should be protected from freezing for at least 24 hr after it is placed and until it has developed a compressive strength of at least 500 psi. Thin sections and high-quality concrete should be protected for longer periods.

4. The lower the temperature, the longer the protection required.

5. The use of a high-early-strength cement, a richer mix, or an accelerator, such as $CaCl_2$, will reduce the period of protection required.

6. In some cases it may be advantageous to use insulated forms instead of heated inclosures.

7. After the required period, protection should be removed in such manner that the drop in temperature will be not more than about 40°F in 24 hr.

The report of this committee contains recommendations on preparation for winter work, heating materials, use of accelerators, curing, protection, and form removal.

8.15 Calcium chloride in concrete during cold weather. A method of accelerating the setting and hardening of the concrete by means of calcium chloride solution incorporated with the mix has been used successfully. The following conclusions are from tests carried out at the National Bureau of Standards [444] on the use of calcium chloride:

1. By addition of 2 percent calcium chloride, the average time of initial set of neat cement at 70°F was reduced from 3 hr 15 min to 1 hr 12 min.

2. Additions of calcium chloride increased the strength of all cements at all ages up to 1 year. The percentage increase was much greater for early ages and for the lower temperatures than for later periods and higher temperatures. Using 2 percent calcium chloride by weight of cement in a 1:2.6 mortar mix, the increases in compressive strength shown in Table 8.2 were obtained. The increases were slightly less for concrete than for mortar. At 1 year, the increase for a 1:2:4 mix cured at 70°F was 8 percent.

3. The most effective amount of calcium chloride was 2 percent at 40 and 70°F and 1½ percent at 90°F.

4. Workability of concrete increased with additions of calcium chloride up to 3 per cent.

The results of tests at 40 and 70°F on concretes containing calcium chloride are shown in Fig. 8.6. For all ages shown (up to 2 years) the strengths are higher when using up to 2 percent of calcium chloride. Other tests have shown that the effect in rich mixes of low water content is much greater than in leaner mixes of higher water content. The specific effect of the use of calcium chloride varies, however, for different brands of cement.

Calcium chloride does not change the general chemical processes of hydration of cements but appears to act solely as a catalyst to free calcium oxide to permit earlier hydration of the cement. Therefore, it can be used

Table 8.2 Increases in compressive strength of mortar by additions of calcium chloride

Age at test, days	Increase in strength at curing temperature indicated, percent of strength of plain mortar		
	40°F	70°F	90°F
1	300	145	90
3	117	68	41
7	75	32	23
28	20	12	15
90	10	14	16

most advantageously in cold weather to reduce the time of protection required or to reduce the time of wet curing at normal temperatures. Dependence should not be placed on calcium chloride to reduce freezing temperatures of concrete, as the amount required for appreciable reduction is so large that it reduces strength and may cause flash set. Calcium chloride should not be used in concrete for cold-weather placement which will be subject to sulfate attack, as it would then be subject to more rapid disintegration; instead, the maximum water-cement ratio should be reduced to 0.45, and protection against freezing should be provided.

There are no tests or other evidence indicating that a small addition of calcium chloride to concrete has any corrosive effect on reinforcement bars of No. 3 size or larger embedded at least ½ in. in dense concrete [888]; although in porous concrete or in prestressed concrete, it may cause serious corrosion and should not be used. Also, it may increase the drying shrinkage considerably, sometimes as much as 50 percent.

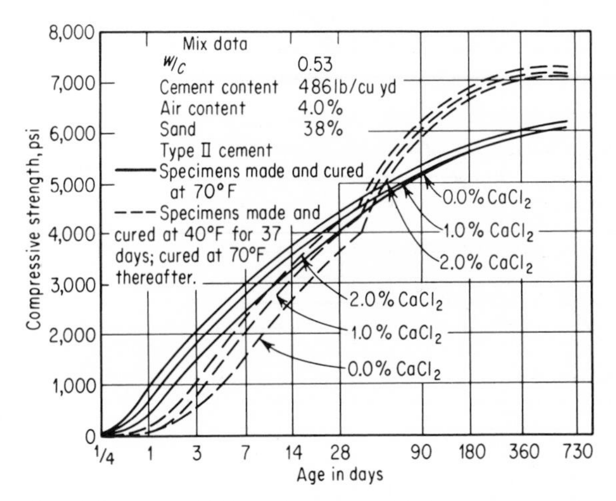

Fig. 8.6. The addition of calcium chloride increases the compressure strength of concrete for two different curing conditions. (*From U.S. Bureau of Reclamation* [106].)

8.16 Concrete work during hot weather. Temperatures above about 90°F serve to accelerate the early hydration of the cement and produce concretes having high strength at early ages. However, these temperatures appear to have harmful influences upon strength at later ages, as shown in Fig. 10.11, even producing retrogression in strength in some cases.

Various difficulties arise when concrete is placed during hot weather. Rapid stiffening of the concrete before compaction because of acceleration of chemical activity and rapid evaporation of mixing water is a source of trouble, but this may be partly avoided by the use of a water-reducing retarder. Also, even though this can be partly offset by the use of a higher water content for the same slump (see Fig. 8.7) and corresponding higher cement factor, the richer mix itself is usually undesirable. Rapid evaporation at exposed surfaces may cause shrinkage and surface checking within the first few hours. If plastic-shrinkage cracks tend to appear, the concrete should be kept moist by means of fog sprays, wet burlap, or evaporation control [897]. Suitable revibration (as late as the concrete will still become plastic under vibration) will do much to prevent plastic-shrinkage cracks or to eliminate those which have already formed. Large thermal contractions of the hardened concrete when it cools later may cause such high tensile stresses that actual rupture will occur. The difficulty of keeping all surfaces moist for the required curing period is also increased. Reliance should not be placed on the protection afforded by forms for curing in hot weather. If practicable, water should be applied to formed surfaces while forms are still in place. Unformed surfaces should be kept moist for at least 24 hr,

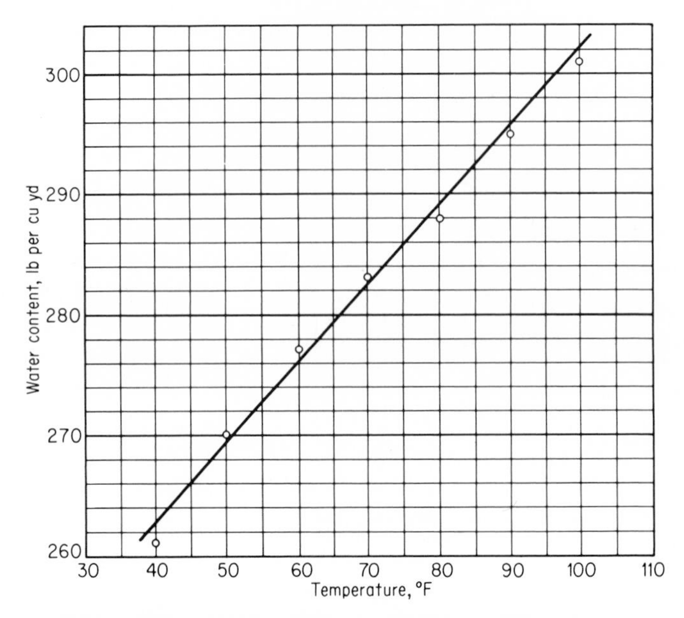

Fig. 8.7. Water requirement for a typical concrete mix (1½-in. maximum aggregate, 3-in. slump) as affected by temperature. The increase of water content accounts in part for greater shrinkage of concrete that is mixed and cured at high temperatures. (*From U.S. Bureau of Reclamation* [106].)

and curing should be started as soon as the concrete has hardened sufficiently to avoid surface damage. As a result of the many difficulties, the placement of concrete in important structures in hot, arid climates is sometimes prohibited during the summer months or is restricted to operations at night when temperatures are lower and evaporation is less.

When concrete is placed during hot weather, various precautions are necessary to obtain good results. The temperature of the concrete as mixed should be kept as low as possible by shading the aggregate piles and the mixer. Sprinkling the aggregates will cool them by evaporation. The mixing water may be kept from heating up by insulating or shading pipelines and tanks, or at least painting them white. In some cases, large quantities of slush ice may be used in the mixing water (see Art. 5.16), but the ice must be completely melted by the time mixing is completed. Cement preferably should not be used at temperatures in excess of 170°F although, other than for its contribution to raising the temperature of concrete, high temperature of cement has no detrimental effects [898]. Certain retarders have been found to delay setting time of concrete, reduce mixing water requirements, and increase strength. Difficulties due to rapid stiffening and drying of the fresh concrete may be reduced by protecting the work from the direct rays of the sun. Sprinkling or covering with wet burlap or other moisture-retaining material is the preferable method of curing, as it is more positive and also has a definite cooling value. Surface-sealing compounds lose some of their efficiency in hot weather; if their use is permitted, they should be preceded by at least one day of water curing; black compounds should be whitewashed promptly to reduce their heat absorbing capacity.

Control of air content in air-entrained concrete is more difficult. This adds to the difficulty of controlling slump. For a given amount of air-entraining agent, warm concrete will entrain less air than cool concrete [898].

8.17 Curing in the laboratory. In laboratory work involving the curing and storage of concrete test specimens, a temperature of 73°F is commonly maintained, and the specimens are stored in an atmosphere of 100 percent relative humidity or under water. Moist curing at 73°F is usually referred to as "standard curing." In investigations of concrete for use in massive structures, such as dams, where the temperature rise due to heat of hydration of the cement may be appreciable, it is desirable to know the properties of concrete under curing conditions similar to those which exist in the mass. Consequently, there is sometimes employed so-called "mass curing," which involves the use of sealed specimens subjected (ordinarily) to temperature conditions which would exist in a mass if no heat were lost (adiabatic conditions); if temperature conditions other than adiabatic are employed, as is sometimes the case, this fact should be made clear.

OPERATIONS AFTER CURING

8.18 Removal of forms. Forms should not be removed until the concrete has hardened sufficiently so that with a factor of safety of 2 it can carry its own weight plus any live loads to which it may be subjected. Many specifications require that the forms be left in place for prescribed intervals of time such as 14 days for slabs and beams, and shorter intervals for wall and column forms. However, as the rate of hardening depends upon the prevailing temperatures, moisture-curing conditions, type of cement, and water-cement ratio of the concrete, it would be better to specify the attainment of some particular strength before permitting removal of the forms. Table 8.3 presents suggested minimum strength requirements for the purpose.

The minimum strengths shown in Table 8.3 are conservative values, as they are intended for application to a wide range of conditions. In some cases where experience shows that lower strengths may be used satisfactorily, some reduction in the tabular values may be considered. On the other hand, especially unfavorable conditions of loading may require higher strengths.

Table 8.3 Strength of concrete for safe removal of forms

Structural classification	Minimum compressive strength, psi
A. Concrete not subject to appreciable bending or direct stress, nor reliant on forms for vertical support, nor liable to injury from form-removal operations or other construction activities	500
Examples:	
Vertical or approximately vertical surfaces of thick sections	
Tops of sloping surfaces	
B. Concrete subject to appreciable bending and/or direct stress and partially reliant on forms for vertical support:	
1. Subject to dead load only	750
Examples:	
Vertical or approximately vertical surfaces of thin sections	
Underside of sloping surfaces steeper than 1:1	
Arch of tunnel lining against solid rock	
2. Subject to dead and live loads	1,500
Examples:	
Columns	
Sidewalls and arch of tunnel lining against unstable material	
C. Concrete subject to high bending stresses and wholly or almost wholly reliant on forms for vertical support	2,000
Examples:	
Roof or floor slabs and beams	
Undersides of sloping surfaces flatter than 1:1	

The determination of an approximate age-strength relationship, so as to know when the forms may be stripped, can be accomplished by testing concrete cylinders cast from the regular concrete mix and cured alongside of, and in the same manner as, the actual structure. New determinations of this relationship should be made whenever any conditions affecting strength, such as type of cement or prevailing temperatures, are materially changed. As it is practically impossible to subject the small test cylinders to conditions identical with those surrounding the concrete in the structure, age-strength relationships determined in this manner serve only as rough indications of the true values. For a simple field test for determining the compressive strength of concrete in place, see Art. 10.10.

When form supports are removed, the operations should permit the concrete to take its share of the load gradually and uniformly.

8.19 Patching. Holes left by tie rods should be hammer-packed with stiff, dry mortar of the same materials as, but somewhat leaner than, that in the concrete. Smoothing the mortar neatly with a wooden block will render the patch inconspicuous. Defective areas should be repaired at once, by approved methods (see Art. 11.22) and not merely by plastering over them. Defective honeycombed areas should be cut out to a depth at which sound concrete is exposed, and filled with concrete matching that of the structure. Spalled areas caused by concrete sticking to the forms when the forms are removed should be chipped back to obtain a good mechanical bond, undercut at the edges, and repaired with mortar matching the concrete. Pitted surfaces caused by dried mortar adhering to the forms at the time of concrete placement should be acceptably repaired and the cause remedied in future work. Throughout the operations of repairing defects and finishing the surface, the surface should not be allowed to become dry, nor should the underlying concrete be damaged by the operations. Rough areas and high spots should be rubbed or ground down to the proper plane and an acceptable degree of smoothness.

8.20 Prevention of damage. Heavy impact on green concrete by construction operations may disrupt the mass and should not be permitted. Floors are subject to surface marring and should be covered. Any projecting ornamentation should be protected from falling material.

Rough handling of projecting reinforcement bars during the early stages of hardening, after the concrete is no longer plastic, may rupture the bond between concrete and steel. Workmen should be prevented from climbing on projecting bars or subjecting them to severe usage. Stripping of forms to which they are attached or through which they protrude should be delayed until there is no longer danger of damage to bond.

Backfilling of earth against concrete walls should be deferred until the concrete is strong enough to carry the load, and even then care must be taken to avoid impact. Damage to waterproof coatings should be prevented by excluding sharp material from the backfill near the concrete, and by puddling the backfill to prevent future settlement.

QUESTIONS

1. What preliminary steps are necessary in preparing the following surfaces upon which fresh concrete will be placed: earth foundations, mass concrete, ordinary structural concrete?
2. List the factors tending to cause segregation of concrete in place.
3. Describe the basic principle causing vibration of a concrete vibrator.
4. Compare the efficiency of internal- and external-type vibrators.
5. How is the efficiency of a vibrator affected by its frequency or speed of operation?
6. How should a mix suitable for hand compaction be modified to be most suitable for vibratory compaction?
7. Why are very dry mixes to be avoided?
8. Why are overwet mixes to be avoided?
9. What is laitance?

10. What conditions are required for the continued hardening or increase in strength of portland-cement concrete?

11. What is meant by curing?

12. How much water is required for reasonably complete hydration of the cement in a concrete mix?

13. Does the amount of water ordinarily used with cement in mortar and concrete greatly exceed that required for completion of the chemical reactions?

14. Will concrete continue to harden or increase in strength if there is no water present?

15. Can concrete which has dried out at the early ages be improved by later moist curing? Explain.

16. What are the disadvantages of using black asphaltic curing compounds?

17. What is the advantage of using calcium chloride as an admixture to concrete?

18. What part has temperature in curing?

19. What temperatures are unfavorable for curing?

20. What is the effect of freezing of fresh concrete?

21. Can concrete which has been frozen develop satisfactory strength?

22. What disadvantages arise because of concreting during hot weather?

23. What precautions should be taken in removing forms from a concrete structure?

24. Why must forms be left longer on beams than on vertical walls?

NINE
FORMS FOR CONCRETE[1]

9.1 Requirements of forms. Concrete formwork must be true to grade and alignment and be braced against displacement, have a surface finish to produce the desired texture of concrete, be strong enough to resist all vertical and horizontal loads acting, and have tight joints to prevent leakage of mortar during placement of the concrete. In general, forms for architectural concrete require even more attention to details than forms for structural concrete, because surface blemishes, sags, bulges, and misalignments of architectural concrete should be reduced to a minimum.

Floor forms must be designed to carry the weight of the concrete and all construction loads (dead and live) without excessive stress, deflection, or settlement. The construction load varies somewhat, but a minimum of 50 psf of floor area is recommended by ACI Committee 622 [900] to provide for weight of workmen, equipment, runways, and impact. Some conditions may justify a smaller load, but many designers use 75 psf for construction with powered concrete buggies.

Deflections are usually limited to $\frac{1}{360}$ of the span. When supporting members rest on the soil, provision should be made to distribute the load over a large enough area to prevent appreciable settlement. Shores supporting successive stories should be placed directly over the shores below.

9.2 Factors affecting form pressures. Wall and column forms must support the lateral pressure of the concrete acting temporarily like a liquid. This pressure is equal to full-fluid head when concrete is placed full height within the period required for its initial set. With slower rates of placing, concrete in the lower parts of the form begins to stiffen, and by the time the form is filled the lateral pressure at the bottom is reduced to less than full-fluid pressure. The effective lateral pressure is dependent primarily on the unit weight, rate of placement and temperature of the con-

[1] Based partly on Ref. 900, which should be consulted for greater details.

crete mix, the use of set-retarding admixtures, and the method of consolidation, whether by hand or by vibration.

The weight of concrete has a direct influence on pressure, since hydrostatic pressure at any point in a fluid is created by the weight of the superimposed fluid. But fresh concrete is a mixture of solids and water whose behavior only approximates that of a liquid and then for a limited time only. The weight of concrete may vary from 80 to 300 pcf, but as most formwork is for ordinary concrete weighing about 150 pcf, that value is ordinarily assumed for form pressure calculations. However, recommended pressures must be adjusted for the unit weight of the mix used.

The average rate of rise of the concrete in the form is referred to as the rate of placing. During the placement of concrete, the lateral pressure of the mass against the forms at any point increases as the concrete depth above this point increases. Later on, as the concrete sets, it tends to support itself, no longer causing lateral pressure on the forms. The rate of placing has an important effect on lateral pressure; the maximum pressure is proportional to the rate of placing up to a limit equal to the full-fluid pressure.

Internal vibration is commonly used to consolidate concrete. As it causes concrete to act as a fluid for the full depth of vibration, it results in temporary lateral pressures locally which are at least 10 to 20 percent greater than those caused by simple spading. Forms must be designed to resist these greater pressures and must be made tighter to prevent leakage.

Revibration of partially set concrete which can be made plastic again and external vibration cause even greater loads on the forms than normal internal vibration and require specially designed forms. The effects of these two conditions have not been adequately investigated to be expressed in a pressure formula; thus the pressures recommended herein are limited to normal internally vibrated concrete.

Temperature of the fresh concrete has an important influence on pressures as it affects the setting time. Low temperatures retard stiffening of the concrete so that a greater depth can be placed before the lower part becomes firm enough to be self-supporting. The resulting greater fluid head causes higher lateral pressures. This is especially important for forms filled during cold weather or when using retarding admixtures.

Other variables having some effect on lateral pressures include consistency of concrete, reinforcement details, ambient temperature, maximum aggregate size, placing procedures, type of cement, depth of placement, cross section of forms, and smoothness and permeability of the forms. However, their effect is generally small and is usually neglected.

9.3 Lateral pressure values. ACI Committee 622 studied all available data on lateral pressures and made recommendations thereon for safe design for forms, considering rate of placement, temperature of concrete mix, and the effect of vibration as the important variables [900].

Columns are usually filled full height in a short time so their vibration extends throughout the full height. The form pressures are high, as the concrete has insufficient time to set and reduce them. They vary uniformly from zero at the top of form to a maximum at the bottom, which is expressed by the formula

$$p = 150 + \frac{9{,}000R}{T} \qquad \text{Maximum} = 3{,}000 \text{ psf or } 150h, \text{ whichever is less} \tag{9.1}$$

where p = maximum lateral pressure, psf
R = rate of placement, ft per hr
T = temperature of concrete in forms, °F
h = maximum height of fresh concrete in form, ft

Although the curve of pressure due to the concrete alone should start at the origin, an initial value of 150 was chosen to include the effects of other construction loads. The equation is not recommended for lifts exceeding 18 ft. Table 9.1 shows the maximum pressure values given by Eq. (9.1).

For walls with R not exceeding 7 ft per hr, the lateral pressure is

Table 9.1 Maximum lateral pressure for design of column forms*

Rate of placement, R, ft/hr	Maximum lateral pressure, psf, for temperature indicated					
	90°F	**80°F**	**70°F**	**60°F**	**50°F**	**40°F**
1	250	262	278	300	330	375
2	350	375	407	450	510	600
3	450	488	536	600	690	825
4	550	600	664	750	870	1050
5	650	712	793	900	1050	1275
6	750	825	921	1050	1230	1500
7	850	938	1050	1200	1410	1725
8	950	1050	1178	1350	1590	1950
9	1050	1163	1307	1500	1770	2175
10	1150	1275	1435	1650	1950	2400
11	1250	1388	1564	1800	2130	2625
12	1350	1500	1693	1950	2310	2850
13	1450	1613	1822	2100	2490	3000
14	1550	1725	1950	2250	2670	
16	1750	1950	2207	2550	3000	
18	1950	2175	2464	2850		

* From Ref. 900.

Table 9.2 Maximum lateral pressure for design of wall forms*

Rate of placement, R, ft/hr	Maximum lateral pressure, psf, for temperature indicated					
	90°F	80°F	70°F	60°F	50°F	40°F
1	250	262	278	300	330	375
2	350	375	407	450	510	600
3	450	488	536	600	690	825
4	550	600	664	750	870	1050
5	650	712	793	900	1050	1275
6	750	825	921	1050	1230	1500
7	850	938	1050	1200	1410	1725
8	881	973	1090	1246	1466	1795
9	912	1008	1130	1293	1522	1865
10	943	1043	1170	1340	1578	1935

* From Ref. 900.

the same as for columns, but with R greater than 7 ft per hr

$$p = 150 + \frac{43{,}400}{T} + \frac{2{,}800R}{T} \tag{9.2}$$

with a maximum of 2,000 psf or $150h$, whichever is less. This equation is limited to rates of placement of not over 10 ft per hr and depth of vibration is limited to 4 ft below the top surface of concrete. Table 9.2 shows the maximum pressure values for walls. Since studs and sheathing are ordinarily uniform throughout their entire height, only the maximum pressure value will be needed for their design. However, wale and tie spacings may be increased near the top of forms where lower lateral pressures occur.

All recommended pressures are for a slump of no more than 4 in. and a unit weight of about 150 pcf; no external vibration will be permitted when using these values. For unit weights of 100 to 200 psf, take a proportionate amount of the pressure value for 150-lb concrete. Recommended pressures may be reduced 10 percent if the concrete is spaded instead of vibrated internally.

When a retarding admixture or a pozzolan is used in hot weather, an effective temperature less than that of the actual concrete should be used in the pressure equation. If either is used in winter, the lateral pressure should be taken as equal to that for a fluid weighing 150 pcf.

9.4 Uplift due to lateral pressure. Freshly placed concrete will cause uplift of the forms, where they slope instead of being vertical, because the pressure

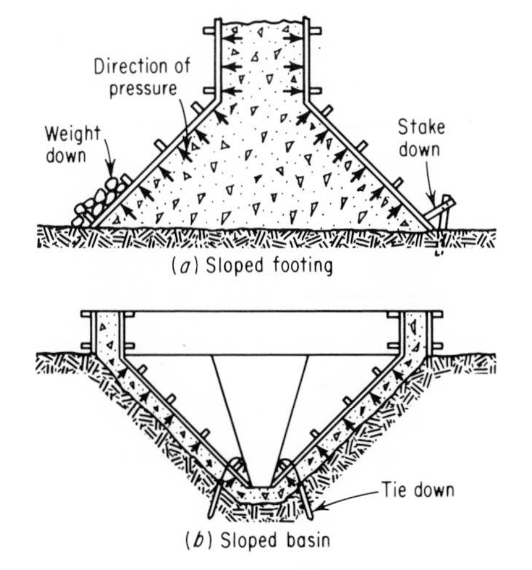

Fig. 9.1. Inclined forms, such as for sloped footing or a basin with sloping walls below grade, require tie-downs, weights, or other means of resisting uplift of the fresh concrete. (*From ACI Committee* 622 [900].)

acts at a right angle to the confining form. Forms for the footing or basin shown in Fig. 9.1 must be weighted or tied down to oppose the uplift force. The effective pressure is calculated from Eq. (9.1) or (9.2) by measuring rate of placing vertically and considering uplift pressure against the horizontal projection, not along the sloping face.

9.5 Other loads. Forms must be designed to resist all possible lateral loads, such as wind, cable ties, inclined supports, and impact from dumping of concrete or from mechanical equipment. Sidesway forces will occur when concrete is placed unsymmetrically on a slab form. For these usual conditions, ACI Committee 622 has recommended that forms be capable of resisting the following minimum loads, acting in any direction:

Slab forms: 100 lb per lin ft of slab edge or 2 percent of total dead load on the form (distributed as a uniform load per linear foot of slab edge), whichever is greater. Consider only the area of slab formed in a single placement.

Wall forms: wind load of 10 psf, or greater, if prescribed by building code; in no case less than 100 lb per lin ft of wall applied at the top of the form, except for walls less than 8 ft high below grade.

9.6 Formwork materials. Materials for forms must be economical and of sufficient strength and suitable quality to produce the desired surface finish of the structure. Most forms are made of lumber, plywood, tempered hardboard, or steel.

Lumber is the most common material for all temporary forms. Partially seasoned stock is best since dry lumber swells excessively and warps when wetted by the concrete, and green lumber may shrink and warp excessively before the forms are filled. Southern yellow pine and Douglas fir are the two most commonly used species, as they have suitable strength and durability, hold nails well, and are economical.

Plywood is often used as sheathing for wall and floor forms. It has the advantage of being available in relatively large sheets and so reduces labor costs. It is made up of several thin plies that may delaminate when used repeatedly; but this condition can be avoided by the use of *exterior*- instead of *interior*-type plywood, as the former is bonded with waterproof glue. When labeled as concrete form grade, it is required to be edge-sealed to protect the glue lines from moisture and to be well oiled. Some plywood has an overlay of resin-impregnated fiber faces on one or both sides to make it more resistant to abrasion and moisture. Some of these are available with special textured surfaces.

Hardboard is a boardlike material manufactured from wood fibers felted under heat and pressure to give it a density of 50 to 80 pfc. Tempered hardboard is given a supplemental treatment involving impregnation with drying oils followed by baking. It should have a factory-applied plastic coating which seals out alkaline water that weakens wood fibers.

Glass-fiber-reinforced plastics are used in some precast concrete operations and especially for architectural concrete because the concrete surfaces which they produce are excellent. They are not limited as to size or shape.

Metal forms are commonly used for tunnels and covered aqueducts as well as for general building construction and many special purpose forms. They are often fairly large sections which are readily removed or collapsed and then moved to be reused many times. Forms for highway and airport pavements, curbs, and gutters also are commonly of metal.

For grain bins, chimneys, dams, and similar vertical structures, the forms are relatively low but are raised from day to day as the work prog-

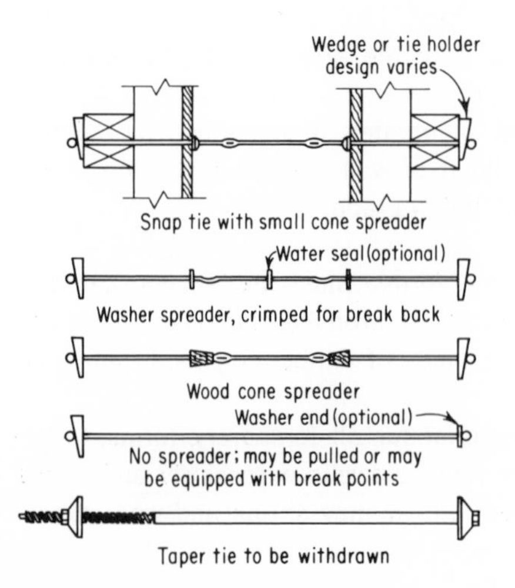

Fig. 9.2. Typical single member ties. The holding device is shown schematically, as many different devices are available. Safe loads range from about 2,500 to 5,000 lb. (*From ACI Committee* 622 [900].)

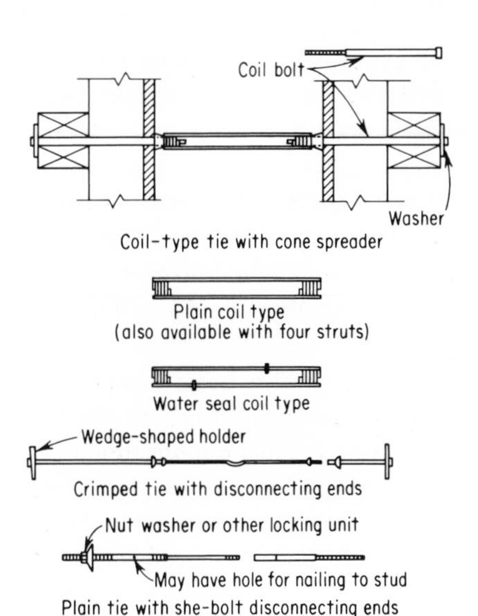

Fig. 9.3. Typical internal disconnecting ties. Safe loads range from about 6,000 to 36,000 lb. (*From ACI Committee* 622 [900].)

resses. For silos and similar round structures which are to be built without construction joints, it is frequently the practice to use sliding metal forms which are kept moving upward by the use of screw jacks, 24 hr a day at a slow and fairly steady rate as concrete is placed, thus preventing the concrete from adhering to them.

Form lining materials including plastics, plastic coated metal, paper, cardboard, fiberboard, and rubber have been used in various thicknesses. Some adhesives are suitable for their attachment to wood or steel forms, or they may be nailed to wood forms.

Form ties are tensile units installed to keep the forms from spreading when subjected to the lateral pressure of the fresh concrete, with or without an arrangement for spacing the forms a definite distance apart and with or without provision for removal of metal to a required distance back from the concrete surfaces. Typical single-member ties are shown in Fig. 9.2 and some internal disconnecting ties are shown in Fig. 9.3. Some of the former may be pulled from the hardened concrete, others are broken back at a weakened section, and some are cut off at the concrete surface. The latter type has an inner part with threaded connections to external members which are removable from the hardened concrete. Water seal and spreader features are available for many ties of both types.

9.7 Form design. The forms must be strong enough to support the anticipated loads safely and stiff enough to keep their shape under those loads. But to keep costs down the forms must not be overbuilt. Some form designs are based on the designer's experience and that what was satisfactory on a previous job can be used safely on the present one. However, with new materials and new systems of formwork and increasing demands for efficiency and economy as well as safety, it is desirable to make a design based on the strength and stiffness of the materials used and the estimated loads to be carried. For highly specialized formwork, for unusually heavy loads, or where there is exceptional danger to life or property, a complete structural design of the formwork may be necessary. To simplify the design work many useful tables for the selection of board sheathing, plywood, joists, wales, and shores have been prepared and presented in Ref. 900.

9.8 Wall forms. Wall forms are constructed using five basic elements as shown in Fig. 9.4. They are (1) sheathing to form the surface of the concrete and to retain it until it hardens, (2) studs to support the sheathing, (3) wales to support the studs and align the forms, (4) braces to give lateral support against construction and wind loads, and (5) separate spreaders and ties or tie-spreader units to hold the forms at the correct spacing under pressure of the fresh concrete.

9.9 Beam forms. Formwork for beams and girders consists of a bottom member and two sides with necessary ties or braces. Figure 9.5 shows a typical form using plywood, but board sheathing may be used. The bottom is commonly made to the exact width of the beam and supported on shore heads. Beam sides overlap the bottom form, rest on the shore heads, and are framed so that they may be removed without disturbing the beam bottom. For fireproofing structural steel the beam forms may be supported by an existing structural frame. A simple wire tie across the top of the beam

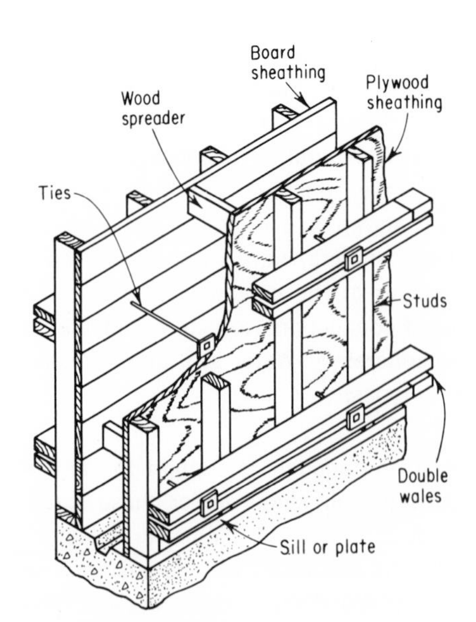

Fig. 9.4. Typical wall form. Alternate sheathing materials are indicated. Wood spreaders are shown, but frequently a combination tie-spreader is used. (*From ACI Committee* 622 [900].)

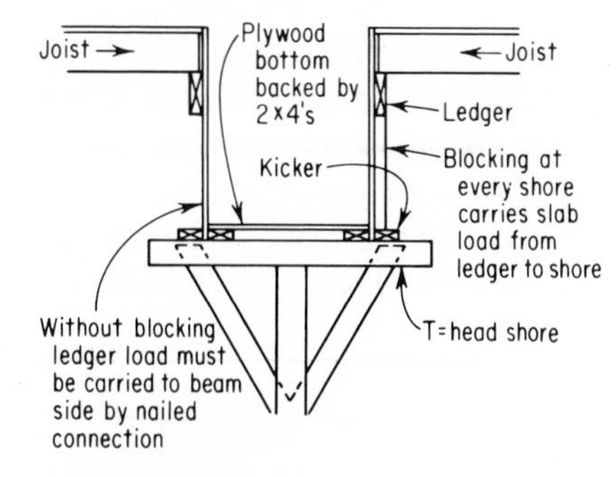

Fig. 9.5. Typical components of beam formwork with slab framing in. (*From ACI Committee* 622 [900].)

is often used, but for deep beams, ties like those used in wall construction are required.

9.10 Column forms. The method of building column forms depends on the materials used as well as the method of clamping or yoking the column. Square or rectangular columns of large size are usually made using plywood sides stiffened by vertical members and yoked by adjustable metal clamps as shown in Fig. 9.6. Lighter columns up to 18 by 18 in. may use horizontal wood battens to stiffen the plywood, these battens being held by long bolts.

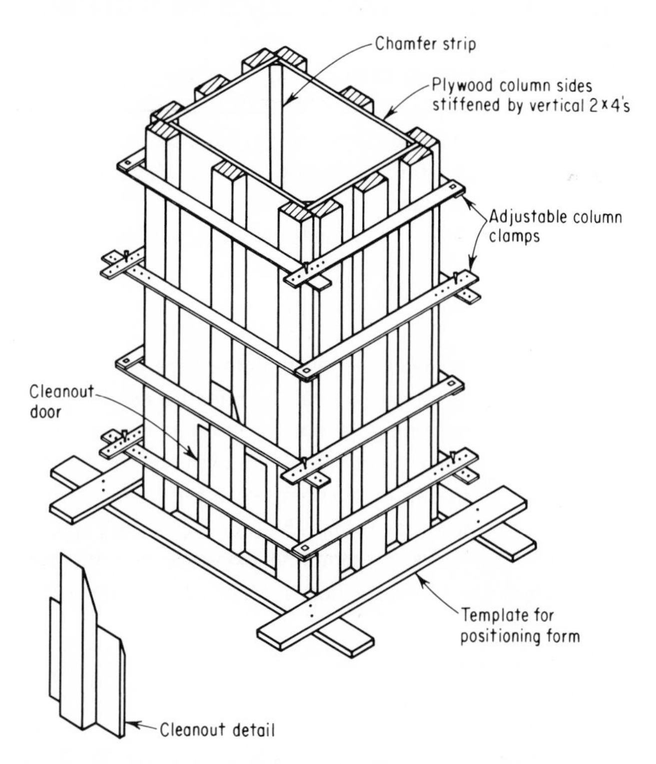

Fig. 9.6. Typical construction of heavy column form. (*From ACI Committee* 622 [900].)

Standard forms for round columns are available in both wood fiber and metal as complete units without need for outside yokes, since lateral pressure is resisted by hoop tension in the forms. The tubular fiber column form is available in diameters ranging from 6 to 48 in. Standard lengths are 12 and 18 ft, but greater lengths may be ordered. No clamps or ties are needed, and no form oil is required, as the tubes are available with polyethylene or wax-impregnated inner surfaces.

9.11 Construction of forms. Forms should be built with tight joints, and holes for tie rods through the forms should be as small as possible, so that water and mortar cannot leak, leaving unsightly sand streaks, rock pockets, and void spaces adjacent to the formed surfaces. This precaution is particularly necessary when the concrete is to be vibrated, as the mix is made unusually fluid during vibration, which causes a greater tendency to leakage than when the concrete is hand tamped. In deep, narrow forms which are inaccessible, holes for inspection and cleaning should be provided at the bottom; these holes are securely closed before the concrete is placed. While the forms are being filled, they should be watched for evidences of displacement and should be adequately secured where necessary, as usually it is impossible to force a form back into position after it has bulged while being filled. When the forms are shifted to a higher level, they should be pulled tight against the concrete already cast.

In some cases where the decorative effect of exposed aggregates is desired, the forms are coated with a cement retarder. This coating prevents the cement in contact with the form from hardening so that after stripping (usually at an early age), the cement paste can be brushed away to expose the coarse aggregate. The same effect can be accomplished on a floor surface by spraying on the retarder or scrubbing the surface at an early age.

9.12 Slip-form construction. Various types of tall buildings are constructed using slip forms, which are frequently raised in small increments as construction progresses continuously. Slip forming is not a new technique; huge grain elevators, silos, chimneys, bridge piers, cores of high-rise buildings, highways, and canal linings have been slip-form constructed for years. However this construction system is now used for 25-story apartment buildings. In such cases the slip-form work of complete exteriors and bearing walls may be done at night and the concrete floors constructed during the day, building one story a day.

The slip forms and decks are custom-made to the requirements of each structure, a typical design being shown in Fig. 9.7. Most forms consist of three distinct elements: (1) the work deck, (2) the slip form, and (3) the finishers' scaffold. Although the form is generally constructed of 1

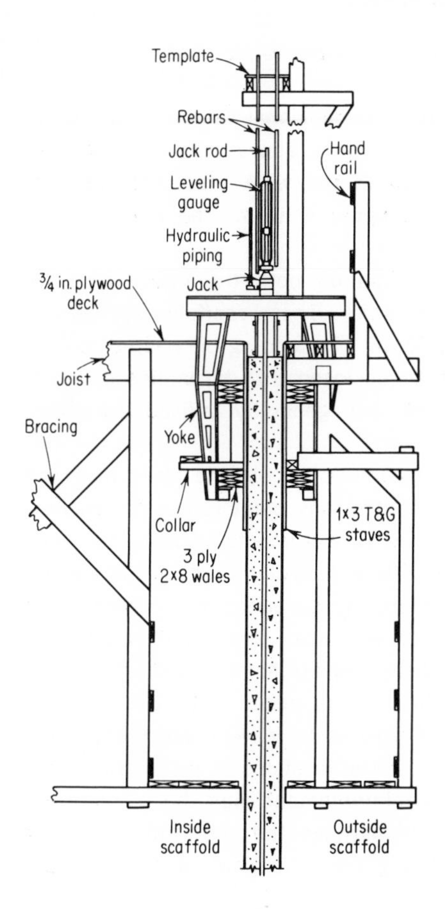

Fig. 9.7. A typical slip form with deck. *(From J. F. Camellerie)*

by 3-in. or 1 by 4-in. tongue-and-groove, straight-grained vertical staves, other materials such as ¾-in. plywood and 10-gauge steel sheets are occasionally used. The form material is attached to 2 or 3-ply timber wales.

The actual height of the form depends on many variables such as thickness of the wall, slump of the concrete, mix design, form design, jacking equipment, and even weather conditions. Most forms, however, are between 3 ft 6 in. and 6 ft high. Form oil is used as a form waterproofing material only—not as a lubricant. Once the slip-forming operation has started no lubricating compounds of any description are used. Usually forms are canted approximately ⅛ in. per ft of height to facilitate movement. Special care must be taken in building forms and arranging jacks so that the forms will draw straight without distortion during the sliding operation.

Yokes, which are spaced approximately 6 ft apart, support the bottom wales, maintain the desired wall thickness, and resist lateral concrete pressure. A jack is mounted on the horizontal arm of each yoke. Care must be exercised in designing and mounting these yokes so that adequate clearance is provided for the installation of horizontal reinforcing rods in the rising concrete.

The deck surface may be plywood or dimensional lumber. On some slip-form jobs, a second deck, placed at a level above the main working deck, is constructed to relieve congestion in the work area. Reinforcing bars for walls are generally stored on the upper platform and passed down as needed.

The construction of finishers' platforms on both the inside and the outside of the wall complete the slip-form unit. These scaffolds are usually suspended 8 ft below the bottom of the form. Depending on weather conditions, these scaffolds may be left open or enclosed with tarpaulins or hardboard. Curing and/or heating devices are generally affixed to these platforms.

Perfect coordination of hydraulic, pneumatic, or electric jacks mounted at each yoke is essential. If the movement at any point is out of synchronization, the structure will be forced out of alighment. Since the jacks are spaced only about 6 ft apart, the lifting capacity of each jack need not be great—3 tons is generally considered to be adequate. These jacks are designed to ride on smooth steel rods which are embedded in the concrete. The jack rods must be braced where they are not encased in concrete (in formed openings or blockouts). The form is raised ⅝ to 1 in. in each jump. General practice in the United States is to leave these rods in the concrete after the structure is completed. The localions of jacks are important to the building designer. Vertical reinforcement must be planned accordingly. This is why close cooperation between the slip-form contractor and the design engineer is essential during the time working drawings are being prepared.

The rate of slide is limited by concrete setting rates. Lifting rates of 18 in. per hr are not unusual if high-early-strength cement or accelerators are used in the concrete. Currently, 9 in. per hr or 16 to 18 ft per day is considered to be a good figure with standard mixes using type I or type II cement.

9.13 Coating of forms. A good form coating should protect and extend the life of the form; it should provide for an easy separation from the concrete and leave forms free of concrete residue; and it should provide a clean concrete surface free from stains or films which mar its appearance or interfere with later operations such as painting, plastering, etc. It should be relatively quick drying and have good weather resistance.

Straight petroleum oils are commonly used. They provide a simple and effective means for impregnating the surface of wooden forms to eliminate local surface variations in absorbency, which could cause differences in the concrete surface. However, such coatings tend to produce more surface blow holes or pocking than other types of coatings [913]. For use on plywood when an exceptionally smooth surface is required, shellac is superior to oil, because it more effectively prevents moisture from raising the grain and causing roughness. For steel forms oils compounded of petroleum oil and other oils of animal or vegetable origin and gums and resins have been found superior to straight petroleum oils in resisting removal of the oil by abrasion by the fresh concrete, and in preventing sticking of concrete to the steel. These compounded oils are heavier in body than straight petroleum oils. One type produces a hard, shellaclike coating which is especially resistant to abrasion. All oiling of forms should be completed before the reinforcement is placed, as otherwise the oil will get on the bars and destroy their bond.

During the entire construction program all form surfaces should be protected against marring. Care should be taken to prevent roughening of forms by operating vibrators against them. If scheduled for reuse after removal from one location, forms should be thoroughly cleaned, reconditioned and reoiled. Metal forms should not be wire-brushed so vigorously as to produce a bright surface, as sticking of the concrete may occur. Where sticking is encountered with steel forms, leaving the cleaned, oiled forms in the sun for a day, or vigorously rubbing the troublesome areas with a solution of paraffin in kerosene will usually improve the condition. It is generally less expensive and more satisfactory to obtain the desired surface effect by proper preparation and reconditioning of the forms than it is to obtain it by working over the concrete after the forms are removed.

9.14 Failure of forms. Many failures of forms have occurred as a result of improper design, construction, or use of the forms. Some of these deficiencies are:

1. Inadequate cross bracing of shores
2. Inadequate horizontal bracing and poor splicing of double-tier multiple-story shores
3. Failure to regulate properly the rate of placement without regard to drop in temperature
4. Failure to regulate properly the rate and order of placing concrete horizontally to prevent unbalanced loadings on the formwork
5. Unstable soil under mudsills
6. Failure to inspect formwork during concrete placement to detect abnormal deflections or other signs of imminent failure
7. Lack of proper inspection by an engineer to ensure the design is properly interpreted
8. Failure to provide for lateral pressures
9. Shoring not plumb

10. Locking devices on metal shoring inoperative, missing, or not locked
11. Overturning by wind
12. Vibration from adjacent moving loads
13. Form damage in excavation because of embankment failure

QUESTIONS

1. List the requirements of concrete formwork.
2. What factors influence the lateral pressure exerted by the concrete upon the forms?
3. Why are some forms subject to uplift pressure?
4. What loads are considered in form design other than those due to pressure of fresh concrete?
5. List the types of plywood used for forms, and their advantages.
6. Why are both very green and thoroughly seasoned lumber less well suited for forms than moderately seasoned lumber?
7. Why is it necessary to oil wooden forms?
8. Are petroleum oils suitable for oiling metal forms? Explain.
9. List the causes of form failures.

TEN
STRENGTH OF CONCRETE

10.1 Properties of hardened concrete. This chapter is the first of four which deal with the more important properties of hardened concrete, such as strength, elasticity, watertightness, resistance to destructive agencies, volume changes, creep, extensibility, and thermal properties. Owing to limitations of space, only a brief summary can be given; references are made to more detailed publications.

It is important to understand that the properties of concrete change continuously with time and ambient conditions. Absolute measures are significant only in so far as they indicate potential qualities. So far as job control is concerned, the results of tests have meaning only to the extent that they measure success in attaining desired qualities. Values employed as the basis for structural design should be safe minima which can be assured with given materials under the conditions surrounding a particular structure, with only an occasional test value below that which is assumed. The strengths specified for job control should be somewhat higher than those assumed in design, the margin of safety depending on the uniformity of job control (see Art. 17.10). Herein, the use of particular test conditions in discussing properties of concrete serves only to illustrate the influence of given variables and to aid in developing judgment in matters of control of job procedures.

Methods of testing are often a large factor in determining quantitative results. The results of tests of concrete have been extensively reported for various conditions of test, but relatively little has been accomplished in the way of comprehensive correlation of the data. In this and the three following chapters will be pointed out only the general range in values for certain common properties for which fairly satisfactory quantitative measures may be given.

SIGNIFICANCE OF STRENGTH

10.2 Resistance to applied forces. A primary function of practically all structures is to carry loads or resist applied forces of whatever nature; other functions such as the retention of fluids (as by a dam or a tank) or the exclusion of weathering or other destructive agencies may be involved also. In order to maintain continuity of structure without undue cracking, tensile strength is of special importance, although its actual magnitude is relatively low; usually (except in most pavements) steel reinforcement is provided to resist tensile forces. In any event, strength is an important direct requisite of practically all concrete.

10.3 Strength as a measure of general quality. The strength of concrete appears to be a good index, whether direct or inverse, of most of the other properties of practical significance. In general, stronger concretes are stiffer, more nearly watertight, and more resistant to weathering and certain destructive agencies; on the other hand, however, stronger concretes usually exhibit higher drying shrinkage and lower extensibility, hence are more liable to cracking. These relationships, together with the fact that strength tests are relatively simple to make, form the basis for the common use of strength in specifying and controlling quality and in evaluating the effects of variable factors such as materials, proportions, manufacturing equipment and methods, and curing conditions.

10.4 Nature of strength. One definition of strength is the ability to resist force; with regard to concrete for structural purposes it is taken, unless stated otherwise, as the unit force (stress) required to cause rupture. Rupture may be caused by applied tensile stress (failure in cohesion), by applied shearing (sliding) stress, or by compressive (crushing) stress. However,

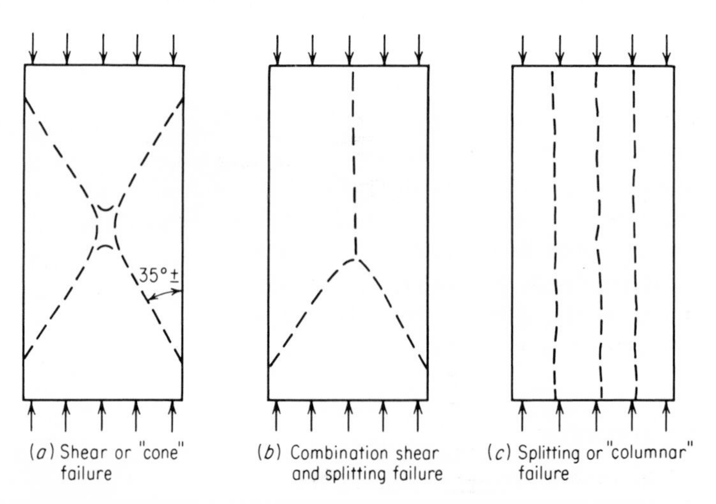

Fig. 10.1. Typical failures of standard concrete test specimens in compression.

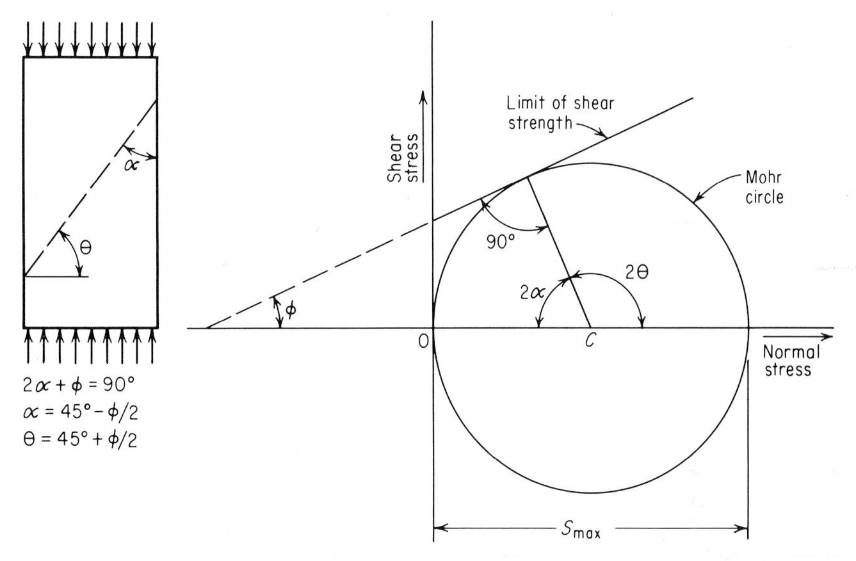

Fig. 10.2. Relation between angle of rupture α and angle of internal friction ϕ.

a brittle material such as concrete is much weaker in tension (Art. 10.6) and in shear (Art. 10.8) than in compression, and failures of concrete specimens under compressive load are essentially shear failures on oblique planes (Fig. 10.1*a*). Since the resistance to failure is due to both cohesion and internal friction, the angle of rupture is not 45° (plane of maximum shear stress) but is a function of the angle of internal friction ϕ; it can be shown by mechanics that the angle α which the plane of failure makes with the axis of loading is equal to $45° - \phi/2$, as illustrated in Fig. 10.2.

Since the angle of internal friction of concrete is of the order of 20°, the angle of inclination of the cone of failure in the conventional test specimen is approximately 35°, as illustrated in Fig. 10.1*a*. Further, the angle of rupture may be caused to deviate somewhat from the theoretical value owing to the complex stress condition induced in the end portions of compression specimens. This deviation results from restraint to lateral expansion under load caused by friction of the bearing plates on the end surfaces. When the strength of concrete is high and lateral expansion at the end-bearing surfaces is relatively unrestrained, as when flow of the bedment occurs, the specimen may separate into columnar fragments, giving what is known as a "splitting" or "columnar" fracture (Fig. 10.1*c*). Often failure occurs through a combination of shear and splitting, as indicated in Fig. 10.1*b*.

Most concrete in structures is subjected simultaneously to some combination of compressive, tensile, and/or shearing stresses in two or more directions, either directly or because of restraint of surrounding portions. If

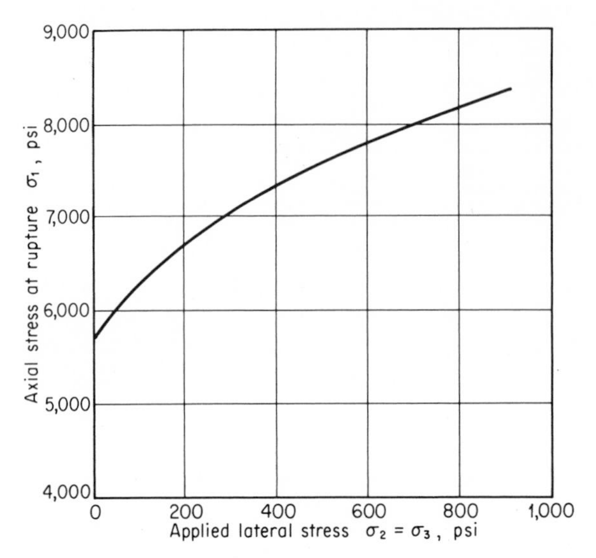

Fig. 10.3. Effect of lateral compression on compressive strength of concrete cylinders [1005].

all loads are compressive, the effect is to increase the principal stress required to produce failure. This relation is illustrated in a specific case by Fig. 10.3, in which the axial stress at rupture is increased by about 3½ times the amount of the imposed lateral stress, which in this case was the same in all lateral directions. In tests at the University of Illinois [1029] it was found that the strength under triaxial compressive loading is equal to the strength under axial loading plus about 4 times the least lateral stress. Price [1028] has shown that a similar relation exists for laterally applied tension, which reduces the axial compressive strength.

KINDS OF STRENGTH

10.5 Compressive strength. Except for highway pavements, most concrete structures are designed under an assumption that the concrete resists compressive stresses but not tensile stresses; hence, for purposes of structural design the compressive strength is the criterion of quality, and working stresses are prescribed by codes in terms of percentages of the compressive strength as determined by standard tests. A further consideration is that compression tests are relatively easy to make. The usual test employs a cylindrical specimen of height equal to twice the diameter, moist-cured at 70 ± 5°F for 28 days and then subjected to slow ("static") loading at a specified rate until rupture occurs; usually loading is completed within 2 or 3 min. Values of strength obtained in this way usually range from

2,000 to 6,000 psi, the most common value being of the order of 3,000 psi—a value which is often used for purposes of general discussion.

Decision as to the appropriate strength for a given structure or member is influenced by considerations not only of resistance to loads (forces) but also of economy, clearance, weight, heat generation (in mass concrete), etc. For example, a building column of 4,000-psi concrete would cost more *per unit of volume* than one of 3,000-psi concrete, but because of the higher permissible working stress and consequent smaller cross section, it might actually cost less overall. Further, the smaller cross section results in a greater percentage of usable floor space. The practical upper limit of strength in this case is either that at which stability becomes a factor or that at which space is needed within the cross section for proper spacing and clearance of reinforcement. Some designers make the lower columns of a tall building of stronger concrete than that in the other members but otherwise make all members of the same concrete in order to simplify construction.

Subject to special considerations for particular jobs, the orders of magnitude of indicated (test) strength for the principal classes of structures are as follows: In buildings, the footings, columns (with the exception just stated), beams, slabs, and walls will average 3,000 to 3,500 psi. Precast structural members and prestressed concrete will average considerably higher, up to twice these values. For pavements, most state-highway specifications call for a flexural strength (modulus of rupture) of about 550 psi, corresponding to a compressive strength of 3,500 to 4,000 psi. For dams, considerations of weight and temperature rise result in relatively low design strengths, usually ranging from 2,000 to 3,000 psi, although considerably lower and considerably higher strengths have been reported. For small and relatively unimportant "nonengineering" work where the concrete is not tested, it is estimated that the usual compressive strength is 2,000 psi or less.

10.6 Tensile strength. Concrete is not usually expected to resist direct tensile forces because of its relatively low tensile strength and brittle nature. However, tension is of importance with regard to cracking, which is a tensile failure; most cracking (aside from that due to load or to settlement of parts of the structure) is due to restraint of contraction induced by drying or by lowering of temperature.

The general relationship between compressive and direct tensile strength of concrete is shown in Fig. 10.4. It is seen that the tensile strength ranges from 7 to 11, and at usual levels averages about 9 percent of the compressive strength; the higher the compressive strength, the lower the relative tensile strength. The maximum value of tensile strength obtained in these tests, and probably nearly the maximum that can be attained,

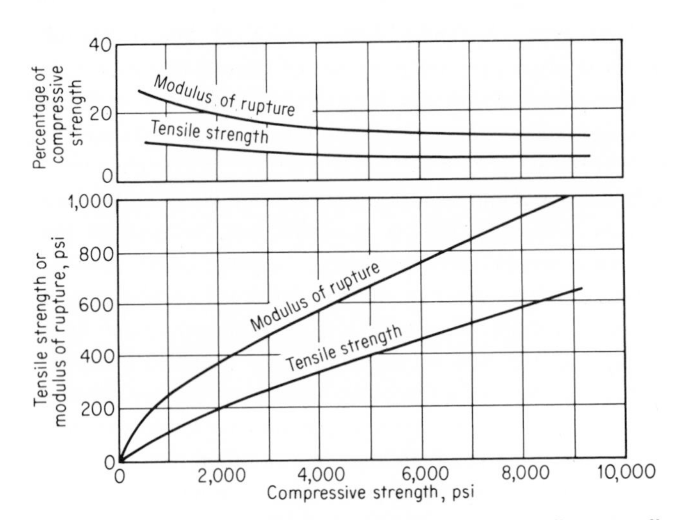

Fig. 10.4. Relations between compressive strength, direct tensile strength, and modulus of rupture [1009].

is 600 psi. Others have found that the type of coarse aggregate has a larger relative effect on tensile strength than on compressive strength.

Direct tension tests of concrete are seldom made, principally because of difficulties in mounting the specimens and uncertainties as to secondary stresses induced by the holding devices. They are never made for purposes of concrete control, and there is no standard test even for research purposes.

Splitting tension. An indirect test for tensile strength of concrete, developed originally in Brazil, has recently come into rather general use and has been standardized (ASTM C496-62T). Some details are given in Test 16 herein. The specimen is the conventional 6 by 12-in. cylinder, made and cured in the same manner as that for compression tests. The cylinder is loaded in compression along two axial lines which are diametrically opposite through bearing strips of ⅛-in. plywood (Fig. 10.5). The plywood

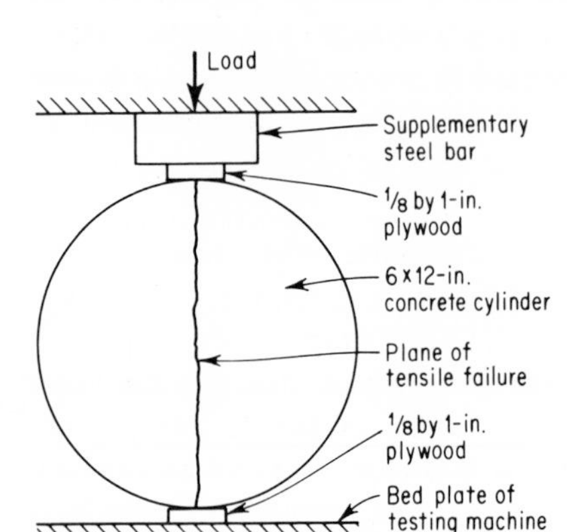

Fig. 10.5. Arrangement for loading concrete specimen for splitting-tension test.

cushion distributes the compressive load over a small width which is sufficient to avoid undue concentration of stress, and it compensates for surface irregularities. The compressive force produces a transverse tensile stress which is practically constant along the vertical diameter. If the compressive strength is at least three times the tensile strength, as it invariably is in concrete, the failure is in tension along the vertical diametral cross section. The splitting tensile strength is computed by means of the following formula, which is derived from the theory of elasticity.

$$\sigma = \frac{2P}{\pi l d}$$

in which σ = splitting tensile strength, psi
P = maximum applied load, lb
l = length of cylinder, in.
d = diameter of cylinder, in.

The tensile strength computed in this manner is apparently about 15 percent higher than that determined by direct tension tests, judging from the relatively few comparative test results available. The splitting tensile strength ranges approximately from 50 to 75 percent of the flexural strength (modulus of rupture) with higher-strength concretes having relatively higher ratios of tensile to flexural strength. It ranges approximately from 8 to 14 percent of the compressive strength with higher-strength concretes having relatively lower ratios, and on the average it is about 10 percent of the compressive strength.

This simple and convenient test is considered to be a useful and reliable index of flexural and compressive strengths.

10.7 Flexural strength. When concrete is subjected to bending, tensile and compressive stresses and in many cases direct shearing stresses are developed. The most common plain-concrete structure subjected to flexure is a highway pavement, and the strength of concrete for pavements is commonly evaluated by means of bending tests on 6 by 6-in. beam specimens. Flexural strength is expressed in terms of "modulus of rupture," which is the maximum tensile (or compressive) stress at rupture computed from the well-known flexure formula

$$\sigma = \frac{Mc}{I}$$

in which σ = stress in the fiber farthest from the neutral axis, psi
M = bending moment at the section, in.-lb
I = moment of inertia of the cross section, in.^4
c = distance from neutral axis to farthest fiber, in.

Since the assumptions on which the flexure formula are based do not hold true at high stresses approaching failure, modulus of rupture is a fiictitious value; however, it is convenient for purposes of evaluation and is commonly used. It ranges from 60 to 100 percent higher than the direct tensile strength, as indicated in Fig. 10.4. The modulus of rupture ranges from 11 to 23 percent of the compressive strength; for concretes of compressive strength of 3,500 to 4,000 psi it is the order of 550 psi, or about 15 percent of the compressive strength. The use of rough-textured or angular aggregates results in relatively high flexural strength as compare to compressive strength.

An approximate relationship between flexural and compressive strength is

$$f'_c = \left(\frac{R}{K}\right)^2$$

in which f'_c = compressive strength, psi
R = modulus of rupture, psi
K = a constant, usually between 8 and 10

10.8 Shear strength. Shear is the action of two equal and opposite parallel forces applied in planes a short distance apart. Shear stress cannot exist without accompanying tensile and compressive stresses. Pure shear can be applied only through torsion of a cylindrical specimen, in which case the stresses are equal in primary shear, secondary tension (maximum at 45° to shear), and secondary compression (maximum at 45° to shear; perpendicular to tension). Since concrete is weaker in tension than in shear, failure in torsion invariably occurs in diagonal tension. Also in members subjected to bending, shear stress is accompanied by tensile and compressive "bending" stresses, and "shear" failures are due to resulting diagonal tension. The use of shear stress in structural design is justified only because it is a convenient measure of diagonal tension.

Tests to determine shearing strength directly are inconclusive because of the effects of bending, friction, cutting, or lateral restraint by the test apparatus. Some investigators have concluded that shear strength is 20 to 30 percent greater than the tensile strength (and thus about 12 percent of the compressive strength), while others have determined the shear strength to be several times the tensile strength—frequently 50 to 90 percent of the compressive strength. Under triaxial compressive loading, the limiting shear stress can be made to exceed the compressive strength under axial loading.

10.9 Bond with reinforcement. A significant property in structural design of flexural members of reinforced concrete is the resistance to slipping of the steel reinforcing bars which are embedded in concrete. Such resistance is provided by adhesion of the hardened paste, by friction, and by bearing of the lugs of deformed bars. In recent practice, round deformed bars are used exclusively (except ¼-in. round, plain bars), and the size and spacing of lugs are such as to produce a high mechanical bond strength. It is customary to specify permissible bond stresses as percentages of the compressive strength of concrete, even though tests have repeatedly shown that there is not a consistent relation between bond strength and compressive strength [1003, 1004, 1024]. It appears that bond strength, at least in the initial stages of failure (slip), depends largely on the magnitude and uniformity of lateral pressures that exist or may be developed between steel and surrounding concrete; thus it may depend not so much on the paste strength and rigidity as on the tendency of the paste to settle or shrink away from the steel during the setting period and subsequently on the tendency of the paste to expand or contract with changes in moisture or temperature.

Bond strength varies considerably with type of cement, type of admixture, and water-cement ratio, all of which influence the quality of the paste. However, it is not reduced significantly by air entrainment. It is increased by delayed vibration, if properly timed and applied, which apparently improves contact after settlement shrinkage has occurred. It is greater for dry concrete than for wet concrete. It is less for horizontal bars than for vertical bars, because of water gain under horizontal bars; this fact is further taken into account in modern codes, which permit higher bond stresses for bars near the bottom of a placement (and thus less subject to water gain) than for bars 1 ft or more from the bottom. Bond strength is reduced by alternations of wetting and drying, of freezing and thawing, or of loading.

Tests to determine bond strength usually consist in pulling a bar out of a concrete prism in which it is embedded. In the standard method, ASTM C234, a bar ¾ in. in diameter is embedded in each of three 6-in. cubes in such a way that one bar is vertical, one is horizontal at 3 in. from the bottom of the specimen, and one is horizontal at 9 in. from the bottom as cast. (Specimens containing the horizontal bars are cast as 6-in. prisms 12 in. deep to develop water gain, and are later divided into two cubes.) During testing, slip at the loaded and unloaded ends of the bar is measured at intervals of load by means of dial gages; and curves of bond stress versus slip are plotted. Loading is continued until the yield point of the reinforcing bar has been reached, or the enclosing concrete splits, or a minimum slippage of 0.01 in. has occurred at the loaded end.

The bond strength of modern deformed bars, which have closely spaced lugs, is much greater than that of the deformed bars formerly used (and therefore to be considered in investigations of old structures). Values vary with the concrete, size and brand of bar, and amount of slip used as a measure of failure. The general order of magnitude of bond stress on deformed bars at free-end initial slip is 100 to 300 psi, and of bond stress at loaded-end slip of 0.01 in. is 300 to 800 psi [1003].

Tests at the University of Texas [1050] indicate that the length of embedment has little influence on the loaded-end slip even when the bars are surrounded by spirals and that usable bond stresses on large high-strength bars are severely limited by the loaded-end specification limit of 0.01 in.

10.10 Nondestructive indications of strength. *Pulse time.* One method of determining the quality of concrete in place involves the use of electronic equipment. An ultrasonic pulse applied at one face of the concrete is received by a sensitive unit at the opposite face. The time taken for the pulse to pass through the concrete, as measured by an electric timing circuit, is used to compute the modulus of elasticity which is taken as a measure of the compressive strength [1015].

Hammer rebound. A simple device for indirectly indicating the compressive strength of concrete is the Schmidt test hammer (Fig. 10.6), which essentially measures the surface hardness. A rather heavy, metal hollow cylinder enclosing a spring-actuated plunger is pressed firmly against the surface of the concrete to be tested, and the spring is released. The rebound of the plunger is indicated on a linear scale graduated in empirical

Fig. 10.6. Schmidt concrete test hammer. *(Pacific Cement and Aggregates.)*

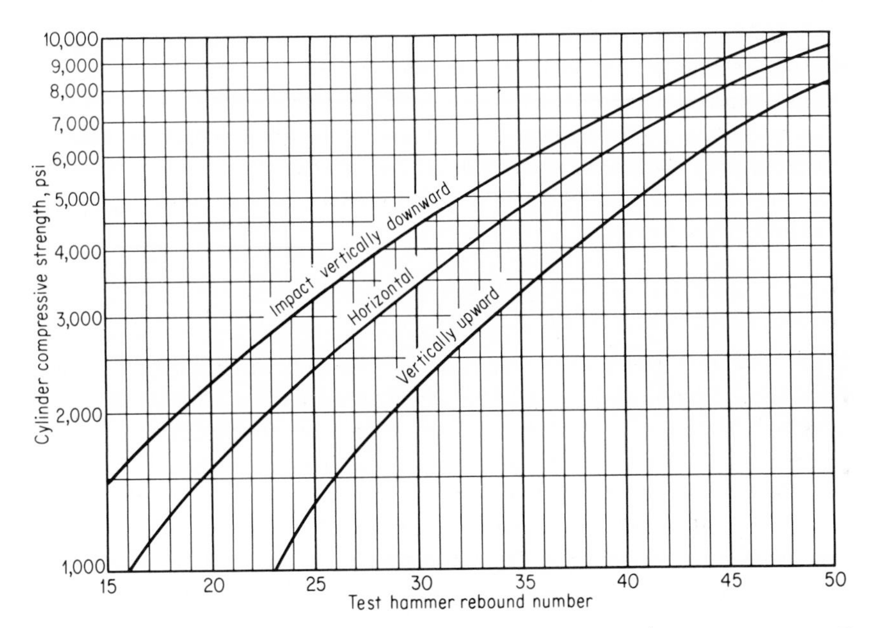

Fig. 10.7. Calibration curves for Schmidt concrete test hammer. (*From Arthur R. Anderson.*)

"rebound numbers." By means of a calibration curve such as one of those shown in Fig. 10.7, the compressive strength is estimated; the higher the rebound number the greater the strength. Several readings are taken, and the average value used.

Although the rebound values are relative rather than absolute and although they are affected by many factors, the test is useful for many purposes. When properly calibrated against compressive-test specimens for the conditions of a particular concrete, it can be expected to indicate the compressive strength within about 15 percent. It is useful to check uniformity—for example, to locate areas of unsatisfactory concrete in a wall. It serves as a convenient substitute for test cylinders cured at the site for evaluating the compressive strength at early ages and thus as a guide to the removal of forms or the discontinuance of heating in cold weather.

Other conditions being equal, rebound numbers are higher for larger masses than for small sections; higher for restrained than for free (loose) specimens; higher for flat surfaces than for convex surfaces, such as the side of a test cylinder; higher for dry surfaces than for wet surfaces after the first day or so; and higher for a horizontal position of the hammer for impact on a vertical surface than for impact vertically downward but lower than for impact vertically upward. They are affected but little by air entrainment or by specific gravity of aggregates.

In the calibration of the hammer, it is recommended that the compression-test cylinder be placed in the testing machine and subjected to a load of approximately 15 percent of the estimated ultimate strength.

The surface against which the plunger is to strike should be smooth; if necessary, it should be smoothed with a carborundum brick. Readings on finished floors or rough-floated surfaces are not likely to be representative of the underlying concrete.

FACTORS AFFECTING STRENGTH

10.11 Effect of component materials. In this and the remaining articles of this section (Arts. 10.12 to 10.15), usually the effects of variable factors—component materials, proportions, curing conditions, temperature, and loading conditions—are evaluated in terms of compressive strength.

Cement. The effect of the cement constituent is manifested through differences in chemical composition and fineness. In general, cements relatively high in lime (as C_3S) gain strength more rapidly than those high

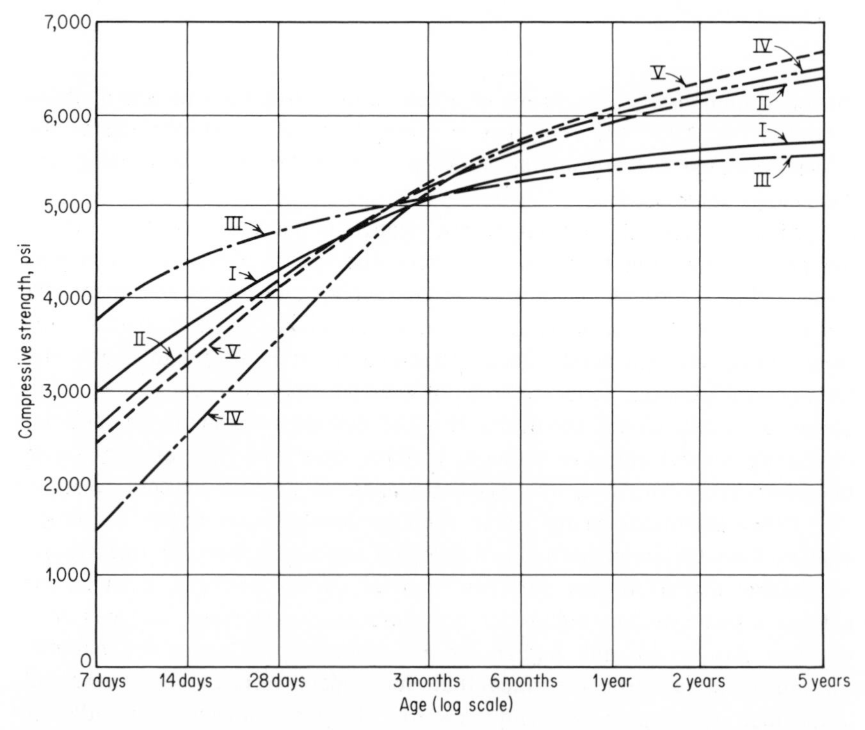

Fig. 10.8. Effect of type of cement on compressive strength of concrete. Aggregate 0 to 1½ in.; 6 sacks cement per cu yd; 6 by 12-in. cylinders; standard curing [1036].

in silica (as C_2S), even though the strengths at later ages (with continued moist curing) may be less; also, alumina (as C_3A) contributes greatly to strength at very early ages (Art. 2.10). Finely ground cements of given chemical composition gain strength more rapidly than coarse cements, but with relatively little effect at later ages; roughly, for a range of specific surface (Wagner basis) of 1,000 to 2,000 sq cm per g, an increase of 1 percent in specific surface results in a 2 percent increase in 7-day strength and a 1 percent increase in 28-day strength [1028, 1032]. Differences in composition and fineness as between brands of cement of a given type sometimes result in wide differences in strength; however, the net effect of these important variables can be shown by consideration of the five standard types as specified by ASTM. In Fig. 10.8 are given typical rates of strength development for concretes containing each of these types. Although the specific surface of these particular cements is not known, the current minimum ASTM requirements (Wagner) were as follows: type I, 1,600 sq cm per g; type II, 1,700; types IV and V, 1,800. For type III (high-early-strength), the specific surface was not specified, but it is characteristically far higher than that of the others. The accompanying tabulation gives the composition of the cements.

	Compound composition, %							
Type of cement	**C_3S**	**C_2S**	**C_3A**	**C_4AF**	**$CaSO_4$**	**Free CaO**	**MgO**	**Igni. loss**
I Normal	49	25	12	8	2.9	0.8	2.4	1.2
II Modified	46	29	6	12	2.8	0.6	3.0	1.0
III High-early-strength	56	15	12	8	3.9	1.3	2.6	1.9
IV Low-heat	30	46	5	13	2.9	0.3	2.7	1.0
V Sulfate-resisting	43	36	4	12	2.7	0.4	1.6	1.0

As shown by Fig. 10.8, the high-early-strength cement exhibits relatively high strength at ages up to 3 months, but thereafter has about the same concrete strength as that for the normal (type I) portland cement. The three cements which are low in C_3A are correspondingly low in early strength, and of these the lowest in C_3S is lowest in early strength. At ages greater than about 3 months, however, all three of these cements exhibit increasingly higher strengths than the normal cement, with little difference among the three.

Aggregate. The effect of type of aggregate of normal weight and given gradation on compressive strength of concrete is minor, probably because all such aggregates are far stronger than the paste. With equal water-cement ratios, concrete containing angular or rough-textured aggregate will usually be somewhat stronger than that containing rounded or smooth aggregate. With equal cement contents, however, somewhat more water is required for workability when angular or rough aggregates are used, and the net effect is that for equal workabilities the concrete strengths are not greatly different.

Differences in maximum size of well-graded aggregate of a given type have two opposing effects on the compressive strength of concrete. With equal cement contents and consistencies, larger aggregates require less mixing water than aggregates of smaller maximum size. On the other hand, with equal water-cement ratios and consistencies, larger aggregates exhibit lower concrete strengths, possibly due in part to some prior cracking of the paste around the larger pieces. The net effect varies with the richness of mix; generally in leaner concretes the larger maximum sizes will result in higher strengths whereas in richer concretes they will result in lower strengths. In any case, the net differences are not large.

Lightweight aggregates are relatively weak, the strength varying with the unit weight, and in some cases the strength of lightweight concrete is governed by that of the aggregate. If lightweight is used as both coarse and fine aggregate, strengths of 5,000 psi are seldom attained, and the practical upper limit appears to be about 7,000 psi.

Water. The effect of type of mixing water is discussed in Art. 4.1. Of principal interest is sea water, which reduces the 28-day compressive strength about 12 percent.

Admixtures. The effect of various admixtures is discussed in Chap. 4. Air-entraining agents (Arts. 4.7 and 11.14) and accelerators (Arts. 4.9 and 8.15) are of practical interest with regard to strength of concrete. When the water-cement ratio is constant, air entrainment reduces strength; but if advantage is taken of air entrainment to reduce the water-cement ratio, the net effect on strength is not large within the range of recommended air contents. The principal accelerator is calcium chloride, which is especially effective, percentagewise, at low temperatures of casting and at early ages (Art. 8.15). Water-reducing admixtures (Art. 4.6) when used in recommended amounts may be expected to increase the compressive strength about 15 percent.

10.12 Effect of proportions and uniformity. With given materials, the influence on strength of proportions of materials in a concrete mix is manifested largely through the effect on the water-cement ratio (or void-cement ratio) required for workability, as discussed in Chap. 6 (also see Art. 2.16). The quantitative effect of variation in water-cement ratio itself is illustrated by Fig. 10.9, which for type I portland cement shows "bands" of strength within which average compressive and flexural strengths will usually fall.

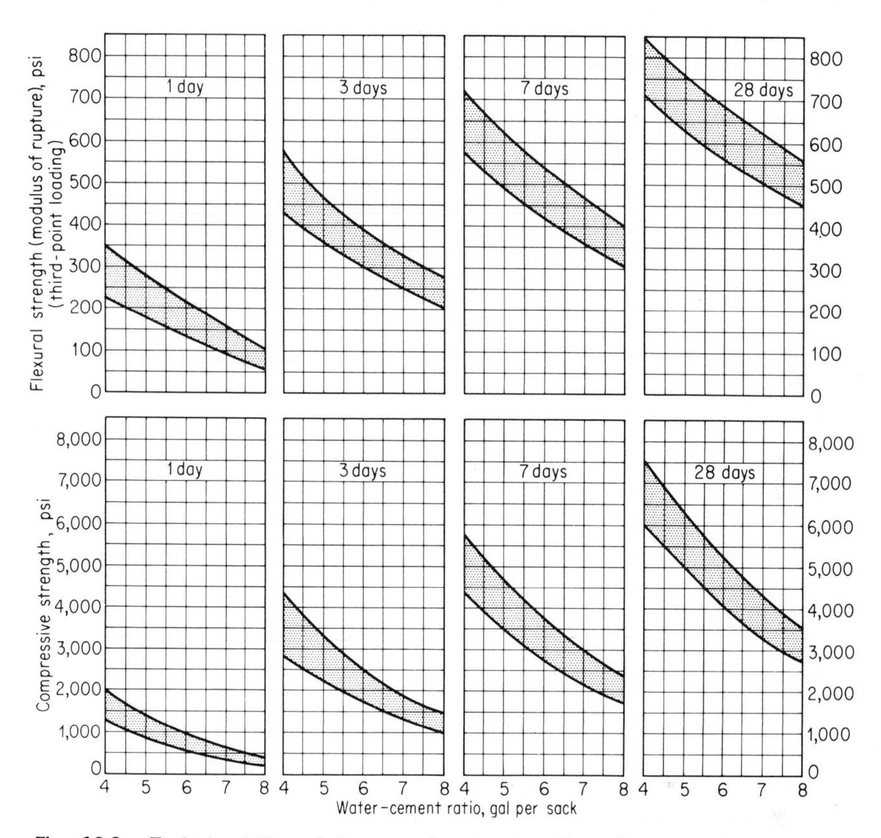

Fig. 10.9. Typical relations between water-cement ratio and compressive and flexural strength of standard-cured concrete containing type I portland cement. (*Portland Cement Association.*)

The lower limit of the 28-day band for compressive strength corresponds roughly to the values (Table 6.6) recommended by ACI as the basis for proportioning concrete mixes; if *minimum* strengths are desired rather than average values as shown in Fig. 10.9 and Table 6.6, correspondingly higher values of average strength must be provided for.

The necessary excess of average strength over desired minimum, or specified, strength depends on the closeness of control of uniformity of concrete. For ordinary jobs it should be of the order of 15 to 25 percent; for jobs under close control it may be as low as 10 percent. If the coefficient of variation of the concrete is known or can be reasonably estimated, the required excess of strength can be determined in accordance with the theory of probability, taking into consideration the proportion of compressive strengths that will be permitted to fall below the specified strength (ACI 214) [1708]. The required average strength can be computed from the following formula:

$$f_{cr} = \frac{f'_c}{1 - tV}$$

in which f_{cr} = required average strength, psi
f'_c = design strength specified, psi
t = a constant depending on the proportion of tests that may fall below f'_c and on the number of samples used to establish the coefficient of variation V
V = forecasted value of the coefficient of variation expressed as a fraction

Values of the constant t are available in various publications. For the common proportions of 1 in 10 and 1 in 5 that are permitted to fall below the nominal minimum and for a large number of samples, $t = 1.282$ and 0.842 respectively. Thus with good field control corresponding to a coefficient of variation of 15 percent, the computed f_{cr} is equal to $1.24f'_c$ and $1.14f'_c$ respectively; if the specified strength is 3,000 psi, the concrete should be proportioned for an average strength of 3,720 and 3,430 psi respectively.

Table 10.1 shows recommended tolerances and requirements for individual and consecutive tests.

10.13 Effect of curing conditions. *Time; moisture.* The curing conditions with respect to time, moisture, and temperature, through their effect on hydration of the cement, exercise an important influence on the strength

Table 10.1 Recommended tolerances and requirements for tests

	ASTM Specification C94 for ready-mixed concrete	ACI building code (ACI 318-63)*	
		Working-stress design	Ultimate-strength design; prestressed concrete
Number of cylinders constituting a "test"	3	2	2
Tolerance for individual tests	1 in 10 may be less than 90% f'_c	1 in 5 may be less than f'_c	1 in 10 may be less than f'_c
Number of consecutive tests the average of which must be equal to or greater than f'_c	5	5	3

* From Ref. 1801.

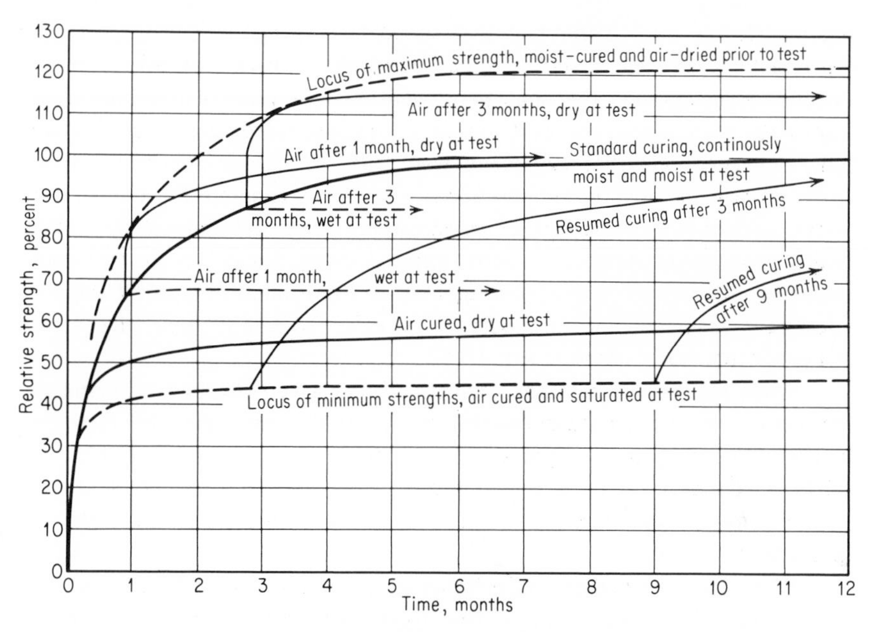

Fig. 10.10. Effect of moist-curing conditions at 70°F and moisture content of concrete at time of test on compressive strength of concrete. (*From H. J. Gilkey* [851].)

of concrete. The longer the period of moist storage, the greater the strength, as indicated in various ways in Figs. 10.8 to 10.13. Exposure to air, with consequent drying, arrests hydration; the rate and extent of drying depend on the mass of concrete relative to the area of exposed surface as well as on the humidity of the surrounding air. The test results shown in Fig. 10.10 are influenced by the moisture content of the concrete at time of test; it is seen that specimens exposed to air and tested in the air-dry condition were one-quarter to one-third stronger than corresponding specimens exposed to air for the same period but saturated just prior to being tested (see also Arts. 2.16 and 10.19). Resumption of moist curing after a period of air-drying results in resumption of hydration, although at a slower rate than that in progress when drying was begun. Long-time tests of concrete cured in water at normal temperature show that hydration continues appreciably (at a decreasing rate) at ages up to 30 years [1041]; these tests were subsequently extended to 50 years with appreciable further increases in strength. Other tests have shown that sealing of concrete to retain the mixing water for curing purposes results in a slighly lower rate of hydration than that of saturated concrete, the minor difference being attributed to self-desiccation during hydration.

Temperature. The influence of temperature of moist curing on concrete strength depends on the time-temperature history. Apparently conflicting results are due to the fact that in some investigations the specimens were cast, cured, and tested at given constant temperatures over a range; in other investigations the specimens were cast at given temperatures and beginning soon thereafter were cured at normal temperature; and in still other investigations the specimens were cast at normal temperature and cured at various constant temperatures. For mass concrete the temperature varies continuously, and some investigations have determined this effect under time-temperature conditions approximating those in the structure under consideration. Steam curing (Art. 8.13) is also of interest with respect to precast concrete products. The effect of temperature of concrete at time of test is discussed in Art. 10.20.

When concrete is cast and maintained at a given constant temperature, the higher the temperature (within limits), the more rapid the hydration and resulting gain in strength at ages up to 28 days, as illustrated in Fig. 10.11*a*. Other tests have shown that at later ages the strengths are not greatly different but that the higher the curing temperature, the lower the strength.

When concrete is cast and maintained at a given temperature for several hours and then cured at 70°F, the higher the initial temperature (within limits), the lower the 28-day strength, as illustrated in Fig. 10.11*b*. The relative strengths at 28 days were maintained at later ages.

When concrete is cast and maintained at 70°F for several hours and then cured at various constant temperatures, the lower the curing temperature (within limits), the lower the strength at ages up to 28 days, as illustrated in Fig. 10.11*c*. At a temperature (33°F) near freezing, the strength is only 47 percent of that of concrete cured continuously at 70°F; at the below-freezing temperature of 16°F, only 9 percent.

From the preceding paragraphs it can be generalized that if the curing temperature is higher than the initial temperature of casting, the resulting 28-day strength will be higher than that for a curing temperature equal to, or lower than, the initial temperature.

Results of more recent and extended tests than those just cited are shown in Figs. 10.12 and 10.13. At temperatures above 73°F the early-age strengths were higher, but the later-age strengths lower, than those at normal temperature. At temperatures below 73°F the early-age strengths were lower, but the later-age strengths higher, than the normal. These and other tests indicate that the optimum temperature of early-age curing is about 55°F for type I and type II cements.

The effect of "mass" curing at continuously changing temperatures, corresponding to those within a large mass of concrete, is influenced greatly by the type of cement, initial temperature, and time-temperature conditions of curing. However, in general it can be said that mass curing accelerates strength development at ages up to about 3 months; thereafter the strength

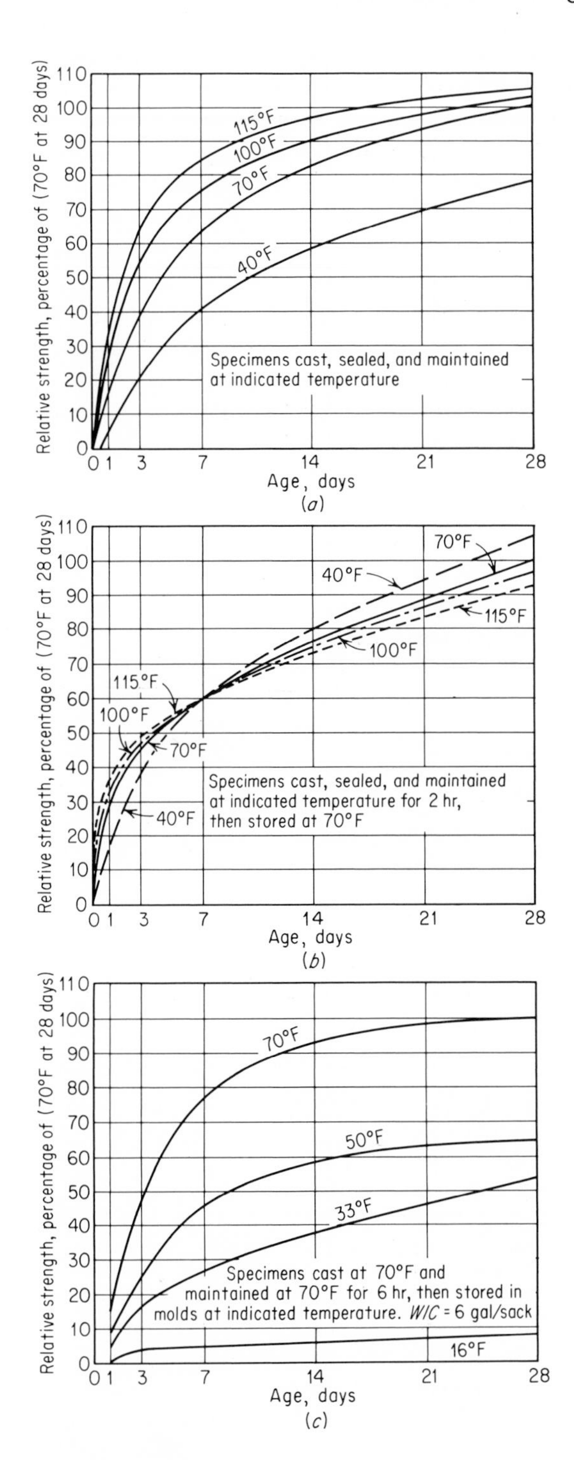

Fig. 10.11. Effect of various temperature conditions of casting and curing on compressive strength of concrete. [(*a*) *and* (*b*) *adapted from Ref.* 1036; (*c*) *adapted from Ref.* 1033.]

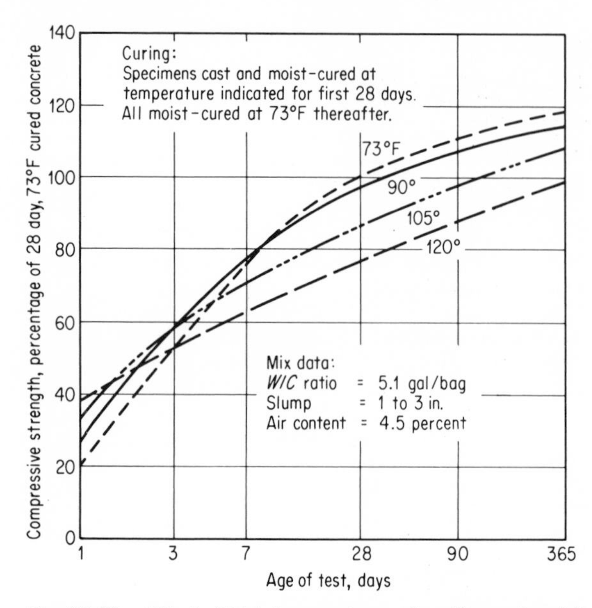

Fig. 10.12. Effect of high temperatures of casting and moist curing on compressive strength of concrete at ages up to 1 year [1027].

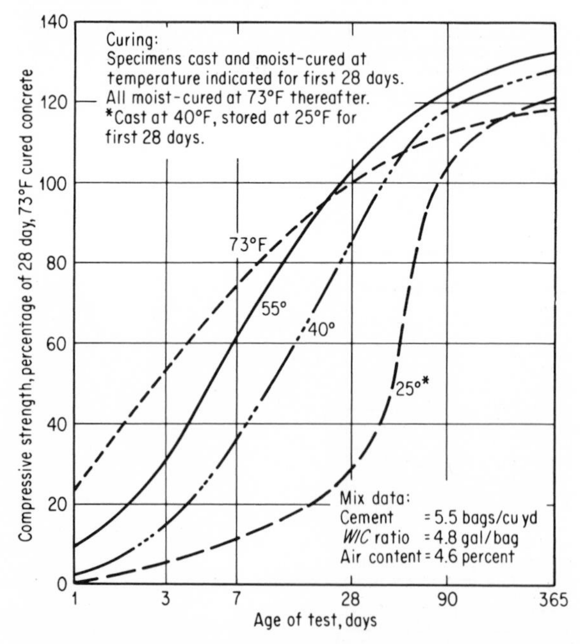

Fig. 10.13. Effect of low temperatures of casting and moist curing on compressive strength of concrete at ages up to 1 year [1026].

of mass-cured concrete is usually somewhat less than that of standard-cured concrete (Art. 14.18) [204]. Roughly, the 28-day strength of mass-cured concrete is about 10 percent higher than that of corresponding standard-cured concrete.

The effect of steam curing at various temperatures on compressive strength is discussed in Art. 8.13.

Destructive agencies. One result of the action of certain destructive agencies, considered as curing conditions, is a reduction in concrete strength through deterioration of the concrete. For example, repeated freezing and thawing reduces strength (Art. 11.15), as does alkali-aggregate reaction (Art. 11.16).

10.14 Effect of loading conditions. Normally, concrete structures are considered as being subject to steady, or "static," loads, and compressive strength is evaluated by means of a test in which load is gradually applied to failure within a few minutes. However, actually most structural members are subjected to long-continued steady loads—at least dead loads—and many are subjected to fluctuations of load or to impact.

Sustained loading. Under steady loading sustained for a number of years, concrete will withstand only about 70 percent of the stress at failure in the conventional test, as indicated by Table 10.2 [1028]. Under a steady load sustained for only 1 day, failure will occur at 90 percent of the strength as normally measured. Long-sustained compressive stresses less than about 70 percent of the strength developed in a short-time loading test have little effect on the strength at the end of the sustained-loading period [1260].

Table 10.2 Effect of duration of load on strength of concrete

Duration of load				Percentage of strength at standard-test loading rate
Min	Hr	Days	Years	
2*				**100**
10				95
30				92
60	1			90
	4	0.17		88
		100		78
		365	1	77
			3	73
			30	69

* Approximate; see Art. 10.22.

Repeated loading. Under a large number of cycles of repeated loading in compression, dry concrete will fail at a stress approximately 50 to 55 percent of the strength under short-time loading [1012]. Even 5,000 repetitions will cause failure at a stress approximately 70 percent of the static strength. In flexure where there is a stress gradient, a somewhat larger repeated maximum stress may be applied before failure occurs [1012*a*]. The endurance ratio, or ratio of fatigue strength to static strength, is lower for saturated concrete than for dry concrete. There is little difference in endurance ratio in flexure as between stresses varying from zero to the maximum and complete reversals of stress.

Impact loading. As would be expected from the fact that concrete is a brittle material, its resistance to impact is relatively low; in practice, reliance is placed on reinforcement or on mass whenever impact is to be expected. In tests to determine the resistance of reinforced concrete slabs [1018] it appeared that the strength under impact was about the same regardless of the compressive strength of the concrete. Resistance to a single impact load causing failure was about three times as great as resistance to a succession of impact loads increasing uniformly in intensity up to the given value.

10.15 Effect of exposure to high temperatures. Certain chemical and physical changes take place when concrete is heated. The uncombined moisture

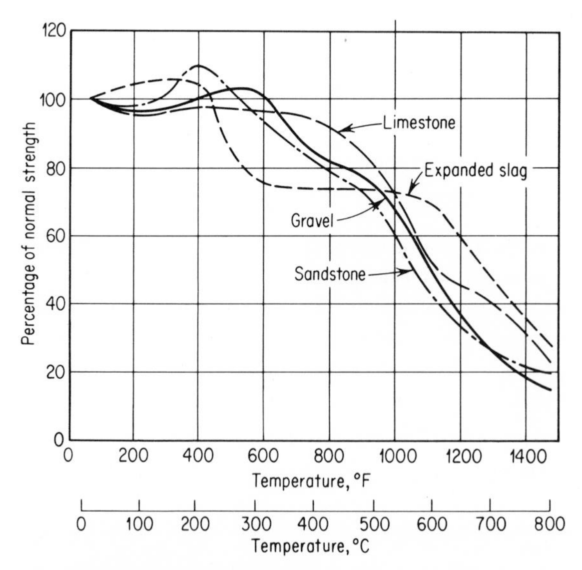

Fig. 10.14. Effect of aggregate and temperature of heating on compressive strength of concrete after cooling [1042]. Maximum aggregate ¾ in.; cement content 5.1 sacks cement per cu yd; 4 by 8-in. cylinders.

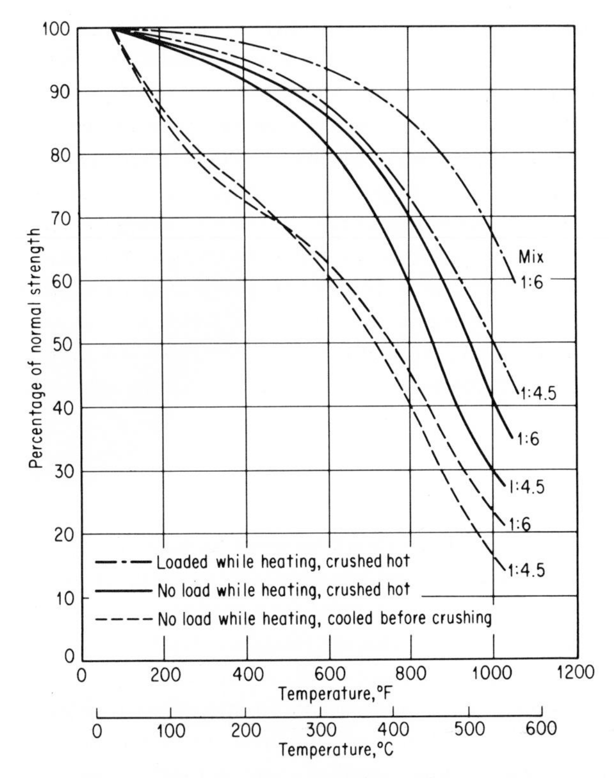

Fig. 10.15. Effect of testing conditions on compressive strength of concrete [1044]. **Maximum aggregate ⅜ in.; 2 by 4-in. cylinders.**

evaporates even at normal temperatures; all of it is lost soon after attaining 212°F, and even the combined moisture begins to be dissipated at this temperature. Also, dehydration of calcium hydroxide occurs when the temperature exceeds about 750°F and causes shrinkage of the cement paste. Various quartzlike aggregates undergo a crystalline transformation at about 1070°F which causes a large expansion and cracking of the concrete.

Various test programs have been conducted to determine the effect of high temperatures on the compressive strength of concrete [1042, 1046]. The test results reported by Zoldners [1042], shown in Fig. 10.14, include four types of aggregate. The test cylinders were heated at the standard rate for fire tests: heat soaked for 1 hour at the desired temperature and slowly cooled in the furnace before testing. For temperatures below about 900°F the expanded-slag concrete had the lowest strength, but for temperatures above 1000°F its strength was the highest. For specimens of all four aggregates which were quenched in water when removed from the hot furnace, the strengths approached a straight line between the values shown in Fig. 10.14 at temperatures of 70 and 1400°F. From the results of a test program on aluminous cement concretes [1043] at temperatures up to 1000°C, it was found that the losses in strength were similar to those indicated above for normal concretes.

Test results reported by Malhotra [1044] for flint concrete shown in Fig. 10.15 indicate that cylinders loaded while heated to 1000°F and crushed while still hot have about double the strength of cylinders cooled before crushing. Specimens which were subjected to the design load while heated were less adversely affected than the corresponding specimens which were not loaded prior to the compression test. The richer mix (1:4.5) was adversely affected somewhat more than the leaner mix (1:6), but it was found that differences in water-cement ratio of a given mix had little effect.

For the effects of normal temperatures upon compressive strength see Art. 10.20.

FACTORS AFFECTING RESULTS OF STRENGTH TESTS

10.16 Specimens vs. structures. Tests of relatively small specimens under standard conditions are admittedly imperfect measures of strength of concrete in structures, which are exposed to a composite variety of stresses and other conditions affecting strength. However, standard tests are probably sufficiently accurate for practical purposes of structural design and job control. In general, load tests on structures and on specimens of hardened concrete from structures indicate higher strengths than those obtained from standard specimens; hence standard methods are considered to be conservative.

Strength tests of separately cast cylinders are seldom used as a direct basis for acceptance or rejection of concrete structures or portions thereof, because of practical and perhaps legal difficulties of enforcement. However, for ready-mixed concrete, alternative specifications on the basis of strength are given in ASTM C94.

Concrete specimens are cured in either of two ways, according to the purpose of the tests. Standard-cured specimens (moist at 70°F) indicate the *potential* strength of the concrete; they are made to check the adequacy of mix proportions or as the basis for the acceptance of concrete. Field-cured specimens indicate the degree to which strength is affected by the actual conditions of moisture and temperature (as nearly as practicable); when used, their purpose is usually to determine when the structure may be put into service.

One factor in testing is the number of samples required to secure a desired accuracy of the average, as discussed in Art. 17.8. Although the criteria given therein are seldom followed, authoritative bodies have recommended the minimum number of tests to secure fairly satisfactory job control. Recommended values are given in Table 10.3.

Table 10.3 Recommended number of specimens and tests for construction

Kind of work	Minimum number of specimens or tests for each class of concrete (for either standard or field curing)	Source of recommendation
Buildings; reinforced concrete in general	(1 test = 2 specimens) 5 tests per class; 1 test for each 150 cu yd; 1 test per day	ACI 318
Ready-mixed concrete (when strength is basis of acceptance)	(1 test = 3 specimens) 3 tests per class; 1 test for each 50 loads of each class	ASTM C94
Pavements	3 field-cured beams and 3 standard-cured beams for each 2,000 sq yd or portion thereof in 1 day	ACI 617
Dams; mass concrete in general	2 tests per shift	Common practice

Interpretation of the results of strength tests should take into account the important influence of the test conditions themselves on the indicated strength. In the following articles of this chapter are discussed the effects of size and shape of specimen, conditions of casting concrete, moisture content and temperature of specimen, position and condition of bearing surfaces, and rate of load application.

10.17 Effect of size and shape of specimen. Most control tests to determine strength are made in compression or in flexure, although the splitting-tension test is increasing in use.

Compression. Standard compression specimens are cylindrical, except that portions of beams broken in the flexure test may be tested in compression being considered as modified cubes (ASTM C116). Standard cylinders are of height equal to twice the diameter; such a specimen is not so short that small variations in length affect the strength significantly and yet is not so long that column action is a significant factor. The common size of cylinder is 6 by 12 in. Methods of making and curing compression specimens are covered by ASTM C31 (field) and C192 (laboratory) and methods of compression testing by ASTM C39 (molded cylinders), C116 (modified cubes), and C42 (cores).

When a given concrete is tested in compression by means of cylinders of like shape but of different sizes, the larger the specimen the lower the strength. Table 10.4 gives, for various sizes of cylinder with height-diameter ratio of 2, the relative strength expressed as a percentage of that for a 6 by 12-in. cylinder, the maximum size of aggregate being ¼ in. throughout. As the size of cylinder is increased above 18 by 36 in., the change in relative compressive strength is small.

The diameter of a cylindrical specimen should be not less than three or four times the maximum size of the aggregate in order to avoid undue influence of boundary conditions and other irregularities. In testing mass concrete it is customary to screen or pick out all pieces of aggregate that would not pass a 1½ or 2-in. sieve and to test this "wet-screened" concrete in a 6 by 12-in. cylinder. Tests by the U.S. Bureau of Reclamation have determined that the compressive strength so indicated is about 30 percent higher than that of the full mass mix containing 0 to 6-in. aggregate and tested in an 18 by 36-in. cylinder [1036]. Cores of diameter 6, 8, or

Table 10.4 Effect of size of compression specimen on indicated strength of concrete*

Size of cylinder, in.	Relative strength, %	Size of cylinder, in.	Relative strength, %
2 by 4	109	12 by 24	91
3 by 6	106	18 by 36	86
6 by 12	**100**	24 by 48	84
8 by 16	96	36 by 72	82

* Adapted from Ref. 1036.

10 in. and $h/d = 2$, drilled from a mix containing 0 to 3 or 6-in. aggregate, are not significantly different in strength from 6 by 12-in. cylinders of wet-screened concrete similarly cured.

The greater the ratio of specimen height to diameter, the lower the strength indicated by the compression test. When cylinders or cores of nonstandard shape are tested, the equivalent strength of the standard specimen ($h/d = 2$) can be computed approximately by multiplying the observed compressive strength by the corresponding factor given in Table 10.5, which is applicable to damp cores of normal-weight concrete within the range 2,000 to 6,000 psi. For example, the correction factor for a cylinder having a height equal to its diameter ($h/d = 1$) is 0.91, indicating that the strength observed by testing such a specimen is about 10 percent greater than the "standard" strength ($1.00/0.91 = 1.10$).

For rough comparisons the factors of Table 10.5 may be applied to the unit compressive strength of rectangular prisms, considering the least lateral dimension of the prism as the diameter of the corresponding cylinder.

However, tests by the U.S. Bureau of Reclamation indicate somewhat lower strengths for the rectangular prisms, as shown in Table 10.6.

As h/d decreases, somewhat lower relative strengths are exhibited by air-entrained concrete and by lightweight concrete as compared with normal concrete, and by moist-cured moist-tested concrete as compared with dry-cured and autoclaved concrete. The effect of h/d is apparently affected but little by the age of the concrete.

Table 10.5 Effect of height-diameter ratio of compression specimen on indicated strength of concrete (ASTM C42)

Ratio of height to diameter of cylinder or core	Strength correction factor
2.00	1.00
1.75	0.98
1.50	0.97
1.25	0.94
1.00	0.91

Compression tests on modified cubes comprising the broken sections of flexure specimens indicate that for high-strength concretes, the strength of modified cubes is about the same as that for standard cylinders but that for low-strength concretes, the modified cubes are up to 25 percent stronger than corresponding standard cylinders [1016].

Relative strengths of mass concrete and of the portion obtained by wet screening through a 1½-in. sieve are given in Table 14.2.

Flexure. Standard flexure specimens are rectangular beams with ratio of breadth to depth of beam not exceeding 1.5; the common size of beam

Table 10.6 Relative strength of concrete cylinders and prisms

Type of specimen	Diameter, in.	Length, in.	h/d	Relative strength, %
Cylinder	6	12	2	100
	6	6	1	115
Prism	6	12	2	93
Cube	6	6	1	113

is 6 by 6 by 21 in. Methods of making and curing flexure specimens are covered by ASTM C31 (field) and C192 (laboratory), and methods of flexure testing by ASTM C78 (third-point loading), C293 (center loading; for specimens smaller than 6 by 6 in.), and C42 (beams sawed from structures).

The effect of size of beam on flexural strength of concrete is similar to that of size of cylinder on compressive strength—in general the larger the specimen the lower the indicated strength.

In accordance with ASTM C78 concrete beams are tested on the side, that is, with the load applied in a plane parallel to the top surface as cast. There is apparently little difference in indicated strength between side loading and loading with the beam in its position as cast.

The span of a beam to be tested under third-point loading is required by ASTM C78 to be, as nearly as practicable, three times its depth as tested. The common 6 by 6 by 21-in. beam is tested on a span of 18 in. The effect of span alone is not significantly large over a range of 18 to 60 in. Center loading and cantilever loading result in somewhat higher indicated strengths than those for third-point loading probably because a greater portion of the beam under third-point loading is subjected to the maximum bending moment, and more points of weakness and incipient failure may be present in this portion.

Deeper beams exhibit lower modulus of rupture than shallower beams of the same width and span.

10.18 Effect of conditions of casting. The method of molding a specimen has an important influence on indicated strength. Insufficient compaction results in pockets and reduces strength; excessive working results in segregation and likewise reduces strength. The standard method of compacting concrete specimens is by hand-rodding with a ⅝-in. round rod; in some cases mechanical vibration is employed instead.

The use of certain types of cardboard cylinder molds results in indicated compressive strengths 3 to 9 percent lower than those obtained when steel molds are used [1002, 1028], perhaps because the relatively rough walls of cardboard molds restrain settlement, or cardboard molds yield during the rodding and a lower density of concrete results, or insufficiently paraffined cardboard absorbs moisture and expands during the setting period, thus subjecting a surface layer of the specimen to tension.

10.19 Effect of moisture content of specimen. A compression specimen tested in the air-dry condition will exhibit 20 to 40 percent higher strength than that of corresponding concrete tested in a saturated condition (Fig. 10.10), probably because of (1) the greater density of dry (and therefore contracted) paste, (2) initial tensile stresses in the paste due to localized restraint of paste shrinkage by pieces of aggregate, and (3) possible development of hydrostatic pressure in saturated paste. Therefore the standard requirement that compression specimens be saturated at time of test is conservative; further, a "dry" condition is variable and indefinite, whereas a saturated condition is uniform. Cores taken from a structure are soaked at least 40 hr immediately prior to the compression test [ASTM C42].

With regard to the moisture content of flexure specimens at time of test, conflicting results have been obtained by various investigators, perhaps largely because uniform drying is difficult to attain. Thoroughly dry concrete exhibits higher flexural strength than that of moist concrete. Partly dry specimens are relatively low in strength; if the exterior of the specimen is more nearly dry than the interior, tensile stresses in the "shell" are developed through restraint of differential shrinkage. Further, localized tensile stresses are developed during drying through restraint of paste shrinkage by the aggregate. Since failure of plain concrete in flexure tests invariably occurs by tensile failure of the extreme fibers, it would be expected that the flexural strength would be reduced by drying the specimen. In any event, ASTM C192 and C42 require that the specimen be moist at time of test. Care must be taken to avoid partial drying of specimens waiting for test.

The tensile strength of moist concrete, as indicated by the indirect splitting test, is somewhat lower than that of thoroughly dry concrete. The tensile strength exhibits a straight-line relationship to the flexural strength but is not directly proportional; it ranges from about 55 percent for lower-strength concretes to about 70 percent for higher-strength concretes. It is less affected than flexural strength by moisture content at time of test.

10.20 Effect of temperature of specimen. The temperature of specimen at time of test has a marked influence on indicated strength; the higher the temperature, the lower the indicated strength. Compression tests of concrete at the University of California indicated in a typical case that the compressive strength at 25°F was 40 percent higher, and at 130°F was 15 percent lower, than that of corresponding specimens tested at 70°F. In tests at the University of Wisconsin [1030] the compressive strength of concrete at 0°F was 40 percent higher, and at 200 to 250°F about 10 percent lower, than that at 70°F. Flexure tests of mortar at the University of Texas [1025] indicated that the modulus of rupture at 40°F was 12 percent higher, and at 100°F was 20 percent lower, than that at 70°F (see also Art. 10.15).

10.21 Effect of bearing conditions. It is important that loading of concrete specimens be applied uniformly over the bearing area; otherwise the stress distribution in the specimen is other than that assumed in computing strength. Further, localized concentrations of bearing due to irregularities, or the use of capping materials which exhibit greater lateral deformation than that of the specimen, tend to produce failure by splitting rather than in compression or flexure as desired.

In order to develop the full strength of a compression or flexure specimen, the following bearing conditions should exist:

1. The specimen should be accurately centered in the testing machine, and the axis of the specimen should be vertical.
2. The bearing surfaces should be perpendicular to the axis of the specimen.
3. A spherically seated bearing block should be used and it should be accurately centered on the specimen with the center of curvature of its spherical surface in the plane of contact with the specimen.
4. The bearing surfaces should be plane.
5. If a capping material is used, it should have strength and elastic properties as near those of the specimen as possible.

However, small variations in certain of the conditions just stated are not of serious consequence. For example, it has been found [1007] that, for 6 by 12-in. cylinders, no significant difference is introduced by an error of ¼ in. in centering of the load or of the bearing block, in inclination of the axis, or in lack of parallelism of the end planes.

Variations in the end surface of a 6 by 12-in. cylinder from a true plane by as little as 0.01 in. give in some instances a reduction in strength of 35 percent [1007]. ASTM C39 requires that bearing faces of the testing equipment be maintained within 0.001 in. of a true plane, and ASTM C31 and C192 require that the ends of all compression specimens that are not plane within 0.002 in. be either capped or ground plane. If full contact between flexure specimens and the load-applying blocks and supports is not obtained, ASTM C78 requires capping, grinding, or shimming with leather strips at these points.

Compression specimens may be capped with portland cement paste at 2 to 4 hr after the time of casting, when the concrete has ceased settling in the molds [ASTM C192]. The cement for capping is mixed to a stiff paste 2 to 4 hr before it is to be used, in order to minimize the tendency of the cap to shrink. The cap is formed by means of a thick plate-glass plate (not window glass) or a machined-metal plate, worked down on the cement paste. Caps formed by this method tend to be slightly concave.

Compression specimens not capped with neat-cement paste at the time of casting, as just described, are capped prior to test. Sulfur caps are permitted regardless of the expected ultimate strength of the specimens; gypsum plaster having a demonstrated compressive strength of 5,000 psi or greater may be used for specimens expected to have an ultimate strength below 5,000 psi; neat alumina cement caps are permitted if given 18 hr to harden; and neat portland cement caps if given 3 days or more to harden. Weak or soft materials such as plaster of paris or cardboard are not permitted by ASTM Specifications.

Approximately equal strengths are obtained by grinding the bearing surfaces of compression specimens to a plane surface (within 0.002 in.) or by capping with neat cement, alumina cement, sulfur (Appendix I), or (within limits) high-strength gypsum plaster (Appendix H). Capping with plaster of paris results in a 5 percent reduction in strength of 3,000-psi concrete and a 15 percent reduction in strength of 8,000-psi concrete [1035]. Bedding the bearing surfaces with shot instead of capping reduces the indicated strength considerably. Prior to capping with sulfur or high-strength gypsum plaster, an inclination, convexity, or concavity of 3/16 in. in both end surfaces of 3 by 6-in. cylinders has no significant or consistent effect on indicated strength [1035].

10.22 Effect of rate of loading. The more rapid the "static" loading of concrete, the higher the observed strength. Within the range of customary testing procedure, however, the effect is not large, as shown in Fig. 10.16 by the results of compression tests on 6 by 12-in. cylinders at the University of Illinois [1014]. As compared with a normal rate of loading (about 35 psi per sec), loading at 1 psi per sec reduces the indicated strength approximately 12 percent, and loading at 1,000 psi per sec increases the indicated strength approximately 12 percent. Tests at the National Bureau of Standards [1039, 1040] tend to confirm this relation and extend it to loading speeds in excess of 10,000,000 psi per sec; at the upper limit of these tests the indicated compressive strength is about 80 percent greater than that obtained from corresponding specimens loaded slowly to failure (in about 30 min). As previously stated, normally a static-loading test is completed within 2 or 3 min.

The results of the tests at the University of Illinois [1014] indicate that the relation between compressive strength and rate of loading is approximately logarithmic, in accordance with the empirical relation

$$S = S_1(1 + k \log R)$$

in which S = strength at a given rate of loading R psi per sec
S_1 = strength at a rate of 1 psi per sec
k = a constant, about 0.08 for 28-day tests

By this relation, the indicated compressive strength of a specimen loaded at 100 psi per sec, thus tripling the rate, would be about 3 percent

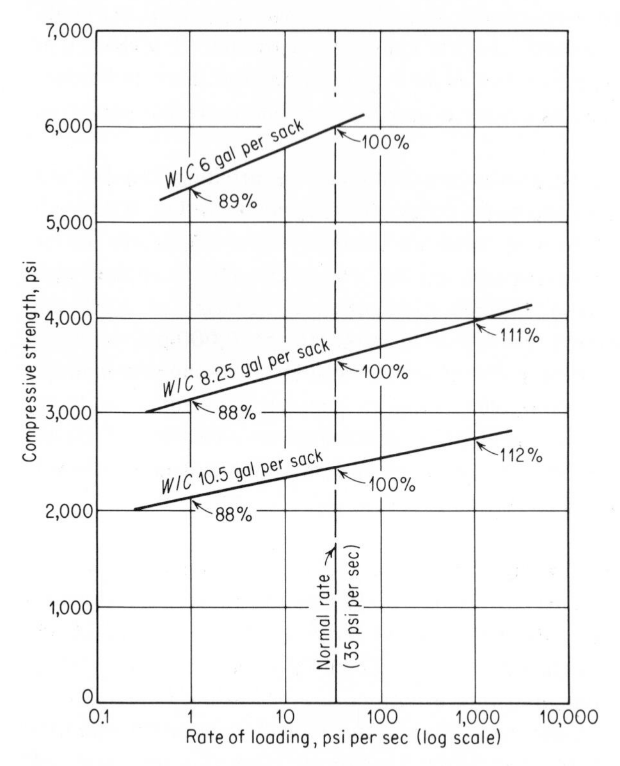

Fig. 10.16. Effect of rate of loading on 28-day compressive strength of 6 by 12-in. concrete cylinders [1014].

greater than that of a specimen loaded at the normal speed of 35 psi per sec.

The effect of rate of loading on flexural strength is somewhat greater in proportion than that for compressive strength. The basis of loading rate differs as between hydraulic and screw-gear testing machines. For hydraulic machines, it is customary to adjust the pumping rate as necessary to increase the total load at a constant rate per unit of time. This cannot be done with screw-gear machines, however. A fixed setting of a screw-gear machine results in a variable increase in load per unit of time; the rate of increase depends on the characteristics of the driving motor, the rigidity of the parts of the testing machine, the size of specimen, and the modulus of elasticity of the concrete. For screw-gear machines it is customary to express testing speed in terms of the rate of travel of the movable head when the machine is running idle.

Based on these considerations, ASTM has prescribed roughly equivalent loading rates for the two types of machine, even though it is not possible to secure exactly comparable rates. In each case the load may be applied more rapidly up to approximately 50 percent of the expected breaking load, since this timesaving procedure has been found not to affect the results appreciably. ASTM C39 requires that the load on compression specimens in hydraulic machines be applied at a constant rate within the range 20 to 50 psi per sec (normally 35 psi per sec is employed), and in screw-gear machines at an idling speed of 0.05 in. per min. For flexure specimens, the prescribed loading rate in hydraulic machines is such that the increase in extreme fiber stress does not exceed 150 psi per min and in screw-gear machines is at an idling speed of 0.05 in. per min [ASTM C293].

In the splitting test for tensile strength of concrete (Art. 10.6), ASTM C496 specifies that the loading be at a constant rate within the range 100 to 200 psi per min splitting tensile stress. For a 6 by 12-in. cylinder the corresponding load range is 11,300 to 22,600 lb per min.

When strain measurements are to be made on a specimen, usually the loading rate is approximately half of that prescribed for tests to determine strength alone.

QUESTIONS

1. With respect to age, how does the strength of concrete differ from that of other construction materials?

2. How does strength correlate with other properties of hardened concrete? Why are strength tests commonly made?

3. What is the nature of "compression" failure of axially loaded concrete?

4. What range of compressive strength of concrete would be considered appropriate for general building construction? Pavements? Large dams?

5. Roughly, what is the ratio of compressive strength of concrete to tensile strength? To flexural strength?

6. What is the principal advantage of the splitting test for tensile strength of concrete?

7. What is the principal factor affecting bond strength at initial slip? How does bond strength correlate with compressive strength?

8. On what relationship between properties is the use of the Schmidt test hammer based?

9. Arrange the five standard types of portland cement in descending order of compressive strength of concrete at the age of (*a*) 7 days and (*b*) 1 year.

10. With given cement content and consistency, will concrete containing crushed coarse aggregate be stronger or weaker than concrete containing rounded coarse aggregate? Explain.

11. What are the pros and cons of using a larger maximum size of aggregate?

12. What factors should be considered in deciding on the excess of average concrete strength specified for the job over the strength assumed in design?

13. What is an "average" value of 28-day strength of concrete (*a*) in compression? (*b*) In flexure?

14. When concrete dries out, what is the immediate effect on strength? The continuing effect? The effect of rewetting?

15. Discuss the effect on strength of curing concrete at constant temperature (*a*) higher than, (*b*) equal to, and (*c*) lower than, that of the temperature at the time of casting.

16. In deciding on a factor of safety for concrete for a given purpose, roughly what quantitative relationship must be taken into account if the load is to be sustained steadily? Repeated many thousands of times?

17. What assumption is commonly made with regard to the relation of strength of concrete in a structure to that of test specimens? Is this assumption justified? Why are the two not directly comparable?

18. What are the purposes of standard-cured specimens? Field-cured specimens?

19. What are the qualitative effects on indicated compressive strength of size of concrete specimen? Slenderness? Moisture content? Temperature at test?

20. What variations in bearing conditions of concrete compression specimens are important? What are minor?

21. Discuss the effect of rate of loading on indicated compressive strength of concrete, with approximate values.

22. Why are rates of loading prescribed differently for hydraulic and screw-gear testing machines? What is the standard rate for 6 by 12-in. concrete cylinders tested in each type? Beams tested in each type?

ELEVEN
PERMEABILITY AND DURABILITY

PERMEABILITY

11.1 Pore structure of concrete. Concrete is inherently porous, as not all the space between the aggregate particles becomes filled with a solid cementing material. To obtain workable mixes it is necessary to use much more water in concrete than is required for hydration of the cement. Furthermore, the absolute volume of the cement and water gradually decreases as chemical combination proceeds, and this makes it impossible for a cement paste of any water-cement ratio to continue to occupy completely the space originally required by the fresh paste; consequenty, the hardened paste develops some voids. In addition, during the mixing of concrete some air is always entrapped in the mass.

As the water and air voids in concrete are generally interconnected, concrete is inherently pervious to water. This is evidenced by its absorption of water by capillary action and by the passage through it of water under pressure. While absorption and permeability may permit disintegrating agencies to damage the concrete or shorten its life, fortunately it is not difficult to obtain concrete that is sufficiently watertight for all practical purposes if materials of good quality are used in proper proportions, if the concrete is well mixed and compacted in place, and if adequate curing is provided.

The porosity of concrete is largely developed during the setting period. Settlement of the solid particles causes the water to rise and form many water channels. Some of the water is trapped below the aggregate particles, and some fills the fine interstices among the cement particles. Hydration of the cement produces a gel which decreases the size of these water voids and increases the watertightness of the concrete (see Art. 2.16) but the voids are never eliminated completely. It is evident that thorough curing is necessary to secure watertight concrete.

11.2 Significance to permeability. The use of concrete in many kinds of hydraulic structures as well as in other construction fields has impressed upon engineers that in some cases watertightness of concrete may be of greater importance than strength. This is not due to any serious loss of water through percolation, but instead to the need for preventing (1) disintegration which results from the freezing of saturated porous concrete, or (2) slow weakening through the dissolving of slowly soluble components. Many structures today bear evidence of the destructive effect of the freezing of permeable concrete or have unsightly surface deposits of calcium carbonate and other compounds resulting from the seepage of water through a defective area.

In general, other conditions being the same, low permeability is associated with high strength and high resistance to weathering.

Permeability tests of concrete are of value (1) to determine the rate of leakage through the walls of a structure such as a concrete pipe; (2) to determine the effects of variations in the cement and aggregate or the effects of various operations in mixing, placing, and curing; (3) to estimate the relative durability and life of concrete as affected by the corrosive action of percolating waters; (4) to determine basic information on the internal pore structure of concrete, which is related directly to such items as absorption, capillarity, resistance to freezing and thawing, shrinkage, uplift, etc.; and (5) to compare the efficiencies of waterproofing materials [110].

11.3 Permeability tests. In determining the permeability under pressure, the inflow method or the outflow method may be used. In either case a cylindrical specimen is sealed at its curved surface inside a suitable metal container, so that water under pressure may be applied to the top flat surface only. The bottom surface is often protected against variations in humidity of the air. A vertical water pipe and water gage is connected to the top of the container, and compressed air is used to maintain a constant pressure on the water column. The inflow is obtained from readings on the water gage.

For the outflow method the apparatus is arranged to collect the effluent from the free surface and to prevent evaporation losses. The outflow will be less than the inflow during the early part of any test, because of absorption, but eventually they will agree.

As water under air pressure readily absorbs some air which would be released inside the concrete specimen as the pressure drops and thus retard the flow of water through the specimen, some provision should be made to replace the column of water as frequently as necessary to prevent the release of air within the concrete.

Tests have shown that the leakage can be determined from D'Arcy's law for viscous flow

$$\frac{Q}{A} = K_c \frac{H}{L}$$

where Q = rate of flow, cu ft/sec

A = area of cross section under pressure, sq ft

$\frac{H}{L}$ = ratio of head of fluid to percolation length

K_c = permeability coefficient for concrete

11.4 Factors affecting watertightness. As shown in Fig. 11.1, the several factors affecting the permeability of concrete fall into three groups:

1. The influence of the constituent materials
2. The effect of methods of preparing the concrete
3. The influence of subsequent treatment of the concrete

In general, other things being equal and for normal concrete, any factor tending to improve the compressive strength of the concrete will have a beneficial effect upon the watertightness. Therefore, the better the quality of the constituent materials, the less permeable the concrete.

11.5 Effect of water and cement. The water-cement ratio and the consistency of concrete are so interrelated that their effects must be considered together. For plastic workable mixes the permeability increases with the water-cement ratio as shown in Fig. 11.2. A water-cement ratio of not more than about 6 gal of water per sack of cement has been recommended for use in thin sections, and one of not more than 7 gal per sack for more massive structures [100]. As dry mixes do not consolidate very readily, more water is required for minimum permeability than for maximum strength. For hand-rodded concrete the permeability increases when the water is reduced below that which will produce a slump of about 2 to 3 in. [1101].

Permeability decreases as the cement-voids ratio increases, and this relationship appears to be more definite than that between permeability and water-cement ratio [1101].

With well-cured concrete and the optimum quantity of mixing water, an increase of cement content above that in a 1:2:4 mix does not materially affect permeability. However, very wet consistencies require a richer mix and tend to produce water gain under the aggregate particles, which increases the permeability.

Greater fineness of the cement improves watertightness in the same way that it improves strength and durability of the concrete.

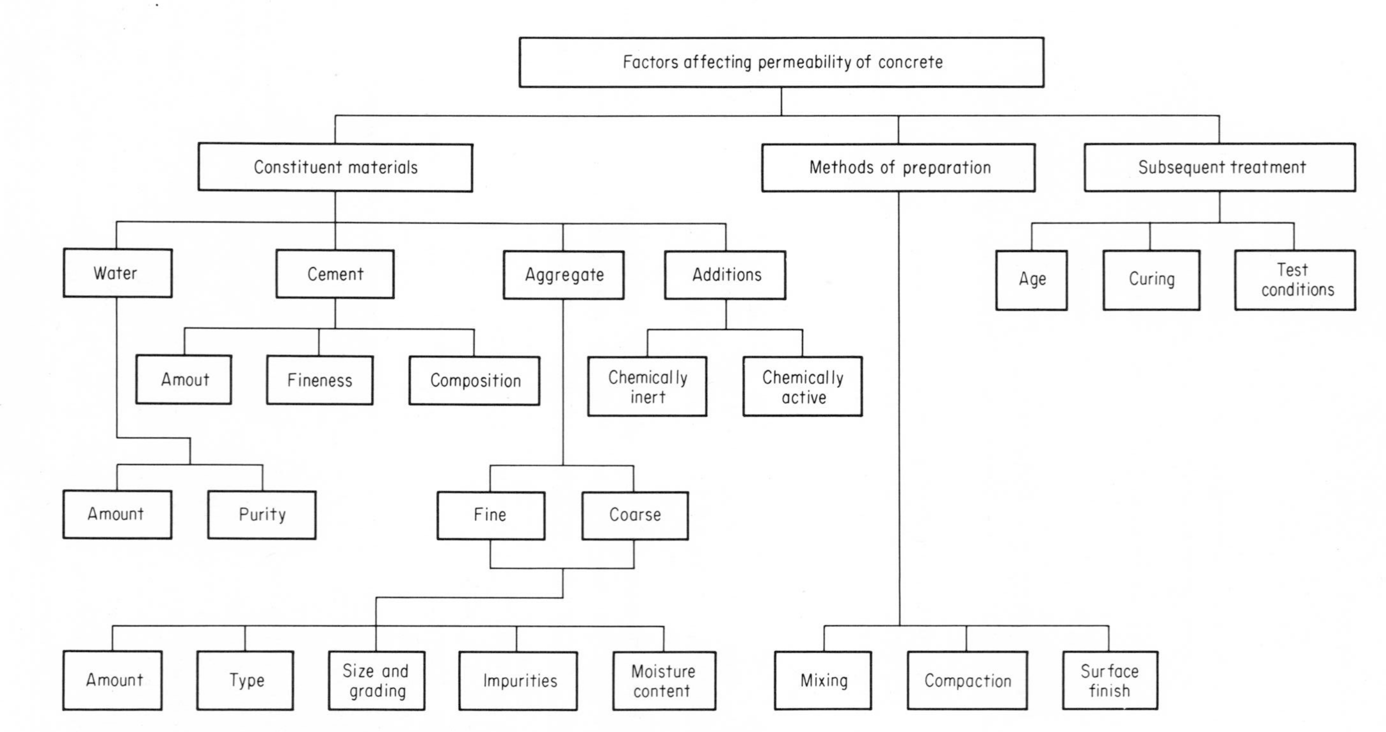

Fig. 11.1. Factors affecting permeability of concrete. (*From Glanville* [1102].)

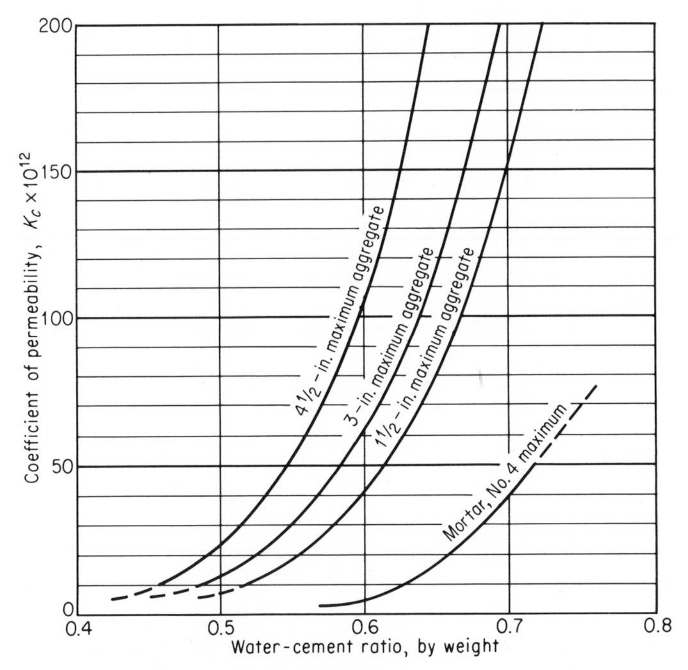

Fig. 11.2. Effect of water-cement ratio upon permeability. (*From Ruettgers, Vidal, and Wing* [1103].)

11.6 Effect of aggregates. As shown in Fig. 11.2, the greater the maximum size of aggregate for a given water-cement ratio, the greater the flow, probably because of the relatively large water voids developed on the underside of the coarser aggregate particles. Aggregates should be sound and of low porosity. A well-graded aggregate is even more important from the standpoint of watertightness than it is from the standpoint of strength, as dense concrete is essential. Sufficient fines must be used, but the mix should not be oversanded.

11.7 Effect of curing. It has been stated that continued hydration of the cement results in gel development which reduces the size of the voids and increases the watertightness of the concrete. Figure 11.3 shows the very great increase in watertightness with curing, the change being even greater than the increase in strength with curing.

11.8 Effect of admixtures and coatings. Various materials are sold for use as admixtures to improve the watertightness of concrete, and some "waterproof" cements are sold for the same purpose. Results of tests on lean-mass concretes containing pozzolans indicate increased resistance to the flow of water when these finely ground materials are used [106]. Al-

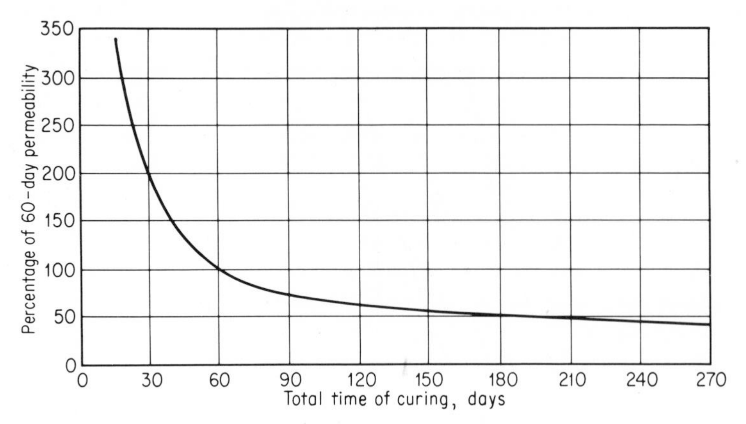

Fig. 11.3. Effect of curing period on permeability. (*From Ruettgers, Vidal, and Wing* [1103].)

though some other materials may be effective, particularly in lean mixes, many produce no beneficial effect. In general, except for pozzolans, the use of extra cement will usually be more effective than the use of other waterproofing admixtures, and the extra cost for an equal effect will usually be less [1100, 1105, 1106].

Water-pressure tests on concrete containing entrained air show that the permeability of concrete is not appreciably affected by entrained air in the percentages ordinarily used on concrete construction if the water-cement ratio remains unchanged [106].

Many surface treatments are effective in reducing leakage through porous concrete when applied to the face subjected to water pressure. The principal ones can be divided into the following classifications:

1. Fabric membranes cemented to the concrete with hot asphalt
2. Asphaltic emulsions
3. Cement plasters, adequately cured
4. Paraffin, silicone compounds, etc. dissolved in volatile solvents
5. Inert fillers in an alkyd resin vehicle

Of these surface treatments the first two are most commonly used when their black color is not objectionable. The other treatments have been most used when a good appearance must be maintained.

11.9 Uniformity of concrete. The permeability test is very sensitive. Minor defects or nonhomogeneous conditions in concrete that would have no appreciable effect on compressive strength influence the leakage through the specimen to a marked degree.

Probably the majority of leaks in concrete structures are due either (1) to defects such as cracks in the structure or (2) to void spaces in the concrete caused by honeycombing or segregation of the constituent materials rather than to inherent porosity of the cement paste or aggregate. To prevent leaks, the mix should be workable so that segregation and honeycomb can be avoided in placing the concrete. The effect of vibratory compaction in reducing the permeability of concrete is shown in Fig .11.4. Care should be taken so that excessive water and laitance do not accumulate at the surfaces of fresh masses and that good bond is obtained between successive lifts. Contraction joints should be designed, if necessary ,with flexible water stops.

11.10 Absorption. Absorption is a physical process by which concrete draws water into its pores and capillaries. In one method of test the concrete is submerged in water for 24 hr, surface-dried, weighed, oven-dried, and weighed again. The difference in the soaked and dry weight is taken as the absorption. It is usually expressed as a percentage of the dry weight. In a second method the concrete is first oven-dried, weighed, boiled in water for 5 hr, cooled in water, and weighed after surface-drying. The percentage absorption is computed as in the first method. The results of these two procedures do not agree, as the loss in weight of a saturated specimen in drying is roughly 5 to 10 percent greater than the absorption after drying [1108].

In both procedures the drying operation withdraws not only the mechanically suspended water but also some of the colloidal water more tenaciously

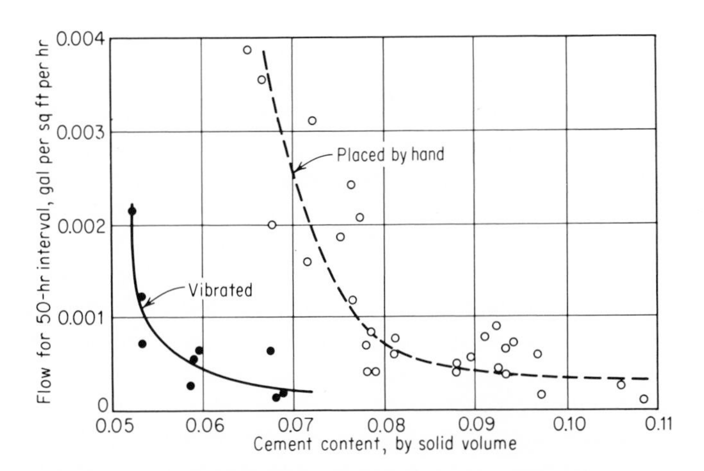

Fig. 11.4. Permeability of hand-placed and vibrated concrete. Slump 2 to 4 in.

held in the cement gel. Hence the absorptions, or porosities indicated by the absorptions, are larger than for the usual temperature-humidity environment of the concrete.

The absorption is considered to be related to the resistance of concrete to weathering, since if no water entered its pores, there would be little or no disintegration caused by freezing or thawing or by aggressive waters. However, actual tests show no reliable correlation of *total* absorption with durability of concrete, but some correlation appears to exist between the *rate* of absorption and durability [110].

DURABILITY

11.11 Deterioration of concrete. The useful life of concrete may be markedly reduced by the disintegrating effects of (1) weathering, including the disruptive action of freezing and thawing and the differential length changes due to temperature variations and alternate wetting and drying, (2) reactive aggregates, (3) aggressive waters in alkali regions, (4) leaching in hydraulic structures, (5) chemical corrosion, and (6) mechanical wear or abrasion. These subjects are covered in some detail in Ref. 1120 by ACI Committee 201.

11.12 Weathering. In severe climates, resistance to weathering is an important consideration. There is, however, no quantitative index or direct laboratory test for weathering resistance. Various accelerated tests to simulate weathering action have been devised; most of these involve alternate cycles of freezing and thawing [1141].

It is considered that the destruction of concrete by freezing is caused by hydraulic pressure developed by the expansion of water when converted to ice, this expansion being about 9 percent. When under pressure, water does not freeze until the temperature is somewhat below the normal freezing temperature of 32°F, but it will freeze even at pressures as great as 30,000 psi provided the temperature drops to about −8°F [1196]. The disruption of concrete by freezing is contingent upon the presence of sufficient water in the void spaces so that high internal pressures will develop, because when the concrete is not sufficiently saturated the expansion of the ice will merely force some of the water into inner voids. Tests have shown that the higher the moisture content of the concrete, the fewer the cycles of freezing and thawing required to rupture the mass, so that when nearly fully saturated an ordinary concrete will disintegrate completely in less than 5 cycles [1141]. But ordinary concrete has many air-filled cavities that cannot readily be filled with water even upon immersion in water. These cavities are entrained air bubbles, pores in the aggregates, and thin fissures under the larger aggregates. The fissures, formed during the period of

Fig. 11.5. Disintegration of concrete at water line. (*From Lamprecht* [1195].)

bleeding, are water-filled in the fresh concrete but may become partially or wholly emptied as hydration proceeds. The pressures developed during freezing are probably more than sufficient to force water into the various cavities, so that after repeated water immersion and freezing a large part of these cavities become filled. A few additional cycles of freezing and thawing, after a critical degree of saturation is reached, cause rapid disintegration of the concrete. The effects of such action are shown in Fig. 11.5.

11.13 Weathering resistance as affected by aggregate, cement, and water. In an average concrete the solid volume of the fine and coarse aggregate constitutes about 70 percent of the total, and the quality of this aggregate plays a very important part in the durability of the concrete.

Aggregates that are readily cleavable and structurally weak or those which are very absorptive and which swell when moistened are subject to disintegration upon exposure to ordinary weathering conditions. Such aggregates should not be used, as they produce an unsound concrete. Examples of the rocks which may be undesirable for this reason are shales, clayey rocks, friable sandstones, various cherts and some micaceous materials.

Physical soundness of the aggregate is commonly tested by soaking it repeatedly in a saturated solution of sodium sulfate or magnesium sulfate, each immersion being followed by thorough drying. The crystallization of the salt carried in solution into the aggregate particle causes an expansive force which simulates, in an accelerated manner, the disruptive force to which the aggregate is subjected in service by freezing of absorbed water. Aggregates are sometimes tested by being subjected to repeated cycles of freezing and thawing (although tests have shown no dependable correlation between the results of soundness tests upon aggregate and the durability of the resulting concrete [1132]), or by preparing concrete specimens and subjecting them to similar cycles. These tests definitely indicate as unsatisfactory certain aggregates that in the past have caused unsatisfactory concrete structures, but they may be unreliable for borderline cases. By far the best information relative to the soundness of an aggregate is obtained from service histories of concrete structures in which the same aggregate was used, taking care to recognize the degree of exposure to which the structures have been subjected.

Temperature changes alone have been found responsible for unsatisfactory service records of some concretes, particularly when the changes are rapid, tending to set up large differential strains between the surface and interior of the mass. Certain aggregates having especially low coefficients of thermal expansion have given poor service at low temperatures, causing high tensile stresses in the matrix of cement paste, resulting in its crazing and cracking [1155].

The resistance to freezing and thawing of concretes hardened from plastic mixes markedly increases as the water-cement ratio is reduced. Table 11.1 shows three series of freezing and thawing tests on 3 by 6-in. mortar cylinders. For series A having constant mix proportions and increasing amounts of mixing water to produce slumps from 0 to 6 in., the durability shows a progressive reduction as the water-cement ratio and unit water content are increased. Although groups 1 and 2 made of nonplastic mortars having low water-cement ratios and high air contents are deficient in strength, as would be expected, they have excellent resistance to freezing and thawing. In series B the three different mix proportions had the same slump and the same unit water content. For these mixes the durability and strength decrease as the water-cement ratio increases. For series C the same mix proportions were used as in series B, but the water-cement ratio was maintained at 0.51, which caused variations in their unit water contents and slumps. This series also shows that the durability is reduced for the higher water contents. Table 6.5 shows that for durable concrete high water-cement ratios can be used only for service at mild temperatures or for massive structures.

In practice there is a tendency for water gain to cause higher water contents in the upper portions of a lift, with the result that disintegration caused by freezing and thawing is usually more serious at the top of a

Table 11.1 Durability of concrete*

Series	Group	Durability	Cycles of freezing and thawing to produce		Volume of air and water, %			Volume ratio, air/water	Mix by wt	Slump, in.	W/C by wt	Cement, pcf	Water, pcf	Compressive strength at 90 days, psi
			0.015-in. expansion	25 percent wt loss	Air	Water	Total							
A	1	Good	360	380	10.5	15.7	26.2	0.67	1:2.75	0	0.29	33.8	9.8	3,300
	2	to	150	240	5.6	21.5	27.1	0.26	1:2.75	0	0.40	33.4	13.4	6,000
	3	poor	104	90	3.0	26.3	29.3	0.11	1:2.75	$1\frac{3}{4}$	0.51	32.2	16.4	6,600
	4		83	70	0.1	30.6	30.7	0.00	1:2.75	$4\frac{5}{8}$	0.60	31.8	19.1	5,700
	5		75	60	0.0	34.1	34.1	0.00	1:2.75	6	0.70	30.3	21.2	4,600
B	6	Good	140	144	3.4	26.3	29.7	0.13	1:2.25	$1\frac{3}{4}$	0.44	37.4	16.4	6,600
	7	to	108	100	3.7	26.3	30.0	0.14	1:2.75	$1\frac{3}{4}$	0.51	32.1	16.4	6,200
	8	poor	82	74	4.3	26.3	30.6	0.16	1:3.25	$1\frac{3}{4}$	0.59	27.9	16.4	4,400
C	9	Good	142	150	6.1	23.2	29.3	0.26	1:3.25	0	0.51	28.4	14.5	5,400
	7	to	108	100	3.7	26.3	30.0	0.14	1:2.75	$1\frac{3}{4}$	0.51	32.1	16.4	6,200
	10	poor	80	80	0.7	30.1	30.8	0.02	1:2.25	$4\frac{7}{8}$	0.51	36.8	18.8	5,900

* From Ref. 106.

The 3 by 6-in. mortar specimens (No. 4 maximum aggregate) were fog-cured 28 days at 70F°, dried 3 days at 120F°, and then soaked 3 days at room temperature (for absorption tests) before undergoing first freezing for 80 min. The specimens, immersed in water-filled rubber bags surrounded by 5°F brine, reached about 15°F at the center. Thawing required another 80 min in 70°F running water. After each 10 cycles of freezing and thawing the 3-day drying and 3-day soaking operations were repeated.

lift than elsewhere. This can be prevented by lowering the water content of the mix (still maintaining adequate workability) as the top of the lift is approached.

The kind of cement used in concrete affects its weathering characteristics largely because of its effect upon the water requirement to produce a given workability of the concrete, and upon its bleeding tendencies. Both these factors influence the porosity of concrete, which leads to its disintegration when water in the voids is frozen.

11.14 Air-entrained concrete. It has been shown [1147, 1148] that the resistance of concrete to disintegration by frost action and by the application of salts for ice and snow removal may be greatly increased by the use of minute quantities of admixtures which will increase the amount of very small disconnected bubbles of air or other gas entrapped in the concrete while it is plastic. These bubbles, having diameters of 0.003 to 0.05 in., provide void spaces which tend to relieve any forces which are developed by the formation of ice and thus prevent disintegration of the concrete.

The beneficial effect of air-entraining agents in greatly improving the resistance of concrete to freezing and thawing is shown in Fig. 11.6. When the air content was increased only a small amount over that for the plain concrete, which generally contained less than 1 percent, a very considerable benefit was obtained. This is indicated by the small reduction in modulus of elasticity, low expansion, and low loss in weight by surface disintegration for air contents only slightly higher than for the plain cements. Concretes containing only about 3 percent of air showed about as good resistance as those with more air.

The lower water-cement ratio obtainable in air-entrained concrete has a beneficial effect upon strength and durability of the concrete and results in less excess water present to produce passageways through the cement paste, thereby reducing permeability and bleeding.

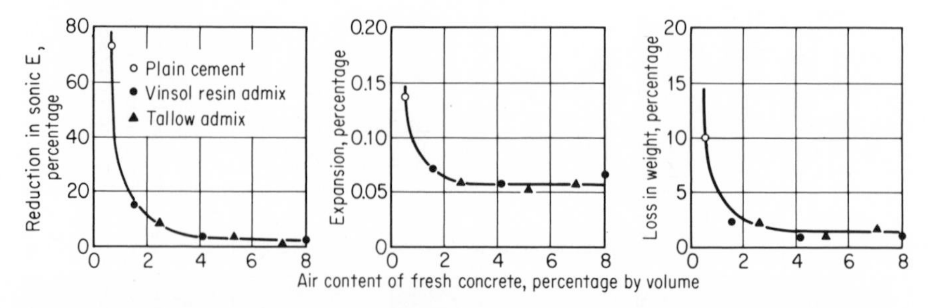

Fig. 11.6. Freezing and thawing test results for concretes containing plain and air-entraining cements. Plotted values are for 225 cycles of freezing and thawing. Concrete contained 6.3 sacks cement per cu yd; slump 2 to 4 in. (*From Gonnerman* [1147].)

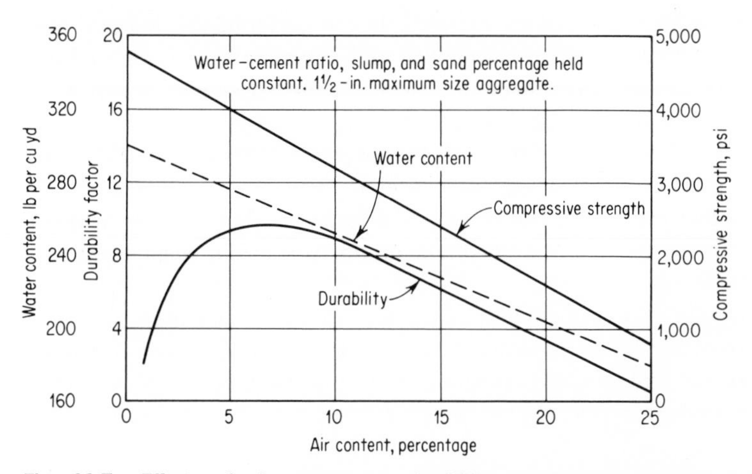

Fig. 11.7. Effects of air content on durability, compressive strength, and required water content of concrete. Durability increases rapidly to a maximum and then decreases as the air content is increased. Compressive strength and water content decrease as the air content is increased [106].

An increase in the entrained air of concrete usually causes reductions in the strength. For each 1 percent increase in the air content the reductions may be 2 to 3 percent in the modulus of rupture, 3 to 5 percent in the compressive strength, and 3 percent in the modulus of elasticity. Considering strength as well as resistance to freezing and thawing, the most satisfactory results will be obtained with about 4 to 5 percent of air when using a maximum aggregate size of 1½ in. With this air content there may be a 7 percent loss in flexural strength and a 12 percent loss in compressive strength. As shown in Fig. 11.7, with higher air contents there will be large losses in strength without any commensating increase in durability [106, 1147].

11.15 Freeze-thaw tests. Accelerated laboratory weathering tests involving freezing and thawing are costly, and are difficult to standardize because of the many possible variables in the test procedure. These variables include (1) type of specimen and condition of concrete at time of test, (2) method of exposure to freezing and thawing, and (3) method of evaluating the results of the test.

The larger the test specimen, the greater the temperature differential between the inner and outer concrete upon exposure to low temperatures. This causes thermal stresses even before freezing begins. However, large specimens are more difficult to saturate, and unless the pores are nearly filled with water, freezing temperatures are not very destructive.

This type of test must be accelerated, but how accelerated it may be and still be significant is not known. In the performance of tests, both the rapidity of freezing and thawing and the actual temperatures of these two conditions are important factors. Whether the specimens are in water or air at time of freezing is significant.

There are four ASTM Standard Procedures [C290–C292, C310] for conducting freezing-and-thawing tests which differ in the rapidity of the freezing-and-thawing cycle (2 to 4 hr, or 18 to 24 hr), the thawing temperature (40°F or 73°F), and the condition of the specimens when frozen (in air, water, or brine).

The results of freezing-and-thawing tests have usually been shown by one or more of the following factors: (1) loss in weight, (2) loss in compressive or flexural strength, (3) expansions of the specimens, (4) condition by visual examination, and (5) reduction in sonic modulus of elasticity (Art. 13.2).

Loss in weight has probably been most commonly used in the past but is now commonly replaced in many laboratories by observations of sonic modulus. Loss in weight is reasonably satisfactory where a gradual loss occurs from the corners and surface of the specimen but is of little value where internal failure occurs as a result of disruption of the coarse aggregate or by the complete disintegration of the mortar mass. The loss in weight is much greater when the freezing specimen is in contact with water than when it is in contact with air.

Loss-in-strength determinations are costly because a large number of specimens are required. Furthermore, compression tests are not well suited to show the degree of internal failures developed, although flexure tests are satisfactory in this respect.

Expansions of durability specimens satisfactorily indicate their condition, but such observations are restricted to specially molded specimens having gage points cast in place. In general, expansions occur more slowly than significant reductions in sonic modulus.

Visual examinations are valuable in conjunction with other types of observations for evaluating the effects of durability tests, but the difficulty of reporting such observations is evident.

Reduction in the sonic modulus of elasticity appears to be the best method of evaluating durability tests. It indicates the effects of disintegration at earlier ages than the other methods, and tests have shown good correlation between changes in flexural strength and changes in the sonic modulus. Figure 11.8 shows the behavior of a good and a poor concrete in such a test. The poor concrete suffers an appreciable reduction of modulus much sooner than the good concrete.

In one ASTM method [C292] the test is continued until the specimens have been subjected to 100 cycles in brine or 200 cycles in water (but to 300 cycles in the other three methods), or until the dynamic modulus

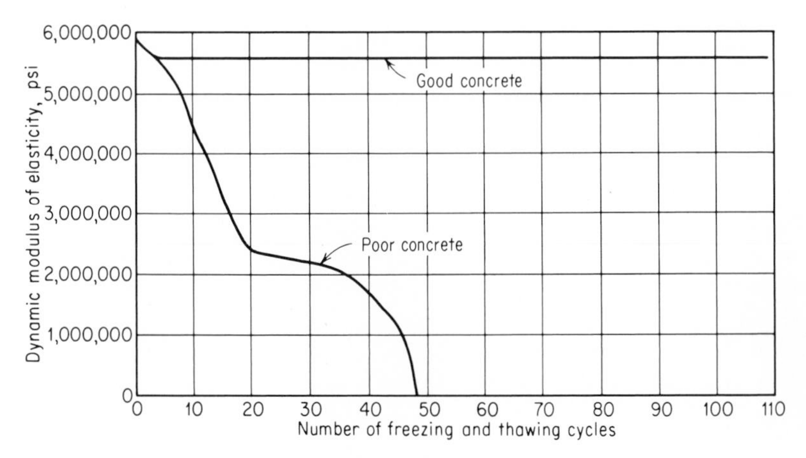

Fig. 11.8. Durability of concrete as determined by its dynamic modulus in a freezing-thawing test. *(From Thomson* [1309]·

of elasticity reaches 60 percent of the initial modulus. The durability factor is then determined from the equation

$$\mathrm{DF} = P\frac{N}{M}$$

where DF = durability factor of specimen
P = relative dynamic modulus of elasticity at N cycles, % of initial modulus
N = number of cycles at which P reaches specified minimum value for discontinuing test or specified number of cycles at which exposure is to be terminated, whichever is less
M = specified number of cycles at which exposure is to be terminated

For example, if $P = 60$, $N = 180$, and $M = 200$, then

$$\mathrm{DF} = 60 \times {}^{180}\!/_{200} = 54$$

An ultrasonic method of studying deterioration and cracking in concrete structures, *in situ,* has been used successfully in the field [1313].

11.16 Reactive aggregates. An expansive reaction between certain types of aggregates and high-alkali cements has been found responsible for the random cracking and disintegration of the concrete in many structures as shown in Fig. 11.9. Not only is the cracking unsightly, but it is indicative of a weakened structure. In dams, the binding of gates and other machinery embedded in concrete have been traced to the excessive expansion which results from such reactions.

Fig. 11.9. Severe cracking of bridge pier caused by reactive aggregates.

Several types of aggregates are known to react with high-alkali cement (see Arts. 2.16 and 3.15). Opaline silica (amorphous, hydrous silica), which is often present in many different kinds of rocks and may form coatings or incrustations on sand or gravel particles, has been responsible for much disintegration. Other reactive aggregates include siliceous limestones and highly siliceous rocks such as chalcedony and some cherts; certain volcanic rocks such as some andesites, dacites and rhyolites; and miscellaneous rocks such as phyllite [1167, 1168]. Some carbonate aggregates and some unstable mineral oxides, sulfates or sulfides have been identified as causing a damaging expansive reaction. Although many kinds of aggregates contain small amounts of undesirable reactive materials, it is not known how much of such materials must be present to produce an adverse reaction with cements high in alkali. Therefore, if field performance records are not available for a particular aggregate, petrographic and chemical examinations and mortar bar expansion tests should be made.

Experience with reactive aggregates in the field and in the laboratory has shown that such materials in combination with high-alkali cements cause deterioration of the concrete, but usually little or no damage is caused when the cements have alkalies, computed as $Na_2O + 0.658K_2O$, of less than 0.5 to 0.6 percent. However, even this limitation may be insufficient to prevent deterioration with some aggregates.

The conditions of exposure are important factors in the development and progress of the reaction. Many structures in which reactive combinations are used show early distress in portions exposed to adverse weathering conditions, whereas protected portions of the same structure fail to develop appreciable distress, if any.

The ASTM has developed two physical tests for determining the potential volume changes resulting from cement-aggregate combinations [C227, C342], and also has developed a chemical test [C289] for determining the potential reactivity of aggregates (see Art. 3.15).

Tests have been made of a number of finely divided pozzolanic minerals for use as correctives by combining with the alkalies while the concrete is still plastic, thus reducing their concentration and preventing later expansive reactions within the hardened concrete [1154]. Some test data indicate that the amount of pozzolanic material required is about 20 g of finely divided reactive silica per gram of alkali in the cement in excess of 0.5 percent [1171]. So far the only materials found to be reliably corrective when of such fineness as to pass the No. 200 sieve and used in practicable amounts are the opaline cherts, fly ash, diatomite, blast-furnace slag, calcined shale (see Table 11.2), Pyrex glass, and certain other active siliceous materials. In some cases pumicite has been found effective in reducing the expansions of reactive combinations, as shown in Fig. 11.10.

But pozzolanic materials may cause increased drying shrinkage in con-

Table 11.2 Effect of calcined Monterey shale on expansions of reactive cement-aggregate combinations*

		Expansion in 40 weeks at 70°F, %	
Cement no.	**Alkali, %**	**No admixture**	**Calcined Monterey shale**
1	0.47	0.067	0.014
2	0.47	0.168	0.015
3	0.51	0.118	0.016
4	0.62	0.255	0.011
5	1.17	0.230	0.086

* From Ref. 1167.

1:2.25 mortar mix; calcined Monterey shale equal to 20 percent by weight of cement substituted for equivalent amount of aggregate.

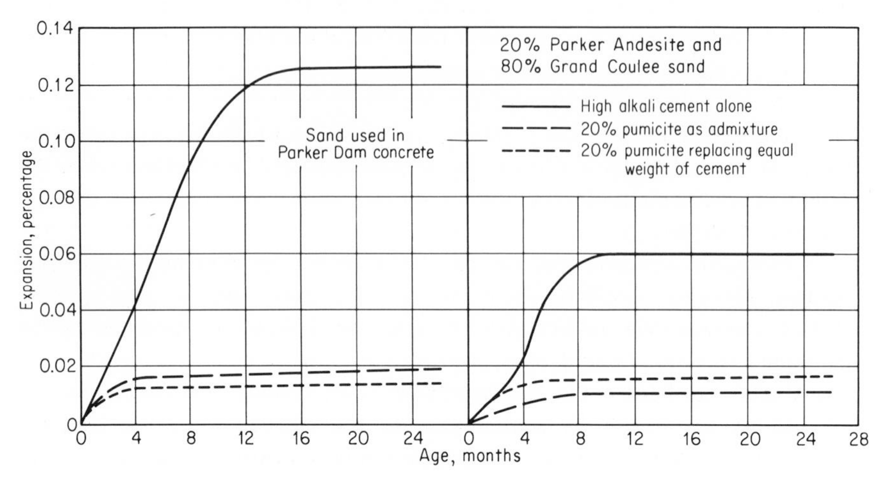

Fig. 11.10. Effect of pumicite on expansions of two reactive cement-aggregate combinations. (*From Blanks and Meissner* [1167].)

crete which is exposed to unusual drying conditions. Also, the rate of development of strength in pozzolanic concrete may be seriously reduced under these conditions.

11.17 Sulfate waters. The sulfates of sodium, potassium, and magnesium, present in alkali soils and waters are known to have caused deterioration of many concrete structures. The sulfates react chemically with the hydrated lime and hydrated calcium aluminate in the cement paste to form calcium sulfate and calcium sulfoaluminate, respectively, these reactions being accompanied by considerable expansion and disruption of the concrete. Alkali solutions in soils increase in strength in dry seasons, as dilution is then at a minimum, and the stronger the concentration of these salts, the more rapid the disintegration. However, continued exposures to concentrations as low as 0.1 percent may be harmful to concrete.

The deposition of sulfate crystals in the pores of concrete also tends to disintegrate concrete. Alkali waters entering concrete may evaporate and deposit their salts. The growing crystals resulting from alternate wetting and drying may eventually fill the pores and develop pressures sufficient to disrupt the concrete.

Disintegration of the first type is best prevented by the use of sulfate-resisting cement (ASTM type V). As shown in Tables 2.3 and 2.7, the C_3A and C_4AF are both relatively low in this cement, the sum of the aluminates being lower than for any of the other types. This combination of low C_3A and C_4AF and accompanying high C_3S and C_2S produces a cement having exceptional resistance to sulfate attack in comparison with other types of cement. A modified cement (ASTM type II) having less than 8 percent C_3A shows the next best resistance to sulfate attack.

The effect of the C_3A content on the resistance to sulfate attack is well shown in Fig. 11.11, which covers the results of an investigation on concrete cylinders partly buried in aggressive alkali soil. All the normal cements developed complete disintegration within 1 to 2 years, but the modified cements having lower C_3A contents resisted disintegration for longer periods [1174a].

The relative degrees of attack on normal concretes by sulfates in soils and ground waters are shown in Table 11.3.

Resistance to disintegration by crystal growth is best obtained by use of a dense, imprevious concrete of relatively low water-cement ratio, and preferably containing entrained air.

For precast products the resistance to sulfate attack, as evaluated in 10 percent $(NH_4)_2\ SO_4$ solution, has been improved by steam curing and further improved by subsequent carbonation for about 24 hr after drying at 100°F for 2 to 8 hr [1179]. The duration of drying and carbonation for effective protection depends upon such variables as mix design and density of the mortar.

Some engineers have contended that if, in concrete exposed to sea water, every care is exercised to use normal cement and sound, nonreactive aggregates, to obtain uniform high density and impermeability of concrete, and to protect reinforcement from corrosion by a minimum cover of 3 in., there need be little danger of disintegration due to sea-water attack. However, since porous concretes with a high C_3A portland cement are subject to disintegration in sea water, it appears necessary to specify at least a moderate sulfate-resistant (type II) cement for structures in sea water. Lean mixes and reactive aggregates should be avoided, segregation and honeycomb

Table 11.3 Attack on concrete by various sulfate concentrations in soils and ground waters*

Relative degree of attack by sulfates	Water-soluble sulfate, as SO_4, in soil samples, %	Sulfate, as SO_4, in water samples, ppm
Negligible	0.00–0.10	0–150
Positive †	0.10–0.20	150–1,000
Considerable ‡	0.20–0.50	1,000–2,000
Severe ‡	Over 0.50	Over 2,000

* From Ref. 106.

† Use type II cement.

‡ Use type V cement.

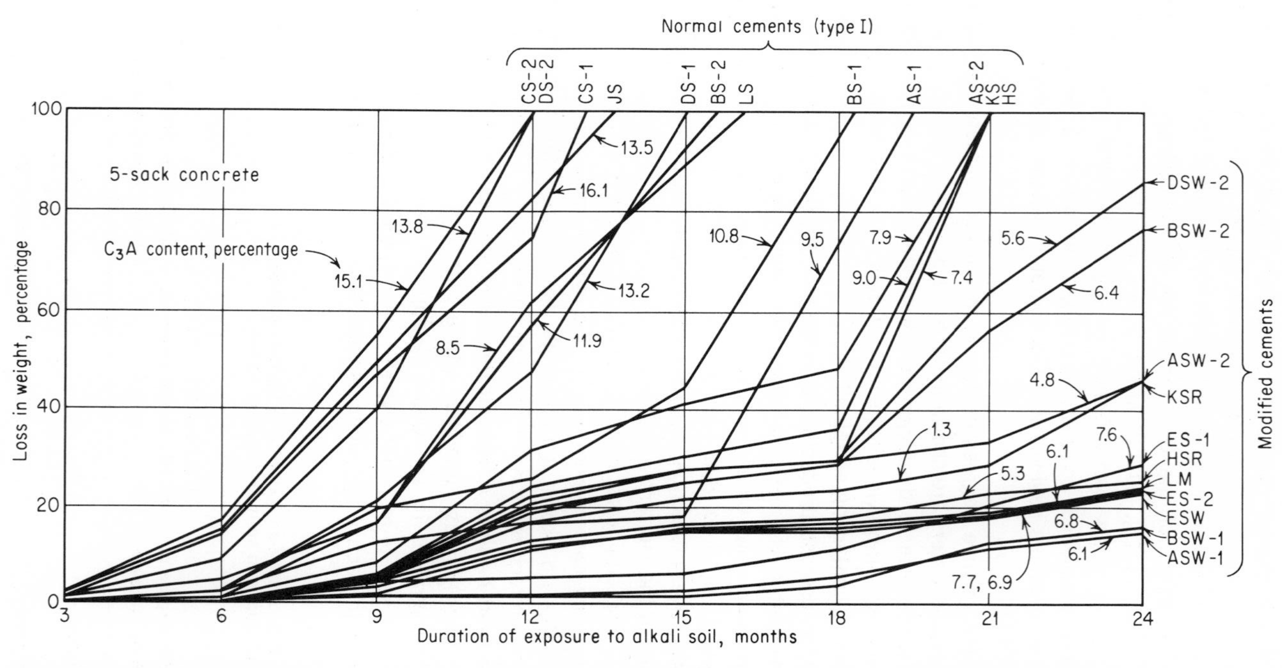

Fig. 11.11. Comparison of performance of normal portland cements with that of modified sulfate-resistant cements. (*From Santon and Meder* [1174*a*].)

should not be permitted to occur, and as noted above, the use of an air-entrained concrete will usually be helpful.

It is of interest that at least one commercial portland-pozzolan cement (containing Monterey shale) has shown considerable resistance to the attack of sulfates [1174b]. Others may be equally effective.

11.18 Leaching and efflorescence. In hydraulic structures water may leak through cracks, along improperly treated construction joints, or through areas of segregated or porous concrete. This water passing through the concrete dissolves some of the readily soluble calcium hydroxide (formed during hydration of the cement) and other solids and in time may cause serious disintegration of the concrete. Many concrete structures bear evidence of this action by the presence of white deposits, or "efflorescence," on their surface. This results from leaching of the calcium hydroxide and subsequent carbonation and evaporation. As problems relating to the dissolving and leaching action of percolating water are, in the main, tied up directly with permeability, they emphasize the need of securing watertight concrete.

In some cases efflorescence is caused by the absorption of rain or ground waters which dissolve calcium hydroxide in the concrete. During the dry season these waters evaporate and deposit the dissolved material on the surface of the concrete. A dense concrete having low absorption will be less susceptible to this condition.

11.19 Chemical attack. Concrete exposed to farm silage, animal wastes, and organic acids from various industries, such as breweries, dairies, and wood pulp mills, is sometimes damaged by surface corrosion which, especially in the case of floors, may cause considerable concern though not be of sufficient magnitude to affect seriously the structural integrity of the concrete. Concrete tanks have been used for the storage of many kinds of materials, some of which are harmful to concrete. Reference 1174, which indicates the chemical effects of various materials on unprotected concrete, also gives suggested protective treatments.

11.20 Corrosion of embedded metal. Portland cement concrete usually provides good protection for embedded reinforcing steel against corrosion caused by the entry of water carrying dissolved oxygen. The protective value of the concrete is caused by its high alkalinity and relatively high-electrical resistivity to atmospheric exposure. The degree to which concrete will provide satisfactory protection for the steel is in most instances a function of the quality of the concrete and the depth of concrete cover over the steel. It is essential to avoid honeycomb and inferior concrete patches and to design against structural cracking and especially any splitting parallel to the steel.

The permeability of concrete is a major factor affecting the corrosion of reinforcing steel. Concrete of low permeability contains less water under a given exposure and hence is more likely to have low-electrical conductivity and better resistance to electrical corrosion. It also resists the absorption of salts and their penetration to the reinforcing steel, as well as providing a barrier against the entry of oxygen.

Low water-cement ratios tend to produce less permeable concrete and thus provide greater assurance against corrosion. A water-cement ratio not in excess of 4.5 gal per sack should be used in all concrete exposed to sea or brackish water or in contact with more than moderate concentrations of sulfates or chlorides at the water or ground line. Air entrainment is recommended as an aid in securing better placement and increased resistance to water penetration.

The corrosion of steel due to the use of sea water for making concrete is discussed in Art. 4.1. Two percent of calcium chloride by weight of cement is usually not significantly corrosive in ordinary reinforced concrete. However, as steel used in prestressed concrete is more sensitive to corrosion, chlorides should not be used in prestressed concrete [1190, 1191]. Portland cement reduces the corrosive effect of a chloride by reacting with a part of it to form a complex calcium chloroaluminate, but sulfate-resisting cement (type V), which has a low tricalcium aluminate content, is less effective in this reaction than regular type I cement.

Aluminum conduit embedded in concrete may corrode, causing concrete to crack and spall. Three causes of the corrosion are (1) galvanic action between aluminum and reinforcing steel, (2) stray electric currents, and (3) alkalies in the concrete. Of these, the first two appear to pose the greatest hazard. Galvanic corrosion of aluminum is speeded if chlorides and moisture are present in the concrete [1191b, 1191c].

11.21 Wear. The wear resistance of concrete is of importance in various types of concrete construction. For floors, pavements, and such hydraulic structures as tunnels and dam spillways, the concrete should withstand destructive wearing forces, which may include abrasion and impact.

Cavitation effects have caused especially severe damage to several hydraulic structures. Under certain conditions of hydraulic flow where disturbances develop, a cavity will occur in the moving water. Where a cavitation pocket occurs adjacent to a concrete surface, the latter will eventually become severely pitted as shown in Fig. 11.12, such damage being caused by extraordinarily high water pressures developed over small areas by the collapse of vacuum pockets. This can be prevented by producing a smooth hydraulic flow.

Fig. 11.12. Cavitation of structural steel angle and concrete at juncture of pier and crest in Stony Gate opening, Parker Dam. (*From ASTM* [110].)

Various types of abrasion and impact tests have been developed. These include rotating discs with silicon carbide grit, the use of loaded wheels, grinding wheel cutters, or steel balls revolved in a fixed path on test slabs or pavements; and fixed charges of grit blasted against the concrete.

In general, the wear resistance of concrete increases as the strength is increased. Lowering of the water-cement ratio through improvement of the aggregate grading and by employing the lowest practicable slump is more effective in improving wear resistance than the same reduction in water-cement ratio resulting from an increase in the cement content. Absorptive cement blankets for floors (see Art. 14.5) and the use of absorptive form linings and the vacuum process (see Art. 14.20) are effective in increasing wear resistance as a result of the lowered water content. Except for absorptive forms there appears to be little difference in the wear resistance resulting from the type of form used. Resistance to wear is increased as the aggregate is harder and tougher and as the percentage of sand is reduced, so long as the required workability is maintained.

Properly applied coats of dry cement and sand or special hard aggregates increase the wear resistance of unformed surfaces, particularly on lean mixes of thick, wet, integrally finished slabs. Quartz, quartzite, and many dense volcanic and siliceous rocks are well adapted to making wear-resistant concrete. The beneficial effect on wear of hard aggregate and of high-strength concrete is shown in Fig. 11.13. Uniformity of concrete is essential to good wear resistance. Pressure applied to the surface by hand troweling after the concrete begins to set increases its density and wear resistance.

Good curing is very beneficial in improving wear resistance. Liquid hardeners such as magnesium or zinc fluosilicate or sodium silicate improve the wearing qualities of moderate- or low-strength concretes, but the floor should be cured and subsequently air dried for at least age 28 days before their application.

As stated in Art. 14.5, floating or troweling a floor surface should be delayed until all surface water has evaporated, for otherwise overly wet mortar with an excess of undesirable fines will develop at the surface to cause a reduction of wear resistance.

When an increase in the percentage of entrained air results in a reduction in strength, the wear resistance of the concrete is reduced. Therefore the air content of concrete should be kept to the minimum required for durability, and should not exceed about 4 percent.

11.22 Restoration of disintegrated concrete. As weathering generally progresses with the age of concrete structures, restoration eventually may be-

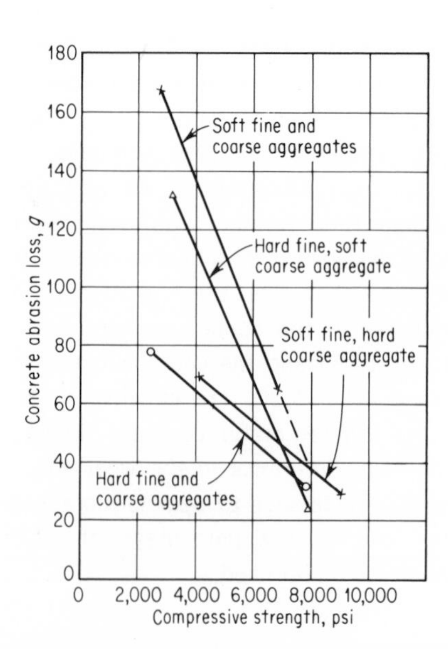

Fig. 11.13. Abrasion resistance increases with strength of concrete and hardness of aggregate. (*From Smith* [1192].)

come necessary for exposed structures. To restore an old structure successfully may require greater engineering skill than to build a new one. Certainly such work calls for a high degree of experience and ingenuity. ACI Committee 201 gives a good discussion of restoration work in Ref. 1199*a*.

For permanency, the restored work must:

1. Be thoroughly anchored and bonded to the old concrete after all disintegrated concrete is cut away
2. By adequately reinforced
3. Be essentially watertight and as dense as possible
4. Provide a new material having volume-change characteristics corresponding to those of the old concrete in its present state
5. Withstand erosion and disintegration due to existing conditions
6. Be thick enough to minimize the number of cycles of freezing and thawing which reach to the depth of the old concrete
7. Not be carried across active cracks or joints, as otherwise the new material will be cracked
8. Be protected from premature loading and against displacement or slumping from inadequate support when thick applications are made on steep surfaces
9. Support all structural loads to which the use of the structure subjects it

For good bond the old surface must be completely cleaned by sandblasting with air or water or both. Steel anchors and a roughened contact surface are essential. A cement grout intimately applied to the previously soaked old surface which has been surface-dried just before placing the new material will help ensure complete contact and improve the bond.

Particular care is required to use a low water-cement ratio for the new concrete to minimize differential movements which might occur and injure the bond, as the older concrete has previously contracted upon setting and drying.

In architectural work, or other work where appearance is important, care should be taken to ensure that the texture and color of the repair will match the surrounding concrete.

Curing must be more carefully done than for the original construction, because the old concrete may absorb moisture from the relatively thin layer of new material and cause excessive drying.

Drains must be installed if there is a possibility of any water tending to collect behind the new concrete, as otherwise hydraulic pressures may develop and cause spalling of the new concrete.

These are the ideal requirements, but they are not fully accomplished in any restoration job, with the result that much restoration work is not very permanent.

The various methods of restoration have included the use of:

1. Wood-floated, steel-trowelled, cement-sand plaster
2. Pneumatically placed mortar (see Arts. 14.8–14.15)
3. Gravity-placed concrete
4. Epoxy compounds
5. Prepacked concrete (see Art. 14.25)

Plaster coats undergo such excessive shrinkage that they invariably crack and soon lose their bond to the old surface. They should not be used.

The advantages of pneumatically placed mortar are ease of placement with minimum need for forms and plant, high strength, and good durability when exposed to freezing and thawing. Its disadvantages include high shrinkage, a wide range in quality dependent upon the skill of the nozzlemen, relatively high porosity and permeability, and a different moisture shrinkage and coefficient of expansion than for the old concrete in the structure. For many kinds of repair the advantages of this method are considered to outweigh the disadvantages [1198]. It has seen extensive application in the past, but many of the jobs have not given perfect service.

Gravity-placed concrete has greater uniformity, density, and impermeability, and lower shrinkage. It readily permits the use of air-entraining agents to increase resistance to freezing and thawing. It is less susceptible to mistreatment because of inferior skill of equipment operators. The concrete materials are more economical than pneumatically placed mortar and,

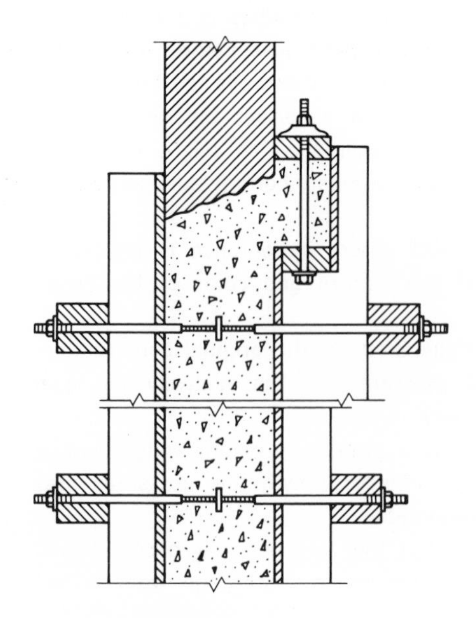

Fig. 11.14. Form for replacement of concrete in walls. By use of anchor bolts, it may be used on surfaces of massive concrete structures [106].)

(*a*)

(*b*)

(*c*)

Fig. 11.15. The prepacked process for repair of concrete. (*From Kelly and Keatts* [1197].) (*a*) **To prepare for restoration with prepacked concrete, the old concrete is trimmed away to sound material;** (*b*) **forms are packed with coarse aggregate, and grout mixture is pumped in;** (*c*) **finished surface, before cleaning.**

having no rebound, may require less cleanup. However, repair work on vertical faces requires forming, and for extensive work the plant required may be considerably more expensive than for pneumatic placing.

As all ordinary concrete shrinks upon setting and drying, it is practically impossible to avoid the development of cracks around a horizontal replacement or at the top of a replacement within a wall. Figure 11.14 shows one arrangement for applying external pressure to the fresh concrete while it is vibrated to reduce this tendency. Note that at the top of the repair work the existing wall is beveled to facilitate a good contact.

Aluminum powder or an expansive cement is sometimes used in the mix to produce an expansive force which tends to make a tighter patch. After the concrete has been placed, the form is closed tight to confine the concrete when it tries to expand. Either of these expansive mixes should be used only under strict control after laboratory tests.

Spalls may be repaired with epoxy resin, an excellent bonding material. Epoxy resin may be used merely as an adhesive or bond coat in place of a portland cement grout, followed by a conventional concrete patch, or it may be used as a binder instead of portland cement to make an epoxy mortar or epoxy concrete (see Ref. 1199b).

As epoxy resins differ appreciably, only a formulation designed for this particular application should be used. Its thermal expansion properties should be checked to be certain that they do not differ too much from those for concrete. Epoxy adhesive should have a film thickness of about 0.03 in. As epoxy concretes are difficult to place and finish, there is need for experienced personnel in their use. These factors, along with their high cost, serve to limit their application.

In prepacked concrete the aggregates are in contact, without the clearance for cement paste which is required for plasticity in ordinary concrete, so that overall drying shrinkage of the concrete is less than that of ordinary concrete of the same cement content, and the tendency to cracking is correspondingly reduced. The bond strength of prepacked concrete to regular concrete is considerably greater than that for regular concrete cast against regular concrete. The general process of restoration with prepacked concrete is illustrated in Fig. 11.15.

QUESTIONS

1. Why does the permeability increase with higher water-cement ratios? With lower cement-voids ratios?

2. How is it possible to produce a reasonably watertight concrete?

3. Discuss the relationship of bleeding to permeability.

4. How does curing affect the permeability of concrete? How does curing produce this effect?

5. What characteristics of the cement and aggregate affect permeability?

6. Of what significance is the absorption and permeability of concrete?

7. What effect may mixing and placing have upon permeability?

8. In general, what are the relative effects upon permeability of various admixtures in comparison with extra cement?

9. What surface treatments are effective in reducing permeability?

10. List the several external factors which cause disintegration of concrete.

11. How does freezing and thawing promote disintegration of concrete?

12. What tests are employed to evaluate the probable resistance of an aggregate to weathering?

13. What effect has water-cement ratio upon weathering resistance of concrete?

14. What amount of air in concrete is usually required for good weathering resistance?

15. List the kinds of observations made in conducting freeze-thaw tests and state the merits and disadvantages of each.

16. What general class of aggregates is known to be reactive with high-alkali cements?

17. How may the harmful action of such aggregates be overcome?

18. What are the essential characteristics of cements resistant to sulfate waters?

19. Discuss the need of sulfate-resistant cements for use in concrete exposed to sea water.

20. What is efflorescence and how is it caused?

21. How does cavitation damage hydraulic structures, and how may cavitation be prevented?

22. How may the wear resistance of concrete be improved?

23. What construction practices have been used to improve the wear resistance of concrete?

24. List the various factors which must be considered to attain permanency for restored concrete.

25. What methods are used in concrete restoration work, and what are the advantages and disadvantages of each?

TWELVE
VOLUME CHANGES AND CREEP

12.1 Types of volume change in concrete. The principal volume changes that may occur in concrete are caused by settlement of the fresh mass, chemical combinations of the cement with water, combinations of high-alkali cements with certain reactive aggregates, changes in moisture content, changes in temperature, and by applied loads.

The effect of chemical combinations between component parts of the mix depends upon whether the final products occupy a lesser or greater volume than the separate constituents. As the chemical combination of cement and water results in a reduction of volume (except for special expansive cements), shrinkage of the concrete occurs, but when reactions occur between the cement and certain aggregates, large expansions and even disruption of the concrete may occur as discussed in Arts. 3.15 and 11.16. When concrete dries, it shrinks, and when it is wetted again, it expands; by common usage, the term "volume changes" often refers to the effects of moisture changes only. In common with most materials, concrete expands or contracts with change in temperature. The effect of rapidly applying a load is to produce deformations which are in large part elastic; the modulus of elasticity and Poisson's ratio are measures of the effects of such loading (Arts. 13.1 and 13.8). If load is sustained over any period, deformation usually continues to take place—this continued deformation with time, due to load, is called "creep."

In discussing volume changes, the movements are usually stated in terms of deformations or of changes in *length*. It is convenient to employ for this purpose a unit called a "millionth" which is simply 0.000,001 per unit length. A change in length of 1,000 millionths corresponds to 0.1 percent or 1.2 in. per 100 ft. The deformations that ordinarily occur in concrete range from a few to about 1,000 millionths. Elastic deformations may be conveniently transformed into equivalent stresses by multiplying strain, in millionths, by the modulus of elasticity in million psi; thus, assuming a modulus of 3 million psi, a strain of 300 millionths corresponds to a stress of 900 psi. If any creep is included in the strain, the sustained modulus of elasticity (Art. 13.7) must be used in place of the ordinary modulus.

12.2 Significance of volume changes and creep. If concrete were free to deform, any volume changes would be of little consequence, but usually it is restrained by foundations, steel reinforcement, or by adjacent concrete subject to different conditions. As the potential movement is thus restrained, stresses will be developed which may rupture the concrete. This is particularly true when tension is developed; thus, contractions causing tensile stresses are more important than expansions which cause compressive stresses. Differences in moisture contents (and thus volume changes) of the exposed and unexposed faces of thin concrete slabs, such as highways and canal linings, may cause curling and eventual cracking. Cracking not only may impair the ability of any structure to carry its designed loads, but it also may affect its durability and damage its appearance. The durability is affected by the entry of water through cracks, which corrodes the steel, leaches out soluble components, and deteriorates the concrete when subjected to freezing and thawing.

Creep, in general, tends to relieve the stress in concrete, especially when reinforced. Thus, when a sustained load is applied to a reinforced-concrete column, creep of the concrete causes a gradual reduction in the load on the concrete and a corresponding increase in the load on the steel. In various structural elements such as continuous beams and slabs, creep relieves some of the stress in the most highly stressed portions and increases the stress in adjacent portions of the concrete, so that finally the stresses are more uniform throughout the member. The relief of high stress as a result of creep is shown in Fig. 12.1. This relieving of the higher stresses serves to reduce the tendency toward cracking. However, creep may cause objectionable sagging of thin, long-span floor slabs or

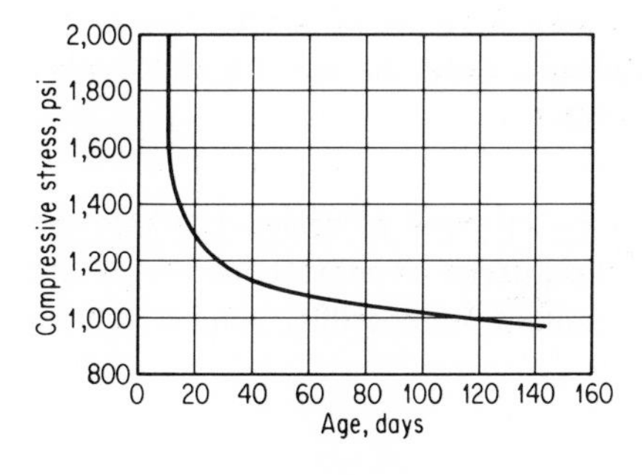

Fig. 12.1. Relaxation of stress under a constant strain of 360 $\times$ 10^{-6} [1275].

other structural elements. Under normal drying conditions, the creep after loading for 2 years may be two to four times the amount of the "elastic" deformation that occurs immediately upon application of load.

12.3 The gel structure as related to volume changes. Cement, after hydration, consists of crystalline material plus a hardened calcium silicate gel resulting from the combination of cement and water. The amount of the gel increases with the age of hydration and is greater for higher water-cement ratios and for finer cements. The amount of gel also depends upon the chemical composition of the cement, as fully hydrated dicalcium silicate is believed to be mostly gel, while hydrated tricalcium silicate is more than half gel. For the water-cement ratios used in average concrete, the gel has a larger volume than the crystalline portions.

The crystalline materials in cement are believed to be unaffected by ordinary drying, but the gel is finely porous and undergoes large volume changes upon wetting and drying (see Art. 2.16). The quantity and characteristics of the calcium silicate gel, therefore, largely determine the potential shrinkage upon drying of hydrated cement.

Water is held in the pores of the gel by such large attractive forces that when it is removed from the pores by evaporation, the forces which formerly attracted the water become effective in compressing and reducing the volume of the gel. All concretes, then, are subject to moisture volume changes in some degree, and the problem involved is so to control conditions that the volume changes have small or practically harmless effects upon the integrity of the structure.

12.4 Shrinkage of fresh concrete. While in the plastic state, fresh concrete settles in the forms and undergoes an appreciable reduction in volume. This occurs as a result of settlement of the solids and the bleeding of free water to the top, where it may be lost by drainage from the forms or by evaporation. Most of this settlement and bleeding usually occurs within an hour or so after placement of the concrete [518]. During this same period some of the water is absorbed by the cement grains, which results in a further reduction of volume. While this total volume change may in extreme cases amount to 1 percent or more, it is not of great importance, as the concrete is in a plastic or semiplastic state, so that no appreciable stresses can result from these volume changes.

12.5 Autogenous volume changes. Volume changes of sealed concrete are of particular importance with regard to the interior of mass concrete, where little or no change in total moisture content is possible. Such volume changes are due to causes other than loss or gain of water, rise or fall in temperature, and external load or restraint, and are called "autogenous" volume changes because they are self-produced by the hydration of the cement.

As the products of hydration have a lesser volume than the sum of the separate volumes of the original components, the system cement plus water contracts as hydration proceeds. Although the volume of cement plus water may decrease, the absolute volume of solids increases, since the newly formed solids are a combination of the original solids and water.

After the cement paste has set, the solids form a structure having channels or pores containing more or less free water. It is thus possible for the hardened solid mass to expand, even though the total volume of cement and water is decreasing, because of the use of internally contained free water in the formation of new solid hydration products and by its absorption causing expansion of the gel.

At later ages, continued hydration of sealed concrete not only may utilize all the free water available in the pores but also may remove some of the more loosely held water from the gel. Loss of water from this gel causes the gel structure to contract, resulting in a decreased volume of the concrete mass.

Of the two competing factors described above, the second is the dominant one at the later ages, so that a contraction finally occurs. In so far as the paste is concerned, the following appear to be the major variables which influence the behavior of a particular concrete: (1) composition of the cement (affecting the nature and rate of chemical reactions and type of reaction products), (2) amount of original mixing water (affecting the rate of early reactions, the porosity of the paste, and the later availability of free pore water), (3) temperature conditions (affecting the rate of reaction), and (4) time (affecting the extent of reaction).

The results of observations on sealed concrete cylinders containing ¾-in. maximum aggregate are shown in Fig. 12.2. All specimens exhibited shrinkage from the time of initial observation at age 2 days. After 4 months, the values for low-heat cement were about 20 percent greater than those for normal cement, but portland-pozzolan cement concretes usually produce a greater autogenous shrinkage than do similar mixes without pozzolan. The change for the rich mix was about 50 percent greater than for the lean mix, and the limestone concrete contracted 20 percent more than the gravel concrete. The effect of consistency of mix was negligible. The magnitude of autogenous shrinkage varies widely, ranging from an insignificant 10 millionths to somewhat in excess of 150 millionths, but the latter value is only about one-fourth the shrinkage of an average concrete due to drying.

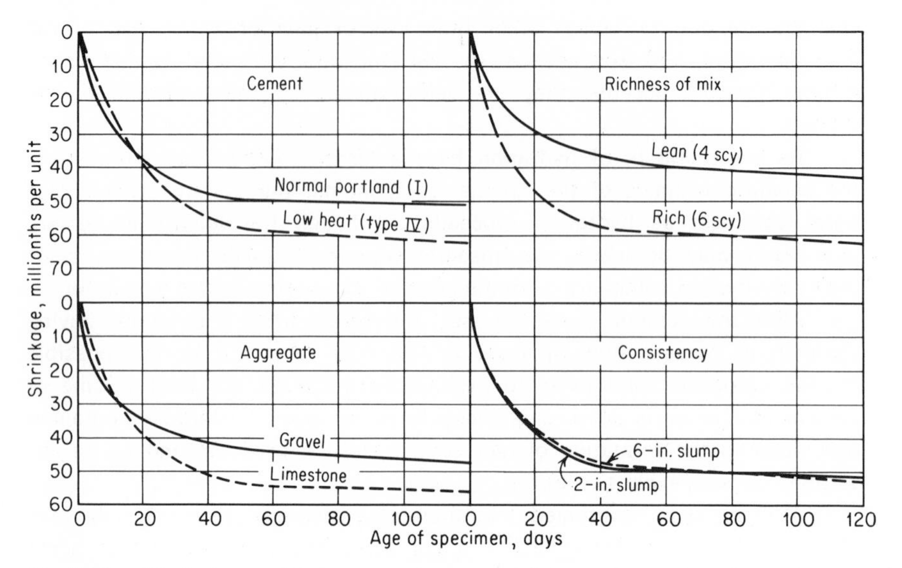

Fig. 12.2. Effect of several factors on autogenous shrinkage of concrete. For each group the graphs show the average for the variables covered by the other three groups. Initial observations at age 2 days.

12.6 Expansive cements. As noted in Art. 2.23 some special cements undergo relatively large expansions at early ages so that if used in concrete which is restrained, compressive stresses will be developed. Later, when drying occurs, the resulting shrinkage is partly offset by expansion which develops as the compression is released. The final result is usually a net contraction somewhat less than that for an ordinary cement [256].

SHRINKAGE AND EXPANSION DUE TO MOISTURE CHANGES

12.7 Factors affecting shrinkage and expansion. Several factors which may be expected to influence the magnitude of volume changes in mortars and concretes caused by variations in moisture conditions, which take place with time and the simultaneous hardening of the cement paste are:

1. Cement and water contents
2. Composition and fineness of the cement
3. Type and gradation of aggregate
4. Admixtures
5. Age at first observation
6. Duration of tests
7. Moisture and temperature conditions
8. Size and shape of specimen
9. Absorptiveness of forms

10. Carbonation
11. Amount and distribution of reinforcement

A test procedure for determining the volume change of cement mortar and concrete is covered by the ASTM [C157].

12.8 Effect of cement and water contents. The water content is probably the largest single factor influencing the shrinkage of cement paste and concrete. Tests have shown that for cements having normal shrinkage characteristics, the shrinkage of the cement paste varies directly with the water-cement ratio [1204].

For concretes, the influence of the water is more pronounced than for pastes, as for any given percentage increase in the amount of water in concrete, the shrinkage is increased by about double that percentage. The explanation of this has been given that for each 1 percent increase in water content, the volume of gel formed is increased about 1 percent, and the shrinkage tendency per unit volume of the wetter gel is increased about 1 percent [1204].

For a given water-cement ratio, concretes of wet consistencies have a high paste content, and thus the shrinkage is greater than for stiffer mixes. For given proportions of cement and aggregate, concretes of wet consistencies have a high water-cement ratio, and again the shrinkage is greater than for stiffer concretes. Hence, in both cases the wet consistencies produce greater shrinkage of the concrete.

In general, a higher cement content increases the shrinkage; the relative shrinkages of neat paste, mortar, and concrete may be of the order of about 5, 2, and 1. However, for given materials and a uniform water content per cubic yard, the shrinkage of concrete varies but little for a wide range of cement contents, because a richer mix will have a lower water-cement ratio and these factors offset each other. For the practical mixes in Table 12.1 the percentages of increase in shrinkage were about half the percentages of increase in cement content [110].

Table 12.1 Effect of cement content on shrinkage*

	Cement content		Shrinkage	
Mix	Sacks per cu yd	Ratio	Millionths	Ratio
Lean	5.0	1.0	560	1.0
Medium	5.8	1.2	640	1.1
Rich	7.2	1.4	700	1.2

* From Ref. 110.

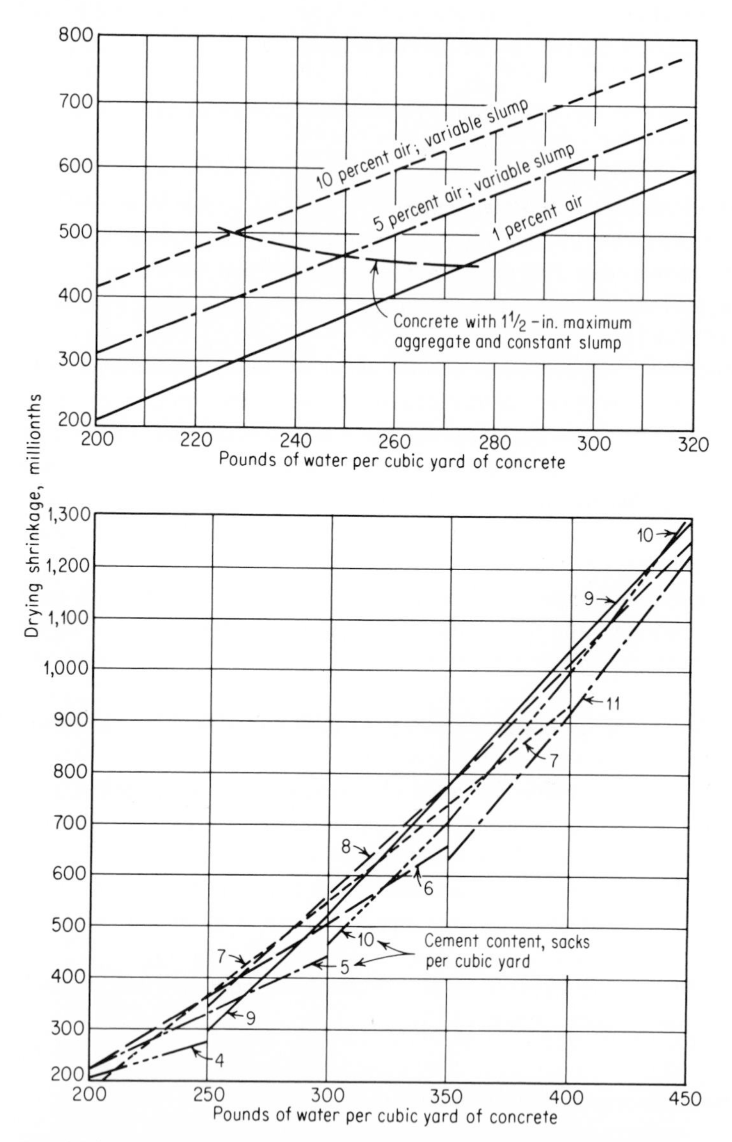

Fig. 12.3. The interrelation of shrinkage, cement content, and water content shows that shrinkage is a direct function of the unit water content of the fresh concrete. For air-entrained concrete of a given mix and consistency, the air content does not affect shrinkage appreciably. (*From U.S. Bureau of Reclamation* [106].)

The interrelation of shrinkage, cement content, water-cement ratio, and water content shown in the lower part of Fig. 12.3 makes clear the all-important effect of water content per cubic yard of concrete and the relatively minor influence of cement content upon shrinkage of concrete.

12.9 Effect of composition and fineness of cement. The shrinkages of several neat cements, all having a water-cement ratio of 0.30 by weight, are shown in Table 12.2. It will be noted that the high-early-strength cement shrinks about 10 percent more than the normal cement. The high shrinkage of the low-heat and portland-pozzolan cements is typical for these types of cement, which are believed to produce large percentages of gel.

The effect of composition and fineness on volume changes of mass-cured mortar bars is shown in Table 12.3. Based upon the more complete data for these tests, the following observations appear to be justified for mass-cured bars, although other tests show the same general effects for standard curing:

1. In general, the finer the cement the greater the expansion under moist conditions, but some cements show a slightly lower expansion with increase in fineness.
2. The contraction in air is not appreciably affected by variations in fineness.
3. The higher the MgO and the C_3A contents, the greater the expansion under moist conditions.
4. The higher the C_2S content, the less the expansion under water but the greater the contraction in air.
5. Contraction of mortar on drying is increased by an increase in ignition loss.

Although these are correct in general, the shrinkage characteristics cannot be predicted reliably from the ordinary analysis. The percentage of C_2S is usually a good guide, but the percentage of C_3A does not bear a consistent relation with the shrinkage performance.

Table 12.2 Shrinkage of neat cements in air at age of 1 year*

Cement	Shrinkage, millionths
Type I, normal, 7% C_3A, 1,400 sq cm per g	2,150
Type III, high-early-strength, 29% minus 5 microns	2,335
Type IV, low-heat, 5% C_3A, 1,900 sq cm per g	2,870
Portland-pozzolan	3,150

* From Ref. 1204.

Table 12.3 Approximate effect of fineness and composition of cement on volume changes of mass-cured mortar*

	Total expansion† under water at age 1 year, millionths	Net contraction† in air of 50 percent R.H. at age 1 year, millionths
Average of 20 cements‡	213	512
Effect of increase in Wagner specific surface of 100 sq cm per g¶	+7	−4
Contribution factor for each % of compound:		
C_3S	+1.6	+4.7
C_2S	+1.1	+7.2
C_3A	+9.4	−1.4
C_4AF	−0.2	+3.7
Effect of 1% increase in MgO content above 2.6%	+53	−19
Effect of increasing SO_3 content from 1.5 to 1.9%	+13	−7
Effect of 1% increase in loss on ignition above 0.9%	−7	+80

* From Ref. 218.
1:3¼ mix; water-cement ratio approximately 0.58 by weight.
† Year includes 28-day mass curing; storage thereafter at 70°F.
‡ Average composition: 43 percent C_3S, 34 percent C_2S, 10 percent C_3A, 10 percent C_4AF. Wagner specific surface, 1,180 sq cm per g.
¶ Within range of 1,000 to 1,600 sq cm per g.

From a large test program on portland-pozzolan cements [451] it has been shown that for a 20 percent replacement of a high-lime portland cement with pozzolan, the average contraction of mortar in air at 1 year is 25 percent greater than for the straight high-lime portland cement mortar, but it is appreciably less than for an ordinary low-heat portland cement, for which on some jobs it has been used as a substitute.

The effect of selected cements upon the shrinkage of concrete is shown in Table 12.4. Even with the same aggregate there is a range of over 2 to 1 in the relative shrinkages exhibited by different cements. When both different cements and different aggregates are considered, the range is still greater. These large differences probably resulted from some internal cracking of the paste, as shrinkages of cement pastes do not vary by such large amounts. In a few cases very fine cracks were barely visible. In other cases cracking was suspected but not visible.

12.10 Effect of type and gradation of aggregate. The drying shrinkage of concrete is but a fraction of that of neat cement as the aggregate particles not only dilute the paste but they reinforce it against contraction. Tests have shown that if the aggregate were readily compressible, as when using porous but nonabsorbent rubber particles, the concrete would shrink as

Table 12.4 Shrinkage of concrete containing different cements and aggregates*

Aggregate	Cement†	Shrinkage characteristic of cement	Water-cement ratio, by wt	6-month shrinkage in air of 50% R.H., millionths: After moist curing 2 days	After moist curing 28 days
Gravel	A	Normal	0.65	820	810
	B	Low	0.63	610	620
	C	Very low	0.62	474	470
Crushed dolomite	A	Normal	0.80	550	496
	B	Low	0.76	400	430
	C	Very low	0.79	400	320
Crushed marble	A	Normal	0.84	650	660
	B	Low	0.80	430	350
	C	Very low	0.81	260	320
Crushed granite	D	High	0.87	550	800
	A	Normal	0.87	730	680

* From Ref. 1204. Specimens were 3 by 6-in. cylinders.
† Potential composition of cements:

Cement	C_3S	C_2S	C_3A	C_4AF
A	51	18	15	7
B	59	13	10	8
C	52	24	6	12
D	38	33	5	15

much as neat cement. The ability of normal aggregates to restrain the shrinkage of a cement paste depends upon (1) extensibility of the paste, (2) degree of cracking of the paste, (3) compressibility of the aggregate, and (4) volume change of aggregate due to drying.

In Table 12.5 is shown the shrinkage of neat cement in comparison

Table 12.5 Effect on shrinkage of adding aggregate to cement paste*

Aggregate	Parts by wt: Cement	Aggregate	Water	Paste, percent by volume	2-year shrinkage in air of 50% R.H., millionths
None (neat paste)	1	0.0	0.40	100	2,705
Gravel	1	2.5	0.40	44	725
Crushed limestone	1	2.6	0.40	44	417

* From Ref. 1204. Specimens were 3 by 6-in. cylinders. Aggregate, No. 4 to 3/8 in.

with the corresponding shrinkages of the same cement diluted with a single sieve size (No. 4 to ⅜ in.) of gravel and crushed limestone, respectively. The reduction in shrinkage due to the aggregate is greater than would be expected considering its relative volume. It is possible that internal cracking of the paste due to the restraint of the aggregate is a factor.

The effect of the size of aggregate upon shrinkage is shown in Table 12.6. For all the smaller sizes of aggregate the shrinkage is fairly uniform, indicating that for these sizes there probably is no internal cracking. However, it is possible that for these concretes the shrinkage is reduced by cracking of the paste when the size of aggregate exceeds about ¼ in.

Table 12.6 Effect of adding various sizes of dolomite aggregate to cement paste*

Aggregate size	1-year shrinkage in air of 50% R.H., millionths
None (neat paste)	2,710
No. 48–No. 28	1,190
No. 28–No. 14	1,240
No. 14–No. 8	1,220
No. 8–No. 4	1,160
No. 4–⅜ in.	940
⅜ in.–¾ in.	690

* From Ref. 1204.
Mix, 1:1 by weight; water-cement ratio 0.40; cylinders 3 by 6 in.

For a given maximum size of aggregate, a wide range of acceptable gradations of aggregate, including continuous and gap gradings, as well as unusual gradings with fines omitted, has very little effect on shrinkage [1204].

Large differences in concrete shrinkage result from the use of different aggregates, as shown in Table 12.7. In general, concretes low in shrinkage often contain quartz, limestone, dolomite, granite, or feldspar, whereas those high in shrinkage often contain sandstone, slate, basalt, trap rock, or other aggregates which shrink considerably of themselves or have low rigidity to the compressive stresses developed by the shrinkage of the paste. Lightweight concretes generally shrink considerably more than normal concretes, and the greater the absorption of the aggregate, the greater the shrinkage of the concrete.

The shrinkage of the aggregates themselves may be of considerable importance in determining the shrinkage of the concrete, as shown in Table 12.8 by the volume changes for various building stones when alternately

Table 12.7 Effect of type of aggregate of a single size on the shrinkage of concrete*

Aggregate	Absorption, %	1-year shrinkage, in air of 50% R.H., millionths
A. Mixed gravel	1.0	560
B. Slate, hand-picked from A	1.3	680
D. Granite, hand-picked from A	0.8	470
E. Quartz, hand-picked from A	0.3	320
F. Sandstone	5.0	1,160
G. Solid glass spheres	0	250
H. Limestone	0.2	410

* From Ref. 1204.
Mix 1:2.5, water-cement ratio 0.40, aggregate size $\frac{3}{16}$ to $\frac{3}{8}$ in., and specimens 3 by 6-in. cylinders.

wetted and dried. The differences between the various grades of sandstone alone are noteworthy.

Various harmful effects of abnormal shrinkage of concretes caused by certain types of aggregate, as observed in actual structures, have included excessive cracking, large deflection of reinforced beams and slabs, and some spalling. The large deflections result from the differential contraction of the reinforced and unreinforced parts of the flexural members. It is very essential that any new source of aggregate be tested to ascertain whether

Table 12.8 Volume changes in building stones*

Stone	Change in 2 weeks, millionths	
	Expansion	Contraction
Sandstone:		
Red, fine-grained	60	180
Red, fine-grained	500	500
Red, very fine-grained	2,060	1780
Limestone:		
White	40	80
Dense, containing clay	260	260
Granite	60	150
Basalt:		
a	260	270
c	480	570

* From Ref. 1201.

its use in concrete would cause excessive shrinkage to develop. Any shrinkage in excess of 800 millionths is taken to indicate an undesirable aggregate. For lightweight concretes see Art. 14.22.

12.11 Effect of admixtures. In general, admixtures that increase the water requirement of concrete increase the shrinkage, and those that decrease the water requirement decrease the shrinkage. Shrinkage is appreciably increased by replacing a part of the portland cement with clays and certain pozzolanic materials such as pumicite or raw diatomaceous earth. However, replacements of fly ash (a pozzolanic material), treated diatomaceous earth, certain water-quenched ground slags, or lime do not change the shrinkage characteristics to an appreciable degree. Of several integral waterproofing compounds tested, none aided in reducing volume changes [1200]. A dispersing agent (calcium lignin sulfonate) and a common wetting agent had little effect on shrinkage [1204]. Calcium chloride added in the amount often used as an accelerator—2 percent by weight of the cement—may increase the drying shrinkage considerably, sometimes as much as 50 percent [110].

The upper part of Fig. 12.3 shows the effect of entrained air in the mix; as the entrained air is increased, the drying shrinkage is increased. However, as the use of entrained air and a given consistency results in a reduction in the water, the shrinkage of air-entrained concrete is not appreciably increased. This is shown by the curve for the mixes using 1½-in.-maximum-size aggregate.

12.12 Effect of age at first observation. The first observation to determine volume changes has usually been made at the age of about 2 days, as by this time the specimen ordinarily is strong enough to be handled without damage. During this period the use of the right moisture conditions will produce little volume change, since mortars and concretes expand under water and contract in air. The curing of specimens in a foggy atmosphere is an attempt to provide suitable conditions, so that no appreciable volume changes will occur. Provided specimens are cured in this manner in rigid molds, which provide frictional restraint against changes in specimen length, it appears probable that volume changes are inappreciable between the time when the mass takes final set and loses its plastic properties, and the time of first observation made at the earliest age that the specimens can be handled safely.

However, if observations for shrinkage are referred to the length after thorough moist curing and just before drying begins, the total shrinkage so determined will be greater than the "net" shrinkage referred to the length upon removal from the forms, by the amount of the expansion during the curing period.

12.13 Effect of moisture and temperature conditions. When ordinary concretes made with normal cement in the form of small laboratory specimens are continuously subject to moist conditions, they undergo an expansion of the order of 200 to 300 millionths. This movement takes place slowly and apparently is not large enough to be a significant factor in the performance of structures. When similar specimens are brought from the completely saturated to the completely dry state, the change in length is of the order of 500 to 800 millionths (roughly equivalent to that caused by a change in temperature of 100°F).

For most structures, normal atmospheric conditions of drying are not as extreme as those to which small laboratory specimens are commonly subjected for the purpose of testing. However, it must not be overlooked that thin walls and slabs, which of all the forms in which concrete can be cast are most susceptible to complete drying, are oftentimes exposed to severe conditions of drying, sun and wind out of doors, and artificial heat in buildings.

As drying proceeds very slowly when hydrating cement is exposed to dry air, some moisture is usually present for a considerable period; thus hydration progresses with the shrinkage; thus, the gel structure grows and becomes more stable. Under these conditions, if water soaking follows a period of drying, the cement paste is not likely to expand to its original volume, because some of the gel was formed and hardened while the paste was of reduced volume. As would be expected, test results show that the original volume is most nearly regained the more complete the hydration before drying began [204, 1201].

The contraction of concrete in dry air will be greatly retarded by coating the concrete with a seal coat such as tar or an impervious lacquer so that the contained moisture cannot escape so readily. Specimens water-cured for 13 days, then air-dried for 1 day, and then coated with tar contracted only 30 millionths at the age of 210 days, whereas similar specimens under similar storage conditions but without the coating contracted about 20 times as much [1201, 1215].

Ordinary variations in atmospheric humidity can markedly alter the shrinkage rate of moist concrete bars, or the expansion of oven-dried bars. After concrete has attained a state of moisture equilibrium, a change in the humidity of the air will produce additional volume changes [1200].

The effect of curing temperatures upon the volume changes of mortar bars may be summarized as follows: (1) Preliminary curing for 28 days at 70°F results in less expansion under moist conditions and greater contraction in air than for curing under mass-concrete conditions, and (2) the higher the preliminary curing temperature, within the range from 70°F to 150°F, the greater the expansion in water and the less the subsequent contraction on drying [204]. High-pressure steam curing at about 350°F, used to some extent in the manufacture of concrete products, practically eliminates subsequent drying shrinkage.

12.14 Effect of duration of tests. Tests on 1 by 1 by 4-in. neat cement bars stored in air showed them to be still contracting very slightly even after 20 years storage, but similar bars stored in water reached their maximum expansion at an age of about 10 years [1202].

Shrinkage is roughly proportional to the logarithm of the age. Mortar and concrete bars about 2 or 3 in. square change their length very slowly after a year or two of uniform conditions of storage, and after three years' storage the volume changes are usually insignificant.

12.15 Effect of size and shape of specimen. The size and shape of a concrete specimen definitely influence the rate of loss or gain of moisture under given storage conditions, and this affects the rate of volume change as well as the total expansion or contraction. Since any large mass is subject to greater variations of moisture content from point to point than is a small specimen, these variations will produce a nonuniform state of volume change within a large mass. Under drying conditions, shrinkage of concrete near the surface will develop tensile stresses there which are in equilibrium with compressive stresses nearer the center. These stresses, which are active over extended periods of time, might cause a plastic yielding of the concrete, permanently elongating the tensile fibers and shortening the compressive fibers, so that the maximum contraction of a large mass might be appreciably less than for a small specimen. Furthermore, for given storage conditions volume changes always continue over a much longer period of time for a large mass than for a small specimen.

It has been found that for the average concrete containing only the original mixing water and cured for a month or more, later shrinkage is

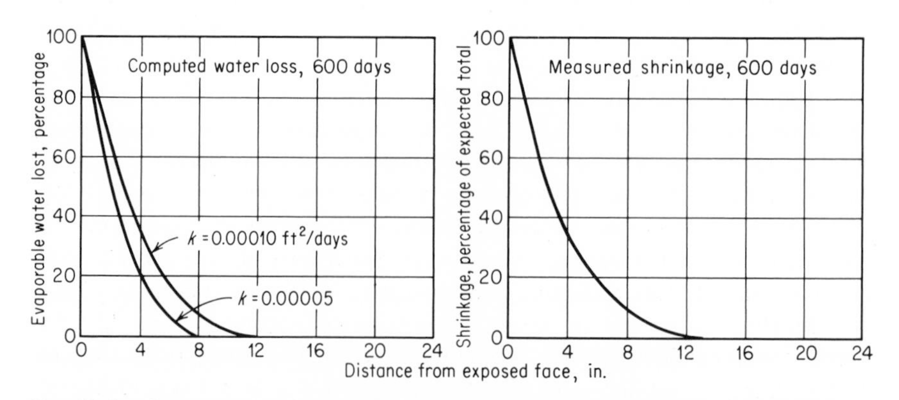

Fig. 12.4. Computed distribution of drying and measured distribution of shrinkage in prisms 2 ft long, drying from one end only. (*From Carlson* [1205].)

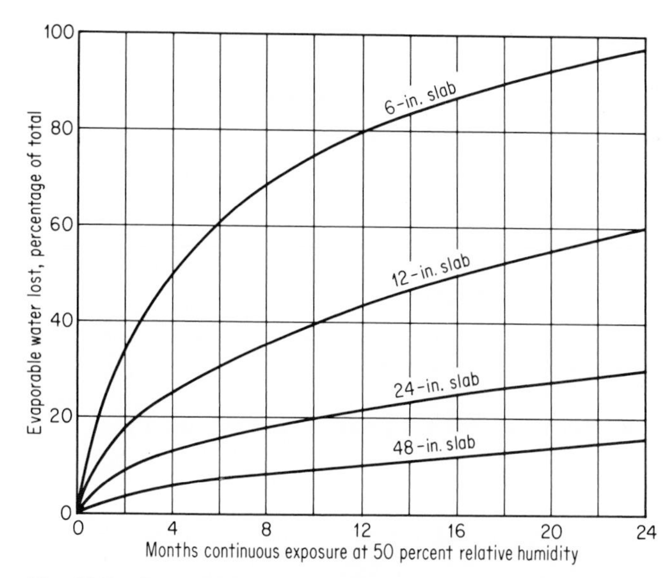

Fig. 12.5. Theoretical rates of drying of concrete slabs. Both faces exposed to 50 percent relative humidity. *(From Carlson* [1205].)

approximately proportional to the amount of water lost. Therefore, when both moisture loss and drying shrinkage are stated in terms of percent of expected total, they should be nearly identical.

Carlson has applied the principles of diffusion of moisture within a mass to the determination of the moisture loss at various distances from a face exposed to air at 50 percent humidity [1205]. A comparison of the computed distribution of moisture losses based on two different moisture-duffusion constants k, and the measured distribution of shrinkage in prisms 2 ft long drying from one end only is shown in Fig. 12.4.

Carrying these studies one step further Carlson has computed the rates of drying of slabs with both faces exposed to dry air, as shown in Fig. 12.5, and the computed distribution of drying in massive concrete structures, as shown in Fig. 12.6. Note that the drying penetrates about 3 in. in 1 month and about 2 ft in 10 years. It is concluded that drying shrinkage in massive concrete is not important unless the surface drying contributes to the beginning of a crack that otherwise might not form.

12.16 Effect of absorptiveness of forms. Tests have shown that for 3 by 3 by 40-in. bars of 1:3 mortar cast in dry, porous, burned-clay molds, the shrinkage is about 15 percent less than for companion bars cast in nonabsorptive molds [1200]. This change in contraction probably results from the lower effective water content of the mix cast in the absorptive molds. Although this condition may prevail for masonry mortars, it does

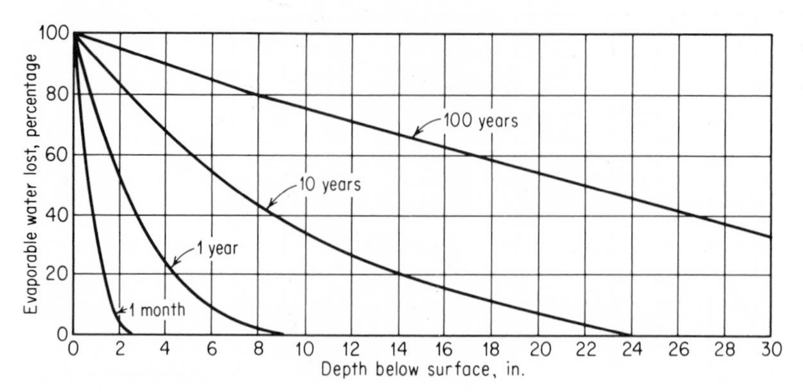

Fig. 12.6. Computed distribution of drying in massive concrete structures. (*From Carlson* [1205].)

not appear feasible to reduce the shrinkage of concrete in any but the thinnest sections by employing absorptive forms.

12.17 Effect of carbonation. It has been shown [1234] that carbonation by CO_2 gas of hydrated portland cement mortars may result in improved strength and hardness and reduced permeability. Irreversible shrinkages and weight gains occur as the various cement hydration products (principally calcium hydroxide) are carbonated, and the carbonated product may possess improved volume stability to subsequent moisture change. Figure 12.7

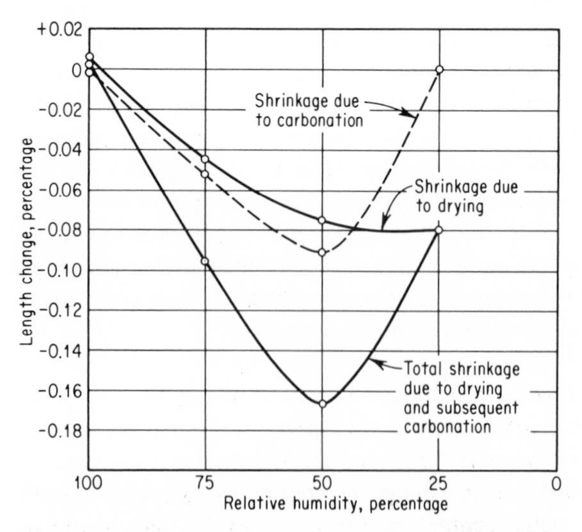

Fig. 12.7. Effect of relative humidity on drying shrinkage and carbonation shrinkage of small mortar prisms [1234].

shows that practically no carbonation shrinkage occurs at a relative humidity of 100 percent, probably because the pores of the cement paste are filled with water so that the CO_2 cannot diffuse readily into the paste. At a relative humidity of 25 percent no carbonation shrinkage occurs in small specimens as water is required for carbonation to occur. But larger specimens carbonate and shrink considerably when exposed at 25 percent relative humidity, presumably because of failure of specimens to maintain equilibrium with this external humidity as water is liberated during the carbonation process. Concentration of carbon dioxide in the exposure atmosphere can also have a considerable influence on carbonation. For a hypothesis on carbonation shrinkage, see Ref. 255 (also see Art. 2.16).

Carbonation is of most importance in connection with precast products, and primarily for concrete blocks, in order to reduce the shrinkage of concrete masonry. Studies have indicated [1235] that a drying and carbonation treatment with hot flue gases can improve greatly the volume stability of concrete block and brick steam cured at atmospheric pressure. The following conclusions were drawn from these studies:

1. The relative humidity in the kiln should be between 15 and 35 percent.
2. The CO_2 content should be as high as possible but not less than 1.5 percent.
3. The temperature should be between 150 and 212°F.
4. Generally a 24-hr treatment will be required, but under optimum conditions a shorter period may provide considerable reduction in shrinkage.
5. Potential shrinkage reduction of more than 30 percent may be obtained under favorable conditions.

12.18 Effect of reinforcement. For a reinforced, unrestrained concrete structure subjected to conditions causing shrinkage, compressive stresses are set up in the steel, and tensile stresses are developed within the concrete, the total compression being equal to the total tensile force. As concrete tends to deform elastically and also to flow plastically under a sustained stress, these tensile stresses cause an elongation of the concrete in tension, but at the same time the loss of moisture causes a large shortening of the concrete. The resultant shrinkage will be less than for plain concrete, the difference being dependent upon the percentage of reinforcement. However, if sufficient reinforcement is used the restraint will be so great that cracking of the concrete will occur.

In a series of tests conducted by Matsumoto [1214] upon 1:2:4 reinforced concrete, the contraction as affected by the steel reinforcement, and the shrinkage stresses as developed in the steel are shown in Table 12.9. Although the contraction is appreciably less for the higher percent-

Table 12.9 Effect of reinforcement upon volume change*

Reinforcement, %	Contraction, millionths	Shrinkage stress, psi	
		Steel, compression	Concrete, tension
0.00	640	0	0
0.55	540	19,000	100
1.23	420	15,500	200
2.18	290	12,000	250

* From Ref. 1214.

ages of reinforcement, these conditions promote high tensile stresses in the concrete. It was noted that when the percentage of reinforcement exceeded 1.5, there was danger of these high tensile stresses causing cracking of the concrete, but that for lower percentages of steel the shrinkage may develop stresses in the steel in excess of allowable values. For another series of restrained shrinkage tests the calculated tension in the concrete was as high as 750 psi, but no cracks were visible [1215]. For these tests the actual stresses must have been materially less than the calculated values as a result of creep within the concrete.

12.19 Prepacked concrete. As stated in Art. 14.25, prepacked concrete is made by filling the forms with aggregates containing no sand and then pumping a cement grout into the void spaces to solidify the mass. This produces a concrete of exceptionally low shrinkage, as the aggregates are in contact and thus not separated by cement paste, which causes most of the shrinkage in ordinary concrete. However, the wet aggregates themselves cause some shrinkage.

12.20 Thermal volume changes. The effect of variation in temperature upon concrete is to produce expansion when the temperature rises and contraction when the temperature falls. An accepted average value of the amount of such change within the ordinary ranges of temperature is 5.5 millionths per °F, although coefficients ranging from 4 to 7 millionths per °F have been observed. The coefficient appears to be influenced principally by the character of the aggregate, being highest for quartz, which by itself has a coefficient of about 7 millionths per °F. Quartz is followed, in order, by sandstone, granite, basalt, and some limestones which themselves may have a coefficient of 3 millionths or less. (However, limestones vary greatly in this respect; some have a coefficient as high as 7 millionths.) Gravel may vary considerably in mineralogical composition but is usually made up of heterogeneous mixtures having a coefficient of about 5 to 7 millionths.

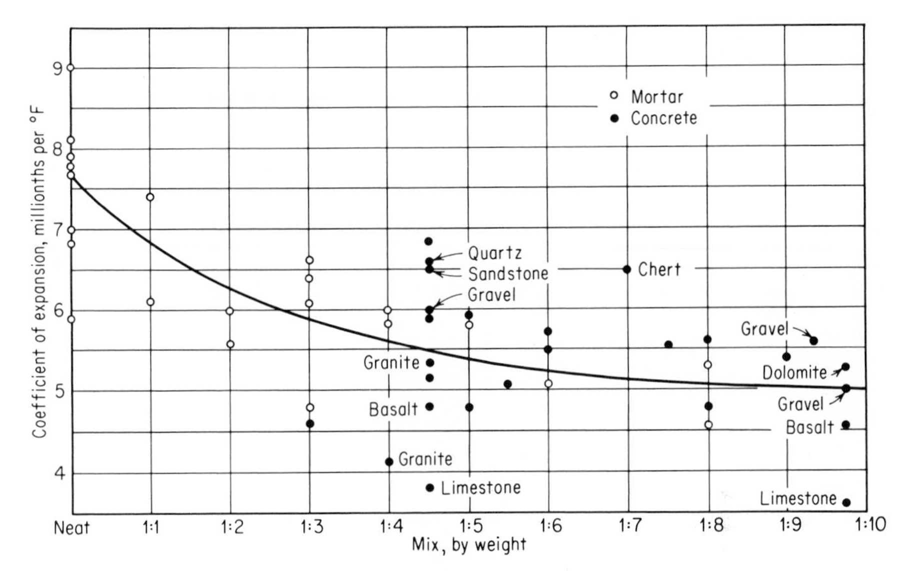

Fig. 12.8. Thermal coefficients of expansion of neat cements, mortars, and concretes. (*From U.S. Bureau of Reclamation* [106].)

Some crystalline rocks are anisotropic; in other words they have different coefficients along each of the various crystalline axes. For example, the coefficient of feldspar is about 0.5 millionths on one axis and 9 millionths on another axis. The coefficient is somewhat greater for rich mixes than for lean ones [110, 1201]. The coefficient for neat pastes has been reported as 6 to 12 millionths per °F. Figure 12.8 shows some experimental values of the thermal coefficient for neat cements and for mortars and concretes containing different types of aggregate.

It has been suggested that damaging internal stresses may develop when the change in volume of aggregate particles due to temperature variations is substantially different from that of cement paste or where there are large differences in the coefficients of expansion among the aggregate particles. However, proof of such failure has not been established.

CREEP OF CONCRETE

12.21 Factors affecting creep. Creep of concrete, resulting from the action of a sustained stress, is a yielding of the concrete with time. It does not include any instantaneous strains caused by loading or any shrinkage or swelling caused by moisture changes. In testing, the latter effects are corrected for by data from unloaded control specimens. Creep may be due partly to viscous flow of the cement-water paste, closure of internal voids, and crystalline flow in aggregates, but it is believed that the major portion is caused by seepage into internal voids of colloidal (adsorbed)

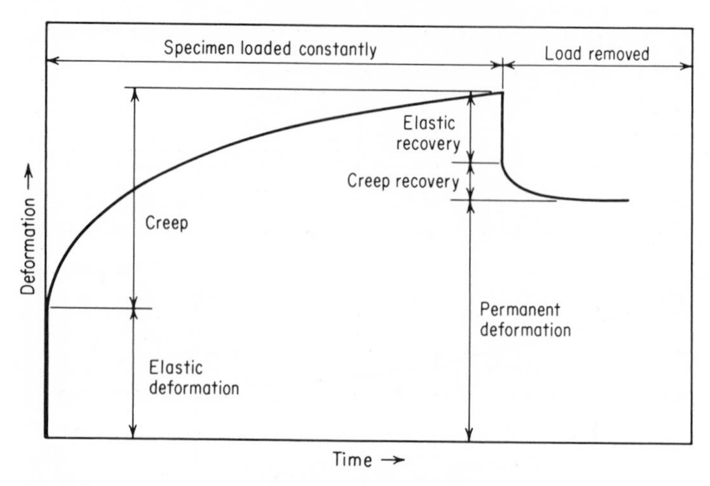

Fig. 12.9. Concrete develops both elastic and creep deformations while under load and shows both elastic and creep recoveries after removal of the load.

water from the gel that is formed by hydration of the cement. Although water may exist in the mass as chemically combined water, and as free water in the pores between the gel particles, neither of these is believed to be involved in creep. The rate of expulsion of the colloidal water is a function of the applied compressive stress and of the friction in the capillary channels. The greater the stress, the steeper the pressure gradient with resulting increase in rate of moisture expulsion and deformation. The phenomenon is closely associated with that of drying shrinkage. Glucklich [1272] has shown that concrete from which all evaporable water has been removed exhibits practically no creep.

Some investigators [1271] prefer to divide creep into two parts: (1) basic creep under conditions of hygrometric equilibrium due to molecular diffusion of the gel and adsorbed water causing a partly viscous (irrecoverable) and partly delayed elastic (partially recoverable) behavior, and (2) drying creep due to a mechanism similar to that involved in free shrinkage due to desiccation, but the applied stress may modify the shrinkage. Drying creep is presumably irrecoverable with respect to stress but may show partial reversibility on restoration of the original moisture content.

The magnitude of the creep depends upon several factors relating to the quality of the concrete such as the aggregate-cement ratio, water-cement ratio, kind of aggregate and its grading, composition and fineness of cement, admixtures, pozzolans, and the age at time of loading. It also depends upon the intensity and duration of stress, moisture content of the concrete, the temperature and humidity of the ambient air, and the size of the mass.

Upon release of the load, both an elastic and a creep recovery occur, so that considering both the loaded and after-unloading periods the deformations are in general as shown in Fig. 12.9.

12.22 Effect of stress and age when first loaded. The greater the degree of hydration of the cement at the time of load application, the lower the rate and total amount of creep. One explanation of this is that the expulsion of moisture from the gel becomes more difficult as the porosity is decreased through hydration. Since the extent of the hydration is indicated by the strength of a given concrete, it can be said that creep varies inversely as the strength.

The creep of concrete cylinders which were kept loaded in dry air for 10 years is shown in Fig. 12.10. Some of these cylinders were loaded at the age of 28 days and others at the age of 3 months. The earlier loading of lower-strength concrete increased both the instantaneous deformations and the total creep by appreciable amounts; still earlier loadings would have produced appreciably greater differences.

For these and many similar tests the approximate proportionality of creep to stress appears to have been maintained for stresses up to about 35 percent of the ultimate, although at very early ages the creep may not be proportional to stress; for example, the early creep for a stress of 600 psi

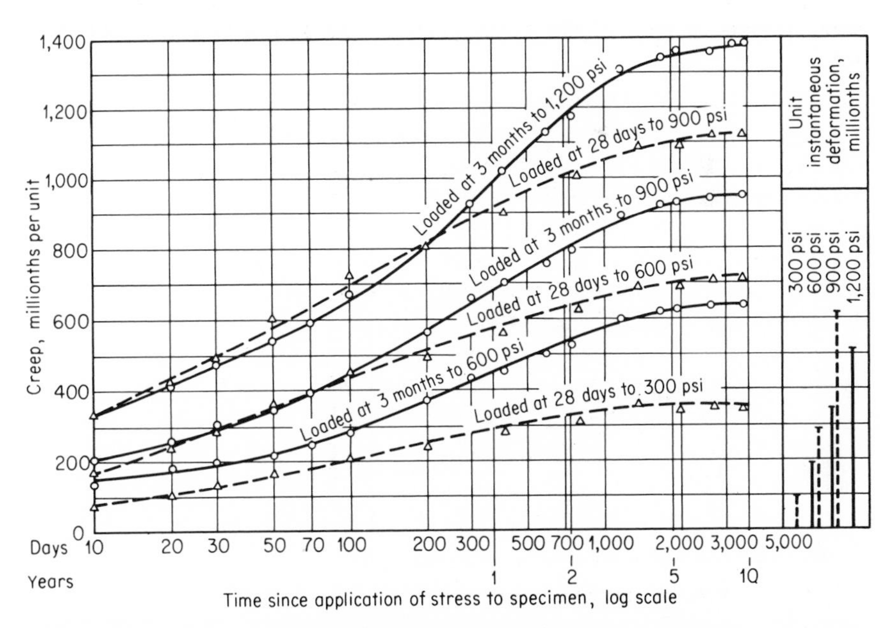

Fig. 12.10. Creep under long-sustained compressive stress. Cylinders 4 by 14 in.; aggregate 0 to 1½-in. crushed granite; aggregate-cement ratio by weight 5.05; water-cement ratio by weight 0.69; cured moist until loaded. Storage after loading, in air at 70°F and 70 percent relative humidity. Creep equals total change corrected for instantaneous elastic deformations and for shrinkage of unloaded control cylinders.

may be three or four times that for a stress of 300 psi. As shown in Fig. 12.10, the cylinders loaded to only 300 psi exhibited maximum creep after about 4 years. Later observations show that those at the higher stresses still appear to be creeping at a very small but measurable rate even after 25 years. For the concrete loaded at the age of 28 days to a stress of 900 psi, the creep after loading for 10 years amounted to 1,130 millionths, or about 1.3 in. per 100 ft. At 25 years the creep was 1,170 millionths.

Previous loading history influences the creep at a later age caused by an increase in sustained load. The later-age creep rates appear to be dependent on the relative load level at early ages; the lower the initially applied load, the greater the creep rate at later ages due to increase of the load [1277].

12.23 Effect of water-cement ratio and mix. A higher water-cement ratio increases the size of the pores in the paste structure, so that water may the more readily escape, and then under a sustained load the water of adsorption may be expelled more readily to cause a high rate of creep as shown in Fig. 12.11.

Tests have shown that the percentage difference in creep of two mixes having the same water-cement ratio was about equal to the percentage difference in paste content of the mixes [1252]. As shown in Fig. 12.11, the

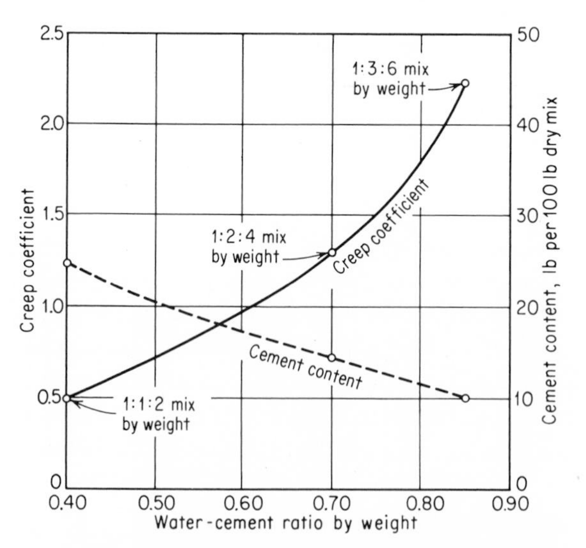

Fig. 12.11. Effect of water-cement ratio on creep. Slump 2 in.; curing 2 days; age at loading 28 days; sustained compressive stress 600 psi; storage condition, 65 percent relative humidity; creep coefficient in millionths per unit length per psi. (*From Lorman* [1254].)

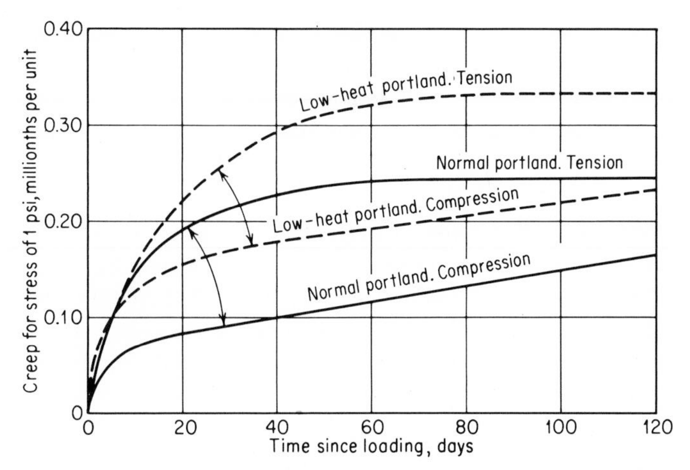

Fig. 12.12. Creep in compression and tension for mass-cured concretes. All specimens sealed in thin sheet-copper containers. Aggregate 0 to $1\frac{1}{2}$ in.; cement content 1.0 bbl per cu yd; slump 1 in.; curing, mass from 40°F. Age at loading 28 days. *(From Davis, Davis, and Brown* [1252].)

creep of a 1:3:6 mix may be over four times that for a 1:1:2 mix of the same 2-in. slump. This may be explained by the fact that, for normal mixes, the increase in creep due to the high water-cement ratio of a lean concrete more than offsets the tendency toward a reduction of creep caused by the low paste content, so that, in general, lean mixes exhibit considerably greater creep than do rich mixes. However, when a constant water-cement ratio is maintained, creep increases as the slump and cement content increase or as the amount of cement paste is increased.

12.24 Effect of composition and fineness of cement. The composition of cement affects the creep primarily through its influence upon the degree of hydration; slow-hardening cements such as low-heat portland and portland-pozzolan cements creep more than cements which hydrate more rapidly. Figure 12.12 indicates that in both tension and compression the creep of concrete made with low-heat cement is about one-third greater than for concrete made with normal cement. Portland-pozzolan cements also show greater creep, and it appears that the creep increases as the percentage of replacement increases. This serves to explain why low-heat portland and portland-pozzolan cements have served so effectively in relieving stresses in large dams as the mass cools and have shown superior resistance to cracking.

Investigations of the influence of fineness of the cement have shown that for concretes containing low-heat cements a reduction of the specific surface from 2,200 to 1,300 sq cm per g increased the creep about one-third, whereas a similar change in fineness for normal cement had no appreciable effect [1252].

Although the use of approved air-entraining agents has no appreciable effect on creep, proprietary compounds should not be used unless their effects have been determined previously.

12.25 Effect of character and grading of aggregate. The effect of mineral character of aggregate is shown in Table 12.10 for six concretes. For any given concrete, the same mineral aggregate was used from fine to coarse; the grading was the same for all mixes. After carrying a sustained stress of 800 psi for about 5 years the maximum creep (1,300 millionths) was

Table 12.10 Effect of mineral character of aggregate upon creep*

Aggregate	Creep after 5 years, millionths†
Sandstone	1,300
Basalt	1,110
Gravel	950
Granite	840
Quartz	790
Limestone	550

* From Ref. 1251.

† 4 by 14-in. cylinders loaded to 800 psi after moist curing to age 14 days. Storage in air of 50% R.H.

exhibited by the sandstone concrete and the minimum (550 millionths) by the limestone.

As all aggregates were batched in a saturated, surface-dry condition, and their absorption factors were generally low, the large variations in creep were not due to the moisture conditions of the aggregates. Neither were they due to seepage from the identical paste used in each mix. It is possible that variations in crystalline slip, particle shape, surface texture, and pore structure of the aggregates may have had some influence. In general, aggregates that are hard, dense, and have low absorption and high modulus of elasticity are desirable when concrete with low creep is needed, as they offer greater restraint to the potential creep of the cement paste. For lightweight concretes see Art. 14.22.

Although the data are somewhat conflicting, it appears that the greater the maximum size of an aggregate graded uniformly from fine to coarse, the less the creep of the concrete. This is particularly true when the water-cement ratio is reduced, as is normally the case when an aggregate of higher fineness modulus is used.

12.26 Effect of moisture conditions of storage. Creep appears to be influenced by the humidity of the air in so far as it affects the seepage of moisture from the concrete. Naturally, an increase in the humidity of the atmosphere reduces the rate of loss of moisture or water vapor to the surrounding atmosphere, slows down the flow of moisture or water vapor to the outer surface, and thus reduces the seepage. Another factor affecting compressive creep is that drying shrinkage at or near the surface results in a reduction of the cross-sectional area remaining in compression and therefore causes higher stresses (and elastic strains) on the central core. The magnitude of creep for various moisture conditions of storage is shown in Table 12.11. Although these values indicate that for a concrete loaded to 800 psi at the age of 28 days the creep in air at 70 percent relative humidity was about double that for water storage, for similar concrete loaded at 3 months to 1,200 psi the creep for the air-storage condition at 70 percent relative humidity was about 2½ times that for water storage.

12.27 Effect of size of mass. The larger the mass subjected to sustained loading, the less the creep. This is probably due to the reduced seepage, as the path travelled by the expelled water is greater with a resulting increase in the frictional resistance to the flow of water from the interior. The general effect of size of specimen upon creep is shown in Fig. 12.13, which includes 6-, 8-, and 10-in.-diameter cylinders stored in fog to eliminate the effect of surface drying. These results show that creep in the 10-in. cylinder is only about one-half that for the 6-in. cylinder. It has been estimated that creep of mass concrete may be about one-fourth that obtained with small specimens stored in moist air.

Table 12.11 Effect of moisture condition of storage upon creep*

Storage condition	Creep after 5 years, millionths
50% relative humidity	850
70% relative humidity	710
100% relative humidity	350
Water	300

* From Ref. 1251.
Aggregate-cement ratio, 5.67 by weight. Water-cement ratio, 0.89 by volume. Age at loading, 28 days. Sustained stress, 800 psi. Specimens, 4 by 14-in. cylinders.

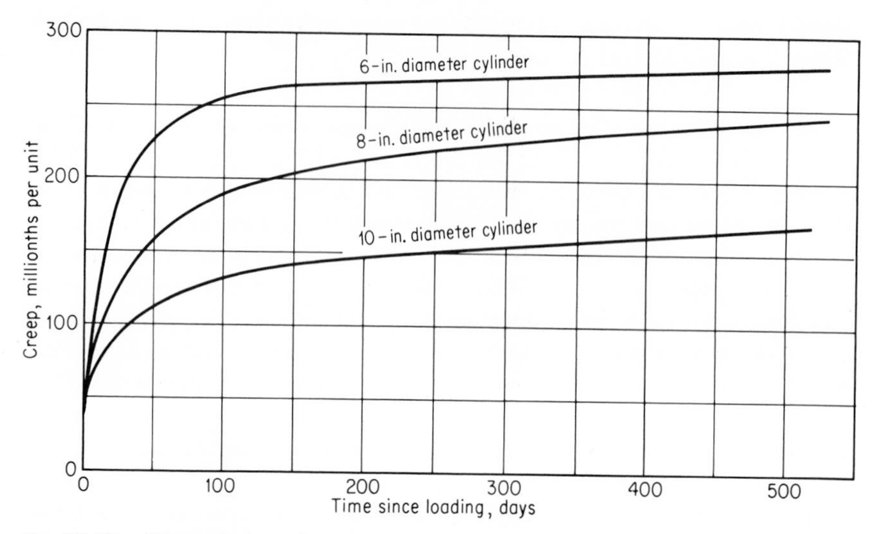

Fig. 12.13. Effect of size of specimens upon creep. Aggregate-cement ratio 6.95 by weight; water-cement ratio 0.61 by weight; age at loading 28 days; sustained stress 800 psi; storage in fog before and after loading. (*From Davis, Davis, and Hamilton* [1251].)

For storage in dry air, seepage may occur much more readily from small specimens, and the effect of size on creep may become more pronounced.

12.28 Creep in axial tension and compression. Test results comparing the creep of concrete under sustained tension and compression are shown in Fig. 12.12. For these and other similar tests the rate of creep in tension for the first few weeks is appreciably greater than the rate of creep in compression, although later the reverse is true. The ultimate value of creep for given conditions is about the same for tension as for compression.

12.29 Estimation of creep. The total creep ϵ per unit stress (exclusive of shrinkage) of a loaded specimen can be expressed by

$$\epsilon = f(K) \ln (t + 1)$$

where $f(K)$ is a function representing the rate of creep deformation with time, and t is time under load, in days. The function $f(K)$ is large when concrete is initially loaded at an early age and small when concrete is loaded later. It varies with the several factors listed above. The function $\ln (t + 1)$ indicates that the rate of creep is relatively rapid at early ages after loading and then decreases gradually, until after a few years it becomes insignificant. Roughly, about one-fourth of the ultimate creep occurs within the first month and three-fourths within the first year. By use of the above equation and a value of K obtained from limited-time creep tests using the actual mix and storage conditions, values of creep may be predicted without performing long-time tests.

An equation giving the approximate creep and creep recovery of concrete based on the modulus of elasticity of the aggregate, the sustained modulus (Art. 13.7) of the matrix, and the proportion of aggregate in the mix has been developed and checked experimentally [1304].

The ultimate magnitude of creep per unit stress (psi) ranges from 0.2 to 2 millionths, but it is usually about 1 millionth per unit of length. This is about three times the elastic deformation of concrete having a modulus of elasticity of 3 million psi (Art. 13.1). The combined effect of elastic plus creep deformations resulting from long sustained loads is represented by a single quantity called the sustained modulus of elasticity (see Art. 13.7).

12.30 Creep recovery. Immediately upon the release of a sustained load, there is an elastic recovery of length followed by a further creep recovery which continues for a period of several days. The magnitude of this latter recovery varies with the previous creep and depends very much upon the period of the sustained load. The longer the loading period, the less the recovery, perhaps because of the hardening of the concrete while in the deformed condition. In general, creep at early ages is largely unrecoverable, but much of that occurring in old or dry concrete is recoverable.

The recovery of one test group after loading for 11 months under stress of 800 psi is shown in Fig. 12.14. The magnitude of creep before release of load is shown in parentheses on the curves. It will be observed that the specimens which were stored in air exhibit about twice the recovery of specimens stored under water. For a group of specimens loaded for only 28 days, the ultimate recovery for air-storage specimens was 39 percent

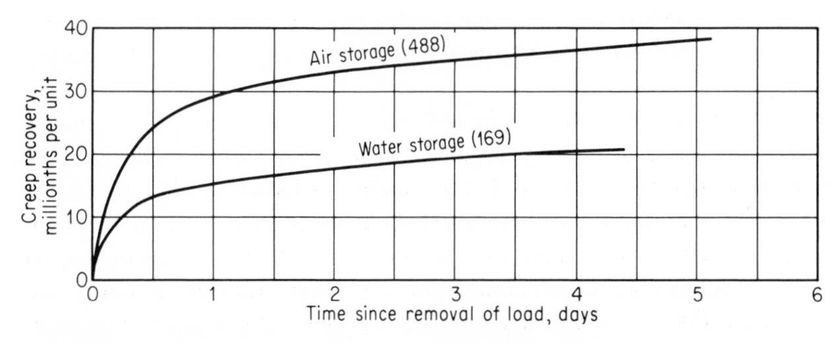

Fig. 12.14. Creep recovery after 11 months under load. Previously sustained stress 800 psi; creep in millionths before release of load indicated in parentheses on each curve. (*From Davis, Davis, and Hamilton* [1251].)

of the creep under stress, while for water-storage specimens it was only 28 percent of the creep.

12.31 Creep of reinforced concrete. Simply supported beams have an unsymmetrical section as they are reinforced only in tension. Also, the concrete in the upper part has a higher water-cement ratio than that of the lower part as a result of water gain, so that it is not so strong. Consequently, even unloaded beams will gradually warp because the top concrete shrinks more than the bottom concrete, and shrinkage is resisted by the steel at the bottom. When subjected to sustained loading these beams will be deflected downward by warping as well as by the load. The deflection-time and compressive strain-time curves reach about 75 percent of the ultimate values within 6 months. The rate and ultimate values are affected by the same factors as those for plain concrete and in addition by such factors as position and amount of reinforcement and ratio of span length to depth of section.

The compressive strains increase much more rapidly than the strains at the tensile-steel level. The reason for this is that the effects of shrinkage and creep are in the same direction in the top but are opposite to each other in the bottom of the beam.

The ratio of span length to depth of beam has a very important effect on creep. An increase in the ratio from 20 to 70 increases the creep deflection of simply supported beams about four to six times.

The use of compressive steel at the section of maximum moment in simply supported reinforced-concrete beams is effective in reducing creep deflection and compressive-creep strains. It has been shown that such steel equal in amount to the tensile steel reduced the creep deflection to about one-half.

At the sections of maximum positive moment in continuous beams having the same section, span length, and load as that for simple span beams, after 1 year of sustained loading, the creep deflection of the continuous beams was between one-third and one-half of the values for the simply supported beams; and the compressive-creep strains of the continuous beams were between 60 and 80 percent of the values for simply supported beams.

The results of tests upon reinforced concrete columns of 5-in. diameter, under load for over 5 years, are summarized in Table 12.12. The stresses in the longitudinal steel and in the concrete, included in this summary, are due to the combined effect of loads, volume changes due to moisture loss or gain, and to hydration of the cement. For the columns stored in air the combined effect of creep and shrinkage transferred more and more load from the concrete to the steel, so that, for 1.9 percent of reinforcement, the stress in the steel increased more than fivefold, and for the col-

Table 12.12 Unit stresses in reinforced-concrete columns*

Nominal strength of concrete at 28 days, psi	Axial steel ratio, %	Total load applied to column, lb	Unit stress, psi							
			At time of application of load†		1 year under load		3 years under load		5½ years under load	
			Steel	Concrete	Steel	Concrete	Steel	Concrete	Steel	Concrete
Columns stored in air of 50 percent relative humidity at 70°F										
2000	5.0	22,300	9,660	875	26,900	−20‡	27,400	−50‡	28,000	−75‡
2000	1.9	14,200	6,540	610	34,800	60	35,700	45	37,100	15
4000	1.9	21,800	7,860	975	37,500	395	40,400	340	41,700	315
Columns stored under water at 70°F										
2000	5.0	19,200	7,200	810	10,050	665	10,890	620	11,400	590
2000	1.9	13,650	5,460	605	7,980	555	9,060	535	9,480	525
4000	1.9	20,600	7,320	925	10,590	865	11,670	840	12,120	835

* From Ref. 1252; diameter of columns, 5 in.
† Age at loading, 28 days.
‡ Minus sign indicates tension.

umns with 5 percent of reinforcement the concrete developed low tensile stresses.

For the columns stored under water the moisture expansion offset a part of the reduced creep occurring under moist conditions with the result that changes in stress with time were much less than for the columns in air storage.

12.32 Creep of prestressed concrete. For pretensioned concrete, where the steel is tensioned before casting the concrete around it, great care should be taken to keep drying shrinkage and creep at a minimum in order to avoid large reductions in the prestress. Consequently, special attention should be given to the choice of materials used, the design and placement of the concrete, the conditions and duration of the curing period, and the age at loading. Under some conditions the loss of prestress of the steel due to shrinkage and creep may be as much as 50,000 psi, although it is usually considered to be about half that much, so that the original prestress should be well over 150,000 psi to keep the percentage loss to a minimum.

In post-tensioned concrete the stressing is done after some shrinkage has occurred. Also, when the steel is not bonded, it is possible to increase the tension at some later time, so that the losses in the prestress force are not so important as those for pretensioned concrete.

QUESTIONS

1. Why is contraction usually of more significance than expansion in a concrete structure?

2. What is the effect of creep of concrete upon the reinforcement steel?

3. How are volume changes related to the gel structure of concrete?

4. Why does the large shrinkage of fresh concrete have little significance?

5. What is the meaning of autogenous volume changes, and what is the final effect of such changes?

6. List the several factors affecting shrinkage and expansion of concrete due to moisture changes.

7. How does the shrinkage of low-heat and portland-pozzolan cements compare with that of normal cement?

8. How is the shrinkage of a cement paste related to its fineness?

9. What effect does the type of aggregate and its maximum size have upon shrinkage of concrete?

10. Two concretes are of the same dry mix, but one has a wetter consistency. How would their shrinkages compare?

11. Two concretes have the same water-cement ratio, but one is a richer mix. How would their shrinkages compare?

12. What is the effect of steam curing upon drying shrinkage?

13. Why is the unit shrinkage less for a large mass than for a small specimen?

14. How do low and high percentages of reinforcement affect shrinkage of concrete?

15. What is the effect of type of aggregate and richness of mix upon thermal volume changes of concrete?

16. How is creep related to the intensity of stress and the age when first loaded?

17. What is the effect of mix and water-cement ratio upon creep?

18. How does the creep of low-heat-cement concrete compare with that for normal concrete?

19. Discuss the effect of character and maximum size of aggregate upon creep.

20. How is creep affected by the moisture condition of the concrete?

21. What is the effect of size of mass upon the creep?

22. In a reinforced-concrete structure, what effect does creep have upon the stresses in the concrete and steel?

THIRTEEN
OTHER PROPERTIES

ELASTIC PROPERTIES OF CONCRETE

13.1 Modulus of elasticity. The modulus of elasticity is defined as the change of stress with respect to elastic strain (deformation) and may be computed from the equation

$$\text{Modulus of elasticity} = \frac{\text{unit stress}}{\text{unit strain}}$$

It is a measure of the stiffness, or of the resistance of the material to deformation.

Although it has been shown in Chap. 12 that concrete exhibits creep under sustained loading, within limits it may be definitely considered an elastic material, and its modulus of elasticity is involved in practically all calculations of structural deformation. It is used in the analysis of reinforced structures to determine the stresses developed in simple elements and to determine the moments, stresses, and deflections in more complicated structures. The change in quality of a given concrete, as shown by its reduction in "dynamic" modulus (Art. 13.2) when subjected to alternate freezing and thawing, is used to determine the relative durability of concrete when exposed to severe climatic conditions.

If in a structure the loads are applied very gradually or are imposed over extended periods of time, it may be more appropriate to use values of the "sustained" modulus (Art. 13.7) instead of that for purely elastic action.

Since the elastic action of concrete is affected by conditions that in practice cannot readily be controlled, engineers concerned with its elastic action usually assume rather general and conservative values for use in design.

13.2 Methods for determining moduli of elasticity. Since concrete, like most other structural materials, is imperfectly elastic, the stress-strain diagram is a curved line. Hence, three methods have been used for computing moduli of elasticity from stress-strain diagrams, as shown for the low-strength concrete in Fig. 13.1.

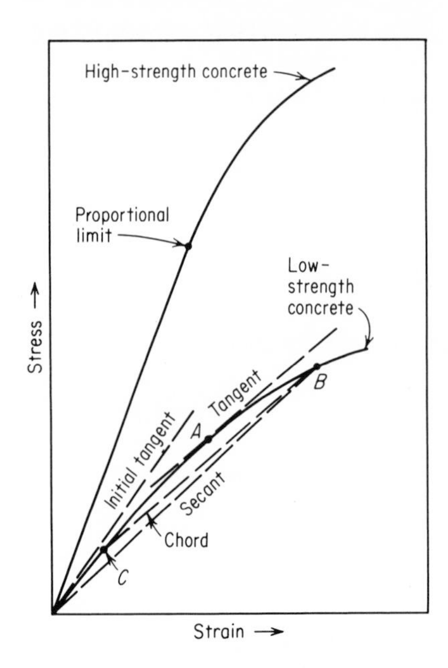

Fig. 13.1. Stress-strain diagrams for concrete, showing methods of determining moduli of elasticity.

The "initial tangent modulus" is represented by the slope of a tangent to the stress-strain curve, drawn through the origin.

The "tangent modulus" is represented by the slope of a line drawn tangent to the stress-strain curve at any point A on the curve.

The "secant modulus" is represented by the slope of a line drawn from the origin to any point B on the curve.

The "chord modulus" is represented by the slope of a line drawn between any points B and C on the curve [C469].

For concrete that has hardened throughly and has been moderately preloaded, the stress-strain diagram may be reasonably straight over the range of stresses allowed in practice (see the curve for high-strength concrete in Fig. 13.1), in which case all the methods described will give nearly the same result. However, many concretes exhibit at least a slight curvilinear relation of stress and strain even at allowable loads.

The initial tangent modulus is of limited value as it has significance only for low stresses. The tangent modulus cannot be determined with accuracy, as the only practicable way to determine its value is to draw a tangent to the curve, by eye, and to compute the slope of the resulting line. The secant modulus is the most practical and is in most general use, because it represents the actual deformation at the selected point, and no uncertainties are involved in its determination, except that any possible slack in the compressometer will require a correction in the first observed strain. The chord modulus avoids the need for such a correction by eliminating the lower strain which may be in error.

The modulus of elasticity in flexure may be determined from the deflection of a loaded beam. For a beam simply supported at the ends and loaded at mid-span, an approximate modulus may be determined from the following expression:

$$E = \frac{PL^3}{48Iy}$$

where E = secant modulus of elasticity, psi
P = applied mid-span load, lb
L = distance between supports, in.
I = moment of inertia of the section
y = mid-span deflection due to load P, in.

The approximation in the above equation arises from neglecting the deflection caused by shear. Considering the effect of both bending moment and shear, the equation becomes [110]

$$E = \frac{PL^3}{48Iy}\left[1 + (2.4 + 1.5\mu)\left(\frac{h}{L}\right)^2 - 0.84\left(\frac{h}{L}\right)^3\right]$$

where μ = Poisson's ratio
h = depth of the beam, in.

As h/L increases, the effect of shear may be appreciable.

A dynamic, "resonance," or "sonic" method for determining the modulus of elasticity uses a prism or cylinder of concrete, which is vibrated and its natural frequency measured [1308–1311]. The apparatus consists of wire supports for the specimen, a driving circuit, and a pickup circuit. The specimen is supported at its nodal points, 0.224L from each end, and caused to vibrate by a radio speaker or driver connected to an oscillator, the frequency of which can be varied. A crystal pickup and output meter attached to one end of the specimen is used to determine the resonance, which causes a maximum deflection when the frequency of the driver is the same as the natural or fundamental frequency of the specimen. This frequency is indicated by the tuning dial on the oscillator.

The dynamic modulus can be calculated from the following equations [C215]:

$$E = CWn^2$$

where E = dynamic modulus of elasticity, psi

$$C \text{ (for prism)} = \frac{0.00245L^3T}{bt^3} \text{ sec}^2/\text{sq in.}$$

$$C \text{ (for cylinder)} = \frac{0.00416L^3T}{d^4} \text{ sec}^2/\text{sq in.}$$

W = weight of specimen, lb
n = fundamental transverse frequency, cycles per second
L = length of specimen, in.
d = diameter of cylinder, in.
t, b = dimensions of cross section of prism, in., t being in direction of vibration
T = correction factor which depends on ratio of radius of gyration K to length of specimen L, and on Poisson's ratio. K for a cylinder is $d/4$ and for a prism is $t/3.464$. Values of T for Poisson's ratio of $\frac{1}{6}$ are shown in Table 13.1

The specimen for this type of test should have a length of about four times its diameter or depth, as for short specimens (i.e., large values of K/L) the corrections that must be applied become large and uncertain.

Table 13.1 Values of correction factor T

K/L	T	K/L	T
0.00	1.00	0.09	1.60
0.01	1.01	0.10	1.73
0.02	1.03	0.12	2.03
0.03	1.07	0.14	2.36
0.04	1.13	0.16	2.73
0.05	1.20	0.18	3.14
0.06	1.28	0.20	3.58
0.07	1.38	0.25	4.78
0.08	1.48	0.30	6.07

The dynamic method is rapid and has the advantage that the modulus of elasticity can be determined without damaging the specimen. This permits the determination of changes in the modulus of a single specimen with time due to curing, durability freeze-thaw tests, or other conditions.

However, this method is not sensitive to small changes in the paste content of concrete and is materially affected by the heterogeneity of the mix, so that determination of the true modulus cannot be assured [1315].

Instruments for determination of the dynamic modulus and the general quality of the concrete, *in situ,* are described in Ref. 1312 and 1313. These instruments measure the velocity of waves through the concrete and are applicable to specimens of any size or shape.

The modulus of elasticity in shear, sometimes called the modulus of rigidity, may be determined dynamically [C215] or by applying a known torque to a cylindrical concrete specimen and observing the angle of twist developed over a given gage length [1318]. Then

$$E_s = \frac{TL}{\theta J}$$

where E_s = modulus of elasticity in shear, psi
T = applied torque, in.-lb
L = length in which angle of twist θ is developed, in.
θ = angle of twist, in radians, in length L
J = polar moment of inertia of cross section

13.3 Effect of method of test on modulus of elasticity. Within limits concrete acts like an elastic material, but also it has been shown in Chap. 12 that concrete has a definite tendency to creep under load. These characteristics are so intimately related that it is difficult to determine where the one leaves off and the other begins, if actually there is a distinction.

During a loading test both these factors are active, with the result that Noble [1302] and Glanville [1250] have shown that the slower the rate of loading, the sharper the curvature of the stress-strain diagram and the lower the computed modulus. From the stress-strain diagrams in Fig. 13.2 it appears that the secant modulus at 1,000 psi for loading over a

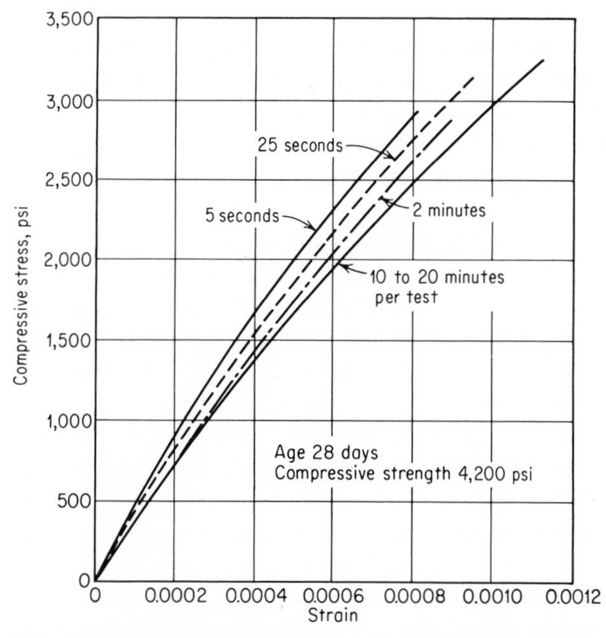

Fig. 13.2. Stress-strain diagrams for concrete determined at different speeds of loading. (*From Noble* [1302].)

10 to 20-min period may be about 15 percent lower than for a 5-sec loading period. All investigations are in agreement that changing the loading time from 2 min to 20 min has no significant effect upon the modulus at allowable stresses. Since some of the loads to which concrete structures are subjected often are of long duration, it appears reasonable to apply the load slowly to permit the rapidly developed early creep to be included in the observed strain. In many cases it may even be desirable to determine a sustained modulus of elasticity, which includes the effect of creep over a considerable period of time and may be defined as the ratio of stress to elastic-plus-creep deformation.

Since the driving mechanism used in the dynamic method produces almost infinitesimal stress in the specimen, the value of Young's modulus computed by this method is apparently the tangent modulus at zero stress, and therefore higher than any secant modulus.

The repeated application of loads within allowable values has no appreciable effect upon the modulus. Although some specimens may show a slightly higher modulus for a second loading than for the first, the reverse has been found for about the same number of tests. In many cases the same values have been obtained for repeated loadings [1302]. However, when the first loading was much beyond allowable values, then subsequent tests showed that the concrete had been damaged and the properties of the material had been permanently changed, resulting in a lowered modulus.

13.4 Effect of characteristics of concrete on modulus of elasticity. The factors which influence the strength of concrete generally influence the modulus of elasticity in similar fashion, although usually to a lesser degree. The principal variables include the richness of mix, water-cement ratio, age and curing conditions, kind and gradation of aggregate, and moisture content at time of test.

The effect of age is shown in Fig. 13.3, which indicates that the modulus increases rapidly during the first few months and continues to increase even at the age of 3 years, particularly for the rich mixes. Comparing the three groups of curves of the figure it is seen that the richer the mix, the greater the relative increase in the modulus with age. Although the 1:4½ mix has the highest modulus at the age of 200 days, the 1:3½ mix has the highest value at the age of 1,000 days. The effect of intensity of stress upon the secant modulus is shown for the several conditions tested. The general reduction in the observed secant modulus with higher stress is due to plastic deformations which cause the curvature of the stress-strain diagram.

The effect of the water-cement ratio is shown in Fig. 13.4, which indicates no particular trend for ratios of about 0.6 to 0.7, possibly because of the less workable character of the drier mixes which had no slump,

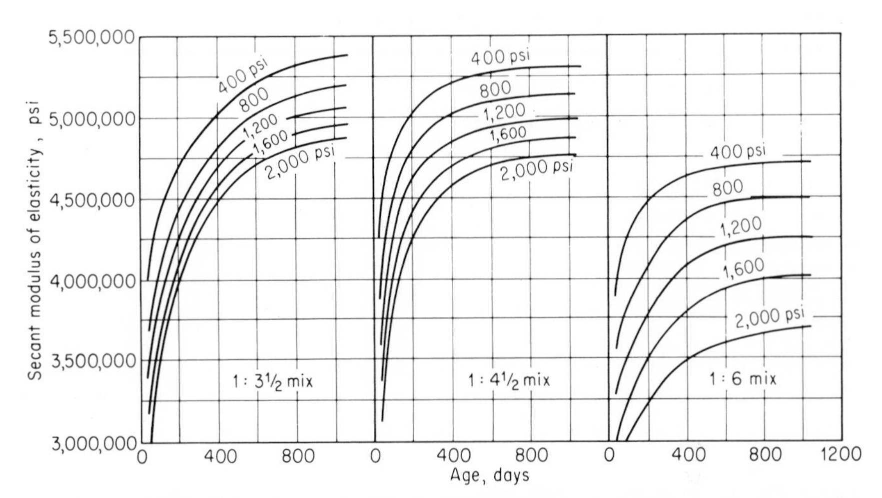

Fig. 13.3. Effect of age, mix, and stress upon modulus of elasticity. All mixes had a 1-in. slump and were stored in damp sand. (*From Davis and Troxell* [1301].)

but clearly shows a general reduction in value of the modulus for the higher water-cement ratios.

The pronounced influence of the type of aggregate also is shown in Fig. 13.4, the hard, flinty aggregates having high values and the soft limestone having the lowest values for the modulus. Since the deformation produced in the concrete is partly an elastic deformation of the aggregate, it is reasonable to expect that the higher values of the modulus would

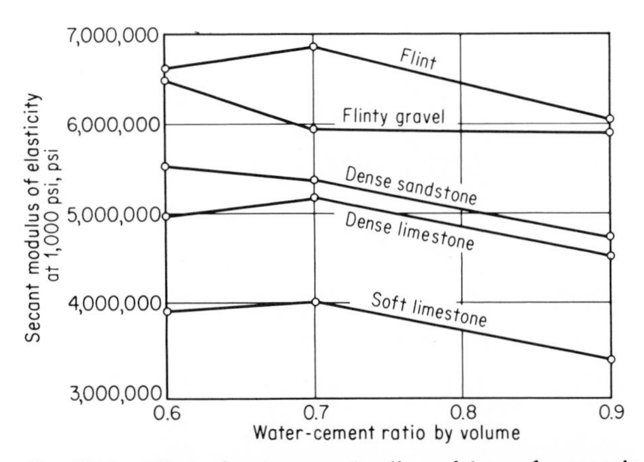

Fig. 13.4. Effect of water-cement ratio and type of aggregate upon modulus of elasticity. All mixes contained 6 sacks cement per cu yd. Age at test, 56 days. (*From Noble* [1302].)

be obtained for the concrete made of stiffer aggregates. In this connection, the following moduli for rock cores are significant:

Type of rock	Modulus of elasticity, psi
Trap	13,300,000
Granite	8,660,000
Sandstone	7,400,000
Limestone	4,000,000

Equations for an approximate evaluation of the modulus of concrete based on the moduli of the aggregate and the matrix, and their proportionate volumes, have been developed and checked experimentally [1304].

The grading of a given type and size of aggregate, as well as the maximum size of a well-graded aggregate, has the same general effect upon the modulus of elasticity as upon strength; if the strength is increased, then the modulus is increased. So long as the mix is not harsh and unworkable, there is a decided tendency for the modulus of elasticity to increase with the fineness modulus [1300]. This effect is due, in large part, to the decrease in water-cement ratio with increase in coarseness of grading.

The influence of free moisture at time of test upon the modulus is shown in Fig. 13.5. The mixes were all of the same consistency and were all cured under water. Seven days prior to testing, one-half of the specimens were placed in dry air at 130°F and then cooled to room temperature

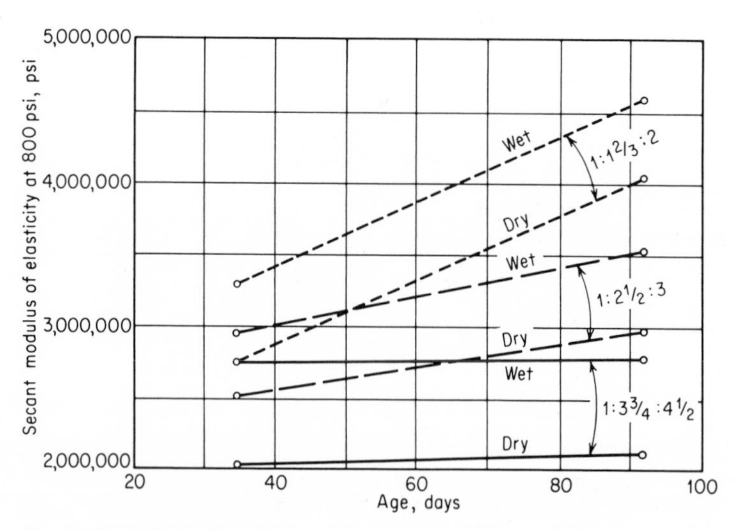

Fig. 13.5. Effect of moisture content on modulus of elasticity. (*From Davis and Troxell* [1301].)

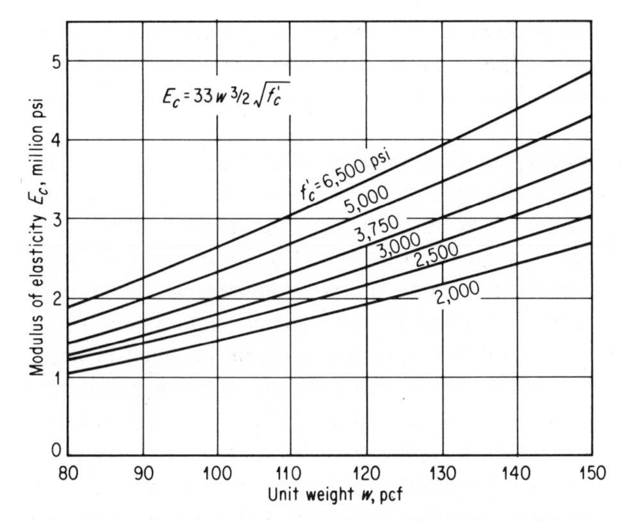

Fig. 13.6. Modulus of elasticity as a function of strength and weight of concrete. (*From Pauw* [1316].)

before loading. Figure 13.5 shows that, regardless of mix or age, the wet specimens exhibited substantially higher moduli than did the corresponding dry specimens, although the ultimate strength was higher for the dry concrete than for the wet by about the same ratios.

In general, both the modulus of elasticity and strength are greater for longer mixing times, but the increase in the modulus is considerably less than the increase in strength. A longer curing period likewise benefits both the strength and the modulus.

13.5 Relationship of modulus of elasticity to strength. Actual tests show that even when dealing with different combinations of the same materials, no universal relationship exists between strength and the modulus of elasticity, and when such variables as different aggregates are introduced, it is possible to produce concretes having the same modulus but having strengths differing by more than 100 percent.

But the static modulus of elasticity E_c of both normal and lightweight structural concrete at age 28 days may be approximately determined by the empirical formula

$$E_c = 33w^{1.5}\sqrt{f'_c}$$

where w = unit weight, pcf
f'_c = 28-day compressive strength, psi

A graphical form of this relationship is shown in Fig. 13.6. It has been accepted by the ACI Code for reinforced-concrete design. The value of the modulus is about as dependent on the weight of the concrete and the method of test used to determine it as it is on the compressive strength of the concrete.

13.6 Effect of type of loading on the modulus of elasticity. For loads up to 50 percent of the ultimate the modulus of elasticity in tension may be slightly smaller than the compressive modulus, but for general design purposes they may be considered equal. The modulus, as determined from flexure tests, is likewise about the same as the compressive modulus.

Studies comparing static compressive and dynamic determinations of the modulus of elasticity show that the static compressive modulus is lower (see Art. 13.3) and that both the age of the concrete and the magnitude of the modulus affect the ratio of the two values [1311]. In general, the older the concrete and the higher the modulus, the better the agreement. For a compressive modulus of 3,000,000 psi at age 28 days the ratio of compressive to dynamic modulus may be about 0.6, but for a compressive modulus of 5,000,000 psi at age 3 years the ratio may approach unity.

The modulus of elasticity in shear varies from about 0.4 to 0.6 of the corresponding modulus in compression [1318]. Its value depends upon the type of aggregate and the characteristics of the concrete, but the conditions producing the highest compressive modulus do not necessarily produce the highest shear modulus.

13.7 Sustained modulus of elasticity. In order to include the elastic and creep effects in one expression and thus to aid in visualizing the net effects of stress-strain conditions up to any given time, a quantity called the sustained modulus of elasticity is occasionally used. The sustained modulus for any conditions is computed by dividing the constant unit sustained stress by the sum of the elastic and creep deformations at any given time. Since most actual structures are not subjected to long-sustained *constant* stresses, the use of the sustained modulus is usually limited to comparisons of concretes in the laboratory. When the sustained stress varies because of a constant dead load and fluctuating live load, a so-called "effective modulus of elasticity" corresponding to the sustained modulus can be determined in the laboratory by reproducing the time-temperature and time-strain conditions observed in any structure [110]. The value of the modulus used in design is generally below that determined in a short-time test, as concrete will show some creep under design loads.

13.8 Significance of Poisson's ratio. For a specimen subjected to simple axial load, the ratio of the lateral strain to the axial strain within the elastic range is called "Poisson's ratio," after the French physicist who determined that its value theoretically should be 0.25 for isotropic elastic materials.

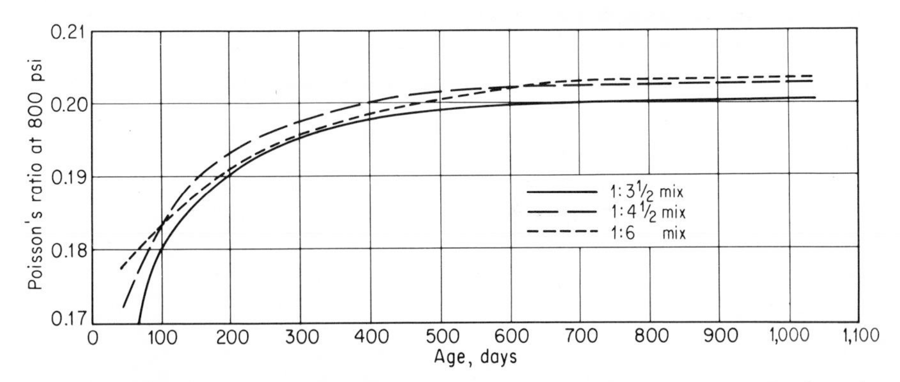

Fig. 13.7. Effect of age on Poisson's ratio for sandstone concrete. (*From Davis and Troxell* [1301].)

Poisson's ratio is not considered in most problems of concrete design, but values for this ratio are generally needed for the structural analysis of flat slabs, arch dams, tunnels, tanks, and other statically indeterminate structures.

13.9 Factors affecting Poisson's ratio. Most determinations of Poisson's ratio for allowable stresses have been within the range 0.15 to 0.20, although values as low as 0.10 and as high as 0.30 have been observed. There appears to be no consistent variation in Poisson's ratio with strength of concrete, richness of mix, or gradation of the aggregate within the range of ordinary concretes. Some test results show it to be slightly greater for low stresses than for high ones, and as shown in Fig. 13.7 it appears to increase with age of the concrete up to 1 or 2 years.

THERMAL PROPERTIES

13.10 Thermal conductivity. Thermal conductivity is of significance in connection with temperature conditions in mass concrete, and also it is important when considering (1) the insulating properties of concrete walls and floors, and (2) the sweating of concrete walls in certain climates. Structural concretes made of normal aggregates conduct heat more readily than lightweight concretes or nonstructural materials commonly used for insulation. The values of thermal conductivity for various types of monolithic concrete walls are shown in Table 13.2. The superior resistance to conduction of heat by the cinder, expanded clay, pumice, and other lightweight concretes is clearly shown by their low conductivities. The coefficients of heat transmission U for various thicknesses of monolithic walls and for 8-in. masonry

Table 13.2 Thermal properties of concrete walls*

Wall type	Conductivity K, Btu/sq ft/hr/°F/in.	Coefficient of heat transmission U, Btu/sq ft/hr/°F		
		4-in. wall	6-in. wall	8-in. wall
Monolithic concrete:				
Gravel or limestone	9.0–12.0	0.90	0.79	0.70
Cinder	5.2	0.65	0.52	0.43
Expanded clay	2.8–4.3	0.54	0.42	0.34
Pumice	2.0	0.36	0.26	0.21
Perlite or vermiculite	0.47–0.85			
Foam concrete, 20 pcf	0.51			
Concrete masory:†				
Gravel or limestone				0.52
Cinder				0.40
Expanded clay				0.36

* From Refs. 1344 and 1345.
† 8 by 8 by 16-in., 3-oval core blocks.

walls are also shown in Table 13.2. This coefficient for a solid wall is expressed as

$$U = \frac{1}{1/f_i + 1/f_o + x/K}$$

where f_i = inside surface coefficient (average value for still air is 1.65)
f_o = outside surface coefficient (average value for wind of 15 mph is 6.00)
x = thickness of wall, in.
K = thermal conductivity, Btu/sq ft/hr/°F/in. thickness

If the wall involves more than one material and if there are air spaces between the component parts, then

$$U = \frac{1}{1/f_i + 1/f_0 + x_1/K_1 + x_2/K_2 + 1/a}$$

where x_1, x_2 = thicknesses of various materials
K_1, K_2 = conductivities of various materials
a = conductivity of air space, which is dependent upon width of space, mean temperature, and enclosing material (for average conditions a value of 1.10 is generally used)

13.11 Condensation as related to thermal conductivity. The condensation of moisture on the inside surface of an exterior wall of a building which is exposed to low outside temperatures t_0 and to warmer inside temperatures t_i may be prevented if the wall has a sufficiently low heat transmission coefficient to maintain the inside surface temperature above the dew point temperature t_d. The maximum heat transmission coefficient which will prevent condensation may be determined from Fig. 13.8, or it may be computed from the equation

$$U = f_i \frac{t_i - t_d}{t_i - t_o}$$

The dew point temperature varies with the room temperature and the relative humidity of the air. Its value for any given conditions can be obtained from hygrometric tables or physics textbooks.

To prevent condensation it may be necessary (1) to decrease the heat-transmission coefficient by using a thicker wall or by furring out with an extra layer of insulation material, (2) to reduce the humidity of the air, or (3) to increase the velocity of air passing over the surface. Of these, the reduction of the heat transmission coefficient is usually the most practical method.

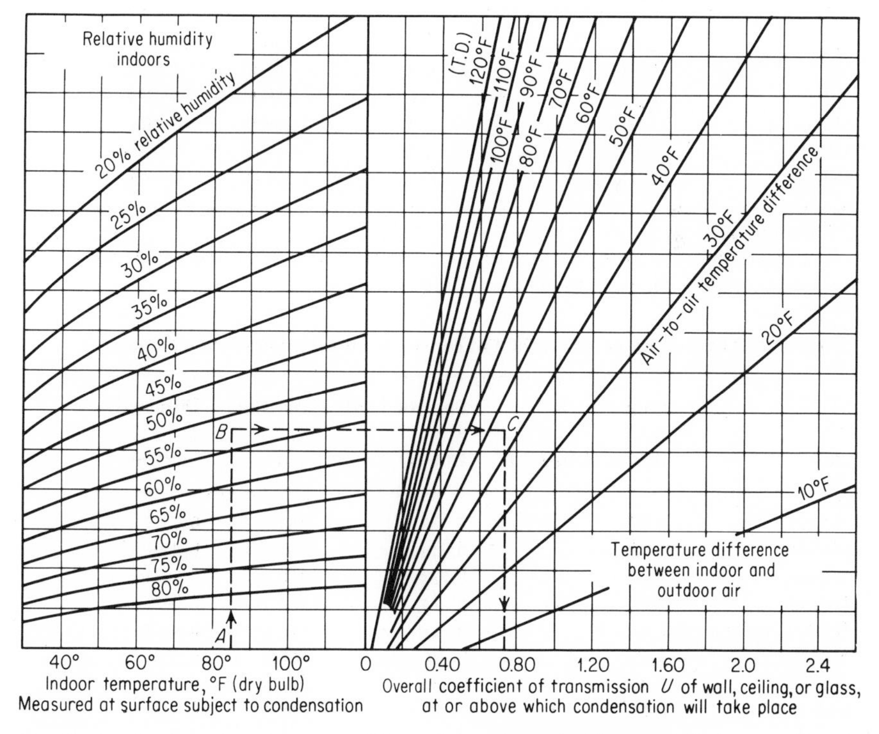

Fig. 13.8. Condensation diagram. (*From "American Architectural Reference Data"* [1347].)

13.12 Thermal properties and their relationships. In calculations involving the flow of heat in concrete masses—such as in connection with studies of thermal volume changes in mass concrete, the extraction of excess heat from concrete, or similar operations involving heat transfer—the thermal conductivity K, specific heat S, and density d must be taken into consideration. This may be done by means of the thermal diffusivity D, which expresses the facility with which the concrete will undergo temperature change upon being heated or cooled. It may be determined from the three properties listed above by use of the equation

$$D = \frac{K}{Sd}$$

or it may be determined directly by observing the flow of heat through a test specimen and computing the average diffusivity that is indicated

Table 13.3 Thermal properties of concrete of Hoover Dam*

Temperatures of concrete, °F	Conductivity K, Btu/sq ft/ft/hr/°F	Specific heat S, Btu/lb/°F	Density d, pcf	Diffusivity D, sq ft/hr
50	1.699	0.212	156.0	0.0514
70	1.688	0.216	156.0	0.0501
90	1.677	0.221	156.0	0.0486
110	1.667	0.229	156.0	0.0467
130	1.657	0.239	156.0	0.0444
150	1.648	0.251	156.0	0.0421

* From Ref. 218.

Mix proportions by weight, 1 : 2.45 : 7.09; 9-in. maximum-size sand and gravel; 1.01 bbl cement per cu yd; water-cement ratio, 0.58 by weight.

Table 13.4 Thermal properties at 70°F of concrete representing various dams*

Dam	Conductivity K, Btu/sq ft/ft/hr/°F	Specific heat S, Btu/lb/°F	Density d, pcf	Diffusivity D, sq ft/hr
Norris	2.105	0.239	160.6	0.0548
Hoover	1.688	0.216	156.0	0.0501
Gibson	1.667	0.222	155.2	0.0484
Owyhee	1.373	0.214	152.1	0.0422
O'Shaughnessy	1.338	0.218	152.8	0.0402
Morris	1.291	0.216	156.9	0.0381
Ariel	0.884	0.235	146.2	0.0257
Bull Run	0.847	0.225	159.1	0.0237

* From Ref. 218.

[1340, 1341]. The former method permits the determination of the variation of the thermal properties and diffusivity with temperature.

Values of the thermal properties of the concrete used in Hoover Dam, showing their variation with temperature, are given in Table 13.3, while values at a temperature of 70°F for concrete representing various dams, as determined by the U.S. Bureau of Reclamation, are given in Table 13.4. These studies have shown that the most important factor affecting thermal conductivity and diffusivity is the mineralogical composition of the aggregate. Listed in the order of increasing diffusivity are the following rock types: basalt, rhyolite, granite, limestone, dolomite, and quartzite. Tests have also shown that diffusivity is increased appreciably by reductions in the water-cement ratio and by a decrease in temperature of the concrete.

13.13 Temperature rise in mass concrete. As cement combines with water, the resulting heat of hydration (Art. 2.13), if not dissipated, causes an appreciable rise in temperature. The rise in temperature in a massive block of concrete, such as a dam, and the subsequent cooling result in stress conditions which tend to produce cracking. Also, the marked changes in temperature influence the properties of the concrete itself.

The rate and total amount of temperature rise in concrete depend upon two principal factors: (1) the rate of heat generation, and (2) the rate at which heat is dissipated. For concrete structures not over a few feet thick, the heat is dissipated relatively fast, so that the *rate* of heat generation is more significant than the total amount. So many variables influence heat generation and its dissipation and therefore affect the temperatures developed in mass concrete that prediction of the actual temperature rises at selected locations is a complicated problem, but a simplified method developed by Carlson serves satisfactorily for many purposes [1364].

Heat generation. The amount of heat generated within a massive concrete structure is dependent upon the type of cement, the temperature of placement, the water-cement ratio, and the cement content.

The effect of type of cement on the generation of heat is shown in Fig. 13.9, where the temperature rise is plotted against the age of concrete protected against loss or gain of heat. These adiabatic time-temperature curves are the basis for all temperature computations for mass concrete. For a given specific heat of concrete, the adiabatic temperature rise may be taken as a measure of the heat generation. In Fig. 13.9 it is evident that the total amount of heat is greatest for the high-early-strength cement, followed in order by the normal portland, modified portland, portland-pozzolan, and low-heat portland cements.

The heat generation is closely related to the compound composition of the cement [218]. In general, heat is generated most rapidly and in greatest total amount per unit of the compound by tricalcium aluminate

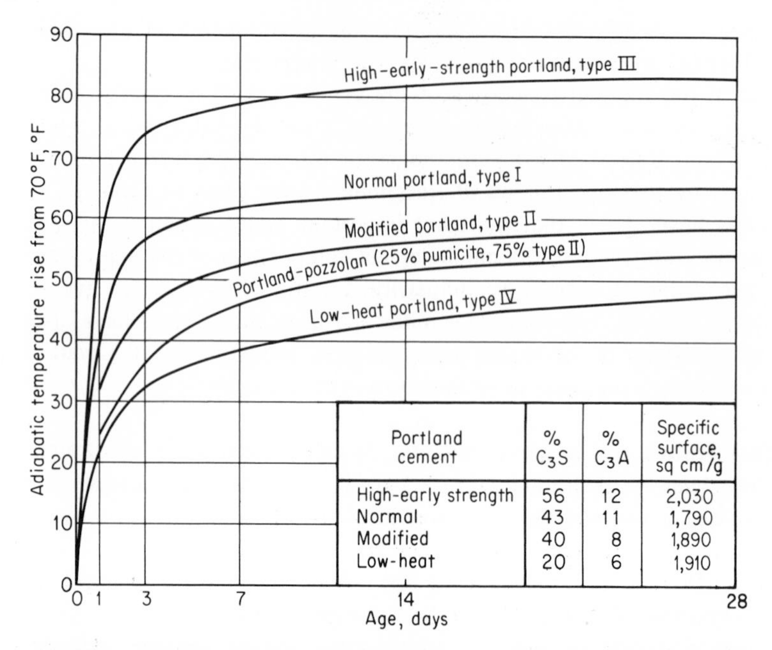

Portland cement	% C_3S	% C_3A	Specific surface, sq cm/g
High-early strength	56	12	2,030
Normal	43	11	1,790
Modified	40	8	1,890
Low-heat	20	6	1,910

Fig. 13.9. Relative effect of type of cement on adiabatic temperature rise of concrete. Cement content 1.0 bbl per cu yd; specific surface of cements by Wagner turbidimeter. (*From Kelly* [1361].)

C_3A, followed in order by tricalcium silicate C_3S, dicalcium silicate C_2S, and tetracalcium aluminoferrite C_4AF. Although fine cements generate heat more rapidly than coarse cements, the ultimate heat generation is practically unaffected by fineness.

The placement temperature of mass concrete and its heat generation (expressed as adiabatic temperature rise) are related as shown in Fig. 13.10. The higher the initial temperature, the more rapid the temperature rise at early ages. For portland-pozzolan cements the ultimate temperature *rise* may be greater for a lower initial temperature, but for low-heat portland cements the temperature rise tends to be higher for the higher initial temperature. In both cases the desirability of placing mass concrete at relatively low temperatures so that lower ultimate temperatures occur is apparent from Fig. 13.10.

Higher water-cement ratios produce greater heat generation. In general, an increase in water-cement ratio of 0.01 by weight increases the heat of hydration at all ages by about 1 cal per g of cement. The corresponding temperature rise in mass concrete containing 1 bbl of cement per cu yd is nearly 1°F.

In general, the heat generated per unit volume of concrete is proportional to the cement content. Although richer mixes tend to have lower water-cement ratios and thus a lower heat generation per pound of cement, the higher temperatures attained by the richer mixes tend to counteract this effect. The effects on the adiabatic temperature rise of the type of cement used and the cement content of the mix are shown in Table 13.5.

Heat dissipation. Heat is dissipated as long as the temperature of the concrete is above that of the surroundings. Several factors tending to reduce the temperature rise by accelerating the dissipation of heat are (1) a large temperature difference between the surroundings and the concrete, (2) free circulation of the surrounding air or water, (3) absence of insulation such as forms or backfill, (4) placement units having a high ratio of exposed surface to volume, (5) small units of placement, (6) long exposure of each unit before covering it with another unit, and (7) concrete of high conductivity. Artifical means, such as the pipe-refrigeration method used for Hoover Dam, may be used to increase the rate of heat dissipation.

Temperature-rise history. When the surface of a 5-ft lift of mass concrete is exposed continuously to the initial temperature, the average temperature rise reaches a maximum about 3 days after placement, as shown in Fig. 13.11.

For a very large dam of concrete having average thermal properties and containing 1 bbl of cement per cu yd, the temperature rise to the 3-day maximum is about 40°F for the normal portland cement and 23°F for low-heat portland cement. When the lift is covered, practically no further heat loss occurs, and the temperature rises again because of the continuing heat generation by the cement. If the surface is exposed for only 3 days, the temperature rise some months later will be about 47°F regardless of the type of cement used, as shown by the solid lines in Fig. 13.11. If the top surface is exposed for 10 days, the final temperature rise above the initial temperature will be about 32°F. Thus the maximum temperature

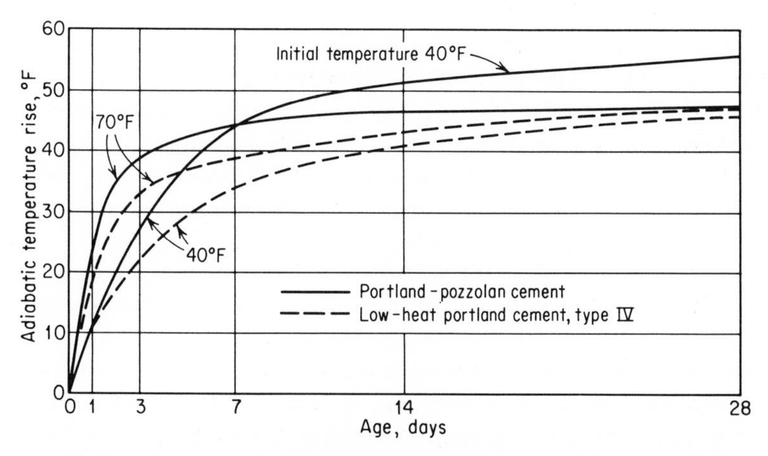

Fig. 13.10. Typical effect of placement temperature on adiabatic temperature rise of concrete. Gravel aggregate 0 to 6 in.; cement content 0.9 bbl per cu yd; W/C = 0.61 by weight. (*From Kelly* [1361].)

Table 13.5 Effect of cement type and cement content on the adiabatic temperature rise of concrete

Cement type	Cement content, sacks per cu yd	Temperature rise in concrete at age indicated, °F		
		3 days	7 days	28 days
I	5	63.4	74.3	81.0
	4	54.0	63.7	69.5
	3	41.8	49.0	53.5
II	5	46.7	58.0	66.0
	4	41.0	49.5	57.5
	3	30.0	39.3	43.5
IV	5	39.4	46.0	56.9
	4	35.6	40.5	44.0
	3	25.9	30.8	37.6

will occur 3 days after placement for normal cement, but with low-heat cement the heat generated after 10 days is sufficient to raise the temperature above the 3-day peak. These temperature-rise histories show that artificial cooling is required if the temperature of the mass is to be reduced to its ultimate stable temperature within a reasonable time. Also, it is essential that artificial cooling be started soon after the concrete is placed, if the maximum temperature is to be reduced appreciably by using a low-heat cement.

Average temperature curves for a concrete dam 50 ft thick, having the same thermal properties and subject to the same exposure conditions as given above, are shown in the lower part of Fig. 13.11. These curves show that the maximum benefit from the use of a low-heat cement is obtained when the rate of heat dissipation can be made, either artificially or naturally, to exceed the rate of heat generation at an early age.

In pavement slabs, say, about 10 in. thick, the temperature rise may be of the order of 10 or 20°F, although with a high-early-strength cement, such slabs under certain conditions may have a temperature rise of as much as 30°F. In massive structures where cement factors higher than those ordinarily employed in dam construction have been used, temperature rises of 100°F and more have been observed [1360].

Only the temperature rise above the initial temperature is shown in Fig. 13.11. For concrete placed during the summer the initial temperature may be much higher than the final stable temperature, so that the total temperature drop—which is the important change, as it is the principal cause of tensile cracking—may be in excess of 100°F. Thus, the influence of the placing temperatures on the temperature drop and consequent contraction may be greater than the effects of a change in the heat-generating characteristics of the cement.

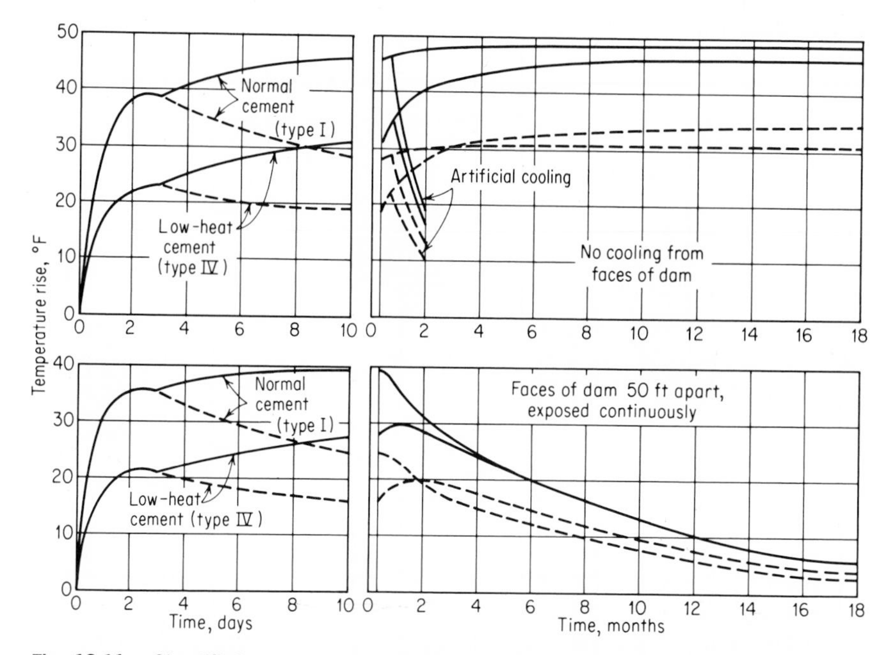

Fig. 13.11. Simplified average temperature-rise history of a very large dam (top) and a moderate-size dam (bottom). Solid lines show temperatures when surface of lift is exposed for 3 days. Dashed lines show temperatures when surface of lift is exposed for 10 days. Concrete placed in 5-ft lifts. *(From Blanks, Meisner, and Rawhouser* [1377].)

Effect of precooling. Since the placing temperature has such an important effect, it has been proposed that concrete for mass structures be precooled. Carlson has shown that for every degree that concrete is reduced in temperature before being cast, approximately 0.6°F reduction in maximum temperature will result in the mass [1362]. This value is based upon the assumption of an ordinary mass concrete and the placement of 5-ft lifts every 4 days.

Precooling of concrete would appear to offer a solution to the temperature problem except that the cost may be excessive. Both the mixing water and the aggregates would need to be cooled to reduce the initial temperature to near freezing (see Art. 8.16). However, it is believed that with cooling of water alone mass concrete often could escape cracking. For each degree that the water of average mass concrete is reduced, the temperature of the fresh concrete is reduced ¼°F.

EXTENSIBILITY AND CRACKING

13.14 Cracking of concrete. Concrete is often subjected to tensile stresses which are not considered in the design of the structure, and being low in tensile strength it cracks if such stresses are appreciable. Many condi-

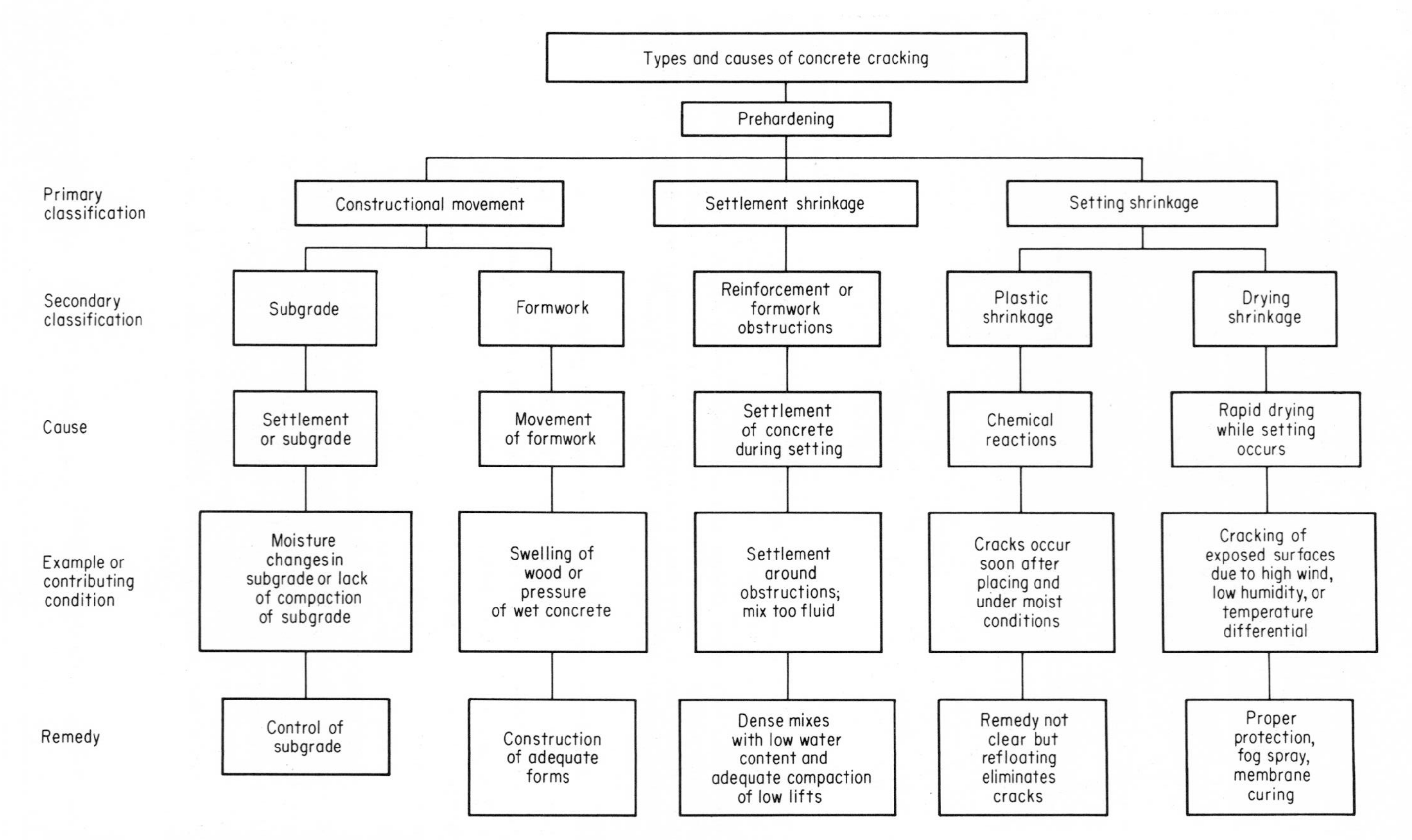

Fig. 13.12. Types and causes of concrete cracking. Prehardening. (*From Mercer* [1375].)

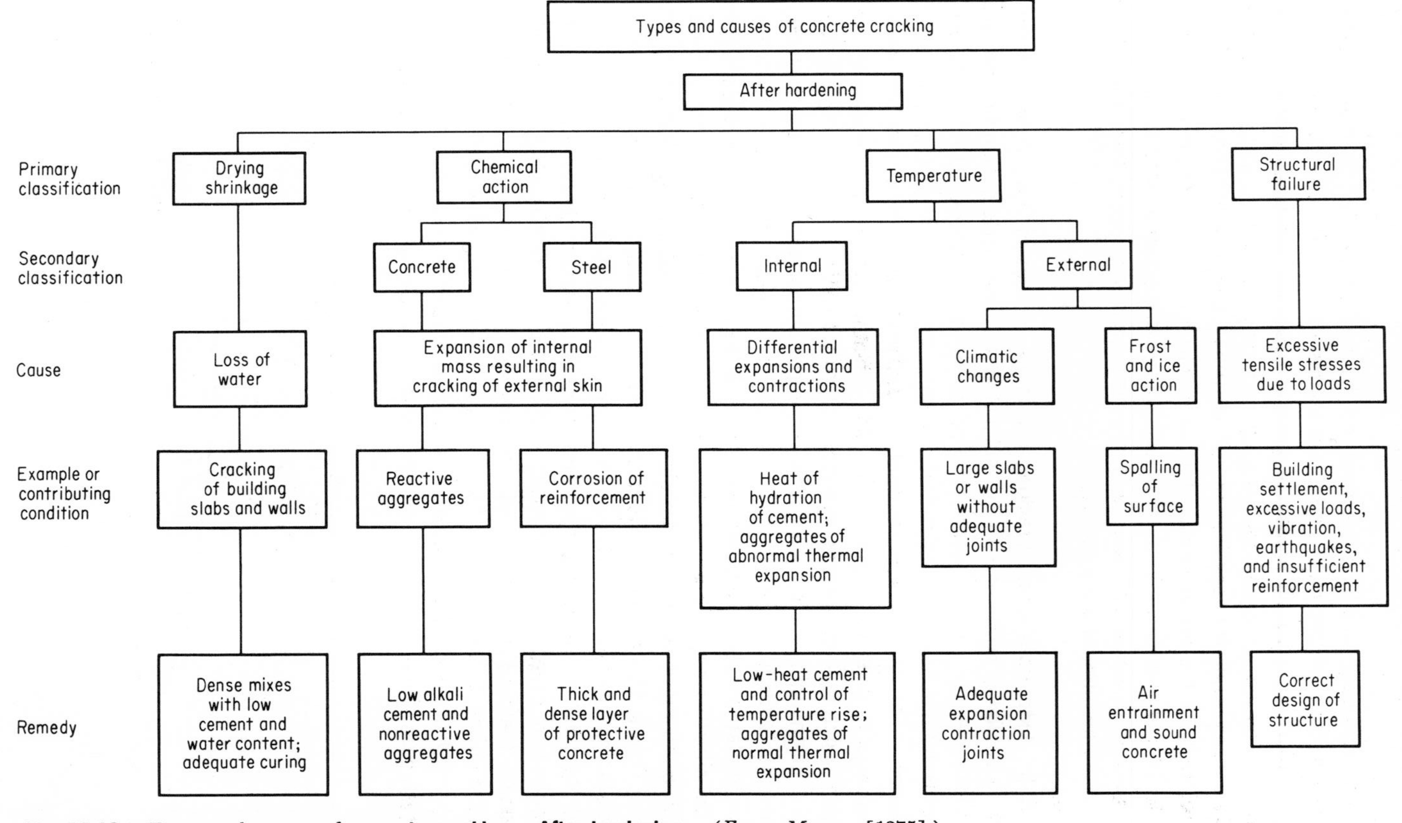

Fig. 13.13. Types and causes of concrete cracking. After hardening. (*From Mercer* [1375].)

tions producing tensile stresses cannot be avoided readily, so that cracking presents one of the most serious problems in concrete construction. Figures 13.12 and 13.13 present a general basis for the classification of the more common examples of cracking and may be helpful in the detection of causes and the application of remedial measures. Although drying shrinkage is the major cause of most cracking, internal temperature effects are the most important in mass concrete, external temperature effects are important in some climates, and chemical reactions have been the cause of much disintegration of concrete.

13.15 Extensibility and cracking. Concrete is said to have a high degree of extensibility when it can be subjected to large extensions without the development of cracks, such as tend to occur in restrained concrete and result from tensile strains caused by shrinkage, thermal changes, or other causes. For minimum cracking the concrete should have low shrinkage, low elastic modulus, high creep, high tensile strength, and—at least for mass concrete—low heat generation.

Extensibility tests have shown the important influence of creep in reducing the tendency of concrete to crack. For sealed specimens loaded in tension at a rate which required 2 to 3 months to produce failure, the extensibility ranged from 80 to 160 millionths, these values being about 1.2 to 2.5 times as great as for similar specimens under rapid loading.

Figure 13.14 shows that if concrete is subjected to drying or cooling conditions tending to cause contraction, an elastic stress will develop if the concrete is restrained, but with time this elastic stress will be partly relieved by creep of the concrete. When the net tensile stress eventually equals the tensile strength of the concrete, a crack occurs. Prolonged moist curing decreases, or at least delays, the cracking of restrained concrete primarily because of the resulting increase in tensile strength. Lean mixes crack less than rich mixes of the same consistency, probably because of

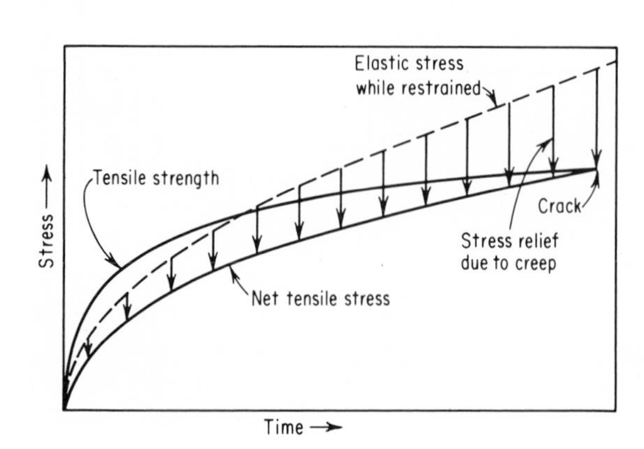

Fig. 13.14. Cracking as affected by restraint and creep. (*From J. W. Kelly.*)

their smaller shrinkage and greater creep. Various aggregates have about the same effect upon cracking as upon drying shrinkage—the greater the shrinkage, the greater the tendency toward cracking. Therefore, it is desirable to use as coarse a gradation of aggregate as possible, consistent with the character of the work. As shrinkage is less the lower the water content of the mix, it is essential that the water content be reduced as much as practicable to reduce cracking to a minimum. Under conditions where drying could not occur, low-heat (type IV) and portland-pozzolan cements have been shown to possess superior extensibility characteristics in comparison with normal (type I) or modified (type II) cements [106].

From the results of large scale tests on continuous reinforced-concrete structures, it has been shown that cracking due to variations in moisture content and temperature cannot be avoided, although through the use of suitable reinforcement it is possible to cause the development of many small cracks close together instead of a few large cracks of the same total width, which are more serious, as they lower the resistance to stress, wear, weathering and penetration of water. In many cases, however, the steel requirement to do this will be uneconomical, and experience has shown the desirability of using suitable contraction joints instead [1216].

Relief joints can be used to eliminate unsightly random cracks and improve the appearance of a structure. In highways they are often formed by sawing a groove about ½ in. wide by 2 in. deep after the concrete has hardened sufficiently to avoid raveling at the edges. The present practice is the use of dummy-groove contraction joints spaced at 15 to 20 ft if the pavement is plain concrete, and from 35 to 80 ft if it is reinforced. Dowel bars across the dummy groove provide for load transfer. These bars are coated with a mastic to permit free movement of the concrete. The grooves are filled with a flexible joint-sealing compound to prevent water penetration into the subbase.

Relief joints can also be used in concrete walls to eliminate unsightly random cracks. They are formed by attaching a wood or metal strip on the inner side of the forms. The reinforcing steel should be placed so that about one-half of it will cross the relief joint. With these planes of weakness, cracks will occur in the relief joints.

13.16 Thermal stress and cracking. The general interrelation of temperature, age, strain, and stress in a typical mass concrete is shown by the diagrams of Fig. 13.15. Since observations at an actual structure do not provide sufficient data to evaluate the elastic and plastic effects directly, the diagrams are for large concrete cylinders stored under mass-concrete conditions. The test cylinders represent concrete in a lift near the base of a large dam where there exists almost complete restraint to horizontal expansion or contraction. Succeeding lifts in the dam are placed at 5-day

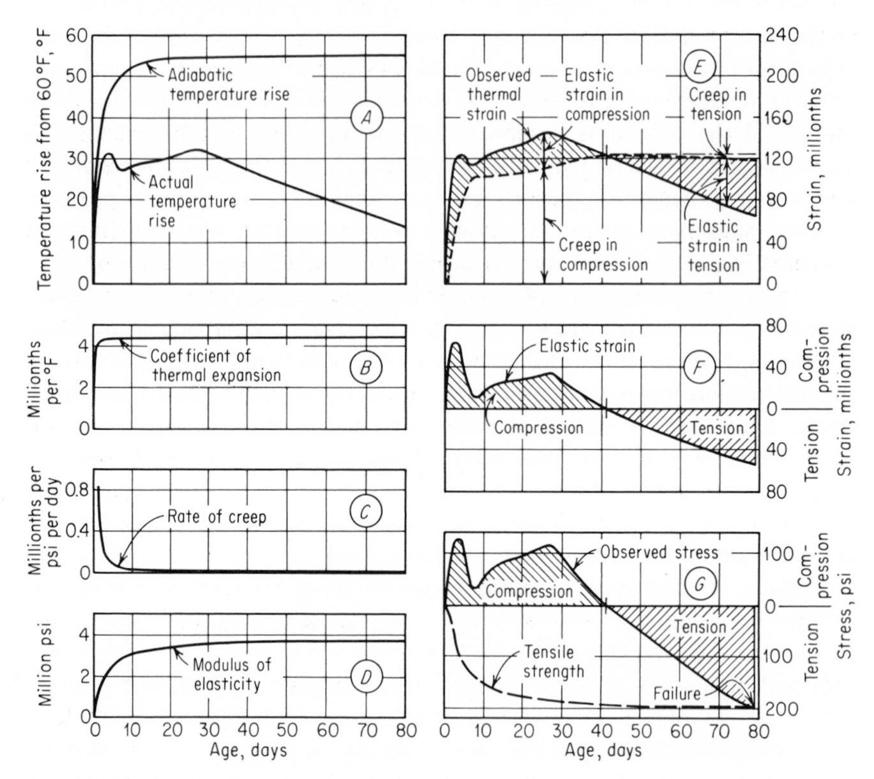

Fig. 13.15. Interrelation of age, temperature, strain, and stress in mass concrete. Portland-pozzolan cement content 1.0 bbl per cu yd; 0 to 3-in. gravel; W/C = 0.63 by weight; initial temperature 57°F; curing in thin copper jackets in accordance with actual temperature rise (diagram A); thermal-strain specimens 12-in. cubes, not restrained; stress specimens 12 by 48-in. cylinders maintained at length observed at age 6 hr, at which time temperature was 60°F. (*From Kelly* [1361].)

intervals, and the resulting temperatures are shown by the lower curve of Fig. 13.15A.

The "observed thermal strains" shown in Fig. 13.15E represent the length changes which would occur in the concrete if it were unrestrained. These strains are almost solely due to thermal expansion, since in mass concrete the net length change caused by cement hydration and moisture change is small. As shown in Fig. 13.15B, the coefficient of thermal expansion is practically constant after the first day or so. Hence the thermal strain (Fig. 13.15E) follows the temperature rise (Fig. 13.15A) quite closely.

Because restraint prevents expansion of the concrete, stresses are developed. Compressive stresses are developed during the early period, when temperatures are rising, and tension occurs soon after the temperature begins to decline. For the condition shown (Fig. 13.15*G* and 13.15*A*) tensile stresses were developed within 15 days after the temperatures began to decline, and at a temperature only 6°F below the peak temperature. As cooling progresses, the tensile stress increases until some equilibrium temperature is reached or until the stress equals the tensile strength, in which case the restraint is relieved by local cracking. In the example cited, the concrete failed in tension at the age of 79 days, even though the temperature was still 15°F above the temperature at placement.

For perfectly elastic and fully restrained concrete the compressive stress due to temperature rise would have been about 300 psi at the age of 3 days and about 500 psi at the peak temperature at the age of 26 days. However, concrete begins to creep as soon as stress is developed, and hence part of the stress is relieved. For the condition of complete restraint, the sum of the creep and the elastic strain (producing stress) is always equal to the thermal strain. Creep occurs rapidly at the early ages but occurs more slowly after the first week or so (Fig. 13.15*C*); at ages of 10 to 30 days the cumulative creep was about 80 percent of the total (thermal) strain, whereas after tension was developed, the creep in tension was only about 10 percent of the total tensile strain (Fig. 13.15*E*). This is influenced by the modulus of elasticity of the concrete, which is low at the early ages but much higher at the later ages (Fig. 13.15*D*); thus the elastic (stress-producing) part of the total strain is high at the later ages (Fig. 13.15*E*).

To summarize, the temperature, restraint, and the properties of mass concrete are so interrelated that the concrete is compressed at the early ages and is stretched at later ages. If the tensile stress resulting from the combined effects of temperature change, moisture change, and load can be kept below the tensile strength, cracking will be avoided. Tensile stress due to temperature change can be reduced through the use of concretes of low heat generation, low temperatures and favorable rates of placement, and concretes having favorable thermal, elastic, and plastic properties.

For mass concrete the most effective single method for reducing cracking is the use of low-heat or portland-pozzolan cement because of their low sustained modulus of elasticity, which includes the effects of both elastic and plastic deformation. Reduced cement content, limitation of thickness of lifts, artificial cooling, and the judicious use of contraction joints are other effective means for controlling the development of cracks in mass concrete [1377].

MISCELLANEOUS PROPERTIES

13.17 Fire resistance. The standard fire resistance of a wall is determined by subjecting a test panel which is at least 9 ft high and which has an area of at least 100 sq ft to a standard fire, the temperature of which follows a specified fire-temperature curve. The fire-resistance period is the time required for a temperature rise of 250°F to develop on the unexposed face while the wall resists its design load during the standard fire test.

The resistance of concrete to the transmission of heat due to a fire depends upon its thickness, type of construction such as solid or hollow-block, type and size of aggregate, and cement content. The load carrying capacity of concrete subjected to fire depends upon the above factors, and in addition is affected by the curing conditions of the concrete.

Tests of wall panels have shown that the thermal fire-resistance period of solid walls is generally doubled when the wall thickness is increased about 40 percent [1386].

The type of wall construction is of considerable importance. For different designs of hollow masonry units of a given concrete mix used in the construction of 8-in. walls, the thermal-resistance period increases with the weight (a measure of net thickness) per square foot of wall. In general, an increase of 50 percent in weight doubles the fire-resistance period [1385].

Comparing solid walls with hollow masonry walls, all made using natural aggregates, tests indicate that pound for pound the relatively dense concrete in solid walls is about 25 percent less effective in retarding temperature rise of the unexposed face than the corresponding dry-tamped concrete in hollow masonry walls. Expanded-shale concrete, however, is about 25 percent more effective in solid slabs than in hollow masonry walls, probably because of tenaciously held moisture within the solid lightweight-concrete slabs [1386].

Natural aggregates may be divided into four groups in regard to their effect upon the fire resistance of concrete. These are (1) calcareous, or lime-bearing, (2) feldspathic, such as basalt, diabase, etc., (3) granites and sandstones, and (4) siliceous aggregates, such as quartz, chert, and flint. These four groups are arranged in descending order of their fire resistance. Certain artificial aggregates, such as expanded shale and slag, are superior in this regard to the best natural aggregates. For 8-in. masonry walls the thermal-resistance period for a wide range of natural and artificial aggregates varied about 25 percent below and above the average for the entire group [1385].

Tests with the above natural and artificial aggregates showed that with a given cement content, the thermal-resistance period increased as the proportion of fine to coarse aggregate in the concrete was increased. For a range of fineness modulus from 2.0 to 4.5, as used in the manufacture of masonry units, the thermal-resistance period with the finest grading was from 15 to 20 percent greater than with the coarsest grading.

These effects of finer grading upon the thermal-resistance period were quite small in comparison with their very large effect in reducing the strength of walls, both before and after fire exposure [1385]. In general, the proportion of fine to coarse aggregate should be governed by the strength requirements rather than by the required thermal resistance.

Although aggregates differ in their fire resistance, tests have shown that by proper selection of the aggregate grading, wall thickness, cement content, etc., any desired requirements can be met with any type of aggregate in common use.

The fire-resistance period of walls, as governed by both heat transmission and strength, increases with the cement content. In general, doubling the cement content will increase the thermal-resistance period at least 10 percent and will increase the load-carrying capacity, both before and after fire exposure, by about 100 percent.

Curing has practically no effect upon the thermal-resistance period but materially increases the strength of walls both before and after fire exposure.

In general, solid 6-in. walls of various ordinary concretes have a fire-resistance period of about 3 hr, whereas corresponding walls of expanded-shale concrete are good for about twice that period. For 8-in. walls of good-quality hollow-masonry units the fire-resistance period varies from 1½ to 3 hr depending upon the type of block and type of aggregate.

The compressive strength of solid bearing walls 6 to 8-in. thick and

Table 13.6 Fire-resistance ratings of reinforced-concrete beams

Coarse aggregate type*	Cover of steel, in.	Fire rating, hr†
1	¾	1
	1	2
	1¼	3
	1½	4
2	¾	½–1‡
	1	1–2‡
	1½	2–3‡
	2	2–4‡

* Type 1 aggregate: Limestone, calcareous gravel, traprock, blast-furnace slag, burned clay or shale, or aggregate containing not more than 30 percent quartz, chert, granite, or similar materials. Type 2 aggregate: Granite, quartzite, siliceous gravel, or aggregate containing more than 30 percent quartz, chert, flint, or similar materials.

† For medium-size beams. Decrease for small beams.

‡ Variable, depending on spalling characteristics of aggregate. The use of mesh to hold cover in place will give ratings about as high as for type 1 aggregates.

Table 13.7 Fire-resistance ratings of reinforced-concrete columns

Round or square size, in.	Aggregate type*	Cover of steel, in.	Fire rating, hr
12 or 14	1	1.5	3
		2.0	4
	2	1.5	1–3†
		2.0	2–4†
16 or larger	1	1.5	4
		2.0	5
		2.5	8
	2	1.5	2–4†
		2.0	3–5†
		2.5	5–8†

* See Table 13.6 for classification of aggregates.
† Variable, depending on spalling characteristic of aggregate. The use of mesh to hold cover in place, or of 1-in. gypsum plaster over concrete, will give ratings about as high as those for type 1 aggregates.

10 ft high may be expected to be at least 400 psi during severe fire exposure [1386]. The compressive strength of 8-in. hollow-masonry walls 6 ft high, after a 3-hr fire exposure, may average about 20 percent of the original strength of the blocks when made of gravel aggregates, or about 35 percent of the original strength of the blocks when made of limestone or expanded shale aggregates [1385]. For the effect of high temperatures on the compressive strength of concrete see Art. 10.15.

The fire resistance of reinforced concrete and of prestressed concrete is dependent primarily on the protective concrete cover of the steel. Table 13.6 shows the fire ratings for reinforced concrete beams, Table 13.7 is for columns, and Table 13.8 is for prestressed concrete. The latter type of construction requires a thicker cover for a given rating as prestressing steel, which is usually cold-drawn to increase its strength, is weakened more by high temperatures than ordinary reinforcement steel is.

13.18 Fatigue strength. When a material fails under many repeated loads which are less than the single static load which would cause failure, it is said to have failed by fatigue. Even for metals, experiments indicate that some crystals in a stressed piece reach their limit of elastic action sooner than others, and eventually after many repetitions of stress some minute element ruptures. Eventually the fracture spreads causing a progressive fracture and complete failure. Somewhat the same condition occurs in concrete.

The results of fatigue tests are commonly shown by an S-N diagram,

Table 13.8 Fire-resistance ratings of prestressed concrete

Type of unit	Restrained or unrestrained	Cross-sectional area, sq in.*	Cover of steel in in. for fire rating shown†			
			1 hr	2 hr	3 hr	4 hr
Girders, beams, and joists	Unrestrained	40 to 150	2	2.5		
		150 to 300	1.5	2.5	3.5‡	
		Over 300	1.5	2.25	3‡	4‡
	Axially restrained	40 to 150	1.5	2		
		150 to 300	1	1.5	2	
		Over 300	1	1.5	1.5	2
Slabs, solid or covered with flat under surface	Unrestrained		1	1.5	2	2.5
	Biaxial restraint		0.75	1.25	1.5	2

* In computing the cross-sectional area of joists, the area of the flange shall be added to the area of the stem, but the total width of the flange so used shall not exceed three times the width of the stem.

† Cover for an individual steel tendon is measured to the nearest exposed surface. Aggregrate to be type 1 except that for expanded shale aggregate the cover may be reduced about 25 percent. For type 2 aggregates, cover may need to be increased. See Table 13.6 for classification of aggregates.

‡ Provide against spalling of cover by means of light 2-in. mesh, U-shaped and having a cover of about 1 in.

in which the stress is plotted against the number of cycles of load to cause failure as shown in Fig. 13.16. If the curve becomes asymptotic to the horizontal axis, the corresponding stress is called the endurance limit or fatigue limit. Most metals have an endurance limit, but it is doubtful that concrete does, at least within 10 million repetitions of load. Thus, the curve continues to slope downward as shown, and the fatigue strength is the stress causing failure after a stated number of cycles of loading. In the following summary of results the fatigue limit of concrete will be considered equal to the fatigue strength at about 1 million to 10 million cycles of loading, as that represents the maximum number of loading cycles for most concrete structures. In some cases the conclusions which follow are

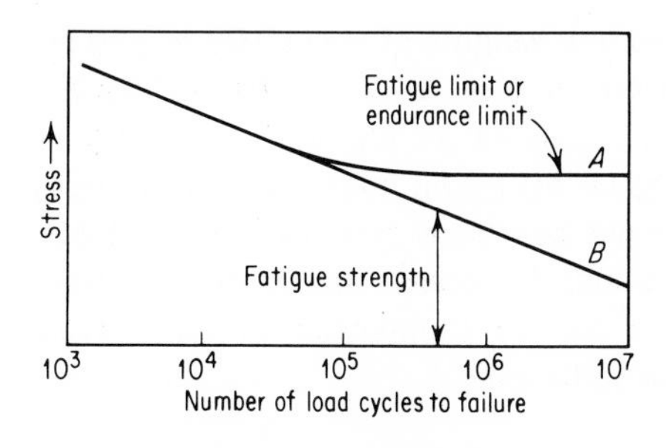

Fig. 13.16. Typical S-N diagram for fatigue tests. Curve A shows an endurance limit; curve B shows no endurance limit.

based on little data, but they give the salient facts available at this time [1398].

1. The fatigue limit for plain concrete, subjected to repeated compressive loads for a range of stress from zero to a maximum compression, is 50 to 55 percent of the ultimate crushing strength.

2. The fatigue limit of plain concrete subjected to repeated flexural loads is about 55 percent of the static ultimate flexural strength, although test results show a range from 33 to 64 percent, depending on such variables as age, curing, moisture content, and aggregate.

3. The fatigue limit for concrete in tension is about 55 percent of the modulus of rupture.

4. The secant modulus of elasticity decreases with repeated loads. The slope of the stress-strain curve decreases in the lower part of the curve but may increase slightly in the upper portion to become concave upward.

5. Rate of testing between 70 and 440 cycles per min has little effect on fatigue strength.

6. Most of the permanent deformation takes place in the early stages of the test, usually the first few thousand cycles. For properly cured and aged concrete the strain stabilizes at fewer load cycles than for concrete at an early age.

7. As the range of stress is decreased, the upper limit of the stress (fatigue strength) is increased substantially.

8. Concerning bond, little can be said except that fatigue failures are possible at loads less than 55 percent of the ultimate static pull-out tests.

13.19 Unit weight. The unit weight of concretes of proportions and materials ordinarily used in building and pavement construction usually ranges from about 148 to 152 pcf, depending upon mix and upon character and maximum size of aggregate. For calculations of an approximate nature, it is probably safe to assume an average value of 150 pcf. For concretes used in massive construction with maximum size of aggregate between 4 and 6 in., the unit weight may be as high as 156 pcf. For mortars and concretes of maximum size ⅜ in. the weight may be as low as 144 pcf. The variation of unit weight with maximum size and specific gravity of aggregate is shown in Table 13.9.

With special lightweight aggregates unit weights of about 100 to 110 pcf are often obtained for structural concretes (Art. 14.22). For lightweight concrete fills having low strengths, unit weights as low as 30 pcf may be obtained. With natural barite aggregate of 1½-in. maximum size, heavy concrete having a unit weight of 225 pcf has been produced (Art. 14.21). With iron ore for sand and steel punchings for coarse aggregate, concretes weighing 270 pcf have been made for use in counterweights for bascule bridges and for the absorption of atomic radiation.

Table 13.9 Average unit weight of fresh concrete*

Maximum size of aggregate, in.	Average values: Air content, percent	Water, lb/cu yd	Cement, sacks per cu yd	Unit weight, pcf† (specific gravity of aggregate‡): 2.55	2.60	2.65	2.70	2.75
3/4	6.0	283	6.02	137	139	141	143	145
1 1/2	4.5	245	5.21	141	143	146	148	150
3	3.5	204	4.34	144	147	149	152	154
6	3.0	164	3.00	147	149	152	154	157

* From Ref. 106.
† Weights are for air-entrained concrete with indicated air content.
‡ Saturated surface-dry basis.

QUESTIONS

1. Why is the secant modulus of elasticity the most practicable one determined from stress-strain diagrams?

2. What effect does the period of loading have upon the modulus as determined?

3. How is the sustained modulus of elasticity determined?

4. Discuss the effect of the following factors upon the modulus; age, mix, water-cement ratio, aggregate, free moisture, and strength.

5. Compare values of the dynamic modulus and the static compressive modulus for the same concrete.

6. Of what value is a knowledge of the thermal properties of concrete?

7. Compare the thermal conductivity of regular and lightweight concrete.

8. What is the cause of sweating on the inside surface of a concrete wall, and how may it be prevented?

9. What is the most important factor affecting the thermal diffusivity of concrete?

10. How does a change in water-cement ratio affect the diffusivity of concrete?

11. Upon what two principal factors does the temperature rise in a concrete mass depend?

12. What is the effect of type of cement, temperature of placement, water-cement ratio, and the cement content upon the amount of heat generated?

13. Will the temperature of a large concrete dam be reduced naturally to its ultimate stable temperature within about 5 years?

14. How significant is the placing temperature of concrete on its later temperature drop and consequent contraction?

15. What advantages may be gained by the precooling of concrete for large masses?

16. What are the principal causes of the cracking of concrete?

17. Will the use of suitable reinforcement prevent cracking resulting from variations in moisture content and temperature?

18. What are the most effective means for controlling the development of cracks in mass concrete?

19. Compare the fire resistance qualities of calcareous and siliceous aggregates.

20. What effect has the cement content on the fire resistance of concrete walls?

21. What is the range of unit weight of various types of concretes?

FOURTEEN
SPECIAL TYPES OF CONCRETE

14.1 Architectural concrete. Concrete for buildings which is left exposed to view is commonly called architectural concrete. To produce an acceptable job using architectural concrete demands attention to details of forms and to the concreting operations which are not ordinarily considered in connection with structural concrete, which is primarily utilitarian. In building construction the structural concrete has usually been concealed, so that its appearance has been considered unimportant. Because the outside facing material protects it from weathering and the design requirements govern its characteristics, the attention of architects and engineers has been focused on compressive strength rather than on durability [1400, 1402].

However, architectural concrete must have a good appearance, and as it may be subject to freezing and thawing, it must be more durable than structural concrete which is protected from the elements. As noted in Art. 11.14 the resistance of concrete to weathering can be controlled in the same way as its compressive strength. It is possible, however, for concrete to be adequately strong to satisfy the structural requirements and yet not be sufficiently durable for the conditions to which it is exposed.

The forms for architectural concrete are usually given especial attention to avoid surface blemishes. Shiplap lumber may be used to prevent the loss of mortar which occurs through the cracks of ordinary forms, but in most cases, large panels of plywood are used instead of narrow boards.

In general a much more workable mixture is required, and the concrete must be placed with greater care, to eliminate the common defects which are permitted in structural concrete. It is necessary to secure a cohesive, plastic mix. This requires a larger proportion of fine material, including sand and cement, than is generally required in structural concrete.

The sand should not be deficient in fine particles nor contain excessive amounts of any particle size. Sands having these objectionable characteristics permit excessive bleeding with consequent water gain and sand streaking.

The maximum size of coarse aggregate should be 1½ in. instead of the ¾-in. or 1-in. maximum size which often is used. This will result in more workable mixtures and the use of less water in mixtures of equal cement content.

The practice of patching and rubbing concrete surfaces to eliminate surface defects is not good. The construction of better forms, the use of a more carefully designed mix, and greater care in placing will usually produce a better job at no greater cost.

As shown in Ref. 1408, many pleasing surfaces are produced by special formwork by lining the forms with textured sheathing, applying a cement retarder to the forms to assist in exposure of the aggregate by brushing and washing, or sandblasting or brush hammering the hardened concrete.

FLOOR SURFACES

14.2 Types and requirements. Two types of concrete floor finishes are used. These are the so-called monolithic or integral finish, which is made an integral part of the base slab, and the separately finished topping. The former has certain inherent disadvantages because of a high water-cement ratio caused by water gain from the deep base slab. This often delays the final finishing until late at night, and produces a low-quality surface having appreciable irregularities and low resistance to wear. Some of these disadvantages may be overcome by the withdrawal of some excess water by the use of vacuum mats. In contrast to the integral finish, the relatively thin concrete topping placed upon a hardened base slab can be made of a low water-cement ratio and can be finished to a true grade [1410–1412].

To render good service a concrete floor should be dense, durable, resistant to wear, and free from cracks or crazing. The surface should be without any appreciable high spots or depressions. In general, it should be troweled to a dense, smooth finish, but in some cases a rough-floated nonskid surface is preferred. Modern high-quality floor surfaces are made of good quality concrete; no longer are they made of wet mortar mixes which are inherently porous, wear readily, and tend to craze. See Ref. 1409 for an excellent paper on concrete floors by ACI Committee 302.

14.3 Preparation of base. If the finish course is to be applied to a hardened base, the latter must be thoroughly cleaned just prior to the placement of the topping. Usually this is accomplished by thorough picking, wire-brushing, and washing of the surface using water under pressure. In some cases, wet sandblasting is used.

If the base course is allowed to dry out before the topping is applied, it is essential that the base be rewetted thoroughly for at least 24 hr prior to placing the topping. Otherwise the bond strength at the joint will be low, and the differential shrinkage of the base and topping may cause them to separate. Any surface water must be eliminated before a preliminary grout coat is applied. The grout should be scrubbed thoroughly into the surface of the base just prior to placement of the topping. This grout should be either neat cement or a 1:1 mix using a fine sand. It should be of a medium consistency and should be about 1/16 in. in average thickness.

14.4 Concrete mix. Ordinary concrete sands can be used in floor topping, although preference should be given to those having not more than 10 percent finer than the No. 50 sieve and not more than 5 percent finer than the No. 100 sieve, as a high percentage of fines may cause dusting and crazing. The gravel should all pass a 3/8-in. sieve. For concrete floors which are to be highly resistant to wear, the aggregates should be hard and tough.

Concrete floor topping mixes are usually about 1:3 by weight of dry materials. The aggregate will generally be about 40 percent sand and 60 percent gravel. The consistency of the mix will depend upon the method of finishing employed, particularly when it is desired to float[1] the surface soon after trimming it to grade. For hand-floating, the slump should be about 1 in., but for power-floating it should have practically no slump. The mix should be workable enough for the power float to fill in and seal all irregularities and level the surface and yet be sufficiently stiff that the float does not sink into or gouge the surface. A mix which is excessively oversanded or overly wet, or concrete which has been overworked in placing or finishing, is likely to have the surface covered with "bled" water or to contain a relatively deep layer of mortar or of paste.

Uniform consistency of the mix is essential, as otherwise wet spots will develop in the slab which interfere with the finishing operations. This requires that the aggregates be reasonably uniform in grading and moisture content and that the mix consistency be carefully controlled. Because of the dryness of the mix, the mixing time should be not less than 2 min.

14.5 Placing and finishing. When the base slab and topping are built monolithically, the floor may be cast in strips, and the elevation of the finished surface can be determined from the tops of the form boards. For

[1] Floating is the first smoothing operation after the topping has been struck off by a straight edge. If done by hand, a wood board or float about 5 by 12 in. and provided with a hand grip is used. If done by power, a steel disk having a diameter of about 2 ft is rapidly rotated over the surface. The floating operation serves to fill in irregularities of the surface and to work a small amount of mortar to the surface to aid the finishing.

topping placed on a hardened base, screeds are used as guides for the straight edges in trimming the concrete to the desired elevation. These screeds are usually strips of ½ by 2-in. steel carefully set to grade; they are removed as soon as the topping is levelled off, and the space they occupied is filled in. The minimum thickness of the topping is usually about 1 in.

The grout mentioned previously should be thoroughly brushed over the base just ahead of placement of the topping. The latter should be spread evenly to a level slightly above grade, thoroughly compacted, and then struck off to grade. In one method, the topping is made to have a slump of about 9 in. to assist in the placing and leveling operations. It is then covered with burlap and a ⅜-in. layer of dry cement, which absorbs the excess water. Upon removal of the burlap and cement drier, the latter is sent to the mixer for use in following batches of the topping mix.

Floating of the surface preferably is done with rotary power equipment, rather than by hand floats, as it permits the use of drier mixes having smaller volume changes with less tendency to check and craze, and permits the use of larger percentages of aggregate, which increases the wear resistance of the slab. The trend is toward still drier and harsher mixes which can be worked effectively with vibratory power floats. The floating operation is begun as soon as the screeded topping has stiffened enough to bear the weight of a man without leaving footprints and is continued until a small amount of mortar is brought to the surface and all irregularities are eliminated. Prolonged floating is undesirable, as the excessive mortar brought to the surface may cause dusting and crazing.

Floating leaves the surface relatively rough-textured. Ordinarily this roughness is eliminated by finishing with steel trowels, either by hand or by power equipment. Neither floating nor troweling should be done too soon while the surface is very soft, or too much in one operation, as in that case excess fines will be brought to the surface resulting in crazing. Any water brought to the surface by the floating or rough-finishing operations should be allowed to evaporate before the surface is again troweled. To produce a hard, smooth surface considerable pressure on the trowel is required, and from three to six separate trowelings with periods between for drying of the surface are necessary. If a hard, glossy finish is desired, final hard troweling should be sufficiently late that all cement paste on the surface will have stiffened sufficiently and that the trowel will produce a shine or gloss on the surface.

14.6 Curing and protection. The floor surface should be protected from drying out as soon as such protection can be applied without damage to the surface. Three common methods of curing involve the use of wet burlap, waterproof paper, or polyethylene sheets. The use of the two latter materials having all edges lapped and sealed has proved to be effective. Sometimes a light fog spray is applied just before they are laid to increase their effectiveness.

If work must be carried on over the floor before it has hardened sufficiently, a suitable layer of cushioning material should be provided. During cold weather it is desirable that the space above and below the slab be enclosed and heated to a suitable temperature, as the floor is usually relatively thin and subject to damage by low temperatures.

14.7 Surface hardeners. For floors which are to be subjected to heavy traffic, aggregates should be selected which are known to be resistant to wear. For some heavy-duty floors it is the practice to sprinkle a slightly dampened mixture of about 1 part cement to 2 parts by weight of iron, alumina, silicon carbide, or hard natural aggregates over the floor surface before floating is begun; this mixture applied at the rate of about ½ to 1½ psf of floor surface is allowed to absorb moisture from the wet topping and is then floated and finished in the normal manner. The heavier applications usually are made in two coats with a floating operation in between.

Floor surfaces that dust and wear rapidly may sometimes be improved by treatment with solutions of certain materials such as fluosilicates of magnesium and zinc, sodium silicate, gums, and waxes. When these materials are absorbed by the floor, they produce crystalline or gummy inclusions which assist in sealing the surface and tend to reduce wear. To remain effective, they must be reapplied periodically.

A nonslip texture may be produced on stairways, ramps, and similar surfaces by application of a dry abrasive grit just previous to the floating operation. The grit, with some cement, is applied to the surface at the rate of about ¼ to ½ psf.

SPRAYED MORTAR AND CONCRETE: GUNITE AND SHOTCRETE

14.8 Use and limitations. Cement mortars and concretes applied by compressed air were formerly designated as "gunite" but are now commonly called "shotcrete." In these processes a mixture of cement and sand or even aggregate up to ¾ in., is transported by air through a hose to a nozzle where water is added and the mixture pneumatically sprayed into place. The coatings so applied can be made hard and strong but are subject to relatively high shrinkage. The results obtained with pneumatically applied mortar and concrete depend to a large extent on the skill of the operator, so the work should be done only by experienced men.

Pneumatically sprayed mortar is used for various types of concrete repair work and as a protective coating over surfaces which would be subject to disintegration or corrosion without such protection. It has been used for the repair of buildings, encasement of structural steel, protective covering of soft rock, building thin curtain walls, and for thin linings of tunnels, reservoirs, swimming pools, and canals.

Although shotcrete is a useful material, it has definite limitations. Usually it is impossible to obtain perfect bond to the base material, so that with time the differential volume changes of the base and the coating, accompanying thermal and moisture variations, may cause separation of the coating from the base. Furthermore, if a massive structure is cracked severely, it is impossible for a thin coating to restrain the structure adequately against future displacements, with the result that the thin coating eventually cracks also. The recommended practice for application of shotcrete is given in Ref. 1416.

14.9 Equipment. The cement gun which represents one type of machine for the shotcrete work is shown in Fig. 14.1, but larger units are used for concrete. It consists of two compression chambers, one above the other. The upper chamber, which is alternately under pressure and free of pressure, serves as a hopper for feeding the lower chamber, which is under constant pressure. By suitable valves the pressure in the upper chamber can be increased to that in the lower chamber and the connecting gate opened; the dry mix then flows into the lower chamber. The connecting gate is then closed, the upper chamber opened up and refilled, and the cycle repeated. The delivery of material from the lower chamber to the hose is regulated by a feed wheel driven by an air motor. A separate

Fig. 14.1. Cement gun used for construction of canal lining by shotcrete method. *(Air Placement Equipment Co.)*

hose carries water to the water ring in the nozzle where radial sprays wet the cement-sand mixture as it flows by.

Shotcrete having up to ¾-in. aggregate may be applied using a wet mix which is pumped into the hose and is then conveyed by pump pressure or by air to the nozzle where a high-velocity air flow shoots it onto the receiving surface. The advantages of applying a premixed mortar or concrete is that mix proportions and water content are established prior to application and can be easily controlled, whereas the water content of the dry mix shotcrete is determined by the nozzleman and may be variable.

14.10 Preparation of base. All surfaces to be coated should be cleaned thoroughly to obtain a good bond; any weak or partly disintegrated material should be removed by chipping followed by sandblasting. This sandblasting can be accomplished with the cement gun equipment using a sandblast nozzle and a low feed rate for the sand. If reinforcement is necessary, it should be anchored firmly to the base material to aid in holding the coating in position.

When the coating is to be applied to a base which has dried out, the base should be moistened thoroughly for a few hours previous to shooting the mix but surface-dried just before shooting begins in order to reduce the shrinkage stresses in the coating. Any absorptive surface to be coated should be moistened to retard any excessive withdrawal of water from the coating, but excess water should be avoided, as it will prevent good adhesion.

14.11 Aggregate. The grading of the sand for dry-mix shotcrete mortar is quite critical. If too fine, it produces a weak coating subject to excessive

Table 14.1 Grading of sand for shotcrete*

Sieve U.S. std.	Percent passing, by weight
⅜ in.	100
No. 4	95–100
No. 8	80–90
No. 16	50–85
No. 30	25–60
No. 50	10–30
No. 100	2–10
Fineness modulus	2.50–3.30

* From Ref. 1416.

shrinkage; if too coarse, the amount of rebound will be excessive, and a rough-textured surface will result. Sand meeting the specifications for good concrete is suitable; that is, it should be graded from coarse to fine within the limits shown in Table 14.1.

The total moisture content of the sand should be about 3 to 6 percent for satisfactory operation of the equipment. If too dry, the sand will not flow uniformly but will come in bursts, and dry patches will result; if too wet, the equipment becomes plugged.

Coarse aggregate No. 4 to ¾ in. in size has been used in amounts up to 40 percent of the combined aggregate in some shotcrete work involving thick applications. For such work a thin mortar layer should be first applied to produce a mortar bed to reduce the amount of rebound. Coarse aggregate is not practical for use when gunning overhead or for thin vertical walls, or when much reinforcement steel is present, as then an excessive amount of aggregate would rebound. Reference 1416 shows some recommended gradings for coarse aggregates.

14.12 Rebound. The material which bounces back from the working face is known as "rebound." It is largely the coarser particles of the mix. The amount of rebound depends upon a variety of factors and so may vary over a considerable range from perhaps 10 to 50 percent of the material handled. The rebound will be greater when applying the shotcrete to vertical or overhanging surfaces than for sloping or level surfaces. Rebound is greater with increased nozzle velocities, and within the range of ordinary consistencies increases with lower water-cement ratios as a result of the less plastic state of the mix. Rebound recovered clean may be used to replace up to 20 percent of total aggregate requirements.

14.13 Mix. Mortar mixes using 1 part cement to from 3 to 5 parts moist sand by weight have been used on various types of work. These values do not represent the mix in place, because the rebound causes the richness of the residual material to increase appreciably. As the rebound varies with the slope of the surface, this must be considered if a given mix in place is desired. In general, the richer mixes shrink more, and high shrinkage may destroy the bond to the base. For this reason overrich mixes should be avoided.

The water content of the mix must be sufficient to impart adequate plasticity to the mortar, so that it will adhere to the surface and so that rebound will not be excessive. On the other hand it must not be so wet as to cause sloughing. Tests indicate that for a sand of average grading the water-cement ratio of the mortar in place was about 0.57 for sloping and 0.54 for overhanging surfaces; these were approximately the maximum ratios that could be used without causing sloughing.

14.14 Mixing and placing. Before introducing the dry materials into the cement gun they should be thoroughly mixed for at least 1½ min. Any material not used within 1 hr after mixing should be wasted.

In placing the shortcrete the hose length should be kept as short as possible, as long hoses require higher air and water pressures and tend to pulverize the sand. Although hose lengths of 350 ft or more have been used, a length not over 100 to 150 ft is preferred. For these shorter hose lengths the air pressure should be about 50 psi at the nozzle, and the water pressure at the nozzle should be at least 15 psi greater than the air pressure. To avoid excessive rebound the air pressure should not exceed 75 psi even for a long hose. Both air and water pressures should be maintained uniform. The volume of air required depends upon the size of nozzle and the air pressure, but for average work, and using a 1¼-in. nozzle, about 250 cu ft of free air per min is required.

The nozzle should be held normal to the surface and usually about 4 ft from it. It should be kept moving to obtain a uniform coating. As thick coatings tend to slough from vertical and overhead surfaces, stiffer mixes should be used for such surfaces and the mortar should be applied in layers of not over ¾-in. thickness. Successive layers should be applied before the first layers have hardened. As there is a tendency for the rebound materials to collect in the forms or to coat the surface of completed work, care must be exercised to clear away all such rebound material.

Wooden templates should be used at corners, edges, and on surfaces where it is necessary to obtain true lines and proper thicknesses. Sometimes the mortar surface is screeded to smooth over the irregularities, but this practice is undesirable, as it may cause cracking and sloughing, particularly on vertical and overhanging surfaces. Steel troweling is an acceptable procedure if a smooth surface is desired, but it should be done not more than 1 hour after placing.

Operations should be suspended when any wind is blowing that forces segregation of the nozzle stream, making it impossible to maintain a proper and uniform mix.

Test cylinders can be made by shooting the mix vertically downward into 6 by 12-in. cylindrical cages of ½-in. mesh hardware cloth mounted on a board. All mortar outside the form should be removed immediately after shooting the specimen so that the wire mesh can be removed before testing.

14.15 Curing. As thin shotcrete coatings tend to dry out readily, the completed surface should be protected from the direct rays of the sun for at least 3 days and should be kept wet for at least 14 days.

MASS CONCRETE

14.16 Characteristics of mass concrete. Various concrete structures such as dams, locks, and bridge piers are massive in size, so that the structure is said to be of mass concrete. This type of concrete is distinguished from ordinary concrete in that it is deposited in large open forms which permit the placement of concrete having large (up to 6-in.) aggregate and a low slump, and the use of powerful internal vibrators for consolidation of the relatively harsh, dry concrete.

Mass-concrete structures insulate themselves so thoroughly against loss of heat generated by hydration of the cement that they may develop very considerable rises of temperature. The heat would be dissipated eventually, but serious contraction cracking would likely occur upon cooling (see Arts. 13.13 and 13.16).

This cracking can be prevented by care in selecting the cement and by care in placing and curing the concrete. In some instances, concrete is placed only during the winter months when initial temperatures of materials are favorable to lower peak temperatures in the concrete. During warm weather the concrete for large dams is commonly placed at temperatures of about 45°F by the use of aggregates precooled in ice water and by the use of chipped ice in the mixing water. The use of large aggregate and stiff consistency permits the use of a low cement content, in some cases as low as about 2 to 3 sacks of cement per cu yd, even including a pozzolan replacement of about 30 percent. The use of a type II cement or a portland-pozzolan cement may reduce the heat generation by one-third. The placement of concrete in low lifts (not over 7½ ft) with several days between placement of lifts to permit dissipation of heat evolved during the early stages is helpful, and the circulation of cold water through pipes buried in the concrete has been used successfully.

14.17 Special treatment of mass concrete. As both loss of moisture and dissipation of heat of hydration in large structures normally requires many years, and as the shrinkage and resultant cracking of such structures as arch dams may endanger their stability, some such structures have been artificially cooled to subnormal temperatures, and then all construction joints have been pressure-grouted. As the temperature returns to normal, or even slightly above because of late heat of hydration, the structure, which is restrained from expanding, is subjected to low compressive stresses. The shrinkage and cooling strains developed with time reduce this initial compression but should not be sufficient to cause tensile cracks.

For determinations of consistency and compressive strength of concrete containing aggregate over 2 in., it is common practice to wet-screen the concrete to extract the larger pieces. Standard slump cones and 6 by 12-in. cylinder molds can then be used satisfactorily with the wet-screened material. The relation between the strength of the wet-screened concrete and that of the regular mass concrete can be determined by comparative tests, or it can be estimated from the results of tests as shown in Table 14.2.

Table 14.2 Relative strengths of mass mix and wet-screened concrete*

Kind of concrete	Curing	Maximum size of aggregate in specimen, in.	Size of test cylinder, in.	Age, days	Relative strength
Full mass mix	Mass	6	18 by 36	365	1.40
Full mass mix	Mass	6	18 by 36	91	1.30
Full mass mix	Mass	6	18 by 36	28	0.97
Wet-screened	Standard	3	9½ by 19	28	0.99
Wet-screened	Standard	3	9½ by 14	28	1.00
Wet-screened	Standard	1½	6 by 12	28	1.10

* From Ref. 1453.
Cement content ranged from 2 to 4 sacks per cu yd, inclusive of 70 lb of pozzolan. Cement was type II; 9½ by 14-in. cylinder was selected as being reasonably representative of mass mix and simpler to use than a 9½ by 19-in. cylinder.

14.18 Effect of temperature and other variables on properties of mass concrete. The thermal-stress conditions described in Art. 13.16 as well as several significant properties of the concrete itself are affected by the temperature of mass concrete during the hardening period. These properties are discussed in the following paragraphs [1361].

Workability and water requirement. With lower mixing temperatures a given mix produces a wetter consistency, or a lower water content is required for a given consitency. Within the range 100 to 40°F, a reduction of 30°F either reduces the water-cement ratio (by weight) by about 0.02 or increases the slump about 1 in. These effects of temperature are similar for both portland and portland-pozzolan cements. The workability obtainable with portland-pozzolan cements is generally higher than when portland cements are used.

Water gain. Water tends to rise from the lower parts of a lift, and this tends to increase the water-cement ratio of the upper portions. Also, water tends to accumulate below the larger aggregates and form areas of weakness and poor watertightness. These effects are more pronounced the lower the temperature; for example, the amount of water collected at the surface of a 6 by 12-in. cylinder was 30 percent greater at 40°F than at 70°F, and 60 percent less at 100°F than at 70°F. The water gain for portland-pozzolan cements usually is far less than for portland cements, but the effect of temperature is similar for both types.

Permeability. The lower the placement temperature of mass concrete, the greater its watertightness provided the effects of water gain are not a controlling factor. At the age of 60 days, a portland-cement concrete mass-cured from 40°F was about 50 percent less permeable, and mass-cured from 100°F was 25 to 50 percent more permeable, than corresponding concrete cured moist at 70°F [1103].

At the age of 28 days, under similar conditions the permeability of mass concrete was twice as great for a low-heat portland cement as for a portland-pozzolan cement, and three times as great for a modified portland cement as for the portland-pozzolan cement.

Strength. Mass concrete cast at normal temperature, with the temperature rising because of the heat of hydration, develops relatively high strength during the first month or so, but at later ages the strength is generally somewhat less than for curing continuously at normal temperature. When cast at about 40°F, with temperature rising thereafter, the compressive strength within a few days is about equal to that for continuous curing at normal temperature, and the ultimate strength is not impaired. The general effects of curing conditions on strength are shown in Art. 10.13.

Relative lean mixes can be used to produce good strengths in mass concrete, as shown in Table 14.3. This table also shows that tests of cores taken from structures almost invariably show greater strength than that of those obtained from crontrol cylinders which are standard cured for 28 days. The extent of such excess strength generally varies with the age of the cores and the conditions allowing continued hydration of the cement.

Elasticity. During the period of cooling of mass concrete a low modulus of elasticity is desirable in order that the stress resulting from a given thermal strain be low. Curing temperatures have the same general effect on the modulus as on strength, and between different cements the moduli vary approximately as the strengths.

Creep. High creep of mass concrete at the later ages is often desirable, so that the tensile stress resulting from the contraction during the cooling period may be relieved so far as possible.

Under moist conditions the direct effect of curing temperature on creep is small. However, curing temperatures probably have an indirect effect upon creep through their influence on the rate of gain in strength—the greater the strength, the less the creep. From this standpoint, slow-hardening cements are advantageous in mass concrete. Concretes containing either low-heat portland cements or portland-pozzolan cements creep considerably more than corresponding concretes containing normal portland cements (see Art. 12.24).

Volume changes. In the interior of mass concrete the tendency to expand because of combination of cement and water is opposed by a tendency to contract because of reduction in the free water, so that little or no volume change results from these factors. Furthermore, no water is lost by evapora-

Table 14.3 Compressive strength of cores and control cylinders of mass concrete*

Dam	Cement type	Average age, months	Mix data			Core strength, psi	Control cyl-inder, psi	Strength ratio core-cylinder, percent
			Cement, bbl/cu yd	Pozzolan, bbl/cu yd	Slump, in.			
Shasta	IV	60	1 15	...	2	5,880	3,760	156
	IV	60	1.00	...	2	5,120	3,170	162
	IV	60	0.85	...	2	3,840	2,370	162
	IV	60	0.70	...	2	3,280	1,660	196
Monticello	II	7.5	0.56	0.24	2	3,980	3,390	117
	II	13.4	0.53	0.24	2	3,800	2,710	140
Friant	IV	60	1.00	...	1	4,520	4,940	92
		60	0.70	0.41	1	3,820	3,110	123
Glen Canyon	II	6	0.57	0.26	$1\frac{3}{4}$	4,200	2,720	154
	II	12	0.52	0.33	$1\frac{3}{4}$	4,120	2,160	191
Canyon Ferry	II	6	0.65	0.28	2	4,150	4,030	103
	II	7	0.47	0.19	2	3,070	2,430	126

* From Ref. 106. Six-inch aggregate, 10-in. cores, 6 by 12-in. wet-screened control cylinders fog cured 28 days.

tion from the interior to cause volume changes there, but some surface checking may result from drying of a shallow surface layer.

MISCELLANEOUS CONCRETES

14.19 Concrete placed under water. Specifications usually prohibit the placing of concrete under water, unless special permission for such operations is granted, as it is difficult to produce as good-quality concrete under water as can be placed in air; in no case can underwater placement be used for small or thin sections. The following comments, chiefly from Ref. 1800, are applicable to this method of placement:

The concrete should contain at least 7 sacks of cement per cu yd, the amount of coarse aggregate should be within about 1½ to 2 times that of the fine aggregate, and the slump should be within 4 to 7 in. The aggregate should be free from loam or other materials which may cause laitance. Coarse aggregate somewhat smaller than that used in open-air concrete work will give best results. Concrete should not be placed in water having a temperature below 35°F, and the temperature of the concrete should not be less than 60°F nor more than 120°F.

Cofferdams, or forms in running water, should be sufficiently tight to reduce the velocity of flow within them to 10 ft per min, and they should be sufficiently tight to prevent loss of mortar through the walls.

The two principal methods used in underwater placement are by bottom-dump bucket and by tremie. Buckets for this work should have doors which open freely downward and outward when tripped. The bucket should be completely filled, covered with a canvas cloth, and lowered slowly to prevent backwash, until it rests on the bottom surface where the concrete is to be deposited. After discharging, the bucket should be withdrawn slowly until well above the concrete.

A tremie is a pipe, having at its upper end a hopper for filling and a bale by means of which the tremie can be handled by a derrick. To start work with a tremie it may be filled by placing the lower end in a box partly filled with concrete, so as to seal the bottom, and then lowering into position, or by plugging the top of the tremie with cloth sacks or other material which will be forced down and clear the pipe of water as it is filled with concrete. The tremie should be kept full at all times while in use to prevent water from entering the pipe, and the lower end of the pipe should be kept below the surface of the layer of concrete to keep water out of the pipe and so that the concrete will not drop through water and cause the cement to be washed out. All flow of concrete should be outward and upward from the lower end of the tremie. The flow can be regulated by a slight vertical movement of the pipe. If a charge is lost, the tremie should be recharged as at the beginning of the operation.

The concrete should be deposited continuously until it is brought to the proper elevation. The top surface should be kept as nearly level as possible to avoid excessive flow within the forms. No attempt should be made to puddle concrete placed under water, as experience has shown that the less it is disturbed after placement, the better it will be. Upon completion of an underwater concrete job, all laitance on the surface should be thoroughly removed before work is resumed.

When other methods of placement are not possible, particularly where flowing water cannot be avoided, the practice has been followed of partially filling coarsely woven cloth bags with the concrete, and after securely tying them, of placing them carefully in header-and-stretcher courses with the aid of divers. Bags used for this purpose should be free from contamination by deleterious materials.

In Art. 14.25 there is described the prepacked method of placement of concrete under water.

14.20 Vacuum concrete. For thin slabs and walls it is usually necessary to place concrete having a relatively high slump to facilitate its consolidation, but unfortunately the wetness of the mix results in concrete structures of relatively low strength and poor resistance to abrasion and delays the time to final finishing of floor slabs. To eliminate these objectionable features of wet concrete mixes, a system has been developed whereby a considerable part of the water and air near the surface is withdrawn by use of vacuum mats which squeeze out excess water as the mat applies pressure to the concrete. For floor slabs the vacuum mats usually have a plywood backing of about 3 by 4 ft. A projecting rim and a rubber flap around the bottom edge of the panel make a tight seal with the wet concrete, and a fabric covering over a wire mesh on the lower face of the mat prevents removal of cement along with the water. Each mat is fitted with a valve-controlled outlet pipe for hose connection to a manifold of the vacuum system, as indicated in Fig. 14.2. The manifolds, which can take care of the hoses from a number of vacuum mats, are connected to a vacuum pump on a truck. These pumps operated by the truck engine usually draw at least a 20-in. vacuum. Appreciable reduction in time to final finishing of floor slabs and to removal of wall forms and an appreciable increase in strength

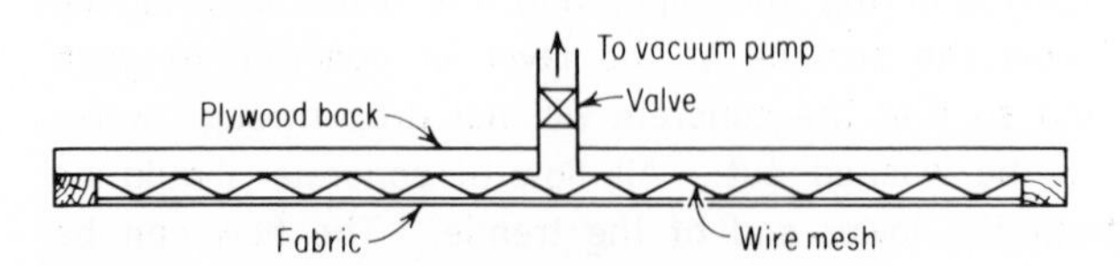

Fig. 14.2. Cross section of vacuum mat.

and general quality of concrete have resulted from the use of the vacuum process.

Vibration of formed concrete during the first few minutes the vacuum is applied is very important. By this vibration, the small openings and channels, created as the vacuum draws out the water, are closed as they are formed. For best results a proper balance must be obtained between the duration and intensity of the vibration and of the vacuum treatment. Too strong a vacuum at the start may stiffen the concrete too rapidly for effective vibration. Too little vibration may not close all the voids developed by the extraction of water.

Vacuum treatment is most effective with coarser sands, minimum practicable percentages of sand, and lower cement contents.

14.21 Concrete for radiation shielding. Penetrating radiation results from the use of nuclear reactors, industrial radiography, and x- and gamma-ray therapy. Radiation shielding is required to reduce the intensity of the radiation to a safe biological level. Density of the shield material is an important factor in the absorption of nuclear radiation. If space is not an important consideration, ordinary concrete in thicker cross section usually produces the most economical shield. If space and access through the shield become more important, as in a hot cell or nuclear reactor, the more costly dense aggregates can be justified. The high density aggregate may consist of a heavy iron ore such as magnetite, limonite or hard hematite, or barium sulfate (barite) rock crushed to give a suitable gradation of concrete aggregate. When very heavy concrete is needed, it may be necessary to grind some of the heavy aggregate to produce fine material, a little of which is required for adequate workability of the concrete mix, and to use steel shot and punchings, but ordinarily it is more economical to use a fine natural sand, even though a slightly lower unit weight results. Natural dense aggregates may cost about 4 to 8 times as much as ordinary sand-gravel aggregates, exclusive of transportation costs, while steel aggregates may cost about 35 times as much.

With fine natural sand and barite aggregate of 1½-in. maximum size, the maximum unit weight of the concrete will be about 215 pcf. With barite sand it will be about 225 pcf [1479]. When average-quality iron ore alone or iron ore with some fine natural sand is used, the maximum unit weight of the concrete will be about 200 to 210 pcf, but if high-density ore is used, the weight may be increased to about 230 pcf. With the addition of iron punchings to the latter concrete the weight may be increased still further to about 270 pcf.

Iron punchings should be free of oil and mill scale. Good-quality iron-ore aggregates are usually sufficiently clean, so that even with 6 sacks of cement per cu yd the 28-day compressive strength for consistencies suitable for hand placement will be well over 2,000 psi, and for stiffer consistencies for vibratory placement the strength will be well over 3,000 psi. With richer mixes, strengths of 6,000 psi or more can be obtained. Similar strengths are obtainable with barite aggregate.

In the case of neutron shields, some compromise in density must be made in order to include sufficient light materials for moderating neutrons to thermal energies. Both hydrogen and oxygen in water are effective for this; thus the water of hydration combined with the cement can supply this need. Proper curing is essential to increase the combined water. Retention of free water at temperatures up to 300°C is improved by incorporating small amounts of lime or about 1 percent of calcium chloride in the concrete.

Boron is one of the most efficient elements in shielding against neutrons and for reducing the production of secondary gamma rays within a shield. But most borate ores retard the set and rate of strength gain of concrete. Also, they frequently contain impurities that tend to reduce the ultimate strength of concrete. However, colemanite and boron frit are two sources of boron suitable for use in concrete, although their cost is very high.

The prepacked method (Art. 14.25) has several advantages for the placing of concrete in shields. Segregation of heavy coarse aggregates, especially steel, can be minimized. For similar materials, greater density and homogeneity can be obtained more consistently by this method than by others. This is important because if segregation occurs the effective concrete thickness is lessened.

The absorption laws governing the attenuation of radiation and the mechanics and design of shielding are given in Ref. 1479.

14.22 Lightweight concrete. Concrete which is considerably lighter than normal concrete is sometimes used for the decks of long-span bridges, for structural concrete, as fire protection for the steelwork of tall buildings, for floor or roof slabs, as thermal insulation, and for building blocks and filler walls. Both natural and artificial aggregates are used in lightweight concrete, but where good strength is essential, only the better artificial aggregates are used. Large deposits of volcanic pumice, lava, and tufa are available in western United States, and in the eastern states much use has been made of expanded blast-furnace slag and cinders as lightweight aggregate, but the principal aggregates for making high-strength lightweight concrete are usually products of the rotary kiln, their vessicular structure resulting from softening the particles by heat and expanding their volume by gases generated within. The oldest of these artificial aggregates is a burned clay known as Haydite, which is produced in large chunks and crushed to size after heating. Most expanded shale and clay aggregates (other than sand sizes) are now processed to the desired size before heating so that they have a hard, sealed, and smoother surface than that of Haydite. As it is difficult to produce a satisfactory lightweight aggregate for good structural concrete that is finer than about No. 16 or No. 30 sieve, it is common practice to use a fine natural sand for the small sizes, although some fine aggregate is made by crushing the coarser material. This latter practice tends to produce a harsh mixture. The irregular surfaces of even the coarser particles of some lightweight aggregates make for harshness and often require a higher paste and mortar factor for workability.

Two lightweight aggregates of recent development are vermiculite and perlite. Vermiculite is made from a lamellar micaceous material, and perlite from a siliceous lava. Both raw materials contain combined water, so that on quick heating to the softening temperature the material puffs, because of the conversion of the contained water into steam. For both these aggregates the fine sizes are produced more readily than the coarse. Neither of them is suitable for high-strength structural concrete, but they have found many other uses, as for concrete partitions and roofs, insulating walls and plaster, etc.

The vessicular structure of these natural and artificial aggregates accounts in part for their light weight. It also gives them high insulating qualities (see Art. 13.10) and results in a concrete of relatively high absorption. In general the absorption of ordinary burned clay aggregates is considerably more than for those which are hard-burned and have a denser shell. As all lightweight aggregates are more porous and absorptive than ordinary aggregates, it is usually necessary to wet them before they are used, or to wet-mix the concrete for a longer period than usual. Otherwise the mix may stiffen unduly before it can be consolidated in the forms. Also, wetting the stockpile reduces its tendency to segregate.

In placing lightweight concrete especial care must be taken to prevent segregation of the coarser particles from the mortar. Because of their lightness, they tend to rise toward the surface and cause segregation more readily than for ordinary aggregates, which tend to move downward in a wet mix. For this reason, highly fluid mixes should not be employed with them. The slump should rarely exceed 3 in.

By use of ⅝ or ¾-in. maximum size of various lightweight aggregates and 7 to 9 sacks of cement per cu yd, compressive strengths of well over 6,000 psi at 28 days and unit weights of 100 to 115 pcf can be obtained. For perlite concretes, strengths of 1,200 down to 300 psi with unit weights of 80 to 30 pcf are obtainable, and for vermiculite concretes strengths of 600 down to 50 psi with unit weights of 70 to 20 pcf can be produced. In general, the lower the weight, the lower the strength for a given mix. The compressive strengths of most lightweight concretes are somewhat lower than for ordinary concretes having the same water-cement ratios and subjected to the same curing conditions. Also, the modulus of elasticity of lightweight structural concretes is lower, probably because of the greater compressibility of the aggregates. The modulus for such concretes ranges from about 1,500,000 to 2,500,000 psi at 28 days, depending on the strength. The shrinkage of lightweight concrete is generally greater than for normal concrete of the same mix and consistency, but some impervious lightweight aggregates produce a concrete having a relatively low shrinkage. The creep of steam-cured lightweight structural concretes loaded at age 1 day to a stress-strength value of 0.5 may not be greater than for some normal-weight concretes, but in general may be from about 20 to 100 percent greater than for normal weight concretes of the same strength [1273]. Reference 1460 covers an extensive investigation of the properties of Haydite concrete, and Ref. 1461 covers other types of lightweight concretes.

Entrained air should be used with practically all lightweight mixes. In addition to the durability it imparts to exposed concrete, it is desirable to promote workability with many lightweight aggregates which tend to produce harsh mixes because of their particle shape and the use of crushed aggregate. Air contents should be determined by the volumetric method. Pressure meters are not satisfactory due to the porous character of the aggregate.

Some types of lightweight concrete can be made without the use of lightweight aggregates. Instead, for one type an admixture (usually containing finely powdered aluminum, magnesium, or zinc) added to a mortar mix generates gas bubbles which cause an appreciable reduction in unit weight to as low as 40 pcf. For another type, an admixture which entrains a large amount of air to produce air-foam or cellular concrete is used. Although these types of concrete may have a unit weight of less than 20 pcf when no aggregate is used, their strength is too low to permit their use as structural concrete. However, they serve effectively as fill material and have excellent insulation qualities—the lighter the material, the better the insulation it provides (see Art. 13.10). The compressive strength of these air-foam concretes varies from 125 psi at 25 pcf to 1,200 psi at 60 pcf, or even 3,000 psi at 115 pcf.

As the physical properties and durability of various lightweight concretes, and their suitability for particular uses, may vary with the different types of lightweight aggregates, these materials should be carefully investigated before approving their use. An economic analysis should also be made. This would include:

1. A study of the savings or extra costs of materials, cement, aggregate, and forms, for both substructure and superstructure, due to reduction in dead load resulting from the use of lightweight concrete

2. A study of the savings or extra costs in the handling of the materials, including mixing, transporting, and placing the concrete, and the excavations and preparation of foundations, due to the use of a lightweight aggregate

14.23 Grouting without pressure. The seating of machinery or structural-steel members on foundations is usually accomplished by grouting with a cement-sand mortar. When the mortar is made sufficiently plastic to flow into place, the later settling and drying shrinkage is likely to be so great that the structure will not be firmly held at the correct elevation. The settling of the cement and sand, leaving a layer of water on the surface is called setting shrinkage, and is the principal objectionable feature of ordinary plastic mortars. The shrinkage upon drying is ordinarily of little significance in comparison with the markedly greater effect of setting shrinkage.

The amount of settlement for a given mix is dependent upon the water content, the sand grading, the cement fineness, and the time that elapses between placement and initial set. A long mixing time or an initial mixing followed by a delayed final mixing is very helpful in reducing the setting shrinkage after placement. The effect of prolonged mixing in reducing the setting shrinkage of grouting mortars is shown in Table 14.4.

As discussed under lightweight concrete in Art. 14.22, certain admixtures can be used to develop gases within a concrete mix. Very small amounts of these admixtures are distinctly helpful in producing fluid grouting mortars which will not settle, because of the expanding influence of the admixture. Such mortars should be used soon after mixing, as the gases do not continue to be developed in sufficient quantities after about 30 to 40 min. For successful results the admixtures should be thoroughly dry-mixed with the cement and sand, as otherwise they tend to float to the surface upon mixing with water.

Table 14.4 Effect of prolonged mixing of grouting mortars upon setting shrinkage*

Mix, cement to sand	W/C by wt	Mixing time, min	Slump, in.	Unit 24-hr setting shrinkage	Mixing time, min	Slump, in.	Unit 24-hr setting shrinkage
1:1	0.40	15	10¼	0.0011	105	9½	0.0005
1:2	0.50	15	10	0.0037	135	9½	0.0005
1:3	0.65	15	9¾	0.0073	150	9½	0.0005

* Tests made using type I cement and sand having a fineness modulus of 2.67. Setting shrinkage is average of three specimens in air for 24 hr. After 4 hr of mixing, the 1:1 mix had a 1-in. slump. From Ref. 106.

A reliable method for eliminating settlement of grouting mortars involves the ramming of a stiff mix into position. If access to the area is to be had on all sides, operations should begin at the center of the bearing area.

The use of admixtures of iron and alkali in a cement mortar produces a grouting mixture which expands slightly because of a chemical reaction between the iron and alkali, thus ensuring a permanent solid filling below machinery bases, or in large cracks which permit leakage of water.

14.24 Pressure grouting. Pressure grouting involves forcing a fluid cement-water paste or mortar into cracks in foundation rock, shrinkage joints between parts of a large dam, or spaces between a tunnel lining and the surrounding rock. The object of such grouting is to plug porous areas and thus stop leakage in hydraulic structures, and to solidify and strengthen fragmented rock adjacent to concrete structures.

Grout mixes for sealing crevices in hydraulic structures consist principally of cement and water. The mixture may vary from thick grout, containing about 4½ gal of water per sack of cement, to very thin grout, containing 20 or more gal of water per sack of cement, depending on conditions. Thick grout is used only when openings in the rock or concrete are large and it is necessary to plug them quickly to prevent useless grouting of large areas or to prevent waste of grout which escapes to the surface. Thin grout can be forced into small openings at lower pressures than thick grout. Although the grout should not be more dilute than necessary to penetrate into thin seams, a thick grout tends to clog the seams and give the false impression that they are filled. Generally, it is advisable to start grouting a hole with dilute grout, then to thicken the grout gradually until no more can be injected. Experience and judgment are essential in determining the consistency to use for best results.

The grout should be mixed by violent agitation for an extended period. It should be used within about 1 hr after the cement is wetted. The cement should not develop rapid stiffening. If narrow crevices are to be grouted, the cement should be fine and free from coarse or flaky particles. In some cases a fine active silica and a small amount of some special lubricant are added to the cement paste. If sand is used, it should be fine; large particles tend to clog the seams.

Some grouting is accomplished at pressures equal to the hydrostatic head of water to be resisted plus 50 percent, but on other projects the pressure may be 2 or 3 times the hydrostatic head. Care must be taken that the grouting pressures do not cause disruption or undue movements of parts of the structure.

The holes for introducing the grouts should be drilled deep enough to secure the specified depth of penetration, or to sound rock as determined by explorations. The sequence of grouting should be such as to avoid compressing air in remote pockets or between adjacent grouting holes; in some locations it may be necessary to install air vents which can be closed when the grout reaches the vent.

One type of grout pump for high-pressure grouting consists of an air-driven reciprocating pump with grout cylinders on the discharge side. Low pressures may be secured by throttling the air supply. Pumping grout into a hole is a continuous operation with this type of equipment.

14.25 Grouted concrete. Although pneumatically placed mortars have been used quite extensively for repair work in the past, their relatively high shrinkage has restricted their use. To overcome this disadvantage a method of grouted concrete known as prepacked concrete has come into use for many types of repair jobs such as tunnel linings, dams and bridges, and even for underwater construction of bridge piers, dry docks, and breakwaters. After filling the forms with compacted aggregate coarser than ¼ in., which is then wetted or preferably inundated with water, the voids in the aggregate are filled by pumping in a grout mix consisting of about 1 part cement, ½ part finely divided active siliceous material, and ½ part fine sand. The flowability of the grout is usually improved by the addition of a pozzolan or other pumpability aids. The resulting concrete is strong and exhibits practically no shrinkage, as the coarse aggregate particles are in direct contact with each other. This latter feature makes it an ideal material for patchwork in hardened concrete (see Art. 12.17). As the forms are subjected to pressure during the grouting operations, they must be made stronger and tighter than is customary for ordinary forms.

14.26 Chemical prestressing using expansive cements. Expansive cements described in Art. 2.23 have been used experimentally to produce expansive concretes for prestressing pipes, slabs, and shells without need of the usual jacking equipment [257a]. The pipes had an inside diameter of 14 in. and a wall thickness of 2 in. The maximum circumferential prestress in the steel was 115,000 psi and the maximum longitudinal prestress was 122,000 psi. However, a part of these stresses was lost as a result of creep and shrinkage of the concrete.

An experimental two-way slab 6 ft 6 in. square and 2 in. thick developed steel tensile stresses of 120,000 psi and concrete compressive stresses of 720 psi before losing a part of these stresses with time. The compressive strength of the concrete after 35 days fog curing was 6,500 psi. When supported at the four corners with clear spans of 5 ft 10 in. each way, it carried an ultimate load of 435 psf.

A one-way slab 8 ft long by 5 ft 4 in. wide and 3 in. thick developed steel tensile stresses of 130,000 psi and concrete compressive stresses of 375 psi, which were decreased to 355 psi as a result of shrinkage on drying.

When tested on a span of 7 ft 6 in., it carried a live load of 262 psf at first crack and an ultimate load of 362 psf.

It should be clear that the composition and manufacture of expansive cements are still in the research and development stage. To use them effectively, data must be available concerning their expansion, shrinkage, strength, elastic, and creep properties.

QUESTIONS

1. What precautions must be considered in connection with architectural concrete?
2. Describe the two types or methods of construction of floor surfaces in common use.
3. What are the disadvantages of the integral-finish type of floor?
4. Why is it undesirable to float or trowel a floor surface while still too soft and wet?
5. Discuss the use and limitations of sprayed mortar.
6. Describe the operation of the cement gun.
7. How is rebound of sprayed mortar affected by the gradation of the sand and by operating conditions?
8. Why are special methods required in placing concrete under water?
9. Describe the operation and use of a tremie in placing underwater concrete.
10. Discuss the factors considered in connection with mass concrete to prevent the development of high temperatures.
11. What is vacuum concrete, and why is it used?
12. How is heavyweight concrete produced?
13. What lightweight materials are used in the production of lightweight concrete?
14. What precautions must be taken in producing lightweight concrete?
15. Discuss the objectives of pressure-grouting.
16. What is prepacked concrete, and what advantageous characteristics does it possess?

FIFTEEN
INSPECTION

15.1 Need for and scope of inspection. In spite of the many advances in knowledge of materials and in mechanized equipment, concrete making is still an art—good organization and good workmanship are essential to the production of quality concrete. The link between the plans and specifications and the finished job is the inspector. Inspection is necessary to ensure that the work is done in accordance with the plans, specifications, and good practice, and to prevent mistakes. Competent inspection is a reliable method of ensuring a structure of high quality.

Inspection is one of several related items which have a direct bearing on the results achieved in concrete construction, as all of the following requirements are considered pertinent to the general problem of obtaining efficiently and satisfactorily, construction that is adequate for a given type of service:

1. Intelligent design. This involves not only the proportioning of parts on the basis of correct mechanical principles, but also the selection of an appropriate grade of concrete.

2. Adequate specifications. This involves appropriate definitions of quality, unambiguous requirements, and enforceable provisions.

3. Reliable construction. This involves the selection of a contractor who is honest and whose plant and equipment are such that control of quality is possible.

4. Competent inspection. This involves the setting up of an organization suitable to the type of construction involved, the selection of capable personnel, and the delegation of sufficient authority.

These requirements are admittedly idealized, but they should serve to point out that inspection is but one step in the production scheme. However, a good design may lose its value, or good specifications their effectiveness, if the inspection is inadequate.

The following items are commonly covered by the inspection of concrete construction [1504]:

1. Sampling, identification, examination, and any field testing of materials
2. Control of concrete proportioning (within specification limits) and the measurement of materials
3. Examination of the foundation, forms, and other work preparatory to concreting
4. Continual inspection of the batching, mixing, conveying, placing, compacting, finishing, and curing of concrete
5. Testing for consistency of concrete, and preparation of any concrete specimens required for laboratory test
6. General observation of contractor's plant and equipment, weather, working conditions, and other items affecting the concrete
7. Preparation of records and reports

The following remarks are pertinent [1501 and earlier editions.]:

Inspection plays an important part in Civil Engineering work. It calls for technical knowledge, tact and judgement. It provides valuable training and is a vital phase in the experience of the young engineer. During the engineer's experience as inspector he is usually getting his first contact with actual construction work and is not only establishing methods and habits of work but is beginning to build his reputation.

Given a good design and proper specifications, it is the inspector's job to apply these to . . . construction . . . in such a manner that in the finished structure the purpose of the design has been faithfully and accurately carried out. He must be able to control the details of fabrication so closely that the desired quality of product will result.

Primarily, the inspector is interested in results rather than methods. But because methods affect results, he must be able to detect improper methods and suggest correct ones in their place. He should have an understanding of the contractor's problems and should be willing at all times to assist in solving them. An appreciation of the contractor's position will do much in making the relations between contractor and inspector friendly and satisfactory.

Lax inspection leads to poor results; careless inspection brings about unsettled conditions and disputes; arbitrary, dictatorial inspection means friction and hard feelings. It is only through firm, intelligent inspection, based on a thorough knowledge of construction principles and an appreciative understanding of the contractor's problems, that good work will be produced.

15.2 Inspection organization. There is a variety of kinds of organization for inspection. In the simplest case the inspection may be performed by one man. However, for a project of any magnitude a department is usually set up specifically to handle the inspection work. Sometimes an inspection department is headed by a chief of inspectors, sometimes by an engineer of materials and tests (or the equivalent) who may perform other functions besides directing inspection. On a construction project the inspection group may sometimes be responsible to the design (administrative) office and sometimes to the construction (field or resident) engineers. On large structures there may be an inspector or inspectors assigned to cover one phase of the work only; a plant inspector is stationed at the batching and mixing plant while a placing inspector is assigned to each concrete gang; in addition to the concrete inspectors there may be general inspectors who act in a supervisory capacity or cover some of the less specific phases of the work.

In the organization of an inspection department the selection of the right man to head it is of considerable importance. He should be given sufficient time and authority so that he can enlist a capable staff and develop a plan of procedure. The funds available for the inspection work should be commensurate with the kind of job to be done, if expected results are to be forthcoming. The authority of the inspection group should be clearly delineated.

15.3 Qualifications of the inspector. Many arguments have taken place on the relative merits of the "practical" and the "technical" type of inspector, the practical type being one who comes up from the ranks of the workmen in the construction industry (and by implication is unhampered by academic training) and the technical type being one who has been trained in a technical school. Probably the ideal inspector would be a combination of the two, and this ideal is approached by many men who have been willing to take the pains to acquire some technical knowledge or to make the effort to obtain practical experience, either through actual apprenticeship or keen observation during a training period under more experienced inspectors.

The general professional qualifications for an inspector are usually summed up by saying that he should have both practical experience and some knowledge of the engineering principles involved in the job to which he is assigned, or as it is sometimes expressed, he should know "the how and the why." This is true for the man who is responsible for the general phases of the inspection work; however, where specialized knowledge or training is the primary requirement, a general engineering education is often entirely unnecessary. It is desirable for any inspector to be acquainted with common practice in the construction field. He should be able to understand the type of specification used in this field; he may be required to interpret contracts and drawings.

The concrete inspector should be fully informed on all phases of concrete technology. It is essential that he understand the bulking influence of moisture in fine aggregate upon volumetric batch quantities, that he be able to recognize whether a mix is well proportioned, and be a good judge of concrete consistency; he should understand the basic principles used in modifying concrete mixes to produce a desired slump or cement factor. He should know whether the forms are sufficiently tight and substantially braced, and during placement of concrete he should know what constitutes satisfactory compaction; if vibration is used, he should be familiar with the undesirable conditions resulting from overvibration as well as undervibration. After placement of the concrete he should be competent to judge whether suitable temperature and moisture conditions are maintained for adequate curing, and should be vigilant to prevent too early removal of the forms or shoring.

Certain traits of character are highly desirable, if not necessary, in an inspector. He should be observant and have a proper sense of proportion that will enable him to give the greater attention to the more important matters. He should have a personality that commands the respect of the workmen and the officers with whom he deals. He should be able to cooperate without having to assume the "good fellow" attitude. He should be straightforward and prompt. He should be possessed of good judgment, tact, a sense of fairness, and the ability to act firmly and impartially. He should understand the extent of his authority and be able to use authority effectively without arousing undue antagonism. He should understand his responsibility and be faithful to it. It is doubtful whether a man possessing all these characteristics actually exists; the degree to which an inspector does possess them, however, may determine his success.

15.4 Responsibility. The inspector's primary responsibility is to see that work assigned to his charge is executed in accordance with the plans and specifications, except as variations are permitted in writing by his superior. He is responsible for thorough knowledge of the specifications, and for the exercise of good judgment. He may be responsible for the use of specified methods of construction as well as materials, because it is often practically impossible from the visual examination of completed work to ascertain whether or not such work has been done in a satisfactory manner and with satisfactory materials; after a structure has been accepted, it may be too late to hold the contractor responsible. The inspector may be responsible for a number of assigned duties, such as the preparation of reports. He should keep a detailed diary of his observations throughout the work, noting particularly all warnings and instructions given to the contractor. He also has certain ethical responsibilities in the way of safeguarding the owner's general interests.

15.5 Inspector training. Inspectors should preferably be given a period of training and apprenticeship under the guidance of an experienced man.

The extent of training will naturally vary with the background of experience the new man brings to his job.

Technical information on concrete construction should be secured and filed accessibly and used by the inspectors in increasing their professional knowledge. In establishing this file, information should be segregated under the subjects to which it pertains. In large inspection organizations such as those maintained by departments of state and Federal government, there may be a number of important cases involving special processes, unusual methods, faulty workmanship, office decisions, or other phases that if filed with the contracts are soon forgotten. Copies of such records belong in the information file.

Inspectors should be urged to improve their technical education by pursuing study or reading courses and attending lectures and meetings of technical societies. When practicable to do so, they should avail themselves of the educational facilities in their district. Technical schools have data, personnel, and facilities that can be utilized to great advantage. Inspectors should also have access to, and avail themselves of, specific information pertaining to their particular problems by calling upon those government agencies best qualified to render aid along the line desired.

Inspectors should be warned that their prestige is lessened when they conduct their inspections by reading from the specifications. The specific details must be studied beforehand, and only a brief reference to a notebook, which may contain pertinent extracts, can be made on the job without "losing face" with the workmen.

15.6 Relations with superior officers. The inspector should explicitly carry out his instructions both as furnished by the specifications and as stated verbally. He should avoid any argument with a superior in the presence of others.

Once assigned to active duty the inspector should have the loyal support of his superior officer. If it is felt that the inspector has exceeded the limits of his authority or failed to exercise good judgment, the superior officer should never argue with or reprimand him before the contractor, his foremen, or any of the latter's workmen. Every effort should be made to maintain the spirit, self-confidence, and morale of the inspector.

15.7 Relations with the contractor. So long as the requirements of the specifications are fulfilled, the contractor is entitled to complete the work at the lowest possible cost. By cooperating with the contractor in any way not inconsistent with the owner's interests, the inspector will aid in reducing the cost of construction. Any change in the specifications that would make possible a saving for both owner and contractor must be made by mutual agreement; the contractor, however, customarily initiates the action by applying to the owner through the inspector. If departures from the letter but not the spirit of the specifications are proposed by the contractor and appear to be necessary, the inspector may arrange to accept the work tentatively and refer the matter to his superior officer.

Inspections should be made promptly when requested. Conditions that will obviously lead to unsatisfactory work should be anticipated whenever possible and in any event should be pointed out at the earliest opportunity in order to avoid waste of materials and labor. The work of inspection should be arranged to cause as little delay and extra work as possible. No demands should be made that are not in accordance with the specifications [1504].

The inspector should familiarize himself with the contractor's methods. Because methods affect results, the inspector should be able to recognize those that will not produce acceptable work and to apprve correct ones in their place. If the specifications permit a choice of methods, the inspector may advise but should not arbitrarily demand that a given method be employed. An appreciation of the contractor's position by the inspector will go far toward making for satisfactory relations between them. The inspector should not attempt to "run the job." Arguments should be avoided and disputed questions referred promptly to the inspector's superior for decision.

The inspector should maintain an impersonal, agreeable, and helpful attitude toward the contractor and his employees, avoiding friction if possible. However, he should avoid familiarity and should accept no personal favors from, or obligate himself to, the contractor. By dealing fairly and by recognizing and commending good work, he can usually secure the friendly cooperation of the workmen. He should take the attitude that changes he suggests are for the benefit of the work and are not made to show his authority. He should not indiscriminately or unjustly criticize the contractor's organization.

Regular and definitely understood methods of communication should be established between the inspection force and responsible representatives of the contractor. Although proper record of all definite action is essential, communication by telephone and by personal conference should be resorted to whenever desirable in the interest of promptness. Oral instructions should be confirmed by memoranda whenever possible, both as a means of established records and of commanding attention. Correspondence and personal contact should be courteous at all times. Instructions should be given only to the authorized representatives of the contractor, usually the superintendent or foreman, except in minor and routine matters and to an extent agreeable to the contractor's organization. The inspector often deals directly with subcontractors, unless his instructions are disregarded;

he should then immediately refer the matters requiring correction to the general contractor who is legally responsible.

Warnings are given in the form of a caution that the faulty work will not be acceptable under the specifications. A good start is important; an incorrect method is more easily corrected the first time it is practiced than after it has been in use. In cases where rejections are made, the contractor should be informed courteously as to why the material is unacceptable. Although he should approach the inspection work without prejudice, the inspector should be watchful of any attempts intentionally made to substitute inferior materials, to hide defects, or to resort to questionable methods, and such attempts should be promptly reported.

Because specifications can often be given various different interpretations, it is always good policy for the inspector and contractor to hold a conference preceding the actual work and come to an understanding on questionable requirements of the specifications.

There are, unfortunately, all classes of contractors, from those who are conscientious and honest to those who are unscrupulous. Until known to the contrary, however, it may be assumed that the contractor will take pride in his work and will endeavor to give satisfaction. Natural dishonesty cannot well be controlled except by experience and refusal to let a contract to one who has already proved dishonest. Even if the contractor is honest, it is exceedingly difficult to secure good work if he is operating at a loss. A contract should only be let when prospects of a reasonable profit are assured. Even then, a foreman, through a desire to cut expenses, may sometimes resort to evasion of specification requirements, the importance of which he does not appreciate.

The inspector should not form habits of procedure which might be anticipated by the workmen. Inspection of the various details and operations should be at irregular intervals, a precaution that is frequently overlooked. The inspector should be on the job to forestall hasty and unsatisfactory work, which is most likely to occur at the beginning or end of a working period. He should always be present when concrete is being placed or other critical work is being done, and in the absence of specific instructions, he must decide whether an operation requires constant or intermittent witnessing [1504].

In any event, test specimens should not be taken except under the direct supervision of the inspector, who will mark them properly for later identification.

Inspectors need not waive their rights for claims against the contractor on account of injuries sustained while performing their duties, provided they have exercised due caution and have strictly observed all job safety regulations.

15.8 Authority of the inspector. Authority should be given the inspector to [1504]:

1. Prohibit concreting until all preliminary conditions (such as completion of forms) have been fulfilled and the work inspected, and until inspection for the concreting has been provided
2. Forbid the use of materials, equipment, or workmanship which do not comply with the specifications
3. Stop any work which is not being done in conformity with the plans and specifications
4. Require the removal or repair of faulty construction or of construction performed without inspection and not susceptible to being inspected later

Ordinarily the inspector is authorized to take direct action in the first three cases above, reporting immediately thereafter to his superior. However, he should stop work only as a last resort, when it is clear that unsatisfactory concrete will result from continuing operations. In the last case, the inspector should obtain the approval of his superior before acting, as the correction of a defect by the complete removal of a member may result in weakness at the junction of the new work with the old.

Ordinarily it will be necessary for the inspector to exercise personal judgment and to make decisions on minor matters not covered by instructions; he should settle as many problems as possible at the job. However, matters of general policy or major items not specifically covered by instructions should be brought immediately to the attention of his superior. An alert inspector will foresee the problems likely to arise, will secure his superior's decision in advance, and thus avoid argument and perhaps delay at the job.

15.9 Specification is inspector's guide. On all work the general and detailed specifications, including all pertinent drawings, prescribe the conditions that must be met by the contractor. The drawings generally prescribe the details of layout, dimensions, tolerances, and certain material requirements for various items; in numerous cases detailed requirements are stated on the drawings.

The specifications regarding the type of concrete must be clear, adequate, rigid enough to secure concrete of the required quality, and yet flexible enough to permit economies to be affected in the selection of local materials. The writer of a specification must have a thorough knowledge of the process of manufacture and an understanding of the effect of the variables on the properties of concrete. Standard specifications for materials and methods of test as published by the ASTM and other bodies are frequently incorporated in job specifications by reference. References 1800 to 1804 contain general specifications for concrete construction.

The inspector is bound by the provisions of the specifications, which include all accompanying documents. Before being used for inspection purposes, copies of the specifications should be examined carefully to ensure that they contain the information essential to a clear understanding of the inspection and test requirements, and for omissions and errors. Specifications should also be checked carefully to ensure that the issue intended for use has been provided. Should any conflict arise between the requirements of the drawings, the detail specifications, and the general specifications, they generally prevail in the order named. A different order of precedence may be specifically indicated in the contract.

15.10 Inspection before concreting. Before concrete is placed in a given section of the work, the specification requirements regarding excavation, forms, reinforcement, embedded fixtures, and joints must be fulfilled and the work inspected. Forms should be of proper size and strength and in their correct location. Careful sighting may assist in detecting irregularities of alignment.

Before concreting begins the cement, aggregate, water, and any other ingredients of the concrete mix should be inspected to see that they are satisfactory for making concrete of good quality (see Arts. 2.17, 3.13, 3.14). If the mix proportions are not given in the specifications, the proper proportions should be determined which will produce concrete of the specified quality (see Art. 6.7). The batching equipment should be checked and adjusted, if necessary, to ensure proper quantities of materials in each batch (see Arts. 7.4–7.7). For a comprehensive checklist for batch-plant inspection, see Ref. 733. No concrete should be mixed until the specification requirements have been met with regard to condition and sufficiency of equipment for mixing, transporting, placing, compacting, finishing, and curing the entire unit of placement.

15.11 Inspection of concreting. Although the inspector must check all concreting operations which involve batching, mixing, conveying, placing, and finishing, as discussed in Chaps. 7 and 8, he should pay particular attention to the batching of materials (especially the cement and water), the time of mixing, the consistency of the concrete, and conditions or methods that might cause segregation in the concrete. He should occasionally test the consistency of concrete at the forms and should more frequently observe the consistency at the mixer, in the conveying devices and at the forms—especially to detect segregation.

After the first floating of floor surfaces, while the surface is still fairly soft, the inspector should check the surface by means of a straightedge, and any high or low areas should be corrected at once.

15.12 Inspection after concreting. One of the very important duties of the inspector is to see that the concrete is properly cured, as covered in Arts. 8.9–8.12. Proper curing exerts such a great influence upon the properties of the hardened concrete and is one stage of concrete making which is so often slighted that the inspector should be watchful of this aspect of producing concrete of good quality.

The inspector should be on hand when forms are removed, to observe the condition of the concrete surface and to advise regarding any finishing or repairing to be done (see Art. 8.19 for the principles involved).

15.13 Concrete samples for tests. To check the strength of the concrete it is necessary to prepare test specimens for shipment to the laboratory. The number of tests (see Table 10.3) depends on the use to be made of the results, but generally not less than two specimens should be made for each 150 cu yd of concrete [1801]. For pavements, it is customary to require at least two specimens for each 1,000 sq yd [1807]. In general, single specimens sampled from different points of the structure provide more useful information than duplicate or triplicate specimens from one point.

The sample of concrete should not be taken from conveying devices, as it may be segregated and not truly representative. If possible, it should be taken from several points within the forms before it has been consolidated; it should be typical of the average concrete being deposited, avoiding batches near the beginning or end of a run. The samples should be taken at irregular times and without prolonged preparations which may provide opportunity for the production of special batches for false sampling purposes. The records should indicate the location in the structure of the concrete batch from which the sample is taken, the general appearance of the mix, and other relevant data such as the source of the aggregates, brand of cement, consistency, temperatures, etc. (see Chap. 16 concerning the inspection records and reports). The general procedure for sampling is covered by ASTM C172.

15.14 Molding specimens. The sample of concrete should be placed in a watertight nonabsorbent container, remixed just enough to make it uniform, and then molded into test specimens. The molds also should be watertight and nonabsorbent to avoid loss of water from the concrete. The entire operation of sampling, remixing, and molding should be completed as promptly as possible; otherwise appreciable evaporation and stiffening may occur.

For compression tests, 6 by 12-in. cylindrical molds are commonly used. The molds are filled in three layers, each layer being consolidated with 25 strokes of a ⅝ by 24-in. round hemispherically tipped steel rod. After the top surface has been leveled, the specimen is covered to prevent evaporation.

Flexure specimens are usually 6 by 6 in. in section. The molds are placed with their long axis horizontal and are filled in two layers, each layer being rodded 50 times for each square foot of area. After rodding each layer, the concrete is spaded along the sides and ends with a trowel. The top surface is finished with a wood float and then covered with wet burlap, which is kept wet until the forms are removed. Details for making and curing concrete specimens in the field are given in ASTM C31.

When compression specimens of mass concrete containing aggregate larger than 1½ in. are to be molded, it is common practice to wet-screen the concrete on a 1½-in. sieve and to mold 6 by 12-in. cylinders from the part passing the sieve. Corrections may be applied to the test results for these cylinders to obtain the compressive strength of the unscreened mix (see Art. 14.17).

15.15 Storing and shipping specimens. Specimens can be damaged by jarring or any movement after set and before sufficiently hardened, and they must be protected against rough handling even after hardening. Specimens made to check the adequacy of the laboratory design for strength of the concrete, or as the basis for acceptance, should be moist-cured at 65 to 75°F and handled very carefully at all times. Specimens for determining the safe time of removal of forms, or when a structure may be put into service, should be removed from the molds at 24 hr and then given the same protection from the elements on all surfaces as is given to the portions of the structure which they represent. However, the small specimens are affected more by freezing or drying conditions than the concrete in the structure, so the storage conditions should be modified to duplicate the treatment given the concrete in the structure.

All specimens should be thoroughly hardened before shipment to the laboratory, and then should be packed in some cushioning material. The standard-cured specimens should be kept damp during shipment. Field-cured specimens should not be shipped until at least three-quarters of the storage period has elapsed. Each specimen should be accompanied by complete identification and relevant data.

15.16 Required strength of test specimens. In Art. 10.12 it is shown that ACI 318 [1801] requires that for designs based on working stresses, the average of any five consecutive strength tests for each class of concrete shall be equal to or greater than the specified ultimate strength, and not more than 20 percent of the strength tests shall have values less than the specified strength.

For designs based on the ultimate strength and for prestressed concrete, the average of any three consecutive strength tests for each class of concrete shall be equal to or greater than the specified ultimate strength, and not more than 10 percent of the strength tests shall have values less than the specified strength.

A method for satisfying these requirements is given in Art. 17.10.

15.17 The field laboratory. The size of the field laboratory and the scope of its work depend upon the type and magnitude of the project. The larger laboratories conduct routine tests on concrete and aggregate, and in addition may conduct investigations of specific conditions peculiar to the project. The smaller laboratories ordinarily make only some tests of the aggregates such as gradation, unit weight, specific gravity, and moisture content. In general, the test data obtained by the laboratory serve as a basis for determining and ensuring compliance with the specifications; for adjusting concrete mix proportions, to secure the maximum value from the materials being used; and for providing a complete record of the concrete placed in every part of the structure.

Specifications usually require the contractor to provide a field laboratory with utilities for the inspector's use. The laboratory should be provided with a desk, work tables, shelves or cabinets, and a sink. A typical floor plan for a small field laboratory is shown in Fig. 15.1. Plans for larger laboratories, with detailed lists of equipment, are given in Ref. 106.

Equipment which may be required for field inspection of concrete is included in the following list [1504]. Several of these items will not be required on some jobs while on others additional equipment will be necessary. All equipment should be kept in good condition.

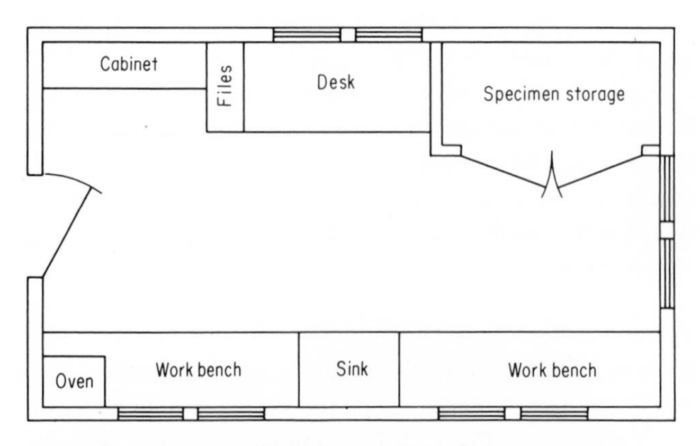

Fig. 15.1. Floor plan of typical small field laboratory.

List of equipment for inspection and field testing

Pocket rule, preferably a 10-ft flexible rule

Light steel tape, 50 ft

Slide rule

Straightedge

Flashlight

Thermometers, 0 to 220°F, armored

Set of standard 8-in. sieves, with cover and pan [C136]

Screen shaker and set of screens with same size of openings as standard sieves, for coarse aggregates

No. 200 sieve, for silt test of aggregate [C117]

Scale, capacity 200 lb or greater, reading to 1/100 lb

Scale, capacity 2,000 g, reading to 1/10 g

Wire basket and container for weighing aggregate under water [C127]

Hot plate or oven for drying aggregates, or other equipment for measuring moisture content of aggregates

Sand sampler

Sand splitter

Metal containers, capacity 1/10, 1/2, and 1 cu ft, for determining unit weight of aggregates and concrete [C29]

Conical mold and tamping rod for preparing sand in a saturated surface-dry condition [C128]

Chapman flask or pycnometer equipment for determining specific gravity and moisture content of sand [C128 and C70]

Slump cone and/or Kelly ball for testing consistency of concrete [C143 and C360]

Vibrator, laboratory model, immersion type

Electric fan

Steel rod, 5/8 in. in diameter by 24 in., hemispherically tipped

Air meter

Bottles and supply of caustic soda for testing sand for organic impurities [C40]

Supply of molds for concrete specimens

Pans, 12-qt pails, shovel, trowel, scoop, brushes, canvas, etc.

First-aid cabinet

Supply of field books, report forms, and stationery

Copy of the various plans and specifications applying to the job

QUESTIONS

1. Why is inspection of concrete construction necessary?
2. What activities are ordinarily covered by the inspection of concrete construction?

3. Discuss the required qualifications of the inspector.

4. Should inspection of details be at regular or at irregular intervals? Explain.

5. What authority should be delegated to the inspector?

6. What are the duties of the inspector before concreting begins?

7. What are the duties of the inspector during concreting?

8. What are the duties of the inspector after concreting?

9. State the precautions to be observed in sampling concrete to be used for casting test specimens.

10. Describe the procedure for molding specimens of concrete sampled in the field.

11. Is it possible for concrete specimens cured in the field for a given period to be truly representative of the concrete in the corresponding structure as the same age? Explain.

SIXTEEN
INSPECTION RECORDS AND REPORTS[1]

16.1 General comments. While the reporting of the inspector's work is secondary to his role as agent to control the quality of construction, records and reports are of sufficient importance to justify his careful and thorough attention. They enable those in authority to know the status of the work in progress, and may be of great value in case of disputes or of future changes in the structure.

The engineering organization and the kind of structure involved have a bearing on the kind and amount of records and reports needed. For small jobs or jobs with which the inspector's superior is in close contact, less extensive reports are needed than for work done for a large organization or at sites which are widely separated.

The resident engineer usually takes care of the general progress reports, quantities of work performed (monthly estimates), records of materials received and on hand, extra work, force-account work, general inspection of contractor's plant and operations, acceptance of parts of structure, and special work; but on some jobs the inspector may be assigned a part of these duties.

The inspector usually makes the necessary mix computations; submits a tabulated daily report regarding the acceptance of materials, the batching, and the placing of concrete; submits weekly or monthly summary reports covering these items; and keeps a diary summarizing the general progress of the work, oral orders given or received, and any outstanding or unusual developments. See Arts. 16.5 and 16.6 for the several items that might be recorded by the inspector.

Tabular forms simplify the work of recording and reporting, and also serve the inspector as a convenient check on the coverage of his daily duties. The order of arrangement of the items on the page will vary with the needs of the organization, but it should be simple and clear.

Field reports to the engineer should be made in duplicate, so that a carbon copy can be retained in the field

[1] Based principally on Refs. 106 and 1504.

office. If more copies are desired for further distribution, they usually will be made in the engineer's office. Some organizations issue a schedule naming the various reports to be made and the persons to whom copies are to be sent.

To facilitate reporting the amount of each kind of concrete placed in each part of the structure, a uniform system of designating the classes of concrete and the parts of the structure should be adopted. Concrete for pavements and canal linings is generally designated by stations, concrete for dams by block numbers or coordinates, concrete for buildings by story and the designations used in the structural design, and concrete for bridges by substructure abutment and pier members, and by superstructure span numbers with designations of their component elements.

When more than one shift is working, the inspector going off duty should advise his relief of the materials and work covered by his report, the serial numbers of samples or specimens taken during the shift, special orders or changes, and any unusual conditions which have developed.

When separate inspectors are needed for inspection of the plant and placing operations, the report forms should be arranged so that each inspector submits a report covering his own work. There will necessarily be some overlapping of items, to identify the various batches with regard to the point of placement in the structure. The two reports may be cross-referenced by batch numbers or the time at which the concrete was mixed.

16.2 Batching and mixing record. The detailed record of batches is kept in the field, and only a summary is included in the daily report. The batching record should include the serial numbers of batches for each kind of concrete, with the time of starting and ending each series; the moisture content of aggregates; the quantities as indicated by the setting of the scales or other measuring devices; the record of the batch or revolution counter of the mixer; the consistency and appearance of the concrete; and the location of placement of the batches in each series. On large installations, automatic recording devices are used to indicate the batch weights, time of mixing, and consistency of the concrete, as described in Ref. 106. The total materials used and the total concrete produced during the shift can be computed from the batch records.

A record of any calibration or checking of measuring devices should be kept, in addition to the record of batches produced.

16.3 Record of materials. The following information should be recorded and reported daily: materials received and on hand, quantity of cement used, materials rejected, disposition of rejected materials. Usually the specifications require that the engineer's representative have access to the contractor's records of materials received, to aid the work of inspection and to check the amounts of materials used on the job.

For the materials (cements, admixtures, aggregates, and steel) of which record is to be kept, convenient tabular forms may be prepared, with a line for each shift and a column for each of the following items: source of material, amount on hand at beginning of shift, amount received during shift, amount received to date, amount on hand at end of shift. The form may contain space for remarks regarding delays in delivery, storage conditions, maximum period of cement storage, damage to, or waste of, materials, and rejection of materials.

16.4 Record of placing and curing. As an aid in preparing the daily report on the various parts of the job, a tabular form somewhat as described below is useful. The tabulation may be provided with columns for the individual parts of the structure worked on during the shift, and with lines as shown in the following list, the spaces in the tabulation being filled in with the appropriate dates: excavation finished or approved, forms inspected or approved, reinforcement and fixtures approved, concrete placed, preliminary curing begun, final curing begun, curing ended, forms removed, defective surfaces repaired, part of structure approved by engineer.

16.5 Daily reports. The daily report should include only those of the items listed below that are significant and appropriate to the particular job [1504]:

Items for possible inclusion in daily report

Identification

Date; shift; type and location of work; contractor's representative; inspector

Materials

Kinds; sources; amounts received, used, wasted, rejected (with reason therefore), and on hand

Field samples shipped to laboratory

Field tests on aggregates: sieve analysis; specific gravity; absorption; moisture content; unit weight and bulking (if on volume basis); deleterious substances

Proportions, batching and mixing (for each class of concrete)

Mix proportions; quantities per batch; computed yield; air content

Period of operation of mixer; number and size of batches; computed total volume

Consistency of concrete at mixer

Parts of structure prepared for placing

Excavation; piling; shoring; forms; reinforcement; construction joints; openings; dowels; embedded fixtures

Placing

Weather (temperature; humidity; wind; sky; times of observation)
Adequacy of organization and equipment
Times of starting and stopping placement; delays; unusual batches
Kinds and volumes of concrete placed; parts of structure completed

Tests of concrete

Consistency at forms; unit weight; air content; temperature
Specimens for strength tests (identification of sample; mix data; curing; age of test; specimens shipped to laboratory)

Curing; form removal (for various parts of structure)

Curing: data and age (in hours) at beginning; date of completion
Forms removed: sides, bottom of member (state age of concrete)
Condition of formed surfaces; defective areas repaired; defective sections replaced

Special work

Grouting; sprayed mortar; ornamental concrete

Force-account work; extra work[1]

Labor; materials; equipment

General

Condition of, or changes in, organization, equipment, or methods
Unusual features

16.6 Diary. A diary should be prepared by each inspector for inclusion in the engineer's records of construction. In general, the entries should include the general progress of the work, delays, unsatisfactory conditions, important understandings or disagreements with the contractor, essence of important conversations, special instructions received, and any data not covered elsewhere which might have a bearing in the future if any question should arise about the work.

The notes should be concise but explicit and should be impersonal in tone; the inspector should keep in mind that at some future date the record may be read in court. Sketches may be useful in amplifying the diary.

16.7 Photographs. Photographs taken at frequent intervals to show the progress of the work form a valuable part of the records of construction.

[1] This is usually covered by a separate daily report.

The inspector should find out whether photographs are desired and, if so, the nature and approximate number to be taken. Usually progress photographs are taken at regular intervals, and additional photographs are taken whenever it is desirable to record such information as the condition of an excavation surface, the condition of forms and reinformcent, faulty methods of placing concrete, or unusual methods of construction.

16.8 Summary report. Some projects require a weekly or monthly summary report to consist of two main sections: (1) a clear, concise, and logically arranged narrative portion in which are described the salient features of the work, and (2) several summarized tabulations recording aggregate gradation, mix properties and concrete strengths, and the results of plant operations.

During the preparatory stages of the work it is desirable to report progress in plant construction, and when a plant is completed, to include a full description, accompanied by photographs and available drawings. After the work is under way, a complete description of the types and methods of forming used, and of the methods and equipment employed in the cleanup preparations for placing and in transporting, placing, consolidating, finishing, and curing the concrete, should be included in the summary report. The reports that follow should contain comments on any changes made in equipment or procedure. In addition to statements of interest concerning normal operations and progress, the reports should contain comments pertaining to special features of the work and difficulties encountered, some of which are illustrated in the following list [106]:

1. Correction of unsatisfactory aggregate gradation, and avoidance of segregation and breakage
2. Means used for heating concrete materials and concrete during cold weather, or for cooling them in hot weather
3. Improvements in batching and mixing procedures, for example, charging and measuring of water, sequence of charging ingredients into mixer, changes in mixing time, alterations in mixer blades
4. Important alterations in plant construction or equipment
5. Difficulties encountered in transporting concrete, such as loss of slump and segregation
6. Problems encountered in making preparations for placing, including unusual form construction and cleanup procedures
7. Unusual placing conditions, for instance, difficult placement, special handling of the concrete into and in the forms, sticking of concrete in hoppers and buckets, premature stiffening, placing in cold weather, segregation, bleeding, harshness, reduction of voids in exposed surfaces, improved practices in lining tunnels and canals

8. Development in finishing procedures, perhaps involving ingenious devices for compacting, screeding, and floating, or methods of removing stains

9. Results of the use of sealing compounds for curing, particularly with reference to changes in the condition of the coating, and to surface defects in the concrete, such as crazing and soft surface; unusual procedures for curing and protecting concrete in cold weather or in hot weather

10. Prolonged delays in the work, with explanation of cause

11. Special tests or investigations made in the field, for example, tests for mixer performance, bond at construction joints, curing with sealing compounds

12. Important cautions, instructions, and concessions to the contractor; protests by the contractor

Monthly reports from some large projects include information in regard to deliveries of bulk and sacked cement, cement handling and blending operations, special sizing of cement for use in grouting contraction joints, weather conditions, temperatures of the concrete and of the ingredients and other routine matters. In some cases there is included a detailed statement of progress of the work with quantities to be used as a basis for payments to the contractor.

QUESTIONS

1. In general, what sort of records and reports is the inspector expected to prepare?
2. What information is generally included in the batching and mixing record?
3. What information is generally included in the placing and curing record?
4. Would you recommend the keeping of a diary by the inspector? If so, what should it include?

SEVENTEEN
ANALYSIS AND PRESENTATION OF DATA

17.1 The problem of transmission of information. It may be taken almost as axiomatic that, at least for engineering purposes, no data are of value until they are put in a form that can be readily understood and utilized. The particular form in which data should be summarized and the extent to which they should be interpreted obviously will depend upon the intended audience. However, there are a number of commonly used procedures for analyzing and reporting data which may be generally applied.

Some of the considerations that may be mentioned in connection with the problem of analysis of data are:

1. Reduction of "raw" data. The units in which the data are recorded are conditioned by the kind of measurements that are made, e.g., loads may be measured in pounds, and temperatures may be measured in terms of electrical resistance. Because most data have meaning only in comparison with similar data, the quantitative measures obtained from a test are reduced to values whose units are acceptable as a basis for comparison; thus loads are reduced to stresses, say, in pounds per square inch, and deformations to strains. Further, the reliability of the data is conditioned by the errors of measurement. In reducing the data, corrections may have to be applied for systematic errors, and in order to express the final data in the appropriate number of significant figures, an estimate should be made of the accidental errors inherent in the measurements and of the effect of these errors on the accuracy of the reduced values.

2. Summary of data. Although in the simple case the summary of test results may merely amount to setting down a few readily comprehensible facts and figures, in connection with large-scale operations there may be accumulated masses of data that are so numerous and variable that it is practically impossible for the mind to digest and evaluate them in unassembled form. Therefore, reports should contain summaries presented in such a manner that as far as possible decisions as to the quality of the material or sufficiency of the process can be made without further reference to the original data. Advantage may then be taken of well-known statistical procedures for summarizing the data.

3. Study of relationships between variables. After the data have been reduced and assembled in manageable form, the final step in the analysis is usually to seek or to develop relationships between the variables involved or between the data obtained from a particular test and previously obtained data or some theory. In many instances the skill with which this is done depends upon the capacity, the ingenuity, and the background of the analyst. Common devices employed in studying relationships are tabulations, graphs, bar charts, and correlation diagrams; the procedure is usually to hold constant (in so far as is known or is possible) all variables except two whose relationship is investigated.

17.2 Variations in data. The primary function of compression tests of field concrete is to insure production of uniform concrete of desired strength and quality, but the quality of concrete is subject to the influence of numerous variables. Each of the ingredients of concrete may cause variations depending upon their uniformity. Variations may also be introduced by practices used in proportioning, mixing, transporting, placing, and curing. In addition to the variations which exist in the concrete itself, strength variations will also be introduced in fabrication, testing, and care of test specimens. Variations in the strength of concrete must be accepted; but consistent concrete of adequate quality can be produced with confidence if proper control is maintained and test results are properly interpreted. For large numbers of data, variations in measurements and measures of properties have been found to coincide closely with variations computed from theoretical considerations. When the data are few, the coincidence is often not so good, but for convenience the concepts developed from the theory of probability (for large numbers) are applied and afford a fairly workable means of summarizing and utilizing data.

As a result of the variations in the test data for concrete, a series of questions arises, such as the following: How may individual test data be condensed to convey a maximum amount of information to the reader of the summary? How many samples must be taken to obtain a fair measure of the quality of the work as a whole? How many tests are necessary to evaluate with a desired accuracy a given property, such as the compressive strength of concrete? The methods of analyses given in the following sections will not give absolute mathematical answers to all these questions, but they will be found useful in guiding judgment.

17.3 Grouping of data. The grouping of most data in concrete testing is according to magnitude. The arrangement of data according to magnitude results in what is technicaly known as a "frequency distribution." In concrete testing the latter usually consists of n measurements of a given characteristic of n different pieces, such as the compressive strength of n test cylinders. A frequency distribution of this type consists of homogeneous data. In statistical parlance they come from a common parent population, and are not a single-valued constant but rather a frequency distribution function. Data on concrete testing may be considered as being predominantly affected by variability of the characteristic observed.

The first operation performed on the raw data is to arrange the items according to magnitude, usually in ascending order. This is sometimes referred to as an array, or called an ungrouped frequency distribution. Merely by inspection, the minimum and the maximum values and the range can be selected, and, by a simple computation, the median or middle value and the average value can be determined.

In some cases the data are separated into groups which are called cells, or step intervals. After the length of the interval has been decided upon, the number of items in each interval, usually called the frequency, is then determined. Often the relative frequency, which is the number in each interval divided by the total number of items, is used. This fraction of the total number is an important characteristic, especially when applied to the percentage below, outside, or beyond a specified limit, thus giving the fraction defective. When there are a large number of items, 13 to 20 class intervals are recommended. Too many intervals may show an irregular distribution. By using larger intervals (fewer divisions) the appearance of the distribution can often be improved. When the total number of items is less than 25, such a presentation ordinarily is of little value.

In the usual graphical presentation the frequencies, actual or relative, are plotted as ordinates to an arithmetical scale on the center line of each interval. When successive plotted points are connected by straight lines,

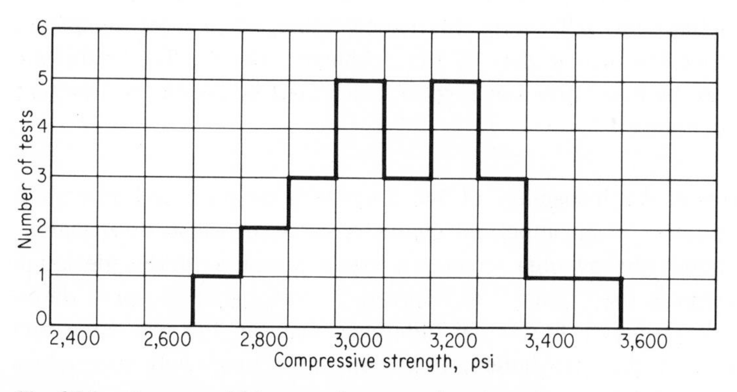

Fig. 17.1. Frequency histogram of compressive strength.

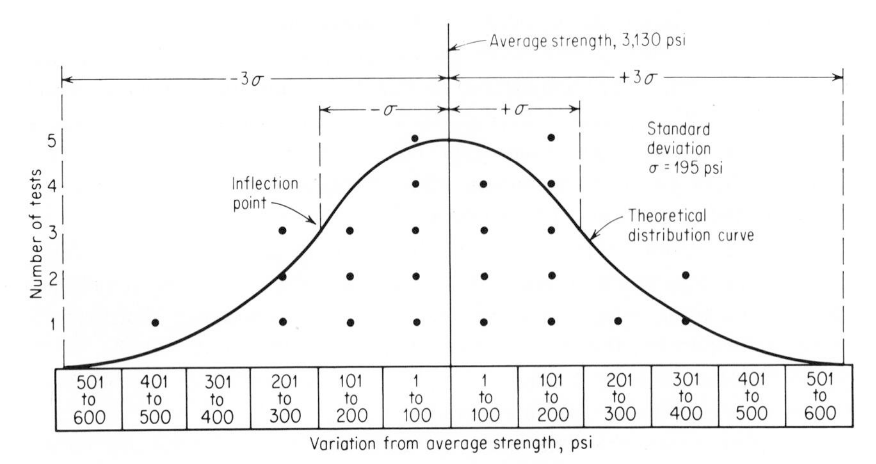

Fig. 17.2. Frequency distribution of compressive strength.

the chart is called a frequency polygon. A different design would result by drawing a wide line or bar along the center line of each interval from the base to the plotted points. The diagram is then known as a frequency bar chart. If instead of filling in the bars they are left open as in Fig. 17.1, which presents the data from Table 17.1, the diagram is designated as a frequency histogram. All these different forms are widely used.

The distribution commonly encountered in materials testing is a bell-shaped curve resembling the theoretical normal frequency curve as shown in Fig. 17.2. If the distribution curve has only one peak, it is said to be unimodal. Many times it lacks bilateral symmetry and is skewed to the left or right (asymmetrical).

17.4 Central tendency. A measure of central tendency, or tendency to be grouped about a central value, is called an "average," which purports to summarize the data by locating this typical value. The most significant average for concrete testing data is the arithmetic mean. The arithmetic mean is affected by every item but is greatly distorted by unusually divergent values at the extremes.

17.5 Dispersion. An inspection of the frequency diagram will furnish a qualitative indication of an important characteristic—the dispersion, scatter, or variation about the average value. A crude measure of this deviation from the average is the "range," or "spread." Although it is easily determined, since it is merely the actual difference between the maximum and minimum items in the distribution, it is dependent upon only two values that are of unusual occurrence (low frequency), particularly when n is large. So the range may give misleading indications of distribution.

A criterion that considers the location of every item, rather than only the two extremes, would be more meaningful. An obvious meaure would be the average distance (disregarding signs) of the items from some measure of central tendency.

The differences between an average $\bar{X}$ and the various items are called deviations and the

$$\text{Average deviation} = \frac{\sum_{i=1}^{n} (X_i - \bar{X})}{n}$$

The quantity $X_i - \bar{X}$ is here taken to be the absolute value of a deviation disregarding signs.

Perhaps the most generally used measure of dispersion is the "standard deviation." It is a special form of average deviation, and it is usually designated by the symbol σ (small Greek letter sigma). The following equation gives first the most general form followed by a simple rearrangement.

$$\sigma = \sqrt{\frac{\sum_{i=1}^{n} (X_i - \bar{X})^2}{n}} = \sqrt{\frac{\sum_{i=1}^{n} X_i^2}{n} - \bar{X}^2}$$

A glance at the general form makes the reason obvious for the often used designation, root-mean-square deviation. In simple words, standard deviation is the square root of the average of the squares of the deviations or differences of the individual test results from their average $\bar{X}$.

As an example of the method of computation of standard deviation, data on the strength of concrete cylinders are presented and summarized in Table 17.1. In this table the standard deviation is computed by the first or more general form of the above equation, which is the easiest to visualize. A short-cut method involving fewer operations and based on the second, or rearranged, form of the above equation is shown in Table 17.2. Tables of squares and square roots of numbers aid in the computation by either method.

The standard deviation, since it uses second powers of all items, places more emphasis on widely dispersed items than does the average deviation. For a normal distribution the average deviation is 0.7979 sigma.

Figure 17.2 shows the data from Table 17.1 plotted according to their variation from the average and illustrates the meaning of standard deviation. Each dot represents a single test result lying within the range indicated at the bottom. The curve approximating the data is plotted symmetrically on each side of the vertical line representing the average. This curve, representing a theoretical distribution, is mathematically related to the stand-

Table 17.1 Computation of standard deviation of concrete strength*

No.	Compressive strength, psi	Deviation, psi	Squared deviation
1	3,180	50	2,500
2	3,320	190	36,100
3	3,260	130	16,900
4	3,510	380	144,400
5	3,360	230	52,900
6	3,110	20	400
7	3,220	90	8,100
8	2,960	170	28,900
9	2,840	290	84,100
10	2,910	220	48,400
11	3,320	190	36,100
12	3,040	90	8,100
13	3,020	110	12,100
14	2,710	420	176,400
15	3,140	10	100
16	2,990	140	19,600
17	3,060	70	4,900
18	3,440	310	96,100
19	3,200	70	4,900
20	3,050	80	6,400
21	3,260	130	16,900
22	2,850	280	78,400
23	3,080	50	2,500
24	3,290	160	25,600
Total	75,120		910,800
Average	3,130		37,950

* Standard deviation, $\sigma = \sqrt{37{,}950} = 195$ psi. Coefficient of variation,

$$V = \frac{195}{3{,}130} = 6.2\%$$

ard deviation. The inflection points of the curve occur at values equal to plus and minus the value of the standard deviation, and within this range 68 percent of all observations will theoretically fall. It is of interest to note that 71 percent of the data of Table 17.1 fall within this range.

Since significant comparisons of dispersions are sometimes difficult to make by using absolute measurements, a variation expressed as a ratio

Table 17.2 Short-cut method of computing standard deviation*

No.	Compressive strength, psi	Squared compressive strength $\times 10^{-4}$
1	3,180	1,011
2	3,320	1,102
3	3,260	1,063
4	3,510	1,232
5	3,360	1,129
6	3,110	967
7	3,220	1,037
8	2,960	876
9	2,840	807
10	2,910	847
11	3,320	1,102
12	3,040	924
13	3,020	912
14	2,710	734
15	3,140	986
16	2,990	894
17	3,060	936
18	3,440	1,183
19	3,200	1,024
20	3,050	930
21	3,260	1,063
22	2,850	812
23	3,080	949
24	3,290	1,082
Total	75,120	23,602
Average	3,130	983.4

* Square of average $= 3130^2 = 9{,}796{,}900$.

$$\text{Standard deviation} = \sqrt{9{,}834{,}000 - 9{,}796{,}900} = 193$$

or percentage must be determined. The one most commonly used is the standard deviation on a percentage basis and is usually called V, the "coefficient of variation."

$$V = \frac{\sigma}{\bar{X}} 100$$

For the data in Table 17.1 the coefficient of variation is 6.2 percent. The coefficients of variation that can be expected on controlled projects are

Table 17.3 Standards of concrete control*

	Coefficients of variation for different control standards			
Class of operation	**Excellent**	**Good**	**Fair**	**Poor**
Overall variations:				
General construction	Below 10.0	10.0–15.0	15.0–20.0	Above 20.0
Laboratory control	Below 5.0	5.0– 7.0	7.0–10.0	Above 10.0
Within-test variations:				
Field control	Below 4.0	4.0– 5.0	5.0– 6.0	Above 6.0
Laboratory control	Below 3.0	3.0– 4.0	4.0– 5.0	Above 5.0

* From Ref. 1708. These standards represent the average for 28-day cylinders computed from a large number of tests.

shown in Table 17.3. The ratings of control are based on experience from a large number of projects and are presented as a general guide in evaluation of concrete control.

A simplified graphical method for the evaluation of test data and the determination of standard deviation, mean strength, and the coefficient of variation is given in Ref. 1709.

17.6 Probable error. Another commonly used criterion of dispersion is the *probable error.* The probable error of a single measurement in a series of measurements (of equal weight) is computed from the expression $0.6745\sqrt{\Sigma(v^2)/(n-1)}$ where the v's are the residuals (or deviations from the mean) and n is the number of observations. The quantity under the square-root sign is seen to be the standard deviation for a limited number of observations. The probable error is *not* the error most likely to occur but is the minimum error which will not be exceeded by 50 percent of the cases. The probable error of the mean of a series of measurements is found by dividing the probable error for a single observation by $\sqrt{n}$. For comparative purposes, the *relative error,* or *precision ratio,* is commonly taken as the ratio of the probable error to the average value of the quantity measured, expressed as a percent; this corresponds to the coefficient of variation.

17.7 Limits of uncertainty of an observed average. The uncertainty of any observed average is less the greater the number of specimens tested under the same conditions. For any observed average the limits of uncertainty with respect to the unknown but true average of a very large number of tests may be determined from a consideration of the law of probability [106].

For an understanding of the limits of uncertainty, consider the following example: Ten groups of five individual tests for compressive strength were made under identical conditions, and an arbitrary strength range for each group established by increasing and decreasing its average strength by 8 percent. Then it was noted that nine of the ten resulting strength ranges included the grand or true average strength for the entire fifty tests. From these results it would appear that if five additional tests were made and a strength range for the group determined as before, there would be a 90 percent probability ($P = 0.90$) that this range would include the previously determined grand average strength. This means that 9 times out of 10

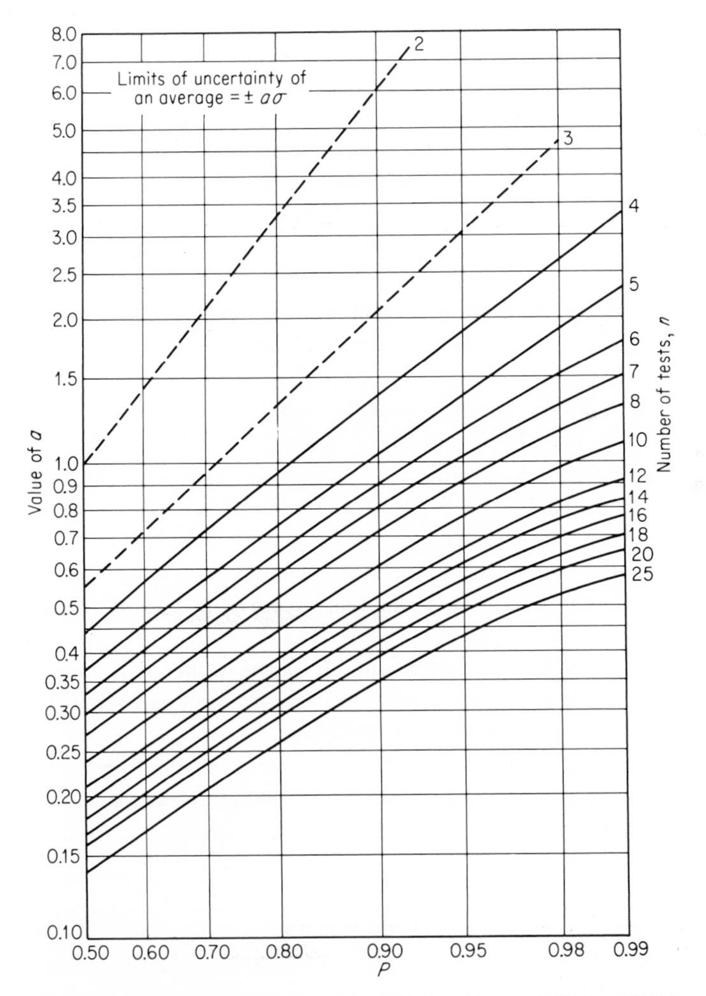

Fig. 17.3. Curves giving values of a and P. *(From ASTM* [1700].)

Table 17.4 Comparison of compressive strength uniformity

Series	Avg compressive strength, psi	No. of tests	Standard deviation, psi	Coefficient of variation, %	Limits of uncertainty, psi	
					9 out of 10	1 out of 2
1	3,130	24	195	6.2	±70	±30
2	3,820	120	270	7.1	±40	±20
3	4,330	40	380	8.8	±100	±40
4	3,510	16	400	11.4	±180	±70

the uncertainty of the group strength average would be less than the ±8 percent, and 1 out of 10 times it would be greater. As most engineering test data follow the law of probability, the limits of uncertainty (value to be added to and subtracted from the test group average) can be determined by multiplying the computed standard deviation, σ, for the test group by a factor a, which depends upon the number of tests, n, in the group and the degree of probability, P, desired. This relationship is graphically shown in Fig. 17.3. A probability of 0.90 is often adequate for engineering work, and a probability of 0.99 is generally regarded as near certainty. The probable error (Art. 17.6) is a special case of limits of uncertainty for a probability of 0.50.

The limits of uncertainty for four series of tests are shown in Table 17.4. For series 4 the average compressive strength is 3,510 psi for 16 tests with a standard deviation of 400 psi. From Fig. 17.3 the value of a for 16 tests for a probability of 0.90 (9 chances out of 10) is about 0.46. Then the limits of uncertainty are $\pm 0.46 \times 400$ or ±180 psi, and the strength range, within which there are 9 chances out of 10 that future averages for the same number of tests will fall, is $3{,}510 \pm 180$, or 3,330 to 3,690 psi.

For a probability of 0.50 (1 chance out of 2) the value of a for 16 tests is about 0.18, and the strength range of the average for this probability is $3{,}510 \pm 0.18 \times 400$, or about 3,440 to 3,580 psi.

These strength ranges may be too wide for the work at hand, in which case either the number of tests must be increased to obtain a more accurate average, or better control must be introduced to eliminate causes of variation, so as to obtain a smaller standard deviation.

17.8 Number of tests to obtain a desired accuracy. As a result of the variability of engineering materials, each test series should include enough tests so that the limits of uncertainty of the average are within the requirements of the tests. For any probability it is apparent from Fig. 17.3 that the limits of uncertainty of an average are narrowed or the reliability is improved as the number of specimens on which the average is based becomes greater.

For an accuracy such that at least 9 times out of 10 the average of the test group will be within, say, 3 percent of the average of a very large group, the required size of the test group may be obtained from Fig. 17.3 after selecting values for P and a. P, the probability factor, is 0.90 for the 9 out of 10 chances that the average of the test group will be within the specified 3 percent of the average of a much larger group. The factor a is determined by dividing the desired percentage limits of uncertainty, 3 percent for this example, by the coefficient of variation V for similar tests, such as 6.2 percent for the corresponding group shown as series 1 in Table 17.3. Then $a = 3.0/6.2 = 0.48$. It will be observed in Fig. 17.3 that the intersection of the values $P = 0.90$ and $a = 0.48$ indicates a need for 15 specimens to meet the specified requirements.

If no data are available for the type of test to be conducted, a preliminary analysis should be made for a group of at least five samples. Then the coefficient of variation V for this group may be used to compute a, and by following the procedure outlined in the foregoing paragraph the number of specimens required for the desired degree of accuracy may be determined.

17.9 Significant figures to retain in presenting test results. The variability of test results for engineering materials does not warrant recording test values in such form as to imply a high degree of accuracy. For instance, reporting compressive strength of concrete to four significant figures may imply an accuracy which rarely exists. The retention of more figures than the precisions of the data warrants causes a wastage of time on computations. However, care must be exercised in rounding the figures to avoid the loss of essential precision.

In general, a significant value is one which has an even chance of being repeated in future tests. As an example, the average strength of the 120 specimens in series 2 of Table 17.4 is 3,820 psi with a standard deviation of 270 and a limit of uncertainty of ± 20 for a probability of 0.50, or one out of two chances. This indicates an even chance that for future tests under the same conditions the average might be in the range $3{,}820 \pm 20$. This shows that the 3 and 8 are really significant, that the 2 being partly significant should be shown even though it could be 0 or 4, and the 0 is not significant but is carried to show the position of the decimal point. Individual test values are not accurate to as many significant figures as the average, but it is desirable to use the same number of figures in the individual values in order that cumulative errors may be avoided. In general, two more places than are known to be significant should be carried in computations, and one more place should be given in the average than is known to be significant. The last column of Table 17.4 covering the limits of uncertainty for average values for one out of two chances, shows that the average for series 2 is significant to the nearest 20 psi and the average for series 4 is significant only to the nearest 70 psi. All averages shown in Table 17.4 should be reported to three significant figures. When properties are more variable than in the example given, fewer significant figures should be shown in the final result.

17.10 Required average strength. As a result of variations in the strength of test cylinders for a given job and the need for having but few test values fall below the specified strength f'_c (see Art. 15.16), it is essential that the required average strength of the concrete, f_{cr}, be in excess of f'_c, the degree of excess strength depending on the expected uniformity of concrete production and the allowable portion of low tests. The required average strength for any design can be approximated from Fig. 17.4 (also see Art. 10.12).

17.11 Statistical summaries. The meaning and significance contained in a mass of measurements is made clear by their interpretation, which in turn is facilitated by a well-arranged presentation. However, the accuracy

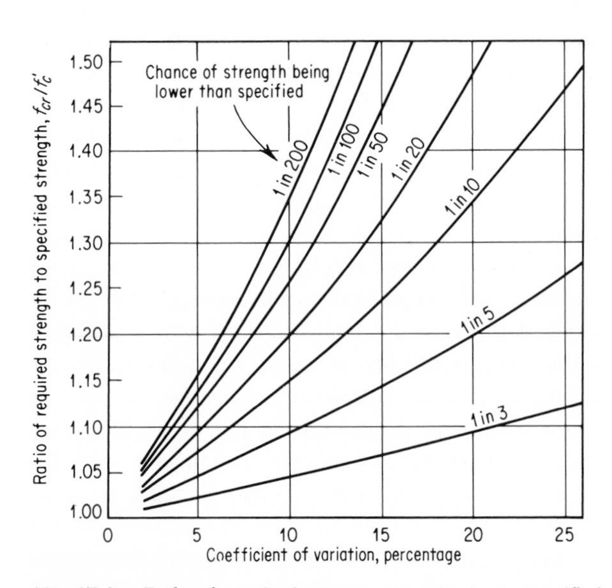

Fig. 17.4. Ratio of required average strength f_{cr} to specified strength f'_c for various coefficients of variation and changes of falling below specified strength [1708].

of the data itself is no way altered by the method of presentation or subsequent study and interpretation.

Since the investigator has no way of knowing what subsequent use may be made of the data, supporting evidence regarding the test specimens themselves should first be presented in brief, clear statements. These should include even more information than at the time may seem relevant, since later developments may prove the value of all collateral evidence presented regarding the history of the specimens, including their selection and preparation, and the grade and character of material.

The next information should deal with the test itself and the conditions under which it was made. This includes the measurements together with adequate description of the methods used to eliminate constant errors and reduce the size of chance or accidental errors as well as observations on the general and specific test conditions and procedures, giving especial attention to the matter of special difficulties and their treatment, so as to present evidence of all precautions taken to establish controlled conditions and secure reliable, trustworthy data.

In the presentation of the data itself, the essential information is generally contained in four statistics, namely, the number of items n, the arithmetic mean $\bar{X}$, the standard deviation σ, and the coefficient of variation V.

17.12 Tables. The presentation of factual material in tabular form economizes space and facilitates the comprehension of the scope and range as well as the significance of the data. Often more varied information can be condensed into a table than can be shown in a chart or diagram, and whether used in the text or in an appendix, tables should have adequate and clearly stated titles, preferably brief, so they can be understood on the first reading. The column headings should be complete and contain units for the values in each bank, or column; these units should not be attached to items in the body of the table. In tables of some length, ease of reading is improved by grouping the lines, such as by the use of horizontal dividing lines or increasing the space between every third or fifth line.

Absolute brevity is important; the use of abbreviations and of keywords with a reference to footnotes aid in securing brevity. As an aid to clarity, the style and arrangement of headings and subheadings should be uniform and consistent. Conditions of test common to all data given in a table should appear as general footnotes. Conditions peculiar only to a subgroup should be shown in special footnotes; superior letters are often employed to call attention to footnotes.

17.13 Figures. Illustrations used in connection with an accompanying text are referred to as figures (abbreviation—Fig.). In reports on concrete these figures may be actual photographs of such items as concreting operations, testing machines, apparatus, and specimens, or line drawings such as a blueprint or black line print reproduced by a direct process, or a print reproduced by some photographic or other process. In the latter case the size of the reproduction may be greater or less than the original, and the size of lettering and weight of line on the original must be proportioned in accordance with any contemplated change in size of the reproduction, so as to make it clear and easily read.

The line drawings most generally used in reports on concrete are usually referred to as graphs, charts, or curve sheets, although the designation diagram is frequently employed, e.g., the stress-strain diagram.

Unless special sizes are required, the 8½ by 11-in. cross section, graph, or coordinate sheets are used. These are available in many varieties of rulings, printed from accurately made plates, and are not to be confused with ordinary quadrille sheets. A widely used ruling is 20 by 20, twenty lines to the inch with every fifth one heavy; another is millimeter paper similarly ruled.

To facilitate direct reproduction such as by blueprinting, select a lightweight paper. Since the printed lines are sufficiently clear when viewed through the paper, it is common practice to draw and letter on the plain side of the sheet. Then, any necessary erasures will not dim the grid lines and thus mar the print.

The choice of scales depends upon the range in the data and the purpose of the graph. The approximate scale is the range in data to be shown, divided by the number of principal divisions, which should preferably be subdivided into five or ten lightly marked smaller divisions. The exact scale should be selected so that the smallest division will be some simpler part of the scale and not an awkward fractional part. Well-chosen scales make the lines on the cross-section paper of assistance in reading values of random points on a curve. Beginners often overlook the rather obvious necessity of placing the principal scale divisions on the heavy lines of the graph paper. The principal values should be lettered in the margins.

The plotted points should be fine pencil dots enclosed in some appropriate symbol such as a circle, triangle, or square. Only the symbols, however, are inked, by use of the drawing instruments, and the inked curve or graph is extended to, and not through, them. If the diagram is to be used for computing (reading values off the chart instead of from tables), the experimental points are not shown, and the curve is drawn as a continuous line of lighter weight than that used when merely a general relationship is shown; in the latter case it should be heavy so as to stand out clearly from the grid, because it is the important part of the illustration.

When the results of loading tests are involved, the loads or stresses should be plotted as ordinates, and the other variables, such as water-cement ratio, should be plotted as abscissas. Very irregular curves should be avoided; smooth curves should be drawn through average values. If several curves are shown on one chart, each curve should be labeled. The title should be appropriate and should indicate the character of the information given by the diagram. If necessary, a few notes may be shown on or below the graph.

If the entire drawing is specially prepared, the ink grid lines are not drawn through any symbols or lettering, and usually only the principal lines, corresponding to the heavy lines on prepared graph paper, are shown.

Avoid possible distortion and unintentional overemphasis which arises by using only a partial scale, i.e., one that does not run continuously from zero. If it is inexpedient to show the entire scale, insert a conspicuous note or break the scale to indicate that it is not complete.

Since the purpose of illustrations is to aid in the presentation of information, the matter of appearance and clearness should be given special attention in the preparation of charts and graphs.

QUESTIONS

1. Should summaries of data be dependent upon reference to the original observations for a complete understanding of their significance? Explain.

2. What devices are commonly employed in the study of relationships between variables.

3. How many variables may be included in a simple graph to show the relationships involved?

4. What is meant by a frequency distribution?

5. What is a frequency polygon? A frequency bar chart? A frequency histogram?

6. What is the common shape of a frequency polygon commonly encountered in materials testing?

7. Name the most significant measure of central tendency in connection with concrete test data.

8. Why does the range or spread as shown by the frequency diagram give a misleading indication of distribution?

9. Give a definition of standard deviation.

10. What is the relationship between the standard deviation and the average deviation for a normal distribution?

11. What is the significance of the standard deviation?

12. What is meant by the coefficient of variation?

13. What is the significance of the probable error as computed from the expression $0.6745\ \sqrt{\Sigma(v^2)/n - 1)}$?

14. Define the relative error or precision ratio.

15. What four statistics are generally employed in summarizing the results of a series of observations?

PART TWO

INSTRUCTIONS FOR LABORATORY WORK

GENERAL INSTRUCTIONS

Prompt and regular attendance is required of every student. If a test is not performed during an assigned laboratory period, special arrangement must be made with the instructor to make up the test.

For the performance of tests, the student will be organized into parties. Before beginning the work of a laboratory period, the student must familiarize himself with the scope and purpose of the test he is to perform and with the laboratory procedure involved. In addition to the particular test instructions, preparatory matter should be read as indicated at the beginning of each test. Lack of preparation may result in exclusion of the student from the laboratory and in any event will make difficult the performance of the work within the assigned period.

Each group shall submit a draft of the proposed data sheet for approval of the instructor in advance of each test, unless one is provided by the instructor. Tabular forms will be arranged whenever possible. All data shall be entered on the data sheet as soon as they are taken.

The following instructions pertain to the performance of work in the laboratory:

1. Unless the materials are laid out on the workbench assigned, each party will consult the instructor at the beginning of the period regarding materials to be used on each particular day.
2. Care shall be taken that small pieces of equipment, such as weights for balances, measuring scales, etc., are not misplaced and lost. Any breakage of apparatus shall be reported immediately. Breakage and loss due to carelessness will be charged to the man responsible for the damage.
3. For later identification, all test specimens must be tagged, showing clearly the party number and specimen designation. Follow supplementary instructions as issued. Use lead pencil for marking all tags, as cement will obliterate ink notations.
4. Upon completion of a test, all apparatus shall be left clean and in order on or near the workbench. All waste material shall be removed from the benches, floor, or testing machines to the waste pile.
5. All expendable material shall be used in an economical manner; this applies particularly to paper toweling and capping materials.

6. In operating testing machines, the instructions given in Appendix B shall be followed; members of a party shall not operate a testing machine for the first time without express permission of the instructor.

Reports. Each student shall submit a separate report on each test, and all reports shall be bound in a regulation folder. On the front of the cover shall be clearly indicated (1) the title of the test, (2) the test number, (3) the student's name, and (4) the party number.

The report should be brief but clear. Advantage should be taken of tabular and graphical methods of presenting data. In addition to subject matter, its clarity, conciseness, method of presentation, legibility, and neatness will receive consideration in the grading of a report. Lack of neatness shall be sufficient cause for rejection of a report.

The following arrangement of the report is suggested:

Title. This should indicate at a glance the nature of the test.

Scope of test. A brief statement of the purpose and significance of the test should be included.

Materials. The materials used or tested should be described.

Apparatus and methods of testing. Special equipment used for the first time should be briefly described. Diagrammatic sketches should be drawn to show the principles of operation of special apparatus. Important features of the testing procedure should be described. Details of standard testing procedures and apparatus may be incorporated by reference.

Data and results of the tests. A complete transcript of all laboratory data shall be submitted in concise tabular form. Incidental observations relating to the behavior of the materials should be included. All equations or formulas used should be clearly stated together with definitions of all symbols employed. Calculations may be made with the slide rule. Steps in numerical work should be clearly indicated. Calculations should be properly checked; this may be done by members of the party working as a group. The results of the tests should be summarized in tabular or graphical form.

Thought and care should be given to the layout of tables and diagrams. A table or diagram should be, as nearly as possible, complete in itself and, in the ideal case, should clearly convey the desired information without the necessity for reference to the text. The instructions given in Arts. 17.12 and 17.13 should be followed in the preparation of tables and diagrams.

Discussion. There should be included a brief discussion in which attention is drawn to the salient facts shown by the tables and diagrams. The test results should be compared with pertinent data given in Part I or in other publications, and definite conclusions should be drawn. Answers to the questions given at the end of each test should be formulated from a study of Part I.

TEST 1 NORMAL CONSISTENCY AND TIME OF SET OF PORTLAND CEMENT

This test is intended to acquaint the student with the principal characteristics of a fresh neat-cement paste and with the determination of the normal consistency and time of set of the cement assigned.

For comparable results it is essential that all tests be made under similar conditions. The standard methods for testing cement developed by the ASTM specify definite procedures, even including temperatures of materials and of storage of specimens, in order to attain as uniform testing conditions as possible.

Test instructions. *Preparatory reading.* Articles 2.4 to 2.6, 2.8 to 2.10, 2.12, 2.14, 2.16, 2.17.

Materials. About 7 lb of portland cement for each group; sufficient for six trials. Thoroughly blend sufficient cement for all groups to ensure uniformity. Note the brand used.

Equipment. Balance (sensitive to 0.1 g) with scoop, set of metric weights, 250 ml graduate, Vicat apparatus, Gillmore apparatus, nonabsorbant mixing plate, small trowel, and three 4-in. square glass plates.

Normal consistency. Weigh out 500 g of cement and place on the mixing plate. From a crater in the center and add a measured quantity of water (take, say, 130 ml for the first trial batch, as the amount required for different cements varies from about 20 to 30 percent by weight). The temperature of the room and dry cement shall be between 68 and 81°F. The temperature of the mixing water shall be between 70 and 76°F. The relative humidity of the laboratory shall be not less than 50 percent. Turn the material at the outer edge into the crater within 30 sec by the aid of a trowel. After an additional interval of 30 sec for the absorption of the water, during which interval the dry cement around the outside of the cone shall be lightly troweled over the remaining mixture to reduce evaporation losses, complete the operation by continuous, vigorous mixing, squeezing and kneading with the hands for 1½ min. With the hands quickly form this paste into a ball, completing the operation by tossing it six times from one hand to the other, keeping the hands about 6 in. apart.

With the ball resting in the palm of one hand, press it into the larger end of the conical ring of the Vicat apparatus held in the other hand, completely filling the ring with paste. Remove the excess at the larger end by a single movement of the palm of the hand. Place the ring with its large end on a glass plate, and slice off the excess paste at the small end by a single oblique stroke of a trowel held at a slight angle with the top of the ring. Smooth the top, if necessary, with a few light touches of the pointed end of the trowel. During these operations take care not to compress the paste.

Place the paste (confined in the ring and resting on the plate) under the rod of the Vicat apparatus. See that the rod of the Vicat apparatus slides easily. Bring the large end of the rod in contact with the surface of the paste, read the scale or set to read zero, and release the rod 30 sec after completion of the mixing. Again read the scale 30 sec after releasing the rod.

Plot the amount of mixing water used and the observed penetration as shown in Fig. 2.3. Make trial pastes with varying amounts of water until sufficient observations are obtained close to the penetration of 10 mm to permit a reliable determination from the graph of the water requirement for normal consistency. A paste is of normal consistency when the rod settles to a point 10 mm ± 1 mm below the original surface in ½ min after being released. Express the amount of water required in percentage by weight of the dry cement.

Keep the apparatus free from all vibrations during the test. Also keep it *clean*, so that the rod will always slide freely. Use 500 g of fresh cement for each trial, and add the entire trial amount of water at one time.

Time of initial set. Mix 500 g of cement with the amount of water required to produce normal consistency, or use the last batch from the preceding operation if it is of normal consistency. Mold part of this paste into the ring of the Vicat apparatus, and place on a glass plate as described under the normal consistency test. From the remainder, on a glass plate, make a pat about 3 in. in diameter, ½ in. thick at the center, and tapering to a thin edge for the determination of time of set by the Gillmore method. Form the pat by drawing the trowel from the outer edge toward the center, then flattening the top. Keep these two specimens in the moist closet in moist air at a temperature of 70 to 76°F. Determine time of initial set by both Vicat and Gillmore methods as described below. Make trials every 15 min beginning 45 min after preparing the specimens. If initial set does not occur within 1½ hr with the Gillmore needle, the test may be discontinued. Report the time required by each method.

Vicat method. Using the paste molded into the Vicat ring, carefully bring the 1-mm needle in contact with the surface of the paste, tighten the set screw, read the scale or set to read zero, and release the rod quickly by releasing the set screw. The initial set is said to have occurred when the 1-mm needle ceases to pass a point 25 mm below the top surface in ½ min after being released. Take care to keep the needle clean as the collection of cement on the sides of the needle will retard the penetration, while cement on the point is likely to increase the penetration. Make no test within ¼ in. of any previous test or within ⅜ in. of the mold.

The time of setting is affected not only by the percentage and temperature of the water used and the amount of kneading the paste receives, but also by the temperature and humidity of the air. Therefore, unless temperature and humidity are maintained at appropriate levels, its determination is only approximate.

Gillmore method. The cement is considered to have acquired its initial set when the pat will bear without appreciable indentation the Gillmore needle 1/12 in. in diameter, weighing 1/4 lb. The final set is considered to have occurred when the pat will bear without appreciable indentation the Gillmore needle 1/24 in. in diameter, weighing 1 lb. In making the test, hold the needles in a vertical position, and apply them lightly to the surface of the pat.

Report. *Report of test results.* Each student shall give test results for his own party and for one other party doing this test on the date assigned. Compare with the ASTM Standard Requirements.

Discussion. 1. Is the time of set of the cement satisfactory?

2. How does fine grinding affect the time of set?

3. What other factors affect the time of set?

4. What is the significance of time of set?

5. What is the difference in meaning of the two words "setting" and "hardening"?

6. What items in the test procedure will affect the results of the normal consistency determination?

7. What percentage of water (by weight of dry materials) should be used with this cement in the preparation of the 1:3 standard sand mortar for tensile strength tests?

8. What is the approximate range in percentage of (*a*) separate C_3S and C_2S, and (*b*) combined C_3S plus C_2S in a normal portland cement?

9. Of what importance is the soundness of cement?

10. What precautions does the manufacturer take to prevent unsound cement?

11. What effect has storage on sound cement?

12. What effect has storage on unsound cement?

13. What is the cause of free lime in cement?

14. What provision is made in the specification for cement should it fail to pass the soundness test as received?

TEST 2 STRENGTH OF TYPE I PORTLAND CEMENT AND TYPE III HIGH-EARLY-STRENGTH CEMENT MORTARS AT VARIOUS AGES

The purpose of this test is to determine the tensile strengths of mortar briquets or the compressive strength of mortar cubes (as directed) using a local sand with both a type I normal portland cement and a type III high-early-strength portland cement. The tests are to be made at the ages of 1, 3, 7, and 14 days, unless otherwise specified by the instructor.

The test results for the two cements will be compared with each other, as well as with the standard requirements for each type of cement, taking into consideration the fact that these tests are made with an aggregate and a water-cement ratio that differ from those specified for standard tests.

Test instructions. *Preparatory reading.* Articles 2.4 to 2.6, 2.8, 2.9, 2.11 to 2.13, 2.16, 2.18; Appendix B.

Materials. About 1 lb of type I portland cement, 1 lb of type III high-early-strength portland cement, and 8 lb of sand will be required for each group. In separate containers, blend a sufficient quantity of each type of cement and of sand to meet the needs of *all groups*. Make note of brands or sources of materials used.

Equipment for molding. Balance (sensitive to 0.1 g) with scoop, set of metric weights, 16 briquet molds with one extra base plate or 16 2-in. cube molds with base plates, 250 ml graduate, two mixing pans, two small trowels.

Equipment for testing. Twelve-inch steel rule, briquet-testing machine, or compression-testing machine for cubes.

Procedure for tensile-strength test. Each group will mold eight briquets of a 1:3 mix by weight of type I portland cement and equivalent saturated surface-dry fine aggregate, using a net water-cement ratio of 0.45 by weight. Consult the instructor concerning any necessary corrections for absorption by the aggregate or for free water. Each briquet requires about 180 g of dry materials. In weighing the aggregates, be sure a representative sample is obtained, taking care to avoid separation of the fine and coarse particles. Place the materials in the mixing pan with the cement on top of the aggregates; mix thoroughly and form a crater in the center. Add the measured amount of water to this crater. Turn the material at the outer edge into the crater within 30 sec. After an additional 30 sec for the absorption of the water, complete the mixing by vigorous and continuous turning and stirring for 2 min.

Immediately after mixing, fill the molds heaping full without compacting while resting on an *unoiled* metal base plate. Press the mortar in firmly, using both thumbs simultaneously and applying a pressure of 15 to 20 lb. Apply pressure 12 times to each briquet, at points to include the entire surface. Then heap the mortar above the mold and smooth off with a trowel. Draw the trowel over the mold in such a manner as to exert a pressure of not over 4 lb. Place an oiled base plate on top of the gang of molds and turn it over. Repeat the operation of heaping, thumbing, heaping, and smoothing. During the final finishing of the briquets, take care that their thickness is exactly equal to the 1-in. depth of the molds. Tag the specimens, giving party number and specimen identification. Store them in the moist closet.

Repeat the above procedure using a type III high-early-strength portland cement.

Test two of the specimens of each group at the ages of 1, 3, 7, and 14 days or as directed. Measure the width and thickness of each specimen to the nearest 0.01 in. before testing. Keep all specimens covered with a damp cloth until tested. Read Appendix B and consult the instructor concerning the operation of the briquet-testing machine. Determine the rate of loading in pounds per minute.

Procedure for compressive-strength test. Each group will mold eight 2-in. cubes of a 1:2.75 mix by weight of type I portland cement and equivalent saturated surface-dry fine aggregate, using a net water-cement ratio of 0.41 by weight. Consult the instructor concerning any necessary corrections for absorption by the aggregate or for free water. Each cube requires about 350 g of dry materials. In weighing the aggregates, be sure a representative sample is obtained, taking care to avoid separation of the fine and coarse particles.

The mixing will be done as described for the tensile-strength test.

In molding the cubes, first half-fill each compartment. Then tamp the mortar in each cube 32 times in about 10 sec using a ½ by 1 by 6-in. wooden tamper. This operation will be done in four rounds, each round at right angles to the previous one and consisting of eight adjoining strokes over the surface of the specimen, as shown in Fig. T2.1. Complete the 32 strokes for one cube before going to the next. When the first layer

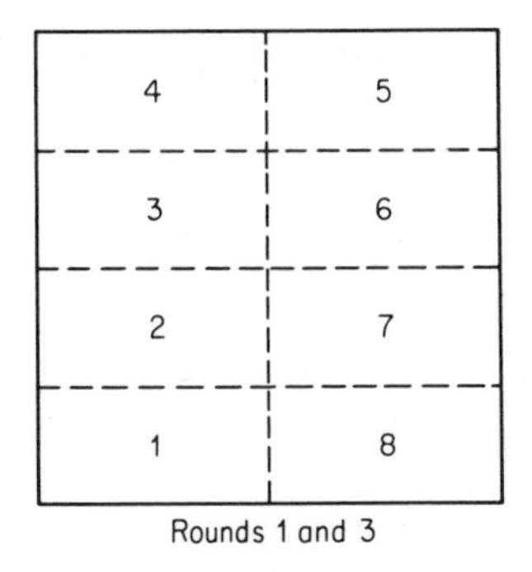

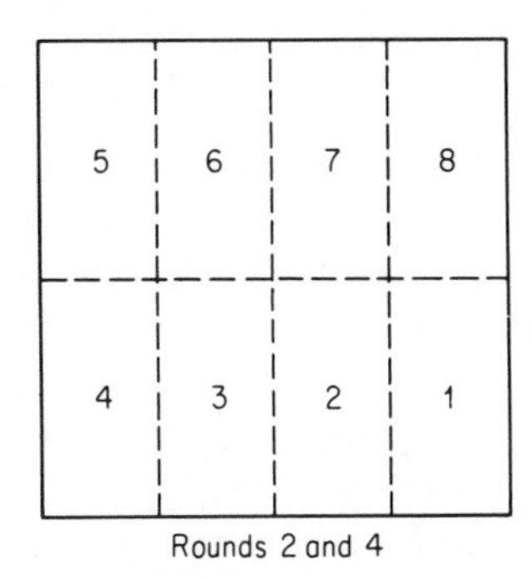

Fig. T2.1. Order of tamping in molding 2-in. cubes.

for all the cubes is completed, fill each compartment with the remaining mortar and tamp as for the first layer. During tamping of the second layer bring in the mortar forced out onto the tops of the molds after each round of tamping, before starting the next round of tamping. Slightly overfill each cube, and smooth off by drawing the flat side of the trowel (with the leading edge slightly raised) once across the top of each cube. Finally, cut the mortar off flush with the top of the mold by drawing the edge of the trowel (held perpendicular to the mold) with a sawing motion over the mold. Tag the specimens, giving party number and specimen identification. Store them in the moist closet.

Repeat the above procedure using a type III high-early-strength portland cement.

Test two of the specimens of each group at the ages of 1, 3, 7, and 14 days or as directed. After removal from storage keep all specimens covered with a damp cloth until time of testing. Clean off the faces to be loaded, which must have been in contact with the true plane surfaces of the mold, and check with a straightedge to see that they are plane. Measure the critical cross-sectional dimensions to the nearest 0.01 in. Read Appendix B and consult the instructor concerning the operation of the testing machine. Without using any capping materials, apply an initial load (at any convenient rate) up to one-half of the expected maximum for specimens having expected maximum loads of more than 3,000 lb. Omit this initial load for specimens having an expected maximum load of less than 3,000 lb. Thereafter, apply load so that the maximum load will be attained within 20 to 80 sec.

Report. *Report of test results.* Compute the unit strength for all specimens. In a neat tabulation submit results of tests made by your own party and one other party making this test on the date assigned. Plot two curves using the same origin of coordinates, to show the average strength of the two types of cement mortars at the various ages included in the test.

Discussion. 1. Compare your results with the standard strength requirements of type I portland and type III high-early-strength portland cement mortars using Ottawa sand.

2. What is (*a*) standard Ottawa sand for mortar-briquet tests and (*b*) graded Ottawa sand for mortar-cube tests?

3. What factors may be responsible for any differences between your results and the standard strength requirements?

4. Discuss the early strength properties of the type III cement tested.

5. For what purposes might the more expensive high-early-strength cement be more economical than type I portland cement?

6. List the factors which are responsible for the strength characteristics of a high-early-strength portland cement.

7. Why is high-early-strength cement unsuited for massive construction?

8. Why is high-early-strength concrete suited for cold-weather construction?

9. Why would the specific surface of cement be a better criterion for predicting variation in strength with fineness than the percent passing the No. 200 sieve?

10. Does the loading rate of the briquet machine used in the tension test comply with ASTM requirements of 600 ± 25 lb per min?

TEST 3 EFFECT OF CURING CONDITIONS UPON COMPRESSIVE STRENGTH OF PORTLAND CEMENT MORTARS

This test is designed to show the effect of various temperature and moisture conditions during the curing period upon the compressive strength of portland cement mortars. Temperature effects will be observed by curing separate groups of specimens at 40, 70, and 100°F. Moisture effects will be shown by curing some specimens moist while others will be stored in dry air.

Test instructions. *Preparatory reading.* Articles 8.9 to 8.17, 10.5 to 10.7, 10.10 to 10.12, 10.19, 10.20, 11.3, 11.7; Appendixes B, H.

Materials. About 5 lb of portland cement and about 15 lb of sand will be required for each party. In separate containers, blend sufficient cement and sand for the use of *all parties* assigned to this test. Make note of brands or sources of materials used.

Equipment for mixing. Balance with scoop, set of metric weights, fifteen 2 by 4-in. cylinder molds or 2 by 2-in. cube molds as assigned, five base plates, 1,000-ml graduate, large mixing pan, two small trowels, and two tamping rods.

Equipment for testing. Twelve-inch steel rule, capping equipment for cylinders, and compression-testing machine.

Procedure. Each party will mold 15 specimens. Use a mix of 1:3 by weight and a net water-cement ratio of 0.45 by weight corrected for the free moisture or effective absorption of the sand. Each cylinder requires about 500 g of dry material and each cube requires about 350 g. In weighing the aggregate be sure a representative sample is obtained, taking care to avoid separation of the fine and coarse particles. Place the materials in the mixing pan with the cement on top of the aggregate; mix thoroughly and form a crater in the center. Add the measured amount of water to this crater. Turn the material at the outer edge into the crater within 30 sec. After an additional 30 sec for the absorption of the water complete the mixing by vigorous and continuous turning and stirring for 2 min.

Immediately after mixing, fill the cylinder molds in three equal layers (two layers for cube molds), tamping each layer 16 times. Carefully finish the top of each specimen. Tag the specimens giving party number and specimen identification. Place nine specimens in the moist closet and leave six specimens in the dry air of the laboratory.

The molds will be removed at the age of 1 day by a laboratory assistant, who will place the specimens in the required storage. Three specimens will be stored continuously under moist conditions at each of the following temperatures: (*a*) 40°F, (*b*) 70°F, and (*c*) 100°F; the remaining 6 specimens will be stored continuously in dry air at 70°F, except that 3 specimens will be transferred to water storage 24 hr before testing.

All specimens will be tested at the age of 14 days. Before testing, measure the lateral dimensions of each specimen to the nearest 0.01 in. Cap the top of each cylinder with casting plaster as instructed in Appendix H. Keep moist specimens covered with a damp cloth. Read Appendix B and consult the instructor concerning the operation of the testing machine. The specimens cured at high or low temperature will be tested last to allow them to come to room temperature before testing. For the report obtain test results from any other party paired with yours, so as to obtain better average strengths.

Report. *Report of test results.* Compute the strength in pounds per square inch for all specimens. In a neat tabulation, submit all results for your party and any other party paired with yours. For the moist-cured specimens plot a graph showing compressive strength of mortar vs. curing temperature. On this same graph plot the strengths of the two groups of specimens stored in dry air at 70°F.

Discussion. 1. Based upon your test data, discuss the influence of temperature and moisture conditions while curing on the strength of cement mortars.

2. Would you expect the compressive strength of concrete in highway slabs and buildings to be affected by similar curing conditions to about the same degree as for your mortar specimens? Explain.

3. Would the effect be the same for massive bridge piers, dams, etc.? Explain.

4. Why does wet curing produce a concrete superior to that kept dry?

5. Why does not later curing at favorable temperatures repair the damage to concrete frozen at an early age?

6. Discuss the relative effect upon strength and permeability of freezing concrete (*a*) immediately after placement, (*b*) after a few hours, and (*c*) after several days of favorable curing.

7. Discuss the effect of moisture condition of the specimen at time of test upon the strength of a concrete specimen (*a*) in compression and (*b*) in flexure.

8. Assuming that the cement you used was a type I portland cement, what 28-day strength would you expect from your specimens under moist curing at 70°F?

9. What routine precautions always should be taken in transferring specimens from the job to the laboratory for test or for storage and test?

TEST 4 SIEVE ANALYSES OF CONCRETE AGGREGATES

The object of this test is to make sieve analyses of typical fine and coarse concrete aggregates, to plot their grading curves, and, using assigned combined grading limits, to determine the required amounts of each of the aggregates to produce a cubic yard of quality concrete.

Test instructions. *Preparatory reading.* Articles 3.8 to 3.12, 6.4.

Materials. About ¼ cu ft of sand, ½ cu ft of ¼ to ¾-in. gravel, and ½ cu ft of ¾ to 1½-in. gravel will be required by the party. These materials will be taken from the storage bins as described below. Make note of sources of materials.

Equipment. Weighing scales sensitive to 0.01 lb, balance with scoop, set of metric weights, large trowel, sand splitter, mechanical shaker, ⅒ and ½-cu ft capacity measuring cylinders, set of standard sieves of the following sizes: Nos. 100, 50, 30, 16, 8, 4, and ⅜, ½, ¾, 1, and 1½ in.

Sieve analysis of sand. From a representative portion of damp bin material select about ¼ cu ft of sand. For sieve analysis, select about a 500-g sample from the ¼ cu ft of sand by use of the sand splitter.[1] Air-dry this sample and weigh to the nearest 0.5 g. Arrange the sand sieves in order by inserting the bottom of one into the top of another with the largest on top and the pan at the bottom. Then place the nest of sieves in the shaker. Place the 500-g sample on the top sieve, cover the sieve, clamp the nest securely, and shake for 2,000 oscillations. Weigh the residue on each sieve and in the pan to 1/1,000 of the weight of the whole test sample. As a check on the sieving operation, continue sieving until not more than 1 percent by weight of the residue passes any sieve during 1 min of hand sieving. If the sum of the weights of material on each sieve does not equal the weight of the original sample to within 1 percent, the test should be repeated. Indicate this check on weighings for all aggregates.

Air-dry the remaining sand, and determine the approximate unit weight in pcf in both the loose and the compact condition. This is done by weighing the ⅒-cu ft cylinder even-full without shaking, and again even-full after shaking the cylinder slightly to cause settlement. In Test 5 the unit weight will be determined accurately by the standard method.

Sieve analysis of ¼ to ¾-in. gravel. From a representative portion of bin material select ½ cu ft of air-dry aggregate by the quartering process. This process involves flattening the blended pile, quartering, and discarding two opposite quarters (see Fig. 3.3). Care must be taken to avoid loss of fine material. This is repeated as many times as necessary to obtain the desired size of sample. Determine its approximate unit weight in pcf in both the loose and in the compact condition by weighing in a ½-cu-ft cylinder.

For sieve analysis select a 15-lb sample from the ½ cu ft of aggregate

[1] If a sand splitter is not available, use the quartering process as described for the gravel.

by the quartering process. Screen the sample by hand successively through all the screens, using one at a time and starting with the largest. Shake on each screen until no more passes. In no case shall fragments in the sample be turned or manipulated through the screen by hand. Weigh the amount held on each screen.

Sieve analysis of ¾ to 1½-in. gravel. Follow the procedure outlined for the ¼ to ¾-in. gravel, but use a 35-lb sample.

Report. *Report of test results.* Submit only the results of tests made by your party.

1. Present the results of the sieve analyses in tabular form, giving for each sieve the total amount coarser and the total amount passing, in percentage of the whole sample. Percentages should be reported to the nearest 1 percent. Indicate the fineness modulus of each material. Provide a column for the proposed combination of the aggregates and its fineness modulus. This is to be filled in from the final grading curve.
2. Prepare a grading chart on 8½ by 11-in. paper, using equal spaces between the standard sieves in the series. On it plot (*a*) the gradings of the aggregates analyzed in the test, (*b*) the grading limits given in Fig. 3.10 and Table 3.9, col. 7, and (*c*) a combined grading curve which will fall within the grading limits just plotted. For a first trial assume about 40 percent sand.
3. Assuming a ratio of cement to total aggregate by weight of 1:6, state your proposed mix in the following forms, always stating the relative amount of cement as unity: (*a*) mix by weight of dry materials, and (*b*) mix by dry, loose volumes. Note that 1 cu ft of cement = 94 lb.
4. Assuming a water-cement ratio by volume of 1.00 for the 1:6 mix in item 3 above, compute the quantity of cement in sacks, the amount of water in gallons, and the quantity of separate aggregates in cu yd (loose) for 1 cu yd of concrete. Assume concrete weighs 150 pcf.

Discussion. 1. In what sizes of particles are the aggregates deficient or oversupplied? How might this be remedied in a practical way?

2. What is meant by "good" grading?
3. What is the practical use of controlling the gradation of concrete aggregates?
4. What precautions should be observed in the use of an "ideal" curve?
5. For maximum economy of cement, how, in general, should the percentage of sand vary with the water-cement ratio of the mix?
6. How does the optimum sand content vary with richness of mix?
7. How is the optimum percentage of sand affected by the use of different maximum sizes of aggregate?

TEST 5 SPECIFIC GRAVITY, UNIT WEIGHT, MOISTURE CONTENT AND ABSORPTION OF CONCRETE AGGREGATES

For the design of a concrete mix, information should be available on the following properties of the aggregates: (1) the bulk specific gravity, which must be known for use in certain types of moisture tests and which is employed in calculating the percentage of voids and the solid volume of aggregates—this latter value in turn is used in computations of yield; (2) the unit weight, which is used in converting proportions by weight into proportions by volume, and in calculating the percentage of voids in aggregates; and (3) the moisture content and absorption of aggregates, which are necessary in order to determine the net water-cement ratio in a batch of concrete made with job aggregates. It is the purpose of this test to acquaint the student with the determination of these properties.

Test instructions. *Preparatory reading.* Articles 3.3 to 3.7.

Materials. For the use of one party: sand—4 lb saturated, 2 lb as stocked, 1⁄10 cu ft air-dry; ¼ to ¾-in. gravel—5 lb saturated, 10 lb as stocked, ½ cu ft air-dry. Make note of sources of material.

Equipment. Platform scale sensitive to 0.01 lb, balance with scoop, metric weights, sample splitter, 1-pt mason-jar pycnometer or 500-ml volumetric flask, conical mold with 1-in.-diameter tamping rod, ⅝-in. diameter tamping rod, 1⁄10 and ½-cu ft measuring cylinders, ring stand, small wire basket, wire-basket container, two 6-in. pans and two 12-in. pans, one large trowel, and fan.

Specific-gravity and absorption capacity of fine aggregate. Screw the pycnometer cap firmly on the jar, and make match marks to designate its position on the jar, as all subsequent settings of the cap must conform to this initial setting. Determine the capacity of the pycnometer by weighing when empty and when filled exactly full with water at room temperature. Gently roll and agitate the jar in an inclined position to eliminate air bubbles, which may tend to accumulate near the top of the jar.[1]

By use of a sand splitter (or by the method of quartering shown in Fig. 3.3), select approximately 1,500 g of the sample of saturated fine aggregate. Spread the sample on a clean, flat surface, expose to a gently moving current of warm air, and stir the *entire* sample frequently to secure *uniform* drying. Continue this operation until the sand approaches a free-flowing condition. Then place the sand loosely in the conical mold, lightly tamp the surface 25 times with the 1-in.-diameter metal rod, and lift the mold vertically. If free moisture is present, the cone of sand will retain its shape. Continue drying with constant stirring, and make tests at frequent intervals, until the cone of sand slumps upon removal of the mold. This indicates that the sand has reached a surface-dry condition.

[1] If preferred, a 500-ml graduated flask may be used in place of the pycnometer.

Note. In the above procedure the first trial determination shall be made with some free water in the sample. If the cone of sand slumps on the first trial, the sand has been dried past the saturated and surface-dry condition. In this case take a new sample of saturated sand or thoroughly mix a few milliliters of water with the dried sand, and permit the sample to stand in a covered container for 30 min. Then resume the process of drying and testing the sand.

Immediately introduce into the pycnometer a 500.0-g sample of saturated surface-dry material, prepared as described above, fill the jar almost to the top with water at room temperature, and screw down the top until the match marks are in line. Gently roll and agitate the jar in an inclined position to eliminate all air bubbles, after which fill it with water to the top. Weigh the pycnometer, and from this weight determine the total weight of water introduced into the jar to the nearest 0.1 g; then empty the jar.

As soon as the above sample has been inundated in the pycnometer, weigh out a second 500-g sample and dry to constant weight at a temperature of 100 to 110°C, cool to room temperature, and weigh. At least 24 hr will be required to oven-dry the sample.

The bulk specific gravity is computed from the formula

$$\text{Bulk sp gr} = \frac{A}{V - W}$$

where A = oven-dry weight of sample, g
V = volume, or capacity, of jar, ml
W = weight, g, or volume, ml, of water *added* to jar

The bulk specific gravity on the basis of weight of saturated surface-dry aggregate is computed from the formula

$$\text{Bulk sp gr (saturated surface-dry basis)} = G = \frac{500}{V - W}$$

The apparent specific gravity is computed from the formula

$$\text{Apparent sp gr} = \frac{A}{(V - W) - (500 - A)}$$

The absorption capacity in percent of the oven-dry weight is computed from the formula

$$\text{Absorption capacity, \%} = \frac{500 - A}{A} \times 100$$

Specific gravity and absorption capacity of coarse aggregate. Select approximately 2 kg (about 4 lb) of the aggregate from the saturated sample by the method of quartering, and reject all material passing a ⅜-in. sieve.

Surface-dry the sample by rolling in a towel until all visible films of water are removed. The larger fragments may be individually wiped. Although the surfaces of the particles will still appear to be damp, this will be taken as the surface-dry condition. Care should be taken to avoid evaporation during the operation of surface-drying. Obtain the weight of the sample in the saturated surface-dry condition. Determine this and all subsequent weights to the nearest 0.5 g.

Immediately after weighing in air, place the saturated surface-dry sample in the wire basket, and determine its weight in water after correcting for the weight in water of the wire basket alone. Then dry the sample to constant weight at a temperature of 100 to 110°C, cool to room temperature, and weigh. At least 24 hr will be required to oven-dry the sample.

The bulk specific gravity is computed from the formula

$$\text{Bulk sp gr} = \frac{A}{B - C}$$

where A = weight of oven-dry sample, g
B = weight of saturated surface-dry sample in air, g
C = weight of saturated sample in water, g

The bulk specific gravity on the basis of weight of saturated surface-dry aggregate is computed from the formula

$$\text{Bulk sp gr (saturated surface-dry basis)} = G = \frac{B}{B - C}$$

The apparent specific gravity is computed from the formula

$$\text{Apparent sp gr} = \frac{A}{A - C}$$

The absorption capacity in percent of the oven-dry weight is computed from the formula

$$\text{Absorption capacity, \%} = \frac{B - A}{A} \times 100$$

Free moisture in fine aggregate. Select from the storage bin about 2 lb of moist sand which is representative of the condition in which it would be used in a concrete batch, particularly as to gradation and dampness. Quickly weigh out about 500 g of the moist sand to the nearest 0.1 g. Place the sample in the pycnometer jar, and partially fill the jar with water. Screw on the cap so that the match marks are in line, and gently roll and agitate the jar to remove all air from the sample. Completely fill the pycnometer with water and weigh.

The free moisture in percent of the saturated surface-dry weight is calculated by the formula

$$M = \frac{V - W - S/G}{S + W - V} \times 100$$

where M = free moisture, %
S = weight of sample of moist sand, g
W = weight, g, or volume, ml, of water *added* to jar
V = volume, or capacity, of the jar, ml
G = bulk specific gravity on a saturated surface-dry basis

The amount of free moisture, by weight, that would be contributed to the mix by the damp aggregate is calculated by the formula

$$m = \frac{MS}{M + 100}$$

where m = weight of free moisture
M = percentage of free moisture
S = weight of damp sample

m is expressed in the same units as S, e.g., in grams. If the aggregate is air-dry, then M is negative and m equals the absorption of water that would occur from the concrete mix.[1]

Absorption of mixing water by coarse aggregate. Select from the storage bin about 10 lb of air-dry, ¼ to ¾-in. aggregate. Weigh out about 3,000 g to the nearest 0.5 g, oven-dry it to constant weight, and when it is cooled, weigh it again; then discard the dried aggregate. The difference in weight referred to the oven-dry weight gives the percentage of internal moisture in the sample. The absorption capacity minus the percentage of internal moisture gives the effective absorption, i.e., the amount of mixing water that can be absorbed by the aggregate, expressed as a percentage of the weight of oven-dry aggregate. A negative result would indicate that the sample is completely saturated and contains free surface moisture.

In correcting the water content of the mix for the effective absorption of the aggregate it is preferable, from a theoretical standpoint, to have

[1] If preferred, the free moisture may be determined by use of a Chapman flask as follows: Fill the flask to the 200-ml mark on the lower neck with water at room temperature. Then slowly pour the 500-g sample of damp aggregate into the flask, and agitate to free any entrained air bubbles. Read the combined volume of water and sand on the scale on the upper neck of the flask. Then

$$M = \frac{V - 500/G - 200}{200 + 500 - V} \times 100$$

where M and G are as before but V equals the combined volume of water and fine aggregate in milliliters.

the effective absorption expressed as a percentage of the saturated surface-dry weight. The effective absorption of mixing water, in percent of the saturated surface-dry weight, is calculated by the following formula

$$E = \frac{WA + 100W - 100S}{WA + 100W} \times 100$$

where E = effective absorption based on saturated surface-dry weight, %
A = absorption capacity based on oven-dry weight, %
S = weight of air-dry sample, g
W = weight of oven-dry sample, g

From a practical viewpoint there is very little difference between the effective absorptions computed for the oven-dry and saturated surface-dry bases. In fact, the difference between two results using the same method may be greater than the theoretical difference between the two effective absorptions mentioned. Therefore, in practice, the simplest, or oven-dry, basis is commonly employed.

Unit weight of fine aggregate. Weigh the standard 1/10-cu ft measuring cylinder to the nearest 0.01 lb. Fill the measure one-third full of air-dry, thoroughly mixed sand, and level the sand with the fingers. Tamp the mass with the rounded end of the standard 5/8-in.-diameter tamping rod, using 25 strokes distributed evenly over the surface. Fill the measure two-thirds full and tamp 25 times as before; then fill the measure to overflowing, tamp 25 times, and strike off the surplus, using the tamping rod as a straightedge.

In tamping the first layer, do not permit the rod forcibly to strike the bottom of the measure. In tamping the second and final layers, use only enough force to cause the rod to penetrate into the layer below.

Weigh the cylinder filled with sand. The unit weight in pounds per cubic foot will be computed from the net weight of the sand and the volume of the cylinder.

Unit weight of coarse aggregate. Use a 1/2-cu ft measuring cylinder, and follow the procedure used for the fine aggregate.

Report. *Report of test results.* Compute and tabulate the following values for both the fine and coarse aggregates tested by your party and any other party paired with yours.

1. Bulk specific gravity
2. Bulk specific gravity, saturated surface-dry basis
3. Apparent specific gravity
4. Absorption capacity, percent of over-dry aggregate
5. Effective absorption of coarse aggregate, percent of oven-dry aggregate

6. Free moisture in fine aggregate, percent of saturated surface-dry aggregate
7. Unit weight, pounds per cubic foot
8. Voids in the dry, compact condition, percent

Compute the water adjustment, in gallons per sack of cement, required to correct for the moisture condition of the aggregate as stocked if 600 lb of saturated surface-dry aggregate (45 percent fine, 55 percent coarse) be used per sack of cement.

Compute the volume of concrete produced by 94 lb of cement (specific gravity = 3.15), 60 lb water, 250 lb fine aggregate, and 350 lb coarse aggregate. Consider the aggregate to be saturated but surface-dry. How many sacks of cement will be required per cubic yard of concrete?

Discussion. 1. What is the distinction between apparent specific gravity and bulk specific gravity?

2. How would the determination of bulk specific gravity of fine aggregate (surface-dry basis) be affected by the 500-g sample's being drier than the surface-dry condition? Explain. Assume that the aggregate becomes saturated during the test.

3. Would the apparent specific gravity be affected in the same manner? Explain.

4. What would be the effect upon the unit weight if the fine and coarse aggregates be combined?

5. What would be the effect upon the unit weight if the aggregate be placed in the cylinder without rodding or shaking?

6. What would be the effect upon the unit weight, if damp fine aggregate be used?

7. Discuss the influence of the fineness of the aggregate upon its bulking characteristics when damp.

8. Discuss the effect of damp aggregate upon the cement content of the mix (computed for saturated, surface-dry aggregates) (*a*) if the materials are batched by weight, and (*b*) if batched by stated bulk volumes.

9. What difficulties arise in the use of aggregates which contain free water or which absorb water? How are they overcome (*a*) in the laboratory and (*b*) on the job?

TEST 6 CHARACTERISTICS OF FRESH CONCRETE

The objects of the test are to observe the characteristic properties of fresh concretes of a given water-cement ratio and given consistency but of variable aggregate gradation.

Test instructions. *Preparatory reading.* Articles 1.3, 5.1 to 5.5, 5.7 to 5.9, 5.13, 6.1 to 6.4, 10.12, Table 6.6, Appendixes C, D, E, F, G.

Organization. Three parties will work as a unit in performing this test.

Materials for three parties. One-half sack of portland cement, ½ cu ft fine sand, 2 cu ft coarse sand, and 2 cu ft ¼ to ¾-in. gravel. Blend a stock pile of each of the aggregates to ensure a uniform gradation and moisture content. Blend the cement and stock it in dry containers. Consult the instructor in regard to the moisture content of the several aggregates and the brands or sources of the materials used.

Equipment for each party. Platform scale sensitive to 0.01 lb, large mixing pan, two large trowels, container for stocking and weighing materials, 1,000-ml graduate, slump cone, ⅝-in. tamping rod, remolding apparatus, two 6 by 12-in. cylinder molds, two base plates, two sheet-metal cover plates, 12-in. steel rule, medicine dropper, and 10-ml graduate.

Procedure. The three parties will make a total of six mixes using a constant water-cement ratio of 0.55 by weight, 3 to 4-in. slump, and ¼ to ¾-in. gravel for coarse aggregate. The kind of sand and percentage by weight of total aggregate will be as follows:

	Sand in total aggregate, %	
Mix No.	**Coarse sand**	**Fine sand**
1	30	
2	35	
3	40	
4	45	
5	50	
6	30	10

For each batch use 30 lb of mixed aggregate, corrected for the effective absorption or free water of the aggregate. In handling and weighing aggregates, take care to obtain representative samples of average moisture content and free of segregation. Place the coarse aggregate on the bottom of the mixing pan, which with the various utensils will have previously been slightly moistened with a damp rag. Place the sand in next. Add 4.5 lb of cement and mix thoroughly while dry. Add $4.5 \times 0.55 = 2.48$ lb (1,123 ml) of water corrected for the absorption or free water of the aggregate and mix wet. Check the slump as instructed in Appendix C. The batch will probably be too dry, so add water and cement in the proper ratio of 0.55, mixing thoroughly after each addition, until a 3 to 4-in. slump is obtained. Keep a record of all quantities of materials added to the mix. Work quickly to avoid stiffening of the mix. If over 30 min elapse before completing the mix, discard it and start again.

All members of the cooperating parties will examine each mix for the following six characteristics, will agree upon the ratings assigned, and will record the results for all mixes in neat tabular form:

1. Cohesiveness. Note whether the concrete tends to hang together well or whether it tends to crumble readily. Rate as high, normal, or low.

2. Proportion of sand. If the pieces of coarse aggregate cannot be imbedded in the mortar without excessive tamping, the mix is undersanded; if they sink into the mortar without tamping, the mix is oversanded (see Fig. 6.4). Rate as undersanded, normal, or oversanded.

3. Troweling workability. Work the concrete with a trowel. If it appears to work smoothly and with little effort, the troweling workability may be called good. Rate as good, fair, or poor.

4. Remolding effort. See Appendix F for method. Record the number of jigs required.

5. Unit weight. Weigh an empty 6 by 12-in. cylinder and base plate combined, measure the diameter and height of cylinder to 0.01 in.,[1] fill the cylinder with concrete in three equal layers, occasionally stirring the concrete in the mixing pan to keep the materials from separating, tamp each layer 25 times to exclude air voids, strike off the top evenly, clean up any concrete on the outside of the forms, and weigh again. The unit weight, in pounds per cubic foot, will be computed from the volume and weight of concrete in the cylinder.

6. Water gain. Make a shallow but broad conical depression in the entire top of the concrete contained in the cylinder to collect any water rising to the surface. Place a sheet-metal cover over the top of the mold to prevent evaporation. At 30 min after filling the mold, measure the accumulation of water on the surface using the dropper and small graduate. Place the concrete remaining from the batch on the debris pile. Wash off all utensils and the mixing pan, and repeat the above procedure for the remaining batches. After the completion of the water gain test smooth off the top of the cylinder and leave on the table.

[1] If available, a 0.1 or 0.2-cu ft measuring cylinder may be used.

Report. *Reduction of data.* Each student will report on all six mixes.

1. From the net weight of concrete in the mold at the time of casting, and the dimensions of the mold, calculate for each mix the unit weight of the fresh concrete in pounds per cubic foot and in pounds per cubic yard.

2. Calculate the percent by weight of water, cement, fine aggregate, and coarse aggregate in each mix. From these percentages and the unit weight of concrete, calculate for each mix the yield in cubic feet of concrete per sack of cement, the cement factor in sacks per cubic yard of concrete, and the amount, in tons, of each of the various aggregates (on a saturated surface-dry basis) required for 1 cu yd of concrete in place.

3. Using the method of solid volumes and assuming 2 percent air voids, compute the cement factor in sacks per cubic yard for each mix. Solid volume = weight ÷ (bulk specific gravity × 62.4). Assume the materials to have the following bulk specific gravities: cement, 3.15; all aggregates, 2.67. Compare with the cement factor computed as specified above on the basis of the unit weight of concrete.

4. Compute the cost of materials per cubic yard of each mix, assuming the following (or other assigned) unit prices of materials: cement, $1.60 per sack: sand, $6.75 per ton; gravel, $6.25 per ton. Neglect the cost of the mixing water.

5. Prepare a table in which are summarized the results of your observations on each of the six mixes and the results of your calculations made as outlined above.

6. Plot a diagram showing percentage of sand (in total aggregate) as abscissas vs. cement factor as ordinates for mixes 1 to 5.

Discussion. 1. Draw conclusions from your test results regarding the optimum percentage of sand for your several concrete mixes of a given water-cement ratio and consistency.

2. Comparing the results for mixes 3 and 6, discuss the effect on workability and yield of using the fine sand.

3. Approximately what differences in strength would you expect between your six mixes?

4. In selecting the best of several mixes of a given water-cement ratio and given consistency, why is it logical to select the one giving the greatest yield?

5. Discuss the general workability (considering remolding effort, cohesiveness, and troweling workability) of the series of six mixes, remembering that all are of about the same consistency.

6. For concretes of a given water-cement ratio and given consistency,

what would be the effect upon yield of overwashing a fine aggregate so that it has few fine particles?

7. What would be the effect upon water gain under similar circumstances?

8. What characteristics of the concrete are measured by (*a*) slump, (*b*) flow, (*c*) ball penetration and (*d*) remolding effort?

9. For an ordinary concrete mix is the water requirement greater (*a*) for lubrication of the mix to impart mobility for placement, or (*b*) for hydration of the cement?

TEST 7 EFFECT OF WATER-CEMENT RATIO UPON COMPRESSIVE STRENGTH AND CONSISTENCY OF CONCRETE OF UNIFORM MIX

This experiment is intended to familiarize the student with the general characteristics of concrete and concreting materials and with laboratory methods of manufacture and test of concrete specimens. In particular, the purpose of the test is to determine the effect of the various water-cement ratios upon the consistency of the fresh concrete and to determine the effect of water-cement ratio upon the strength of the hardened material. Thus the student may develop for himself a curve expressing the relationship between water-cement ratio and strength.

To decrease the work of the student, the present program involves the casting and testing of 3 by 6-in. cylinders using a maximum size of aggregate of ¾ in., although tests are ordinarily made on 6 by 12-in. cylinders with concrete containing up to 1½-in. maximum size of aggregate.

For convenience only, there is employed in this experiment the procedure of fixing arbitrary proportions and obtaining desired consistencies by varying the quantity of mixing water. For convenience also, in this experiment, the compression tests of the cylinders will be made at the age of 14 days, while the standard age of test is 28 days. For normal materials, similar results may be expected at the early age as at the later one.

Test instructions. *Preparatory reading.* Articles 5.1 to 5.5, 513, 514, 6.1 to 6.4, 6.12, 10.11 to 10.22, 14.20; Table 6.2, 6.5, 6.6; Appendixes B, C, G, H.

Organization. Two parties will work as a unit in making this test.

Material for two parties. About ½ sack of cement, 2 cu ft of fine aggregate, and 2 cu ft of ¼ to ¾-in. aggregate as assigned by the instructor. Blend a stock pile of each of the required aggregates to ensure a uniform gradation and moisture content. Blend the cement and stock it in dry containers. Consult the instructor concerning the moisture content of the aggregates and the brand or source of all materials.

Equipment for molding specimens (one party). Six 3 by 6-in. cylindrical molds with three base plates, containers for stocking and weighing materials, large mixing pan, two trowels, ⅝-in.-diameter tamping rod, 2-qt. measure or 1,000-ml graduated cylinder, slump cone, 12-in. steel rule, 0.1 or 0.2-cu ft measuring cylinder, and platform scale sensitive to 0.01 lb.

Equipment for testing specimens (one party). Testing machine as assigned with spherical bearing block and plain block, 6-in. calipers, 12-in. steel rule, capping plates, small mixing pan, and trowel.

Procedure for preparing specimens. The two parties working together will make a total of six batches of concrete having slumps of ¼ to ½, 1 to 2, 2 to 3, 4 to 5, 6 to 7, and 8 to 9 in. Of these, mixes 1, 3, and 5 will be made by the odd-numbered party. The proportions for each mix will be 1:5.5 by weight, on a saturated surface-dry basis. The total aggregate will contain 45 percent fine and 55 percent coarse aggregate, or as assigned.

A separate batch, containing 35 lb of dry materials, will be made up for each mix. Follow the instructions in Appendix G for batching and mixing the concrete and filling the molds.

Each student will observe the general characteristics of each mix before it is molded into test cylinders and will make note of the *cohesiveness* and *troweling workability* and *unit weight* of the mix as outlined in Test 6.

Procedure for testing specimens. Test in compression at the age of 14 days (or as assigned) each of the 12 cylinders previously molded.

Determine the mean diameter and the length of each specimen to the nearest 0.01 in. Weigh each specimen to the nearest 0.01 lb. Cap the specimens as outlined in Appendix H. Consult Appendix B for directions for operating the testing machine. Observe and sketch the characteristic fractures of the specimens.

Report. *Reduction of data.* Each student will report on all six mixes.

1. Calculate the water-cement ratio for each mix on the basis of *saturated surface-dry* aggregates. To do this, correction must be made for the effective absorption of aggregates which are less than fully saturated, and for the free moisture (in excess of the amount required just to saturate the particles) in the wet aggregates. Express the water-cement ratio in terms of weight, bulk volumes, and gallons per sack.
2. From the net weight of concrete in the measuring cylinder, calculate for each mix the unit weight of the fresh concrete in pounds per cubic foot and in pounds per cubic yard.
3. Assuming the proportions of the various materials in the freshly molded concrete to be the same as those which were weighed for the batch, calculate the percent (by weight) of water, cement, fine aggregate, and coarse aggregate in each freshly placed mix. From these percentages, calculate for each mix the yield in cubic feet of concrete per sack of cement, the cement factor in sacks per cubic yard of concrete, and the amounts of each of the various aggregates required for 1 cubic yard of concrete in place, expressed in tons and in cubic yards (bulk volume). Assume that the materials in a loose, damp condition, as purchased on the market, have the following unit weights: fine aggregate, 90 pcf; ¼ to ¾-in. aggregate, 95 pcf.
4. Using the method of solid volumes, determine the cement factor in sacks per cubic yard for each mix. Solid volume = weight ÷ (bulk specific gravity × 62.4). (Note. Assume the materials to have the following bulk

specific gravities: cement, 3.15; all aggregates, 2.67.) Compare with the cement factor computed above.

5. Defining as the solidity ratio the ratio of the sum of the solid volumes of cement and each of the aggregates to the volume of the resulting concrete, calculate this ratio for each of the mixes.

6. Compute the unit weight of hardened concrete at time of test in pounds per cubic feet and the compressive strength of concrete in pounds per square inch for each of the cylinders.

7. Prepare a table in which are summarized in a clear manner the results of your calculations made above and the results of your observations on the fresh batches.

8. Prepare a chart to show the relation between strength and water-cement ratio. On this chart draw (in dotted lines) the water-cement ratio–strength curves for the ages of 7 and 28 days shown in Fig. 10.9. Using the results of your tests, locate a "job curve" for your materials. Note that the tests of your cylinders having the ¼ to ½-in. slump may give a result which has no relation whatsoever to the curve you are trying to draw.

Discussion. 1. State the "water-cement ratio" law, or principle. Does it apply to all mixes? Does it account for extreme variations in some of your results?

2. What is the function of the aggregates in concrete?

3. Discuss the strength of concrete as developed in your tests in relation to water-cement ratio and to density or solidity ratio. Explain why, in general, the solidity ratio is not a criterion of strength.

4. What factors, in addition to those mentioned above, affect the strength of concrete?

5. What considerations besides strength affect the selection of a water-cement ratio?

6. What consistency (slump) would you recommend for concrete for use in rather heavy footings and walls?

7. Why is it that a rich mix may be weaker than a leaner mix, both mixes being of the same lots of cement and aggregate?

8. What is the effect of age of the concrete upon the water-cement ratio–strength curve?

9. Approximately what loss in compressive strength would result if 2 percent free moisture in the combined aggregate, for a mix intended to have a water-cement ratio of 0.65 by weight, were ignored? Assume the mix to be approximately 1:6 by weight.

10. If you were in charge of construction and found that certain portions of the structure required a wetter consistency for proper placement, how would you modify the mix to maintain a uniform quality of hardened concrete?

11. For a given water-cement ratio, which is the more economical—a stiff mix or a fluid mix? Why is the more economical mix not always used on the job?

12. Is it ever the practice on the job to extract water from concrete during or immediately following placement? Explain.

13. What variations in the relative amounts of fine and coarse aggregates would normally be made for use in lean and for use in rich mixes?

14. What relative compressive strengths would you have obtained in your tests if 6 by 12-in. cylinders had been used in place of 3 by 6-in. cylinders?

15. What relative compressive strengths would you have obtained in your tests if 6-in. cubes had been used?

16. Discuss the characteristic fracture of your test specimens.

TEST 8 EFFECT OF WATER-CEMENT RATIO UPON COMPRESSIVE STRENGTH, CEMENT FACTOR, AND COST OF CONCRETE OF UNIFORM CONSISTENCY

This test is designed to show the effect of various water-cement ratios on the compressive strength of concretes of uniform consistency. From the quantities of materials required per cubic yard of concrete, the student will compute the effect of the various water-cement ratios upon the richness of mix and the cost of the several concretes produced.

Test instructions. *Organization.* Two parties will work as a unit in performing this test.

Preparatory reading. Articles 5.1 to 5.4, 5.13, 5.14, 6.1 to 6.4, 10.11 to 10.22, Appendixes B, C, G, H.

Materials for two parties. One-half sack portland cement, 2 cu ft fine aggregate, and 2 cu ft ¼ to ¾-in. gravel as assigned. Blend a stock pile of each of the required aggregates to ensure a uniform gradation and moisture content. Blend the cement and stock it in dry containers. Consult the instructor concerning the moisture content of the aggregates and the brand or source of all materials.

Equipment for molding specimens (one party). Platform scale sensitive to 0.01 lb, large mixing pan, two large trowels, containers for stocking and weighing materials, 1,000-ml graduate, slump cone, ⅝-in. tamping rod, six 3 by 6-in. cylinder molds, three base plates, 0.1 or 0.2-cu ft measuring cylinder, 12-in. steel rule.

Equipment for testing specimens (one party). Testing machine as assigned with spherical bearing block and plain block, 6-in. calipers, 12-in. steel rule, capping plates, small mixing pan, and trowel.

Procedure for molding specimens. The two parties will make a total of six mixes using the following net water-cement ratios by weight: 0.40, 0.48, 0.55, 0.62, 0.70, and 0.80 by weight. Each mix will have a 3 to 4-in. slump.

For each batch use an amount of mixed aggregate (45 percent fine and 55 percent coarse or as assigned), corrected for the effective absorption or free water of the aggregate, as shown in the following tabulation:

Mix No.	Water-cement ratio, wt	Combined aggregate (saturated surface-dry), lb	Cement for first trial, lb
1	0.40	26	8.2
2	0.48	28	7.0
3	0.55	30	6.0
4	0.62	30	4.6
5	0.70	30	4.0
6	0.80	30	3.4

Follow the instructions in Appendix G for batching and mixing the concrete and filling the molds.

The proper amount of cement to use must be determined by trial. For the various mixes, use the weights of cement shown in the above tabulation as a first trial. Add the water (corrected for the effective absorption or free water of the aggregate) required to produce the desired net water-cement ratio (1 lb water = 453 ml). Thoroughly mix the batch; check the slump as outlined in Appendix C. The batch will probably be too dry, so add water and cement in the proper ratio, mixing thoroughly after each addition, until a 3 to 4-in. slump is obtained. Keep a record of all quantities of materials added to the mix.

All members of both parties will examine each mix for the characteristics listed below (see Test 6 for procedure), will agree upon the ratings assigned, and will record results for all mixes in neat tabular form.

1. Cohesiveness
2. Troweling workability
3. Remolding effort
4. Unit weight

Two 3 by 6-in concrete cylinders will be cast for compression tests at the age of 14 days.

Procedure for testing specimens. Test in compression at the age of 14 days (or as assigned) each of the 12 cylinders previously molded.

Determine the mean diameter and the length of each specimen to the nearest 0.01 in. Weigh each specimen to the nearest 0.01 lb. Cap the specimens as outlined in Appendix H. Consult Appendix B for directions for operating the testing machine. Observe and sketch the characteristic fractures of the specimens.

Report. *Reduction of data.* Each student will report on all six mixes.

1. Calculate the water-cement ratio for each mix on the basis of *saturated surface-dry* aggregates. To do this, correction must be made for the effective absorption of aggregates which are less than fully saturated, and for the free moisture (in excess of the amount required just to saturate the particles) in the wet aggregates. Express the water-cement ratio in terms of weight, and gallons per sack.

2. From the net weight of concrete in the measuring cylinder, calculate for each mix the unit weight of the fresh concrete in pounds per cubic foot and in pounds per cubic yard.

3. Assuming the proportions of the various materials in the freshly molded concrete to be the same as those which were weighed for the batch,

calculate the percent (by weight) of water, cement, fine aggregate, and coarse aggregate in each freshly placed mix. From these percentages, calculate for each mix the yield in cubic feet of concrete per sack of cement, the cement factors in sacks per cubic yard of concrete, and the amounts of each of the various aggregates required for 1 cubic yard of concrete in place, expressed in tons and in cubic yards (bulk volume). Assume that the materials in a loose, damp condition, as purchased on the market, have the following unit weights: fine aggregate, 90 pcf; ¼ to ¾-in. aggregate, 95 pcf.

4. Compute the unit weight of hardened concrete at time of test in pounds per cubic foot and the compressive strength of concrete in pounds per square inch for each of the cylinders.

5. Prepare a table in which are summarized in a clear manner the results of your calculations made above and the results of your observations on the fresh batches.

6. Prepare a chart to show the relation between strength and water-cement ratio. On this chart, in dotted lines, draw the water-cement ratio–strength curves for the ages of 7 and 28 days shown in Fig. 10.9. Using the results of your tests, locate a "job curve" for your materials.

Discussion. 1. Discuss the net water content per cubic yard to produce a given consistency, in relation to the richness of the mix.

2. Discuss the strength of concrete in relation to its water-cement ratio and richness of mix (percentage of cement).

3. Discuss the cost of concrete as affected by richness of mix.

4. For concretes of equal consistency, what effect does the variation in richness of mix and in water-cement ratio have on (*a*) cohesiveness and (*b*) workability as measured by troweling and the remolding effort?

5. Discuss the effect upon the voids in the concrete as the mass hardens.

6. For maximum economy in practical mixes, how would the percentage of sand be varied with the richness of the mix?

7. Discuss the effect upon compressive and flexural strength of the moisture content of the specimen at time of test.

8. Discuss the effect upon compressive strength of the speed of testing.

9. Discuss the effect upon compressive strength of the end conditions of the specimen.

TEST 9 TRIAL-MIX PROPORTIONING OF CONCRETE

Of all the methods in use for the design of concrete mixes the trial-mix method is the one most readily understood by the beginner. This test is intended to give the student some experience with the application of this method, so that he may appreciate its advantages and limitations.

Test instruction. *Preparatory reading.* Articles 5.1 to 5.4, 5.13, 6.5, 6.6. Appendixes C, G.

Organization. Several parties should be assigned to this test, so that a series of different mixes may be made during the assigned period.

Materials for each party. Six pounds of portland cement, ½ cu ft fine aggregate, and ½ cu ft ¼ to ¾-in. gravel, all aggregates to be saturated and surface-dry or of known moisture content.

Equipment. One 6 by 12-in. cylinder mold with cover and base plate, containers for stocking and weighing materials, large mixing pan, two large trowels, 1,000-ml graduate, slump cone, ⅝-in. tamping rod, 12-in. steel rule, platform scale sensitive to 0.01 lb, 0.1 or 0.2-cu ft measuring cylinder, medicine dropper, and 10-ml graduate.

Procedure. Each of the several parties working on this problem will make a trial batch using the net water-cement ratio and slump assigned by the instructor. The following values are suggested:

W/C, by wt	Slump, in.
0.50	1–2
0.50	4–5
0.60	1–2
0.60	4–5

If more than four parties are available, a given set of conditions should be assigned to more than one party so that the parties working on a given assignment may check one another. Each trial batch will be made using 6 lb cement and the proper amounts of fine and coarse aggregates for best combination of workability and yield.

If the aggregate is not saturated and surface-dry, but of known moisture content, it will be necessary to make a preliminary trial mix to determine the approximate amount of aggregates to be used so that a correction may be made for free water or water absorbed by the aggregates. For this work weigh out 20 lb of each aggregate in a separate covered container, then deposit about half of each weighed lot into the mixing pan. Follow the procedure of Appendix G for mixing a batch. Based upon your studies and previous experience with concrete mixes, use a suitable trial ratio of fine to coarse aggregate, and add in sufficient amounts to produce the desired slump. Check the weights of all materials used.

Determine (1) cohesiveness, (2) proportion of sand, (3) troweling workability, (4) unit weight, and (5) water gain as in Test 6. Clean all equipment. Exchange the following values for your mix for corresponding items obtained by all other parties from their mixes:[1]

1. Water-cement ratio, by weight
2. Actual slump, in inches
3. Mix, parts by weight
4. Percent sand in aggregate
5. Sacks of cement per cubic yard of concrete
6. Gallons of water per cubic yard of concrete
7. Remarks on bleeding, cohesiveness, etc.

Report. Each student will report on all mixes.

1. Make comparisons and draw conclusions from these tests regarding the influence of the several variables upon the characteristics of the mixes.

2. Would large batches of concrete made in the field be expected to have the same workability as the small batches having the same proportions prepared in the laboratory?

3. Why is it desirable to use saturated surface-dry aggregates in trial mix designs? Would it be possible to use aggregates which are not saturated surface-dry? Explain.

[1] It is suggested that, if possible, these items be entered in a tabulation on a bulletin board or blackboard for ease of exchange and review.

TEST 10 CONCRETE-MIX PROPORTIONING BY ACI CALCULATION METHOD

Before mixing operations begin on an important job, a tentative mix should be determined from the results of trial batches or by the calculation method (see Arts. 6.5 to 6.8). This mix can be modified slightly, as required, when actual placement of concrete begins.

As an example of mix design and the determination of amounts of materials, assume the following conditions:

A precast reinforced concrete beam for use in an ordinary exposed building in a mild climate is 14 ft long. Its width is 10 in., overall depth is 18 in., depth to steel is 16 in., reinforcement is three No. 7 (7/8-in. diameter) bars, the outer bars having their centers at 2 in. from the side form. The concrete is to have a slump of 5 in. and a compressive strength of 3,000 psi at 28 days. Coarse aggregate is natural gravel having a dry-rodded weight of 100 pcf, an absorption of 0.5 percent, and a specific gravity of 2.70. The natural sand has a fineness modulus of 2.70, a specific gravity of 2.65, and 2 percent free moisture. The concrete will be assumed to contain 1 percent of entrapped air. A type I cement will be used.

Following the ACI method of concrete mix design (Art. 6.7), select or compute:

1. Maximum stock size of aggregate (e.g. 3/4, 1, 1½ in. etc.) based on bar spacings
2. W/C as governed by strength or exposure
3. Water content, pounds per cubic yard
4. Cement content, sacks per cubic yard
5. Weight of saturated surface-dry sand and gravel per cubic yard of concrete
6. Weight of sand and gravel, as stocked, per cubic yard of concrete
7. Weight of materials required to precast the beam: cement, water, sand, and gravel (as stocked).

TEST 11 ADJUSTMENT OF CONCRETE MIX TO GIVE DESIRED CEMENT FACTOR OR WATER-CEMENT RATIO AT CONSTANT CONSISTENCY

The object of this test is to acquaint the student with the general procedure for adjusting a concrete mix of given cement factor and consistency to produce a mix having the same consistency but a different cement factor. This problem arises during the trial-mix method of design of concrete mixes, particularly when the aggregates are not saturated surface-dry. In this method a mix is first produced which has the desired slump and a suitable ratio of fine to coarse aggregate, but the desired water-cement ratio or desired cement factor may not be obtained. By a simple method described in Art. 6.6, the mix can be adjusted as desired.

Test instructions. *Preparatory reading.* Articles 5.1 to 5.4, 5.13, 5.14, 6.6; Appendixes C, G.

Materials. About ⅛ sack of cement, ½ cu ft of fine aggregate, and ½ cu ft of ¼ to ¾-in. aggregate as stocked.

Equipment. Containers for stocking and weighing materials, large mixing pan, two trowels, ⅝-in.-diameter tamping rod, 1,000-ml graduate, slump cone, 12-in. steel rule, and platform scale sensitive to 0.01 lb.

Procedure. Assume that a 1:2.40:3.30 mix by weight on a saturated surface-dry basis produces the desired slump and contains 5.83 sacks of cement per cu yd and 1 percent air voids. The net water content of the concrete is 37.5 gal per cu yd, giving a water-cement ratio of 0.57 by weight. The specific gravity of the cement and aggregate will be taken as 3.10 and 2.65 respectively, or as assigned.

Prior to the laboratory period compute the modified or adjusted mix having the same consistency but a cement factor of 5.50 sacks of cement per cu yd, the adjustment being based upon the fact that the water content in gallons per cubic yard is practically constant for given materials and a given consistency, at least for a limited range of mixes, and therefore a certain solid volume of fine aggregate can be substituted for an equal solid volume of cement without an appreciable change in the consistency (Art. 6.6). Report the adjusted mix proportions and water-cement ratio to the instructor before beginning any laboratory work.

In the laboratory prepare a 1:2.40:3.30 mix having a net water-cement ratio of 0.57 by weight and using 30 lb of aggregate on a saturated surface-dry basis. Follow the instructions in Appendix G for batching and mixing the concrete. Make three determinations of the slump to obtain a reliable average value.

Also make a batch of the modified mix using the same aggregates. Make certain that the moisture contents of the aggregates are the same as for the first batch, so as to avoid errors in the desired net water-cement ratio. Make three determinations of the slump. Note the general character of the modified mix in comparison with the original mix as regards workability, and note any tendency of the modified mix to be undersanded or oversanded.

Dispose of all concrete on the debris pile, and clean all equipment used.

Report. *Computations.* Show all computations on the determination of the modified mix.

Discussion. 1. Which may be expected to stiffen a given concrete mix most if added in equal solid volumes—cement, fine aggregate, or coarse aggregate?

2. Is the rule for altering strengths at constant consistency by substituting aggregate for an equal solid volume of cement a basic principle or simply a convenient approximation?

3. How might the validity of this rule be checked readily by simple tests at limiting values of substitution?

4. Should the valid range of application of this rule be the same regardless of the aggregate used?

5. Did the application of this rule appear to work satisfactorily for the two mixes used in this test? How nearly constant did the workability remain as judged by the slumps of the two mixes?

6. If the application of this rule did not work satisfactorily in this test, endeavor to give the reasons therefore.

TEST 12 ADJUSTMENT OF CONCRETE MIX TO PRODUCE A GIVEN CHANGE IN CONSISTENCY

Two methods are available for adjusting the consistency of concrete: For constant proportions of cement and aggregate, an increase in the water-cement ratio increases the slump; and for a constant water-cement ratio, an increase in the paste content increases the slump. The object of this test is to determine quantitatively the effect of these two adjustments on consistency to give the student an understanding of some of the factors used in adjusting concrete mixes.

Test instructions. *Preparatory reading.* Articles 5.1 to 5.4, 5.13, 5.14, 6.6; Appendixes C, G.

Organization. Two parties will work as a unit in making this test.

Materials for two parties. About 1/5 sack of cement, 1/2 cu ft of fine aggregate, and 1/2 cu ft of 1/4 to 3/4-in. aggregate as assigned by the instructor.

Equipment for one party. Containers for stocking and weighing materials, large mixing pan, two large trowels, 1,000-ml graduate, slump cone, 5/8-in. tamping rod, 12-in. steel rule, and platform scale sensitive to 0.01 lb.

Procedure. Each party will prepare one batch of concrete. The mix will be 1:5 by weight, on a saturated surface-dry basis. The total aggregate will contain 45 percent fine and 55 percent coarse aggregate or as assigned. The total weight of aggregate per batch will be 30 lb. The initial slump of the concrete will be about 2 in. Follow the instructions in Appendix G for batching and mixing the concrete. Make three determinations of the initial slump to obtain a reliable average value. Record the amount of water used.

One party will determine the effect of increasing the water-cement ratio upon consistency as follows: Add water to the mix until the slump is about 5 in., and then add more water until the slump is about 8 in. Measure the additional water in each case. Make three determinations of slump for each consistency.

The second party will successively increase the slump of their mix to about 5 in. and then to about 8 in. by adding cement paste having the same water-cement ratio as required to produce the 2-in. slump. Make three determinations of slump for each consistency. Record the amount of additional cement required.

As a mix tends to stiffen with time, the work should be done as rapidly as possible to avoid the introduction of appreciable errors from this source.

Report. *Reduction of data.* Each student will report on all six mixes.

1. From the data for the three mixes having a constant cement-aggregate ratio, determine the net water-cement ratio for each mix. Also, deter-

mine the change in water-cement ratio corresponding to each change in slump. Compute the average change in water-cement ratio (or percent change in water content) for a 1-in. change in slump.

2. From the data for the second group of three mixes having a constant water-cement ratio, determine the mix proportions by weight for each mix. Assuming that 1 cu ft of these concretes weighs 148 lb, compute the cement content in barrels per cubic yard for each of the three mixes of the second group. Determine the change in cement content corresponding to each change in slump. Compute the average change in cement content in barrels per cubic yard for a 1-in. change in slump.

Discussion. 1. Compare your test result with any corresponding ones reported in Art. 6.6.

2. If concrete of a given strength for placement by hand tamping must have a 5-in. slump, whereas that for placement by vibration need have a slump of only 2 in., what saving in material costs per cubic yard of concrete may be realized by the use of vibrators for compacting the concrete. Assume that cement costs $1.60 per sack and aggregate costs $6.00 per ton.

3. How can the consistency of a mix be modified without appreciably varying the cement factor or the general texture? Should this alter the strength?

4. How can the consistency be varied without altering the water-cement ratio? Should this affect the strength or texture?

5. For given aggregates how can the yield be increased without appreciable loss of workability? What will be the effect upon strength?

6. For given aggregates how can yield be increased without altering the strength?

7. Can yield be altered without appreciably altering either consistency or strength? Explain.

TEST 13 EFFECT OF CAPPING MATERIALS AND END CONDITIONS BEFORE CAPPING UPON COMPRESSIVE STRENGTH OF CONCRETE CYLINDERS

Concrete cylinders cast in the field as well as many cylinders cast in the laboratory have irregular end surfaces. To provide flat-end bearing surfaces for compression tests the ends are usually capped with a rapid-hardening compound. This experiment is designed to acquaint the student with the effectiveness of certain capping materials, and to show that certain materials which may be effective for capping cylinders having slight irregularities are unsatisfactory for capping cylinders with very irregular end surfaces.

Test instructions. *Preparatory reading.* Article 10.21; Appendixes B, C, G to I.

Organization. Two parties will work as a unit in performing this test.

Materials for two parties. About ⅕ sack of cement, ½ cu ft of fine aggregate, and ½ cu ft of ¼ to ¾-in. aggregate as assigned by the instructor.

Equipment for molding specimens (one party). Eighteen 3 by 6-in. cylindrical molds with base plates, 12 circular plates for preparing concave-ended specimens, 12 circular plates for preparing convex-ended specimens, containers for stocking and weighing materials, large mixing pan, two trowels, ⅝-in.-diameter tamping rod, 1,000-ml graduate, slump cone, 12-in. steel rule, and platform scale sensitive to 0.01 lb.

Equipment for testing specimens (one party). Testing machine as assigned, spherically seated bearing block, plain bearing block, 6-in. calipers, 12-in. steel rule; small mixing pan, trowel, and capping plates for gypsum caps; and sulfur-capping equipment.

Procedure for preparing specimens. The two parties working together will make a total of thirty-six 3 by 6-in. concrete cylinders in four batches. The proportions for the mix will be 1:5 by weight, on a saturated surface-dry basis. The total aggregate will contain 45 percent fine and 55 percent coarse aggregate or as assigned. The water-cement ratio will be the same for all batches and should give a slump of about 2 in.

A total of 30 lb of aggregate will be required for each of the four batches. Follow the instructions in Appendix G for batching and mixing the concrete and molding into compression-test cylinders. Determine the amount of water required with the first batch to produce the desired 2-in. slump; then use the same amount of water in the remaining three batches.

From each batch mold three cylinders with plane normal ends, three with both ends convex 3/16 in., and three with both ends concave 3/16 in., the convex and concave ends being produced by circular metal plates inserted in the bottom and top of each mold. Care must be taken in mixing and placing the concrete that all test cylinders will be of equal quality. Number the specimens so that each batch of nine cylinders may be identified.

Procedure for testing specimens. Each of the 36 cylinders will be tested in compression at the age of 14 days or as assigned. First, determine the mean diameter of each specimen to the nearest 0.01 in. Cap one specimen of each type of end condition, from each of the four batches, with each of the three capping materials, i.e., casting plaster, Hydrostone (a high-strength gypsum compound), and a sulfur compound. For capping with casting plaster and Hydrostone, follow the instructions in Appendix H. For capping with the sulfur compound follow the instructions in Appendix I.

Consult Appendix B for directions for operating the testing machine. Observe and sketch the characteristic fractures of the specimens for each type of capping material and for each type of end condition.

Report. *Reduction of data.* Each student will report on all 36 cylinders.

1. Calculate the unit compressive strength for each cylinder and the average compressive strength for each group of four cylinders for a given end condition and type of capping material.
2. Present the results of the experiment on a diagram with compressive strengths as ordinates, the three types of end conditions being shown at equal distances apart along the axis of abscissas. Show a separate curve for each kind of capping material.

Discussion. 1. Does the end condition *before* capping have any effect on the compressive strength?

2. What effect would the end condition *after* capping have on the compressive strength?
3. Are the three types of capping materials equally effective in developing the potential compressive strength of the concrete?
4. Do the characteristic fractures for each group of specimens give any indication of the influence of end condition before capping or of the type of capping material upon the method of failure?
5. What are the requirements of a good capping material?
6. How do inaccuracies in centering the spherically seated bearing block affect the indicated compressive strength of the concrete?

TEST 14 EFFECT OF SHAPE OF TEST SPECIMEN UPON INDICATED COMPRESSIVE STRENGTH OF CONCRETE

This test is designed to show the student the compressive strengths of concrete cubes in comparison with cylinders having various ratios of height to diameter, all being made of the same concrete mix.

Test instructions. *Preparatory reading.* Articles 3.11, 10.17; Table 6.3; Appendixes B, C, G, H.

Organization. Two parties will work as a unit in performing this test.

Materials for two parties. About ¼ sack of cement, ½ cu ft of fine aggregate, and ½ cu ft of ¼ to ¾-in. aggregate as assigned by instructor.

Equipment for molding specimens (one party). Twelve 3 by 6-in. cylindrical molds with base plates, three 3-in. cube molds with base plates, containers for stocking and weighing materials, large mixing pan, two trowels, ⅝-in.-diameter tamping rod, 2⅞-in.-diameter rod for smoothing top surface of 3-in. cylinders, 2-qt measure or 1,000-ml graduated cylinder, slump cone, 12-in. steel rule, and platform scale sensitive to 0.01 lb.

Equipment for testing specimens (one party). Testing machine as assigned, spherically seated bearing block, plain bearing block, 6-in. calipers, 12-in. steel rule, small mixing pan, trowel, and capping plates.

Procedure for preparing specimens. Each party will prepare one batch of concrete. The mix will be 1:5 by weight, on a saturated surface-dry basis. The total aggregate will contain 45 percent fine and 55 percent coarse aggregate or as assigned. The total weight of aggregate per batch will be 35 lb. The slump of the concrete will be about 3 in. Follow the instructions in Appendix G for batching and mixing the concrete and molding into test specimens.

From each batch of concrete mold three 3-in. cubes, three 3 by 6-in. cylinders, three 3 by 4-in. cylinders, three 3 by 3-in. cylinders, and three 3 by 2-in. disks. Smooth the top surface of all specimens, using the 2⅞-in.-diameter rod for the 3-in. cylinders and disks.

Procedure for testing specimens. Test in compression at age 14 days, or as assigned, each of the specimens previously molded. Determine the lateral and axial dimensions of each specimen to the nearest 0.01 in. Cap all specimens with casting plaster following the instructions in Appendix H. Consult Appendix B concerning directions for operating the testing machine. Observe and sketch the characteristic fractures for each shape of specimen.

Report. *Reduction of data.* Each student will report on all 30 specimens tested by the two parties assigned to this test.

1. Calculate the unit compressive strength for each specimen and the average unit compressive strength for each group of identical specimens.

2. Taking the strength of the 3 by 6-in. cylinders as 100, determine the relative strength of the other shapes of specimens.

3. Present the relative strengths and ratios of height to diameter of the several shapes in tabular form.

Discussion. 1. Why and how does the ratio of height to diameter for a test specimen affect its compressive strength?

2. Why are the unit compressive strengths of 6-in. concrete cubes usually higher than for 6 by 12-in. cylinders?

3. What is the general effect of size of test cylinder upon the unit compressive strength?

4. What are the usual limitations on *minimum* dimension of test specimens with respect to the *maximum* size of aggregate?

5. What is the effect upon compressive strength when concrete is wet-screened to reduce the maximum size of aggregate remaining?

6. Some concrete cores drilled from a structure are to be tested for strength. They are of constant diameter but of variable height, are relatively dry, and the ends are not plane. What measures should be taken to make the reported strengths as significant as possible?

7. If a 6-in.-diameter core cut from a wall 4 in. thick has a compressive strength of 3,000 psi, what would be the probable strength of a 6 by 12-in. cylinder of the same quality of concrete?

TEST 15 PHYSICAL AND MECHANICAL PROPERTIES OF CONCRETE

The objects of the test are to observe the behavior of concrete under compressive loading and to determine the following physical and mechanical properties:

1. Proportional limit
2. Compressive strength
3. Initial tangent modulus of elasticity
4. Secant moduli of elasticity at stresses of 500, 1,000, 1,500 and 2,000 psi
5. Chord modulus of elasticity between a lower stress corresponding to a strain of 50 millionths and a stress at 40 percent of the ultimate [C469]
6. Weight per cubic foot

Test instructions. *Preparatory reading.* Articles 13.1 to 13.7; Appendixes B, H, I.

Specimen. Concrete cylinder, 6 by 12 in.

Equipment. Caliper, 12-in. steel scale, capping equipment, compressometer and compression testing machine.

Procedure. 1. From the instructor obtain data regarding the kinds and and proportions of the constituent materials, the water-cement ratio and the consistency of the mix, the curing and storage conditions, and the age of the specimen. From these data predict the ultimate strength. However, if scheduled by the instructor, the student may cast concrete cylinders for testing at various ages to develop the relationship of age, compressive strength, and modulus of elasticity.

2. Determine the mean diameter of the cylinder at its midsection, and determine its average length, making measurements to 0.01 in. Weigh the specimen to 0.01 lb.

3. Cap each end of the specimen with the material provided. If Hydrostone is used, mix sufficient material with water to produce a mixture of fairly stiff consistency. Place a trowelful of the paste in the center of the capping plate or cylinder and work it out to the edge by rotating the cylinder against the plate, or vice versa. Allow the cap to set ½ hr before testing the specimen with plates attached.

4. Study the action of the compressometer, note its gage length and multiplication ratio, and determine the strain corresponding to the least reading of the dial. Attach the compressometer to the central portion of the specimen and remove the spacer bars.

5. After the caps have hardened, center the specimen in the testing machine and center the spherical bearing block on top of the specimen. Centering operations should be carried out by actual measurements.

6. Adjust the compressometer dial to read zero and make sure that most of its range is available.

7. Apply a preliminary load up to about two-fifths of the estimated ultimate load for the specimen, at a speed of about 35 psi per sec, and then release the load. This is done primarily to seat the gages. Reset the compressometer gage and then apply load continuously at a speed of about 10 psi per sec, reading the compressometer after each load increment of about one-twentieth of the estimated ultimate load. Consult the instructor regarding this value. At a load of three-quarters of the estimated ultimate load remove, or loosen, the compressometer. Thereafter, apply load continuously and record the maximum load.

8. Draw a sketch to show the type of failure.

Report. 1. Plot a stress-strain diagram. Draw a smooth curve through the plotted points. Note that the curve may not pass through the origin of the graph. Mark the proportional limit on the diagram. Determine the several moduli of elasticity specified.

2. Compute the compressive strength and the unit weight of the concrete.

3. Tabulate the test results in a suitable form.

Discussion. 1. Compare your results with the range of values indicated in Chaps. 10, 13.

2. Discuss briefly the important facts disclosed by the test.

3. What factors affect the development of strength of concrete?

4. How is the compressive strength affected by moisture content at time of test?

5. How is the modulus of elasticity affected by age and by moisture content at time of test?

6. Why is the compression test the one most frequently made for concrete?

7. What was the purpose of using the spherical bearing block in this test?

8. List the various precautions which should be taken in positioning the spherical bearing block.

9. Are the strength correction factors given in Table 10.5 for concrete specimens having height-diameter ratios below two rational or empirical values?

TEST 16 SPLITTING TENSILE STRENGTH OF CONCRETE CYLINDERS

The purpose of this test is to determine the tensile strength of concrete by splitting cylinders of the concrete in a compression testing machine. A direct axial tension test of concrete is not made ordinarily, as adequate gripping of the ends of the specimen is very difficult to accomplish.

Test instructions. *Preparatory reading.* Articles 10.6, 10.10.

Test specimen. A 6 by 12-in. cylinder will have been prepared and cured previously, or one will be provided by the instructor who will give full information on the materials and mix used and the curing history.

Equipment for testing. Large calipers, 12-in. steel rule, device for making diametral lines on ends of specimen, two plain bearing blocks 12 in. long, two plywood strips ⅛ by 1 by 12 in., and compression testing machine equipped with spherical bearing block.

Procedure. Draw diametral lines on each end of the specimen using a device that will ensure that they are on the same axial plane. Determine the diameter of the specimen to the nearest 0.01 in. by averaging three diameters measured near the ends and the middle of the specimen and lying on the plane containing the diametral lines marked on the ends. Determine the length of the specimen to the nearest 0.1 in. by averaging the two lengths measured between the diametral lines on the ends.

To position the specimen, center one of the plywood strips along the center of the lower bearing block. Place the specimen on the plywood strip and align so that the lines marked on the ends of the specimen are vertical and centered over the plywood strip. Place a second plywood strip lengthwise on the cylinder, centered on the lines marked on the ends of the cylinder. Check to see that the projection of the plane of the two lines marked on the ends of the specimen intersects the center of the upper bearing plate and that the bearing plates and the specimen are centered on the spherical bearing block.

Apply the load continuously and without shock, at a rate of 100 to 200 psi per min splitting tensile stress (corresponding to applied total load in the range of 11,300 to 22,600 lb per min for 6 by 12-in. cylinders) until failure. Record the maximum applied load, the type of fracture, the appearance of the concrete, any defects, and the estimated proportion of coarse aggregate fractured during the test.

Report. *Report of test results.* Compute the splitting tensile strength of the concrete to the nearest 5 psi. In the report include all available information on the concrete, including its moisture condition at the time of test and the test results obtained.

Discussion. 1. What is the probable axial tensile strength of the concrete tested?

2. What is the effect of moisture at time of test on the tensile strength?

3. What is the effect of age (with moist curing) on the tensile strength?

4. Discuss any conditions in concrete structures where the tensile strength of the concrete is significant.

TEST 17 DEMONSTRATION OF ENTRAINED AIR IN CONCRETE

Experience has shown that the resistance of concrete to disintegration by freezing and thawing is much improved by the entrainment of air in the concrete. Entrained air also promotes greater workability of the concrete. The range in air content which is effective to increase durability and workability and yet not to lower strength appreciably is not great. Hence, success in the application of this principle requires especially careful measurement of the ingredients and control of the concrete-making process to regulate the amount of entrained air.

Various mehods for measurement of the volume of entrained air have been developed, and three have been standardized by the ASTM. Two of the latter methods involve a comparison of the volume of a batch of air-entrained concrete with the solid volume of the same batch as computed from the summation of the solid volumes of the several components of the mix [C138] or as determined by displacement of the mix under water [C173]. Both these methods are subject to appreciable errors unless controlled very carefully.

A third method [C231] which has been developed is one in which the reduction in volume of the air voids in a sample of the fresh concrete (and hence a reduction in volume of the concrete) is effected by application of water pressure to the sample, the quantity of air being determined from a consideration of Boyle's law. In general, this pressure method gives percentages of air about one higher than the others and is the method most commonly used.

The purpose of this test is to acquaint the student with (1) the characteristics of air-entrained concrete in comparison with plain concrete, (2) the pressure method for the measurement of air content, and (3) the direct volumetric method for the measurement of air content.

Test instructions. *Preparatory reading.* Articles 4.4 to 4.7, 5.11, 5.12.

Preparation of concrete mixes. In this demonstration the instructor will make two concrete mixes using the same nominal cement content, one without entrained air and the other with entrained air. Also, a preliminary half-batch mix of concrete without air should be made to coat the inside of the drum mixer, as otherwise some of the first batch would be lost. For the air-entrained concrete a slightly lower sand content can be used. An effort will be made to have the slump of the two mixes as nearly equal as possible.

The instructor will give the students full information on each mix, including kind and amounts of materials used and the moisture condition of the aggregates.

Preliminary determinations on the mixes will include their slump, Kelly ball penetration, and weight of a given volume of concrete.

Air content by pressure method. This method depends on an application of Boyle's law. Pressure is applied to a known volume of concrete, and the reduction in volume measured. Since the air entrained in the concrete is the only significantly compressible ingredient, the observed reduction in volume is due to compression of the air. The amount of air is shown on the calibrated scale of the apparatus.

The procedure for the test, as conducted by the instructor, will be as follows: Fill the calibrated container (lower part of the assembly) in three equal layers, rodding each layer with 25 strokes of the steel tamping rod. Follow the rodding of each layer by tapping the sides of the bowl smartly 15 times with the mallet. Slightly overfill the bowl, and strike off any excess concrete by sliding the strike-off bar across the top flange with a sawing motion until the bowl is just level full.

Thoroughly clean the flanges of the bowl and the conical cover. Assemble the apparatus, making a watertight connection. Introduce water into the head and graduated tube assembly to about the halfway mark on the tube. Avoid turbulence and the inclusion of air in this operation. Incline the assembly about 30° from vertical, and, using the bottom of the bowl as a pivot, describe several complete circles with the upper end, simultaneously tapping the conical cover lightly to remove any entrapped air bubbles above the concrete. Fill the water column to the zero mark. Close the tube at the top and apply the required pressure. Read the air content. Release the pressure and read the scale again. The actual air content is the difference between the two readings on the scale.

Air content by volumetric method. This method involves releasing the entrained air by agitation of the concrete in water and then determining its amount by observing the reduced volume of the combined concrete and water. The procedure for this test will be as follows: fill the bowl of the Rollimeter apparatus in the same manner as for the pressure method above; strike off any excess concrete; and clean the flanges of the apparatus and assemble it, making a watertight connection. Insert the funnel for filling the apparatus with water, adding water until it appears in the window in the neck. Remove the funnel and adjust the water level, using a rubber syringe, until the bottom of the meniscus is level with the zero mark. The graduations marked on the window correspond to percentages of the volume of the bowl. Attach and tighten the screw cap.

Invert and agitate the unit until the concrete settles free from the base; and then, with the neck elevated, roll and rock the unit until the air appears to have been removed from the concrete. Set the apparatus upright, jar it lightly, and allow it to stand until the air rises to the top. Repeat the operation until no further drop in the water column is observed; then remove the screw cap. Add, in small increments, one measuring cupful (equal to 1 percent of the volume of the bowl of the apparatus) of isopropyl alcohol, using the syringe to dispel the foamy mass on the surface of the water. Read the level of the liquid in the neck, estimating to the nearest 0.1 percent. The air content is the reading plus 1 percent for the amount of alcohol used.

Compression tests. After measuring the air content, four 3 by 6-in. cylinders will be cast for each mix and tested at the age of 14 days.

Report. *Reduction of data.* The report on this test will include all observed data and the following items for each mix:

1. Air content by each method.
2. Unit weight.
3. Percent sand, by weight of total aggregate and by solid volume of total mix. Use specific gravity of 2.60 for sand and 2.70 for gravel, or as assigned.
4. Net W/C by weight.
5. Net W/C by volume, in gallons per sack.
6. Water content, pounds per cubic yard.
7. Yield, cubic feet per sack.
8. Cement factor, sacks per cubic yard.
9. Compressive strength.

Discussion. Write a one-page discussion on air entrainment based on the demonstration and reading assignment.

APPENDIX A
SUMMARY OF USEFUL VALUES

Weight, measures, temperatures

1 in. = 2.540 cm	1 cm = 0.3937 in.
1 lb = 0.4536 kg	1 kg = 2.2046 lb
1°F = 0.5556°C	1°C = 1.80°F

Cement and water equivalent

1 sack cement = 1 cu ft = 94 lb = 0.25 bbl
1 bbl cement = 4 cu ft = 376 lb = 4 sacks
1 cu ft water = 62.4 lb = 7.48 gal
1 gal water = 8.33 lb
1 lb water = 454 ml

Comparative water-cement ratios. On various bases:

Volume	Weight	Gal/sack
1.0	0.664	7.48
1.505	1.0	11.25
0.134	0.089	1.0

Example: W/C by volume = 1.505 × W/C by wt

Solid volume. Equals (wt of given quantity)/(bulk sp gr × 62.4).

Specific gravity of portland cement. Varies from about 3.05 to 3.20. Assume 3.15 when value is unknown.

Specific gravity of aggregates. For siliceous sands and gravels, use average value of 2.65.

Effective absorption of mixing water by air-dry aggregates. For siliceous sands and gravels (excluding sandstones) the effective absorption commonly ranges from about 0.2 to 0.5 percent by weight. See Table 3.4 for absorption capacities.

Free moisture in aggregates. See Table 3.5.

APPENDIX B
INSTRUCTIONS ON OPERATION OF TESTING MACHINES

Screw-gear testing machine. The following instructions will be observed in the operation of all testing machines of the screw-gear type:

1. Do not operate any machine for the first time without the assistance of the instructor.
2. Do not start a machine without determining beforehand the direction and speed with which it will move.
3. Do not reverse a machine before it comes to a complete stop.
4. Accurately center the specimen on a smooth base plate which is centered in the testing machine. Balance the scale beam with the poise set at zero load. See that the spherically seated bearing block operates easily and is centered on top of the specimen, with the center of curvature of its spherical surface in the top surface of the specimen. Lower the head until it just clears the bearing block, shift to slow speed, and continue to move the head downward at this speed.
5. Apply the load using a speed such that the scale beam may be kept balanced with ease. Use a maximum speed of the movable head of about 0.05 or 0.06 in. per min.
6. Keep the scale beam balanced continuously while the load is being applied. Exercise particular care in keeping it balanced as the maximum load is approached, so that the ultimate strength of the test specimen may be accurately determined. As soon as failure occurs, stop the motion of the movable head; then, before shifting the poise, record the indicated reading.
7. When the test is completed, disengage the clutch and stop the motor. If a testing machine is left running, parts may be broken.
8. Draw a sketch of the characteristic fracture of the specimens tested. Remove the tested specimens to the waste pile, and clean up the debris around the testing machine.

Hydraulic testing machine. The following regulations will be observed in the operation of all testing machines of the hydraulic type:

1. Do not operate any machine for the first time without the assistance of the instructor.
2. Accurately center the specimen on a smooth base plate which is centered in the testing machine. Adjust the load pointer to indicate zero load. See that the spherically seated bearing block operates easily, then center it on top of the specimen, so that the center of curvature of the spherical surface will be in the top surface of the specimen.
3. Reduce the clearance between the bearing block and the head of the machine, then apply load at a rate of about 2,000 psi per min. Carry the maximum-load pointer along with the regular-load pointer, so that the maximum load will be indicated. Be careful to shut off any low-load dials which may be damaged by exceeding their range.
4. As soon as the specimen fails, close the loading valve and open the unloading valve. When the test is complete, stop the motor. Record the maximum load.
5. Draw a sketch of the characteristic fracture of the specimens tested. Remove the tested specimens to the waste pile, and clean up the debris around the testing machine.

Briquet-testing machine. The following regulations will be observed in the operation of briquet-testing machines having a graduated beam for determining the load. Consult the instructor for operating instructions for other types.

1. Place the poise weight at zero, put the shot bucket on the hook on the right end of the graduated beam, and balance the beam by the counterpoise weights at the left.
2. Place shot bucket on the hook at the left end and the bucket counterweight on the right hook. If machine is no longer balanced, determine error by moving poise weight on the graduated beam. If a positive load is indicated, it must be subtracted from all future readings.
3. See that roller clips which bear on briquet rotate freely.
4. Clean off any fins or sand grains on the sides of the briquet which would prevent even bearing of the roller clips on the specimen.
5. Insert briquet in the testing machine.
6. Engage the hand crank by moving to the left, and insert the pin to lock crank in position.
7. Open shot spout, and keep beam balanced by turning crank counterclockwise. Observe time of loading, so that rate of loading may be checked. ASTM Specification C190 requires a loading rate of 600 ± 25 lb per min.

8. Make certain that flow of shot stops instantly when the specimen breaks.

9. Measure load by placing bucket on left hook and balancing with bucket counterweight on right hook, and beam poise weight supplemented by additional weights if necessary. Correct for any error observed in step 2.

10. Return all shot to the reservoir, taking care to avoid spilling or contamination.

APPENDIX C
PROCEDURE FOR MAKING THE SLUMP TEST[1]

For the slump test, place the dampened truncated cone-shaped metal mold, 12 in. high by 8 in. in diameter at the base and 4 in. in diameter at the top, on a smooth, moist rigid base. The bottom of the mixing pan may be used if reasonably flat and if care is taken to avoid jarring the pan after removing the slump cone.

Place the newly mixed concrete in the mold in three layers, each approximately one-third the volume of the mold. In placing each scoopful of concrete, move the scoop around the top edge of the mold as the concrete slides from it, in order to ensure symmetrical distribution of concrete within the mold. Rod each layer with 25 strokes of a ⅝-in. rod, 24 in. in length and rounded at the lower end. Distribute the strokes in a uniform manner over the cross section of the mold, each stroke just penetrating into the underlying layer.

After rodding the top layer, strike off the surface of the concrete with a trowel, leaving the mold exactly filled. Clean the surface of the base outside the cone of any excess concrete. Immediately remove the mold from the concrete by raising it slowly in a vertical direction. If the pile topples sideways, it indicates that the materials have not been uniformly distributed in the mold, and the test should be remade. The entire operation should be completed within an elapsed time of 1½ min.

Measure the slump immediately by determining the difference between the height of the mold and the height of the vertical axis (not the maximum height) of the specimen. Clean the mold thoroughly immediately after using.

[1] From ASTM C143.

APPENDIX D

PROCEDURE FOR MAKING THE BALL-PENETRATION TEST[1]

The ball-penetration test may be conducted on the concrete as placed in the forms prior to any manipulation or in a suitable container. The minimum depth of the concrete shall be at least three times the maximum-size aggregate but never less than 8 in. The minimum horizontal distance from the center line of the handle to the nearest edge of the level surface on which the test is to be made shall be 9 in.

Level the surface of the concrete, working it as little as possible. Set the base of the apparatus on the concrete, lower the weight to the concrete surface, and release slowly. Read the penetration to the nearest ¼ in. Take a minimum of three readings from a batch or location. These readings shall not be taken with the feet of the stirrup within 6 in. of a point where the feet rested in a previous test. No correction shall be made for any slight settlement of the feet of the stirrup.

[1] From ASTM C360.

APPENDIX E
PROCEDURE FOR MAKING THE FLOW TEST[1]

Check height of drop of table to see that it is ½ in. instead of ¼ in. as used in the remolding test.

For the flow-table test, first wet and clean the round metal table top of all gritty material, and remove all excess water with a rubber squeegee or wet cloth. Place the flow-table mold, having a height of 5 in. and top and bottom diameters of 6¾ and 10 in., respectively, in a central position on the table.

Hold the mold firmly in place and fill in two layers, each approximately one-half the volume of the mold. Rod each layer with 25 strokes of the ⅝ by 24-in. hemispherically-tipped rod, each stroke just penetrating into the underlying layer.

Strike off the top surface with a trowel. Remove all excess concrete outside the mold, and again clean the table top with the squeegee. Immediately remove the mold from the concrete by a steady, upward pull; then raise and drop the table ½ in., 15 times in about 15 sec. Take the diameter of the spread concrete as the average of six symmetrically distributed measurements read to the nearest ¼ in.

$$\text{Flow } \% = \frac{\text{spread diameter} - 10 \text{ in.}}{10} \times 100$$

[1] From ASTM C124.

APPENDIX F
PROCEDURE FOR MAKING THE REMOLDING TEST[1]

Before using the remolding apparatus, check height of drop of table to see that it is ¼ in. instead of ½ in. as used in the flow test.

The Powers remolding apparatus consists essentially of a base plate and an outer shell 12 in. in diameter and 8 in. high, which is mounted centrally on a flow table set for a ¼-in. drop. Inside the outer shell is an inner ring 8¼ in. in diameter and 5 in. in height, mounted with its bottom 2¾ in. above the top of the base plate as shown in Fig. F.1. To make a test, set a standard slump cone on the base plate in the center of the inner ring. Fill the slump cone and remove it as instructed in Appendix C. Place the 8⅛-in.-diameter horizontal disk of the remolding apparatus on top of the slumped concrete, and jig the flow table at the rate of one drop per second until the disk settles down to 3³⁄₁₆ in. above the base plate, which is indicated when mark on vertical rod reaches top of guide. During the jigging operations, the concrete moves downward in the inner ring and outward and upward into the annular space between the inner and outer rings.

This test measures the relative effort required to change a mass of concrete from one definite shape to another by means of jigging. The amount of effort, called "remolding effort," is taken as the number of jigs required to complete the change. The results of the remolding test are of value in studying the mobility of masses of concrete made with varying amounts of water or cement and with various types and gradations of aggregates.

[1] From Ref. 515.

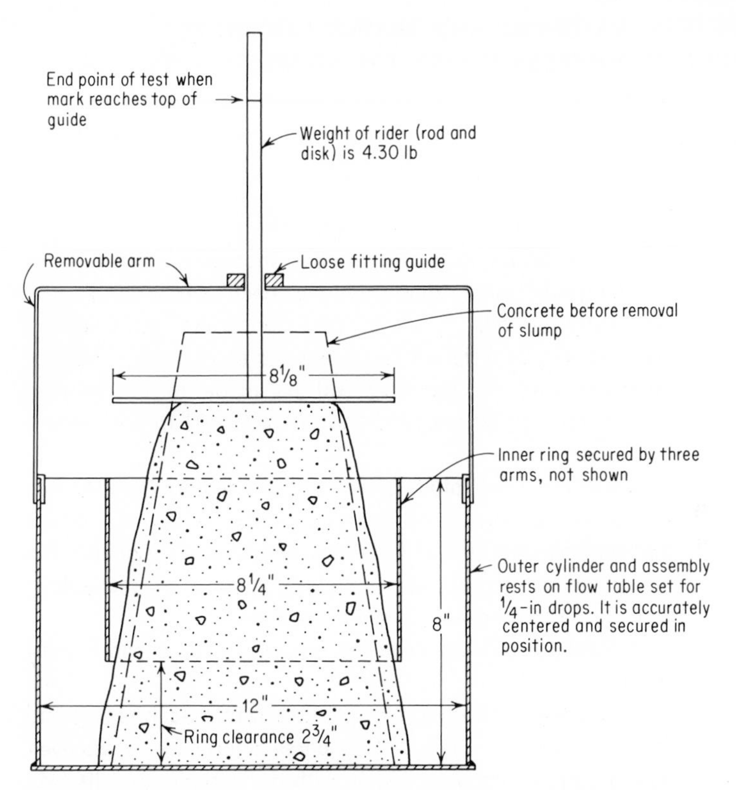

Fig. F.1. Remolding apparatus for determining remolding effort of concrete.

APPENDIX G

PROCEDURE FOR BATCHING AND MIXING CONCRETE AND MOLDING COMPRESSION-TEST CYLINDERS

To expedite the work in the laboratory, the students will calculate batch quantities for each mix prior to the laboratory period. Each of the component sizes of aggregates will be weighed separately for each batch. Weight measurements for batching will be made to the nearest 0.05 lb for aggregates, and the nearest 0.01 lb for cement and water.

Blend a small stock of each of the required aggregates to ensure a uniform gradation and moisture content. Blend the cement, and stock it in dry containers. Consult the instructor concerning the moisture content of the aggregates and the brand or source of all materials.

Measure out the required quantities for a batch (corrected for effective absorption or free moisture). Take care that segregation of the materials within each size group does not occur. Spread the coarse aggregate in an even layer in the mixing pan, which, with the various utensils, will have previously been moistened or wiped with a wet cloth. Place the fine aggregate in next and the cement on top. Turn the dry mixture several times with the trowel to obtain a uniform color. Take care not to spill any of the ingredients, because any loss of materials will change the mix and cause errors in the calculated results. Add water gradually and mix thoroughly. The weight of water used must be accurately known and may be determined by weighing the water container and its contents before and after mixing, or by measuring in a graduated glass cylinder. Add water until the desired water-cement ratio or slump is obtained. Determine the consistency of the mixture by the slump test, as outlined in Appendix C.

Metal molds should be sealed to their base plates to prevent loss of water. Then molds and base plate should be coated very lightly with oil. (Cardboard molds should be heavily paraffined.) Fill each mold with concrete in three layers, tamping each layer 25 times with the ⅝-in. steel tamping rod before adding the next layer. Take special care to consolidate the bottom layer. While filling the molds, occasionally stir and scrape together the concrete remaining in the mixing pan to keep the materials from separating. Fill the molds completely, smooth off the tops evenly, and clean up any concrete outside the cylinders.

If yield determinations are required the procedure above must be modified slightly as follows: Before filling the forms, determine their mean height and mean diameter to the nearest 0.01 in. Weigh the mold (or set of molds) with its base plate, both empty and when filled with concrete.[1]

After filling the required number of molds, place the remaining concrete on the waste pile. Thoroughly wash off all utensils and the mixing pan.

When each specimen has been molded, place on the top thereof a slip of paper on which is written with *lead pencil* the party number and specimen identification. The students will leave the molded specimens on the work table or as instructed. The following day a laboratory assistant will strip off the molds, and mark and place the specimens in the curing room for the remainder of the period until time of test.

[1] If 0.1 or 0.2-cu. ft. standard measuring cylinders are available, they may be used for yield determinations.

APPENDIX H
PROCEDURE FOR CAPPING COMPRESSION CYLINDERS WITH GYPSUM COMPOUNDS

Except as noted, casting plaster or Hydrostone will be used for capping specimens made in the tests outlined in this book. In mixing the plaster, use a small (6 or 8-in.) pan and a small trowel specifically provided for the purpose.

The caps will be formed on machined-metal base plates, which have been well oiled before use. A template or guide will be used to aid in keeping the axis of the specimen perpendicular to the base plate.

All measurements and weighings should be completed and all pertinent equipment assembled before the plaster is mixed with water.

Mix three heaping trowelfuls of plaster with water to a plastic (not wet) consistency. With the trowel, place on the base plate a small heap of plaster, sufficient to form a cap about 1/16 in. thick, having a diameter equal to that of the specimen. Quickly press the end of the cylinder to be capped into the plaster, using the template to guide the cylinder. Press firmly, so that some plaster is forced out at the edge of the cylinder all around.

The amount of plaster specified above will cap six or eight small cylinders. Do not mix more than this at one time, because it will stiffen in the pan before it can be used. Thoroughly clean the pan and trowel after each mixing operation. As soon as the plaster caps have hardened sufficiently (usually 1/2 hr or more), the cylinders can be slipped off the baseplate and tested.

APPENDIX I

PROCEDURE FOR CAPPING COMPRESSION CYLINDERS WITH SULFUR COMPOUND

Composition and properties of the compound. Sulfur capping compounds commonly consist of a mixture of about two parts sulfur to one part powdered silica or fire clay. When subjected to changes in temperature, the physical changes which take place in the compound are the same as for pure sulfur. At normal temperatures the material has a hard crystalline form, and as the temperature is raised, the compound goes through the following stages in the order mentioned: a viscous liquid, a free-flowing liquid, a viscous liquid again, and when the temperature is raised excessively, the mixture becomes stringy and gas bubbles form. When the material is cooled, the above stages are reversed.

The second, or free-flowing, stage is the desirable state of the material to use in forming caps. Under no conditions should gas bubbles be present, as voids in the hardened cap seriously reduce the strength of the cap.

Method of melting compound. Heat the empty melting pot electrically or over a very low flame in order to obtain a uniform distribution of heat. When the thin coating of sulfur adhering to the pot begins to melt, add about a pound of crushed material, which must be stirred continuously until it has melted to the free-flowing stage. Slowly add sufficient crushed material to do the capping at hand. Vigorously stir any added material to prevent the formation of large lumps.

The material is a poor conductor of heat; therefore a slow heating rate must be used. The melting cannot be forced by using excessive heat, because a sulfur fire will inevitably result.

Method of forming caps. Lightly oil the machined capping plate and forming ring and then assemble them, as shown in Fig. I.1, so that the end of the concrete cylinder will slide vertically into position concentric with the ring. From a single ladleful, pour enough compound into the ring so that

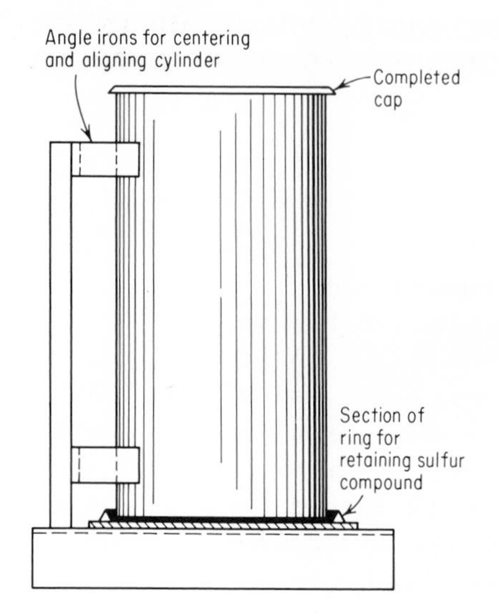

Fig. I.1. Equipment for capping concrete cylinder with sulfur compound. Some devices cap both ends of a horizontal cylinder in one operation.

when the specimen is forced down into it the compound will flow over the edges. Within about 10 sec place the specimen in the jig and force it down into the molten compound. When the cap has solidified, slide the specimen from the plate, invert it, break off the surplus material, and remove the ring.

If a specimen has a very irregular end, it will be necessary to cast a preliminary cap by wrapping paper around the specimen with the rough end on top and allowing the paper to extend above the specimen. With the specimen in a vertical position, pour a little molten compound on top of the specimen, filling all low spots. Add a final cap following the procedure outlined above.

During the capping operation the material must be stirred continuously, in order to prevent a crust from forming on the top surface.

Care of equipment and compound. When capping is finished, ladle all the unused compound out of the pot onto a smooth, cold, oiled, level metal surface, and as soon as it sets, remove it and place it in the container provided. Remove all surplus material from the ladles before they are allowed to cool. Reclaim the sulfur caps on broken cylinders, unless it is impossible to remove them or unless the compound has become too badly contaminated with broken fragments of concrete.

APPENDIX J
SELECTED REFERENCES AND SPECIFICATIONS PERTAINING TO PLAIN CONCRETE

Purpose of references. The discussion of plain concrete in Part I serves only to point out the most essential facts concerning concrete and its constituent materials. Throughout Part I references have been made to many publications, which are here grouped into various general classifications, to indicate the sources of material presented in the text and to serve as a guide to the student for further study.

Sources of information. Information concerning concrete is available in (1) various published texts or separate summaries, (2) professional papers presented before, or specifications adopted by, various technical societies such as the ASTM, the ACI, the Highway Research Board, or the American Association of State Highway Officials, (3) reports from governmental scientific organizations such as the National Bureau of Standards, and from various universities, (4) articles in technical and trade journals including *Engineering News-Record, Concrete,* and *Rock Products,* and (5) bulletins and reports issued by trade associations such as the Portland Cement Association, National Sand and Gravel Association, and the National Crushed Stone Association.

American society for testing materials. Much information on concrete is made available through the publications of the ASTM. This society is composed of three classes of members: (1) those producing materials, (2) those using materials, and (3) the general-interest group, including members of engineering faculties and of scientific bureaus of the government. Technical papers on materials are published in the *Proceedings of the American Society for Testing Materials* and in *Materials Research and Standards.* Committees of the society develop specifications for materials and the testing of materials, which are submitted to the membership for approval.

American concrete institute. The ACI is, in this country, the most important agency devoting all its energies to the dissemination of knowledge concerning concrete and concrete materials. Its membership may be grouped into the same three classes (producers, consumers, and general-interest groups) as for the ASTM. The *Journal of the American Concrete Institute* is published monthly; it presents many worthwhile technical papers. The annual *Proceedings of the American Concrete Institute* include these same papers with all discussions which have been presented.

Portland cement association. The Portland Cement Association (PCA) is a nonprofit trade association in which most of the cement companies in the United States have membership. Its functions are fourfold, including (1) the study and solution of manufacturing problems which the cement manufacturers present to it, (2) the promotion of concrete construction, (3) researches in concrete and concrete materials to learn their characteristics and to improve the quality of concrete structures, and (4) the dissemination of knowledge concerning concrete and concrete materials by means of public lectures, the free distribution of pamphlets, and by papers presented in the proceedings of the technical societies.

SECTION 100 GENERAL

100. *Design and Control of Concrete Mixtures,* 10th ed., Portland Cement Association, Chicago, 1952, 68 pp. An excellent brief consideration of the problem of making quality concrete. Contains reprints of several of the most used ASTM Specifications for aggregates and concrete.

101. McMillan, F. R.: *Basic Principles of Concrete Making,* McGraw-Hill, New York, 1929, 99 pp. A well-planned and very understandable book covering what the title indicates. Summarizes in concise form a wealth of data and develops a usable "philosophy" of concrete mixtures.

102. Bauer, E. E.: *Plain Concrete,* 3d ed., McGraw-Hill, New York, 1949, 441 pp. A textbook on concrete containing a fairly detailed discussion of the selection of materials and the production of concrete.

103. Neville, A. M.: *Properties of Concrete,* Wiley, New York, 1963, 532 pp.

104. Taylor, W. H.: *Concrete Technology and Practice,* Elsevier, New York, 1965, 639 pp.

105. Blanks, R. F., and H. L. Kennedy: *The Technology of Concrete Construction,* Wiley, New York, 1955.

106. *Concrete Manual,* 7th ed., U.S. Bureau of Reclamation, Denver, Colo., 1963, 642 pp. A handbook of information, advice, and instruction relating to the control of concrete construction.

107. *Handbook for Concrete and Cement,* U.S. Army Corps of Engineers, Waterways Experiment Station, Vicksburg, Miss., 1949 (with later loose-leaf revisions), 920 pp. A handbook of test and inspection procedures for concrete and portland cement.

108. "Progress with Concrete, 1923–1948," *Proc. ACI,* vol. 44 (1948), pp. 693–741. A symposium. See also pp. 345–348.

109. *Control of Quality of Ready Mixed Concrete,* National Ready-mixed Concrete Association, Munsey Building, Washington, 1945, 75 pp.

109a. "Cement and Concrete Technology," *ACI Comm. 216,* ACI SP-19, 1967, 146 pp.

110. *Significance of Tests and Properties of Concrete and Concrete-making Materials,"* *ASTM Spec. Tech. Pub.* 169*A* (1966), 571 pp. An important summary by many authorities.

111. *Symposium on Mix Design and Quality Control of Concrete,* Cement and Concrete Association, London, 1955, 548 pp.

112. Van Vlack, L. H.: *Elements of Material Science,* Addison-Wesley, Reading, Mass., 1959, 528 pp.

113. Witt, J. C.: *Portland Cement Technology,* 2d ed., Chemical Publishing, New York, 1966, 346 pp.

114. *Review of Lectures and Discussions—Ohio River Concrete School,* Pittsburgh Engineer District, U.S. Army Corps of Engineers, Pittsburgh, Pa., January, 1935, 158 pp.

115. *ASTM Standards,* pts. 9 and 10 (Cement, Concrete, etc.), American Society for Testing Materials, Philadelphia, 1966, or later supplements or editions. NOTE. A numerical listing of current ASTM Specifications pertaining to concrete is given at the end of this bibliography. References in the text which carry a letter prefix designate particular ASTM Specifications which may be found in the book of Standards.

SECTION 200 CEMENT

General references

201. Bogue, R. H.: *The Chemistry of Portland Cement,* 2d ed., Reinhold, New York, 1955, 793 pp.

202. Czernin, W.: *Cement Chemistry and Physics for Civil Engineers,* Chemical Publishing, New York, 1962, 139 pp.

203. Lea, F. M.: *The Chemistry of Cement and Concrete,* 2d ed., St. Martin's, New York, 1956, 637 pp.

204. Taylor, H. F. W.: *The Chemistry of Cements,* vol. 1, Academic, New York, 1964, 460 pp.

205. Neville, A. M.: *Properties of Concrete,* Wiley, New York, 1963, 532 pp. Includes bilbliographies.

206. Witt, J. C.: *Portland Cement Technology,* Chemical Publishing, New York, 1947.

207. Davis, A. C.: *Portland Cements,* 2d ed., Concrete Publications, London, 1943.

Historical information

210. Davis, A. C.: *A Hundred Years of Portland Cement, 1824–1924,* Concrete Publications, London, 1924.

211. Hadley, E. J.: *The Magic Powder,* Putnam, New York, 1945.

212. Lesley, R. W.: *History of the Portland Cement Industry in the United States,* International Trade Press, Chicago, 1924.

213. Redgrave, G. R., and C. Spackman: *Calcareous Cements,* 3d ed., Griffin, London, 1924.

214. Steinour, H. H.: "Who Invented Portland Cement?" *J. Portland Cement Assoc. Res. Develop. Labs.,* vol. 2, no. 2 (May, 1960), pp. 4–10.

215. Clausen, C. F.: "The Evolution of the Cement Kiln—A Historical Sketch," *J. Portland Cement Assoc. Res. Develop. Labs.,* vol. 4, no. 1 (January, 1962), pp. 33–45.

216. Bogue, R. H., and H. H. Steinour: "Origin of the Special Chemical Symbols Used by Cement Chemists," *J. Portland Cement Assoc. Res. Develop. Labs.,* vol. 3, no. 3 (September, 1961), pp. 20–21.

217. Gonnerman, H. F.: "Development of Cement Performance Tests and Requirements," *Res. Develop. Lab. Portland Cement Assoc. Res. Dep. Bull.* 93 (March, 1958).

218. Savage, J. L.: *Special Cements for Mass Concrete,* U.S. Bureau of Reclamation, Denver, Colo., 1936, 230 pp. Prepared for consideration of the Second Congress of the International Commission on Large Dams. An excellent summary of the evolution of special cements in the United States up to the mid 1930s with particular reference to mass-concrete construction. A comprehensive summary of several important laboratory investigations on cements for mass concrete and a summary of field experience in special problems on the construction of concrete dams. Good annotated bibliography.

Important collections and summaries

220. *Setting and Hardening of Plasters,* First International Symposium on Chemistry of Cements, Faraday Society, London, 1919.

221. *Proc. Intern. Symp. Chem. Cements, 2nd, Stockholm, 1938,* Ingeniörsvetenskaps Academian, Stockholm, 1939.

222. *Proc. Intern. Symp. Chem. Cements, 3rd, London, 1952,* Cement and Concrete Association, London, 1954.

223. *Proc. Intern. Symp. Chem. Cements, 4th, Washington, 1960,* 2 vols., Natl. Bur. Standards Monograph 43, 1962, 576 pp. and 564 pp.

224. Taylor, H. F. W.: *The Chemistry of Cements,* Lecture Series No. 2, 1966, Royal Institute of Chemistry, London, 1966, 27 pp.

225. Brunauer, S.: "Tobermorite Gel—The Heart of Concrete," *Am. Scientist,* vol. 50, no. 1 (March, 1962), pp. 210–229.

226. Brunauer, S., and L. E. Copeland: "The Chemistry of Concrete," *Sci. Am.,* vol. 210, no. 4 (April, 1964), pp. 81–92.

Papers and monographs

230. Verbeck, G.: "Pore Structure," in "Concrete and Concrete-making Materials," *ASTM, Spec. Tech. Pub.* 169-*A,* 1966, pp. 211–219. Also *Res. Develop. Lab. Portland Cement Assoc. Res. Dep. Bull.* 197.

231. Powers, T. C.: "The Nature of Concrete," in "Concrete and Concrete-making Materials," *ASTM Spec. Tech. Pub.* 169-*A,* 1966, pp. 61–72. Also *Res. Develop. Lab. Portland Cement Assoc. Res. Dep. Bull.* 196.

232. Carlson, E. T.: "Some Properties of the Calcium Aluminoferrite Hydrates," Nat. Bur. Standards, *Building Science, Series 6,* June, 1966.

233. Verbeck, G.: "Cement Hydration Reactions at Early Ages," *J. Portland Cement Res. Develop. Labs.,* vol. 7, no. 3 (September, 1965), pp. 57–63. Also *Res. Develop. Lab. Portland Cement Assoc. Res. Dep. Bull.* 189.

234. Seligmann, P., and H. R. Greening: "Studies of Early Hydration Reactions of Portland Cement by X-ray Diffraction," *Highway Research Record* 62 (1964), pp. 80–105. Also *Res. Develop. Lab. Portland Cement Assoc. Res. Dep. Bull.* 185.

235. Kantro, D. L., L. E. Copeland, C. H. Weise, and S. Brunauer: "Quantitative Determination of the Major Phases in Portland Cements by X-ray Diffraction Methods," *J. Portland Cement Assoc. Res. Develop. Labs.,* vol. 6, no. 1 (January, 1964), pp. 20–40. Also *Res. Develop. Lab. Portland Cement Assoc. Res. Dep. Bull.* 166 and *Bull.* 108, 113.

236. Powers, T. C.: "Physical Properties of Cement Paste," in Ref. 223, pp. 577–609. Also *Res. Develop. Lab. Portland Cement Assoc. Res. Dep. Bull.* 154 (1963).

236a. Mikhail, R. S., L. E. Copeland and S. Brunauer: "Pore Structures and Surface Areas of Hardened Cement Pastes by Nitrogen Adsorption," *Res. Develop. Lab. Portland Cement Assoc. Res. Dep. Bull.* 167 (1964).

237. Copeland, L. E., D. L. Kantro, and G. Verbeck: "Chemistry of Hydration of Portland Cement," in Ref. 223, pp. 429–465. Also *Res. Develop. Lab. Portland Cement Assoc. Res. Dep. Bull.* 153 (1963).

238. Brunauer, S., and S. A. Greenberg: "The Hydration of Tricalcium Silicate and β-Dicalcium Silicate at Room Temperature," in Ref. 223, pp. 135–165. Also *Res. Develop. Lab. Portland Cement Assoc. Res. Dep. Bull.* 152 (1963).

239. Kantro, D. L., S. Brunauer, and C. H. Weise: "Development of Surface in the Hydration of Calcium Silicates," *Advan. in Chemistry Ser.* 33, American Chemical Society, 1961, pp. 199–219. Also *Res. Develop. Lab. Portland Cement Assoc. Res. Dep. Bull.* 140 (see also *Bull.* 151).

240. Steinour, H. H.: "Progress in the Chemistry of Cement, 1887–1960," *J. Portland Cement Assoc. Res. Develop. Labs.*, vol. 3, no. 2 (May, 1961), pp. 2–11. Also *Res. Develop. Lab. Portland Cement Assoc. Res. Dep. Bull.* 130.

241. Kantro, D. L., L. E. Copeland, and E. R. Anderson: "An X-ray Diffraction Investigation of Hydrated Portland Cement Pastes," *Proc. ASTM*, vol. 60 (1960), pp. 1020–1035. Also *Res. Develop. Lab. Portland Cement Assoc. Res. Dep. Bull.* 128.

242. Powers, T. C.: "Some Physical Aspects of the Hydration of Portland Cement," *J. Portland Cement Assoc. Res. Develop. Labs.*, vol. 3, no. 1 (January, 1961), pp. 47–56. Also *Res. Develop. Lab. Portland Cement Assoc. Res. Dep. Bull.* 125.

243. Brunauer, S., D. L. Kantro, and C. H. Weise: "The Surface Energy of Tobermorite," *Can. J. Chem.*, vol. 37 (1959), pp. 714–724. Also *Res. Develop. Lab. Portland Cement Assoc. Res. Dep. Bull.* 105.

244. Steinour, H. H.: "The Setting of Portland Cement," *Res. Develop. Lab. Portland Cement Assoc. Res. Dep. Bull.* 98 (November, 1958), 124 pp.

245. Powers, T. C.: "Structure and Physical Properties of Hardened Portland Cement Paste," *J. Am. Ceram. Soc.*, vol. 41, no. 1 (January, 1958), pp. 1–6. Also *Res. Develop. Lab. Portland Cement Assoc. Res. Dep. Bull.* 94 (see also *Bull.* 90).

246. Hansen, W. C.: "The Properties of Gypsum and the Role of Calcium Sulfate in Portland Cement," *ASTM Bull.* 212 (February, 1956), p. 66. See also Ref. 222, p. 318; *ASTM Spec. Tech. Pub.* 266 (1959), p. 3; and *Proc. ASTM*, vol. 61 (1961), pp. 1029 and 1038.

247. Lerch, W.: "The Influence of Gypsum on the Hydration and Properties of Portland Cement Pastes," *Proc. ASTM*, vol. 46 (1946), pp. 1252–1292. Also *Research and Develop. Lab. Portland Cement Assoc. Res. Dep. Bull.* 12.

248. Powers, T. C., and T. L. Brownyard: "Studies of the Physical Properties of Hardened Portland Cement Paste," *Proc. ACI*, vol. 43 (1947). Also *Research and Develop. Lab. Portland Cement Assoc. Res. Dep. Bull.* 22.

249. Meissner, H. S.: "The Optimum Gypsum Content of Portland Cement," *ASTM Bull.* no. 169 (October, 1950), p. 39.

250. "Significance of Selected ASTM Tests for Cement," *Proc. ASTM*, vol. 61 (1961), pp. 1029–1058.

251. Pressler, E. E., S. Brunauer, D. L. Kantro, and C. H. Weise: "Determination of Free Calcium Hydroxide Contents of Hydrated Portland Cements and Calcium Silicates," *Anal. Chem.*, vol. 33, no. 7 (June, 1961), pp. 877–882. Also *Res. Develop. Lab. Portland Cement Assoc. Res. Dep. Bull.* 127.

252. Helmuth, R. A.: "Capillary Size Restrictions on Ice Formation in Hardened Portland Cement Pastes," in Ref. 223, pp. 855–869. Also *Res. Develop. Lab. Portland Cement Assoc. Res. Dep. Bull.* 156 (April, 1963).

253. Powers, T. C., L. E. Copeland, and H. M. Mann: "Capillary Continuity or Discontinuity in Cement Pastes," *J. Portland Cement Assoc. Res. Develop. Labs.*, vol. 1, no. 2 (May, 1959). Also *Res. Develop. Lab. Portland Cement Assoc. Res. Dep. Bull.* 110.

254. Powers, T. C., H. M. Mann, and L. E. Copeland: "The Flow of Water in Hardened Portland Cement Paste," *Highway Research Board Special Report 40* (July, 1959), pp. 308–323. Also *Res. Develop. Lab. Portland Cement Assoc. Res. Dep. Bull.* 106.

255. Powers, T. C.: "A Hypothesis on Carbonation Shrinkage," *J. Portland Cement Assoc. Res. Develop. Labs.*, vol. 4, no. 2 (May, 1962), pp. 40–50. Also *Res. Develop. Lab. Portland Cement Assoc. Res. Dep. Bull.* 146 (see also *Bull.* 187).

256. Klein, A., T. Karby, and M. Polivka: "Properties of an Expansive Cement for Prestressing," *Proc. ACI*, vol. 58 (1961), pp. 59–82.

257. Klein, A., and G. E. Troxell: "Studies of Calcium Sulfoaluminate Admixtures for Expansive Cements," *Proc. ASTM*, Vol. 58 (1958), pp. 986–1008.

257a. Lin, T. Y., and A. Klein: "Chemical Prestressing of Concrete Elements Using Expanding Cements," *J. ACI* (September, 1963), pp. 1187–1218.

258. Monfore, G. E.: "Properties of Expansive Cement Made with Portland Cement, Gypsum, and Calcium Aluminate Cement," *J. Portland Cement Assoc. Res. Develop. Labs.*, vol. 6, no. 2 (May, 1964), pp. 2–9. Also *Res. Develop. Lab. Portland Cement Assoc. Res. Dep. Bull* 170.

259. Gustaferro, A. H., H. Greening, and P. Klieger: "Expansive Concrete–Laboratory Tests of Freeze-Thaw and Surface Scaling Resistance," *J. Portland Cement Assoc. Res. Develop. Labs.*, vol. 8, no. 1 (January, 1966), pp. 10–36. Also *Res. Develop. Lab. Portland Cement Assoc. Res. Dep. Bull.* 190.

260. Lossier, H., and A. Caquot: "Expanding Cements and Their Application—Self-stressed Concrete," (in French), *Génie Civil*, vol. 121 (1944), pp. 61–65, 69–71. Abstr. *Proc. ACI*, vol. 41 (1945), p. 238.

261. Lafuma, H.: "Expansive Cements," in Ref. 222, pp. 581–592.

262. Copeland, L. E., E. Boder, T. H. Chang, and C. H. Weise: "Reactions of Tobermorite Gel with Aluminates, Ferrites and Sulfates," *J. Portland Cement Assoc. Res. Develop. Labs.*, vol. 9, no. 1 (January, 1967), pp. 61–74.

263. Carlson, R. W.: "Development of Low-heat Cement for Mass Concrete," *Eng. News-Record*, vol. 109 (1932), pp. 461–463. Contains a good summary of the characteristics of the major compounds in cement.

264. Davis, R. E., R. W. Carlson, J. W. Kelly, and G. E. Troxell: "Properties of Mortars and Concretes Containing High-silica Cements," *Proc. ACI*, vol. 30 (1934), pp. 369–389; see vol. 32 (1936), pp. 80–114 for later report on same investigation. Bibliography.

265. Kalousek, G. L., and C. H. Jumper: "Some Properties of Portland Pozzolana Cements," *Proc. ACI*, vol. 40 (1944), pp. 145–164.

266. States, M. N.: "Specific Surface and Particle Size Distribution of Finely Divided Materials," *Proc. ASTM*, vol. 39 (1939), pp. 795–807.

267. Travis, P. M.: "Measurement of Average Particle Size by Sedimentation and Other Physical Means," *ASTM Bull.* 102 (January, 1940), p. 29.

268. Wagner, L. A.: "A Rapid Method for the Determination of the Specific Surface of Portland Cement," *Proc. ASTM*, vol. 33, pt. II (1933), pp. 553–570.

269. Klein, A.: "A Suspension Turbidimeter for Determination of Specific Surface of Granular Materials," *Proc. ASTM*, vol. 34, pt. II (1934), pp. 303–314.

270. Biddle, S. B., and A. Klein: "A Hydrometer Method for Determining the Fineness of Portland Puzzolan Cements," *Proc. ASTM*, vol. 36, pt. II (1936), pp. 310–324.

271. Klein, A.: "An Improved Hydrometer for Use in Fineness Determinations," *Proc. ASTM*, vol. 41 (1941), pp. 953–966.

272. Roller, P. S.: "Fineness and Particle Size Distribution of Portland Cement," *Proc. ASTM,* vol. 32, pt. II (1932), pp. 607–625.

273. Blaine, R. L.: "Studies of the Measurement of Specific Surface by Air Permeability," *ASTM Bull.* 108 (January, 1941), p. 17. See also *ASTM Bull.* 118 (October, 1942), p. 31, and 123 (August, 1943), p. 51 (bibliography).

274. "Proposed Method of Test for Fineness of Portland Cement by Blaine Air-permeability Apparatus," report of ASTM Committee C1, *Proc. ASTM,* vol. 45 (1945), pp. 195–199.

275. Young, Roy N.: "The Autoclave Test and Interpretations," *Proc. ACI,* vol. 34 (1938), pp. 13–24.

276. McCoy, W. J.: "Significance of Tests for Heat of Hydration of Cement," *Proc. ASTM,* vol. 63 (1963), pp. 861–865.

277. Kester, K. E.: "Significance of Tests for Fineness," *Proc. ASTM,* vol. 63 (1963), pp. 866–879.

278. Gilliland, J. L.: "Significance of False Set Tests," *Proc. ASTM,* vol. 63 (1963), pp. 880–885.

279. Sawyer, J. L.: "Control of False Set by the Use of Anhydrite and Gypsum Blends," *Proc. ASTM,* vol. 63 (1963), pp. 918–931.

280. Berger, E. E.: "Significant Time of Set Studies and Their Possible Relation to Current Time of Set Problems," *Proc. ASTM,* vol. 63 (1963), pp. 886–898.

281. Mather, B.: "Laboratory Tests of Portland Blast-furnace Slag Cements," *J. ACI* (September, 1957), pp. 205–232.

290. *ASTM Standards,* pt. 9 (Cement, Lime, Gypsum), American Society for Testing Materials, Philadelphia, published annually. Contains standard and tentative specifications for cements and methods of sampling and testing cement. Especially of value is the appended "Manual of Cement Testing," the purpose of which is to emphasize factors which may affect results of tests and to call attention to less apparent influences which are important in cement testing. It also contains a bilbiography on cement testing.

SECTION 300 AGGREGATES

300. Griffith, J. H.: "Physical Properties of Typical American Rocks," *Iowa State Coll. Eng. Expt. Sta. Bull.* 131 (March, 1937), 56 pp. Summary of tests to determine hardness, thermal expansion, strength, specific gravity, absorption, and porosity of a number of different rocks. References.

301. "Symposium on Mineral Aggregates," *Proc. ASTM,* vol. 29, pt. II (1929), pp. 740–901. A group of twelve papers on the properties, testing, and inspection of aggregates. Also see *ASTM Spec. Tech. Pub.* 83 (1948), 233 pp.

302. Walker, S.: *Effect of Grading of Gravel and Sand on Voids and Weights,* National Sand and Gravel Association Circular 8, Washington, 1930.

303. Levison, A. A.: "The Bulking of Moist Sands," *Public Roads,* vol. 5, no. 5 (July, 1924), pp. 21–23.

304. Johnson, W. R.: "Comparison of Methods for Determining Moisture in Sands," *Proc. ACI,* vol. 25 (1929), pp. 261–279.

305. Graf, S. H., and R. H. Johnson: "Study of Methods for Determining Moisture in Sand," *Proc. ASTM,* vol. 30, pt. I (1930), pp. 578–590.

306. Woolf, D. O.: "The Cone Method for Determining the Absorption by Sand," *Proc. ASTM,* vol. 36, pt. II (1936), pp. 411–422.

307. "Abrams-Harder Field Test for Organic Impurities in Sands," report of ASTM Committee C9, *Proc. ASTM,* vol. 19, pt. I (1919), pp. 321–324.

308. Paul, Ira: "New Laboratory Method for Determining the Organic Matter in Washed Fine Aggregates," *Proc. ASTM,* vol. 39 (1939), pp. 892–898.

309. "Selection and Use of Aggregates for Concrete," report of ACI Committee 621, *J. ACI* (1961), pp. 513–542.

310. Swenson, E. G., and V. Chaly: "Basis for Classifying Deleterious Characteristics of Concrete Aggregate Materials," *J. ACI* (1956), pp. 987–1002.

311. Woolf, D. O., and D. G. Runner: "The Los Angeles Abrasion Machine for Determining the Quality of Coarse Aggregate," *Proc. ASTM,* vol. 35, pt. II (1935), pp. 511–529.

312. Jackson, F. H.: "The Correlation of the Los Angeles Rattler Test with Service Behavior," *The Crushed Stone J.,* vol. 13, no. 2 (March, April, 1938), pp. 13–17.

313. Woolf, D. O.: "The Relation between Los Angeles Abrasion Test Results and the Service Records of Coarse Aggregates," *Highway Research Board Proc.,* vol. 17 (1937); also National Sand and Gravel Association Circular 18, Washington, May, 1938, 10 pp.

314. Woolf, D. O.: "Methods for the Determination of Soft Pieces in Aggregate," *Proc. ASTM,* vol. 47 (1947), pp. 967–985.

319. Price, W. H., and D. G. Kretsinger: "Aggregates Tested by Accelerated Freezing and Thawing of Concrete," *Proc. ASTM,* vol. 51 (1951), pp. 1108–1119.

320. Lang, F. C., and C. A. Hughes: "Accelerated Freezing and Thawing as a Quality Test for Concrete Aggregate," *Proc. ASTM,* vol. 31, pt. II (1931), pp. 435–452.

321. Wray, F. N., and H. J. Lichtefeld: "The Influence of Test Methods on Moisture Absorption and Resistance of Coarse Aggregate to Freezing and Thawing," *Proc. ASTM,* vol. 40 (1940), pp. 1007–1020.

322. Wuerpel, C. E.: "Modified Procedure for Testing Concrete Aggregate Soundness by Use of Magnesium Sulfate," *Proc. ASTM,* vol. 39 (1939), pp. 882–889.

323. Bean, L., and J. J. Tregoning: "Reactivity of Aggregate Constituents in Alkaline Solutions," *Proc. ACI,* vol. 41 (1945), pp. 37–52.

325. Walker, S., and C. E. Proudley: "Studies of Sodium and Magnesium Sulfate Soundness Tests," *Proc. ASTM,* vol. 36, pt. I (1936), pp. 327–335.

327. McCown, V.: "Significance of Sodium Sulfate and Freezing and Thawing Tests of Mineral Aggregates," *Highway Research Board, Proc.,* vol. 11 (1931), pp. 312–337.

330. Weymouth, C. A. G.: "Effects of Particle Interference in Mortar and Concrete," *Rock Products* (Feb. 25, 1933), pp. 26–30.

331. Weymouth, C. A. G.: "A Study of Fine Aggregate in Freshly Mixed Mortars and Concretes," *Proc. ASTM,* vol. 38, pt. II (1938), pp. 354–373.

332. Connor, C. C.: "Some Effects of the Grading of Sand on Masonry Mortar," *Proc. ASTM,* vol. 53 (1953), pp. 933–948.

333. Buslik, D.: "Mixing and Sampling with Special Reference to Multi-sized Granular Material," *ASTM Bull.* 165 (April, 1950), pp. 66–73. Probability considerations applied to sampling.

335. Moore, W.: "Prospecting for Gravel Deposits by Resistivity Methods Described," *Pit and Quarry,* vol. 38, no. 3 (September, 1945), pp. 77–80.

336. Parsons, W. H., and W. H. Johnson: "Factors Affecting the Thermal Expansion of Concrete Aggregate Materials," *Proc. ACI,* vol. 40 (1944), pp. 457–468.

337. Williams, G. L.: "The Effect of Belt Transportation on Concrete Aggregate Grading," *Proc. ACI,* vol. 38 (1942), pp. 329–332.

340. Tuthill, L. H.: "Developments in Methods of Testing and Specifying Coarse Aggregates," *Proc. ACI,* vol. 39 (1943), pp. 31–32.

341. Hubbard, F., and H. T. Williams: "Strength of Concrete as Related to Abrasion of the Blast-furnace Slag Used as Coarse Aggregate," *Proc. ASTM,* vol. 43 (1943), pp. 1088–1094.

342. Wuerpel, C. E.: "Aggregates for Concrete," *Concrete,* vol. 52, no. 9 (September, 1944), p. 14.

345. Collins, A. R.: "Use of All-in Aggregates in Concrete," *Concrete and Constr. Eng.,* vol. 40, no. 1 (January, 1945), pp. 15–19.

350. Litehiser, R. R.: *The Effect of Deleterious Materials in Aggregate for Concrete,* National Sand and Gravel Association Circular 16, Washington, 1938, 8 pp.

355. Markwick, A. H. D.: "Aggregate Crushing Test for Evaluating Mechanical Strength of Coarse Aggregates," *J. Inst. Civil Engrs.,* vol. 24, no. 6 (April, 1945), pp. 125–133.

360. Berkey, C. P.: "The Nature of the Processes Leading to the Disintegration of Concrete, with Special Reference to Excess Alkalies," *Proc. ACI,* vol. 37 (1941), pp. 689–692.

361. Hansen, W. C.: "Studies Relating to the Mechanism by Which the Alkali-aggregate Reaction Produces Expansion in Concrete," *Proc. ACI,* vol. 40 (1944), pp. 213–228.

362. Meissner, H. S.: "Cracking in Concrete Due to Expansive Reaction between Aggregate and High-alkali Cement as Evidenced in Parker Dam," *Proc. ACI,* vol. 37 (1941), pp. 549–568.

363. Stanton, T. E., et al.: "California Experience with the Expansion of Concrete through Reaction between Cement and Aggregate," *Proc. ACI,* vol. 38 (1942) pp. 209–236. See also paper by Stanton and discussions in *Trans. ASCE,* vol. 107 (1942), pp. 54–126.

364. Mielenz, R. C., K. T. Greene, and E. J. Brenton: "Chemical Test for the Reactivity of Aggregates with Cement Alkalies; Chemical Processes in Cement Aggregate Reaction," *Proc. ACI,* vol. 44 (1948), pp. 193–221 (see ASTM C289).

365. "Methods and Procedures Used in Identifying Reactive Materials in Concrete," *Proc. ASTM,* vol. 38 (1948), pp. 1055–1127. A symposium.

370. Walker, S.: *Effect of Characteristics of Coarse Aggregates on the Quality of Concrete,* National Sand and Gravel Association Bulletin 5, Washington, 1930.

380. Walker, S.: "Production of Sand and Gravel," *Proc. ACI,* vol. 51 (1955), pp. 165–180.

390. *Bibliography on Mineral Aggregates,* Highway Research Board Bibliography 6, Washington, 1949.

See also ASTM Specifications C29, C30, C33, C40, C70, C87, C88, C117, C125, C127, C128, C131, C136, C142, C144, C227, C235, C289, C294, C295, C330–C332, C342, C535, C566, C586, D2419, E12, E13.

See also Refs. 100–107, 110, 1132, 1135, 1155, 1159–1171, 1200, 1201, 1240, 1303, 1460–1479.

SECTION 400 WATER AND ADMIXTURES

Water (mixing, washing, curing)

400. Abrams, D. A.: "Effect of Impure Waters When Used in Mixing Concrete," *Proc. ACI,* vol. 20 (1924), pp. 442–486.

400a. Steinour, Harold H.: "Concrete Mix Water—How Impure Can It Be," *J. Portland Cement Assoc. Res. Develop. Labs.,* vol. 2, no. 3 (September, 1960), pp. 32–50.

401. "Sea Water for Mixing Concrete," *Proc. ACI,* vol. 36 (1940), pp. 313–314.

402. "Water for Making Concrete," *Proc. ACI,* vol. 44 (1948), pp. 414–416.

403. "Requirements for Water for Use in Mixing and/or Curing Concrete," in *Handbook* *for Concrete and Cement,* serial CRD-C400-48, U.S. Army Corps of Engineers, Waterways Experiment Station, Vicksburg, Miss., 1949.

404. Dempsey, J. G.: "Coral and Salt Water as Concrete Materials," *Proc. ACI,* vol. 48 (1952), pp. 157–168.

See also Refs. 106, 1188, 1189, 1504; ASTM C87; AASHO T26.

Admixtures, general

405. "Admixtures for Concrete," report of ACI Committee 212, *J. ACI* (November, 1963), pp. 1481–1524. Bibliography.

406. "Admixtures in Concrete," *Proc. ACI,* vol. 47 (1951), pp. 25–52. A symposium.

410. McMillan, F. R., and T. C. Powers: "A Method of Evaluating Admixtures," *Proc. ACI,* vol. 30 (1934), pp. 325–344.

412. Williams, G. M.: "Admixtures and Workability of Concrete," *Proc. ACI,* vol. 27 (1931), pp. 647–653.

413. Smith, G. A.: "Effect of Celite on the Modulus of Elasticity of Concrete," *Proc. ACI,* vol. 28 (1932), pp. 613–626. See also vol. 26 (1930), pp. 184–201.

414. Voss, W. C.: "Lime in Concrete," *Rock Products* (June, 1937), pp. 39–40.

See also ASTM C494.

Air-entraining agents

420. Gonnerman, H. F.: "Tests of Concretes Containing Air-entraining Portland Cements or Air-entraining Materials Added to Batch at Mixer," *Proc. ACI,* vol. 40 (1944), pp. 477–508. Bibliography also published as *Research and Develop. Labs. Portland Cement Assoc. Res. Dep. Bull.* 13, Chicago.

421. Wuerpel, C. E.: "Laboratory Studies of Concrete Containing Air-entraining Admixtures," *Proc. ACI,* vol. 42 (1946), pp. 305–359. Extensive bibliography. See also *U.S. Army Corps of Engineers Bulletin* 30, Waterways Experiment Station, Vicksburg, Miss., 1947. Also *Civil Eng.,* vol. 16, no. 11 (November, 1946), pp. 496–498.

422. Wuerpel, C. E., and H. Weiner: "The Reaction of Vinsol Resin as It Affects the Air-entrainment of Portland Cement Concrete," *ASTM Bull.* 130 (October, 1944), pp. 41–42.

423. Axon, E. O., et al.: "Effect of Air-entrapping Portland Cement on the Resistance to Freezing and Thawing of Concrete Containing Inferior Coarse Aggregate," *Proc. ASTM,* vol. 43 (1943), pp. 981–994.

424. Bean, L. and A. Litvin: "Effect of Heat on Portland Cements Containing Vinsol Resin," *Proc. ASTM,* vol. 45 (1945), pp. 766–770.

425. Blaine, R. L., J. C. Yates, and J. R. Dwyer: "The Testing of Portland Cements Containing Interground Vinsol Resin," *Proc. ASTM,* vol. 45 (1945), pp. 732–752.

426. Higginson, E. C.: "Some Effects of Vibration and Handling of Concrete Containing Entrained Air," *Proc. ACI,* vol. 49 (1953), pp. 1–12.

See also Refs. 522–534, 1144–1148; ASTM C175, C226, C233, C260; AASHO M134.

Gas-forming agents

430. Carlson, R. W.: "Powdered Aluminum as an Admixture in Concretes," *Proc. ASTM,* vol. 42 (1942), pp. 808–818.

431. Menzel, C. A.: "Some Factors Influencing the Strength of Concrete Containing Admixtures of Powdered Aluminum," *Proc. ACI,* vol. 39 (1943), pp. 165–184.

Water-reducing retarders

435. Symposium on "Effect of Water-reducing Admixtures and Set-retarding Admixtures on Properties of Concrete," *ASTM Spec. Tech. Pub.* 266 (1966), 246 pp.

436. Grieb, W. E., George Werner, and D. O. Woolf: "Water-reducing Retarders for Concrete–Physical Tests," *Public Roads*, vol. 31, no. 6 (1961), pp. 136–152. Also *Highway Research Board Bull.* 310 (1962), pp. 1–32.

437. Gaynor, R. D.: *Tests of Water-reducing Retarders*, National Sand and Gravel Association Publication 12 (August, 1962), 14 pp.

See also Ref. 405.

Accelerators and retarders

440. Rapp, P.: "Effect of Calcium Chloride on Portland Cements and Concretes," *Highway Research Board, Proc.*, vol. 14 (1934), pp. 341–381.

441. Yates, J. C.: "Effect of Calcium Chloride on the Strength Development of Concrete Stored at Low Temperatures," *Highway Research Board, Proc.*, vol. 21 (1941), p. 288. See also vol. 21, pp. 288–304, and vol. 23 (1943), pp. 296–300.

442. Newman, A. J.: "Effects of the Addition of Calcium Chloride to Portland Cements and Concretes," *Concrete and Constr. Eng.*, vol. 38, no. 5 (1943), pp. 159–167. Discusses shrinkage.

443. Roseberg, A. M.: "Study of the Mechanism through Which Calcium Chloride Accelerates the Set of Portland Cement," *J. ACI* (October, 1964), pp. 1261–1270.

444. Rapp, P.: "Effect of Calcium Chloride on Portland Cements and Concretes," *Natl. Bur. Standards Research Paper* 782 (April, 1935).

445. "The Use of Calcium Chloride as an Admixture in Concretes for High Early Strength and Low Temperature Protection," *Proc. ACI*, vol. 38 (1942), pp. 400–401.

446. Newman, A. J.: "Effects of the Addition of Calcium Chloride to Portland Cements and Concretes," *Concrete and Constr. Eng.*, vol. 38, no. 5 (May, 1943), pp. 159–167.

447. Forbrich, L. R.: "The Effect of Various Reagents on the Heat Liberation Characteristics of Portland Cement," *Proc. ACI*, vol. 37 (1941), pp. 161–184.

448. Newman, E. S., et al.: "Effects of Added Materials on Some Properties of Hydrating Portland Cement Clinkers," *J. Research Natl. Bur. Standards*, vol. 30 (April, 1943), pp. 281–301 (R.P. 1533).

449. Shideler, J. J.: "Calcium Chloride in Concrete," *Proc. ACI*, vol. 48 (1952), pp. 537–560.

See also Refs. 109, pp. 47–48; 405; 872; ASTM D98, D345; AASHO M144.

Pozzolanic materials

450. McMillan, F. R., and T. C. Powers: "Classification of Admixtures as to Pozzolanic Effect by Means of Compressive Strength of Concrete," *Proc. ACI*, vol. 34 (1938), pp. 129–144.

450a. Ruiz, A. L.: "Strength Contribution of a Pozzolan to Concretes," *J. ACI* (March, 1965), pp. 315–326.

450b. Abdun-Nur, E. A.: "Fly Ash in Concrete: An Evaluation," *Highway Research Board Bull.* 284 (1961), 138 pp. A review of the literature from 1934 to 1939.

451. Davis, R. E., et al.: "Properties of Mortars and Concretes Containing Portland Pozzolan Cements," *Proc. ACI*, vol. 32 (1936), pp. 80–114. Also *Proc. ACI*, vol. 46 (1950), pp. 377–384.

452. Kalousek, G. L., and C. H. Jumper: "Some Properties of Portland Pozzolanic Cements," *Proc. ACI*, vol. 40 (1944), pp. 145–163.

453. Davis, R. E., et al.: "Properties of Cements and Concretes Containing Fly Ash," *Proc. ACI*, vol. 33 (1937), pp. 577–612.

454. Frederick, H. A.: "Application of Fly Ash for Lean Concrete Mixtures," *Proc. ASTM*, vol. 44 (1944), pp. 808–818.

455. Stanton, T. E., and L. C. Meder: "Resistance of Cement to Attack by Sea Water and Alkali Soils," *Proc. ACI,* vol. 34 (1938), pp. 433–464.

456. "Symposium on the Use of Pozzolanic Materials in Mortars and Concretes," *ASTM Spec. Tech. Pub.* 99 (August, 1950). See especially paper by R. E. Davis, "Pozzolanic Materials and Their Use in Concretes: A General Review."

457. Blanks, R. F.: "The Use of Portland-Pozzolan Cement by the Bureau of Reclamation," *Proc. ACI,* vol. 46 (1950), pp. 89–108.

458. Blanks, R. F.: "Fly Ash as a Pozzolan," *Proc. ACI,* vol. 46 (1950), pp. 701–707.

459. Washa, G. W., and N. H. Withey: "Strength and Durability of Concrete Containing Fly Ash," *Proc. ACI,* vol. 49 (1953), pp. 701–712.

459a. Scholer, C. H., and G. M. Smith: "Use of Chicago Fly Ash in Reducing Cement-Aggregate Reaction," *Proc. ACI,* vol. 48 (1952), pp. 457–464.

See also Refs. 106, 405; ASTM C311, C350; C402; U.S. Federal Specifications SS-C-208c.

Curing aids

460. Jackson, F. H., and W. F. Kellerman: "Tests of Concrete Curing Materials," *Proc. ACI,* vol. 35 (1939), pp. 481–500.

461. *Curing Concrete Pavements,* Highway Research Board Wartime Road Problems Pamphlet 1 (now designated as *Current Road Problems*), Washington, 1942.

See also Refs. 440, 442, 854–858; ASTM C156, C171, C309, D98; AASHO M73, M139, M144, M148.

Water-repelling agents

470. Dunagin, W. M., and G. C. Ernst: "A Study of the Permeability of a Few Integrally Waterproofed Concretes," *Proc. ASTM,* vol. 34, pt. I (1934), pp. 383–392.

471. Jumper, C. H.: "Tests of Integral and Surface Waterproofings for Concrete," *Proc. ACI,* vol. 28 (1932), pp. 209–242.

473. "Exterior Waterproofers for Masonry," *Natl. Bur. Standards Tech. Information on Building Materials,* no. 5 (1936), 3 pp.

474. Parker, W. E.: "Integral Waterproofing Materials for Concrete," *Proc. ACI,* vol. 44 (1948), pp. 77–79; also additional note by E. H. Logan, p. 329.

475. Anderegg, F. O.: "Testing Surface Waterproofers," *ASTM Bull.* 156 (January, 1949), pp. 71–77.

See also Ref. 405; AASHO M117.

Coloring agents

480. "The Use of Color in Concrete," report of ACI Committee 408, *Proc. ACI,* vol. 27 (1931), pp. 975–999.

481. *Colored Concrete,* Portland Cement Association, Chicago.

See also Ref. 1800, sec. 727.

Paint for concrete surfaces

483. "Painting on Concrete Surfaces," report of ACI Committee 407, *Proc. ACI,* vol. 29 (1933), pp. 1–7.

484. *Painting Concrete,* Portland Cement Association Information Sheet ST1, Chicago, 2 pp.

485. Burnett, G. E., and A. L. Fowler: "Painting Exterior Concrete Surfaces with Special Reference to Pretreatment," *Proc. ACI,* vol. 43 (1947), pp. 1077–1086.

See also Ref. 1800, secs. 729–731.

Surface hardeners

486. "Concrete Floor Treatments," *Natl. Bur. Standards Tech. Information on Building Materials,* no. 9 (1936), 6 pp.

Agents for "nonshrink" or expanding mortars

490. "The Action of Embeco in Concrete and Mortar," *Embeco, Non-shrink Method of Grouting,* Master Builders Co., Cleveland, Ohio. See also other trade literature of this company.

Miscellaneous

493. Hornibrook, F. B.: "The Effectiveness of Various Treatments and Coatings for Concrete in Reducing the Penetration of Kerosene," *Proc. ACI,* vol. 41 (1945), pp. 13–20.

494. "Guide for Use of Epoxy Compounds with Concrete," report of ACI Committee 403, *J. ACI* (September, 1962), pp. 1121–1142.

Cement grinding aids

495. Rockwood, N. C.: "Aids to Clinker Grinding," *Rock Products,* vol. 42, no. 5 (May, 1939), pp. 38–39.

496. Kennedy, H. L.: "Practical Application of Catalysis and Dispersion to Cement and Concrete," *J. Boston Soc. Civil Engrs.,* vol. 24, no. 1 (1937), pp. 28–45.

Steel reinforcement

499. *Code of Standard Practice and Specifications for Placing Reinforcement,* Concrete Reinforcing Steel Institute, Chicago, 1937, 14 pp. (Or later editions.)

See also ASTM A15, A16, A82, A184, A185; AASHO M31, M32, M42, M53, M54, M55, M137.

SECTION 500 PROPERTIES OF FRESH CONCRETE

Workability: consistency

500. Bahrner, V.: "Report on Consistency Tests on Concrete Made by Means of the Vebe Consistometer," *Report No.* 1, Joint Research Group on Vibration of Concrete, *Swed. Cement Concrete Inst. Roy. Inst. Technol. Stockholm Bull.* (March, 1940).

501. Burmister, D. M.: "The Concrete Flow Trough," *Proc. ASTM,* vol. 31, pt. II (1931), pp. 554–569.

502. Davis, H. E., and J. W. Kelly: "Rating the Characteristics of Fresh Concrete," *Proc. ASTM,* vol. 36, pt. II (1936), pp. 372–379. Visual rating of various factors of workability.

503. Glanville, W. H., A. R. Collins, and D. D. Mathews: "The Grading of Aggregates and Workability of Concrete," 2d ed., *Dep. Sci. Ind. Res. Brit. Govt.,* Road Research Technical Paper 5 (1947). Illuminating discussion of nature of workability.

504. Herschel, W. H., and E. A. Pisapia: "Factors for Workability of Portland Cement Concrete," *Proc. ACI,* vol. 32 (1936), pp. 641–658.

505. Kelly, J. W., and Milos Polivka: "Ball Test for Field Control of Concrete Consistency," *Proc. ACI,* vol. 51 (1955), pp. 881–888.

506. Mather, Bryant: "Partially Compacted Weight of Concrete as a Measure of Workability," *Proc. ACI,* vol. 63 (1966), pp. 441–449.

507. Patch, O. G.: "Grand Coulee Consistency Meters," *Proc. ACI,* vol. 35 (1939), pp. 204–206. Load on dumping arm of tilting mixer recorded automatically.

508. Pearson, J. C.: "A Study of Slump and Flow of Concrete," *Proc. ACI,* vol. 27 (1931), pp. 1137–1142. A discussion of papers on this subject.

509. Plowman, J. M.: "Measuring the Workability of Concrete," *Engineer,* vol. 209, no. 5447, pp. 1007ff (1960); no. 5458, pp. 392ff. Use of a sloping channel with vibrator attached.

510. Polatty, J. M.: "New Type of Consistency Meter Tested at Allatoona Dam," *Proc. ACI,* vol. 46 (1950), pp. 129–136. Blade on arm in mixer, with electrical indicator.

511. Powers, T. C.: "Studies of Workability of Concrete," *Proc. ACI,* vol. 28 (1932), pp. 419–448. Includes description of remolding test.

512. Purrington, W. F., and H. C. Loring: "The Determination of the Workability of Concrete," *Proc. ASTM,* vol. 28, pt. II (1928), pp. 499–504. Measurement of power consumption during mixing of concrete.

513. Thaulow, Sven: "Field Testing of Concrete" (in Norwegian; abridged English text), Norwegian Portland Cement Assoc., Oslo (1952). Résumé in *J. ACI* (March, 1954), Newsletter, pp. 10–11, 24–26.

514. Vollick, C. A.: "Uniformity and Workability of Freshly Mixed Concrete," pp. 73–89 in *ASTM Spec. Tech. Pub.* 169-*A* (1966). Significance of tests and properties.

515. "The Wigmore Consistometer," *Roads and Road Construction* (England), vol. 27, no. 315 (March, 1949). Abstrs. in *Highway Research Abstr.* (April, 1949) and *J. ACI* (June, 1949), p. 752.

Bleeding; stiffening; setting

516. Kelly, T. M.: "Setting Time of Concrete," pp. 102–115 in *ASTM Spec. Tech. Pub.* 169-*A* (1966). Significance of tests and properties.

517. Lerch, William: "Plastic Shrinkage," *Proc. ACI,* vol. 53 (1957), pp. 797–802.

518. Powers, T. C.: "The Bleeding of Portland Cement Paste, Mortar, and Concrete," *Proc. ACI,* vol. 35 (1939), pp. 465–480. Reprinted as *Research and Develop. Lab. Portland Cement Assoc. Res. Dep. Bull.* **2** (July, 1939).

519. Steinour, H. H.: "Further Studies of the Bleeding of Portland Cement Paste," *Research and Develop. Lab. Portland Cement Assoc. Res. Dep. Bull.* 4 (December, 1945), 88 pp.

520. Tuthill, L. H., and W. A. Cordon: "Properties and Uses of Initially Retarded Concrete," *Proc. ACI,* vol. 52 (1956), pp. 273–286.

521. Valore, R. C., J. E. Bowling, and R. L. Blaine: "The Direct and Continuous Measurement of Bleeding in Portland Cement-water Mixtures," *Proc. ASTM,* vol. 49 (1949), pp. 891–908.

Air entrainment

522. "Concretes Containing Air-entraining Agents," *Proc. ACI,* vol. 40 (1944), pp. 509–572. A symposium.

523. "Entrained Air in Concrete," *Proc. ACI,* vol. 42 (1946), pp. 601–609. A symposium.

524. "Measurement of Entrained Air in Concrete," *Proc. ASTM,* vol. 47 (1947), pp. 832–913. A symposium.

525. Bartel, F. F.: "Air Content and Unit Weight of Freshly Mixed Concrete," pp. 116–124 in *ASTM Spec. Tech. Pub.* 169-*A* (1966). Significance of tests and properties of concrete and concrete-making materials.

526. Cordon, W. A.: "Entrained Air—A Factor in the Design of Concrete Mixes," *Proc. ACI,* vol. 42 (1946), pp. 605–620.

527. Crawley, Walter O.: "Effect o Vibration on Air Content of Mass Concrete," *Proc. ACI,* vol. 49 (1953), pp. 909–920.
528. Klein, W. H., and Stanton Walker: "A Method for the Direct Measurement of Entrained Air in Concrete," *Proc. ACI,* vol. 42 (1946), pp. 657–668.
529. Klieger, Paul: "Effect of Entrained Air on Concretes Made with So-called 'Sand-Gravel' Aggregates," *Proc. ACI,* vol. 45 (1949), pp. 149–164.
530. Menzel, Carl A.: "Procedures for Determining the Air Content of Freshly Mixed Concrete by the Rolling and Pressure Methods," *Proc. ASTM,* vol. 47 (1947), pp. 833–864.
531. Mielenz, Richard C., Vladimir E. Wolkodoff, James E. Backstrom, and Harry L. Flock: "Entrained Air in Unhardened Concrete," *Proc. ACI,* vol. 55 (1959), pp. 95–121.
532. Portland Cement Association: "Air-entrained Concrete," ST89 (1962), 7 pp.
533. Powers, T. C.: "Void Spacing as a Basis for Producing Air-entrained Concrete," *Proc. ACI,* vol. 50 (1954), pp. 741–760.
534. Walker, Stanton, and D. L. Bloem: *Control of Quantity of Entrained Air in Concrete,* National Ready Mixed Concrete Association, 1950.

Composition; properties

535. Bloem, D. L., R. D. Gaynor, and J. R. Wilson: "Testing Uniformity of Large Batches of Concrete," *Proc. ASTM,* vol. 61 (1961), pp. 1119–1140. Reprinted as *Natl. Ready Mixed Concrete Assoc. Pub.* 100.
536. Dunagan, W. M.: "A Study of the Analysis of Fresh Concrete," *Proc. ASTM,* vol. 31, pt. II (1931), pp. 362–386. See also *Proc. ACI,* vol. 26 (1930), pp. 202–210.
537. Helms, S. B.: "Air Content and Unit Weight of Hardened Concrete," pp. 309–325 in *ASTM Spec. Tech. Pub.* 169-*A* (1966). Significance of tests and properties.
538. Hime, W. G., and R. A. Willis: "A Method for the Determination of the Cement Content of Plastic Concrete," *ASTM Bull.* 209 (October, 1955), pp. 37ff.
539. Lorman, W. R.: "Verifying the Quality of Freshly Mixed Concrete," *Proc. ASTM,* vol. 62 (1962), pp. 944–959.
540. Vollick, C. A.: "Uniformity and Workability of Freshly Mixed Concrete," pp. 73–89 in *ASTM Spec. Tech. Pub.* 169-*A* (1966). Significance of tests and properties.
541. Walker, Stanton: "Ready Mixed Concrete," pp. 340–358 in *ASTM Spec. Tech. Pub.* 169-*A* (1966). Significance of tests and properties.
542. Walker, Stanton, Delmar L. Bloem, Richard D. Gaynor, and John R. Wilson: "A Study of the Centrifuge Test for Determining the Cement Content of Fresh Concrete," *Mater. Res. Std. (ASTM),* (June, 1961). Reprinted as *Natl. Ready Mixed Concrete Assoc. Pub.* 95, 12 pp.

SECTION 600 PROPORTIONING OF CONCRETE MIXES

600. Abrams, Duff A.: "Design of Concrete Mixtures," *Structural Materials. Research Lab. Lewis Inst. Bull.* 1, Chicago, 1918.
601. American Concrete Institute: "Recommended Practice for Selecting Proportions for Concrete (ACI 613-54)," *Proc. ACI,* vol. 51 (1955), pp. 49–64.
602. American Concrete Institute: "Recommended Practice for Selecting Proportions for Structural Lightweight Concrete (ACI 613A-59)," *Proc. ACI,* vol. 55 (1959), pp. 535–565.
603. American Concrete Institute: "Proposed Recommended Practice for Selecting Proportions for No-slump Concrete," *Proc. ACI,* vol. 62 (1965), pp. 1–22, 1125–1138.
604. American Concrete Institute: *ACI Manual of Concrete Inspection,* chaps. 3, 5, 4th ed., 1957, 240 pp.

605. Bloem, Delmar L., and Stanton Walker: *Proportioning Ready Mixed Concrete,* National Sand and Gravel Association Circular 91 and *National Ready Mixed Concrete Assoc. Pub.* 114 (October, 1963), 44 pp.

606. Edwards, L. N.: "Proportioning the Materials of Mortars and Concretes by Surface Area of Aggregates," *Proc. ASTM,* vol. 18, pt. II (1918), pp. 235–283.

607. Fuller, W. B., and S. E. Thompson: "The Laws of Proportioning Concrete," *Trans. Am. Soc. Civil Engrs.,* vol. 59 (1907), pp. 67–143.

608. Glanville, W. H.: "Grading and Workability," *Proc. ACI,* vol. 33 (1937), pp. 319–326.

609. Goldbeck, A. T., and J. E. Gray: *A Method of Proportioning Concrete for Strength, Workability, and Durability,* National Crushed Stone Association, *Bull.* 11 (1953), 31 pp.

610. Kellerman, W. F.: "Design Concrete Mixtures for Pavements," *Proc. ASTM,* vol. 40 (1940), pp. 1055–1065.

611. Kennedy, Charles T.: "The Design of Concrete Mixes," *Proc. ACI,* vol. 36 (1940), pp. 373–400.

612. Kitts, Joseph A.: "Coordination of Basic Principles of Concrete Mixtures," series of articles in *Concrete,* 1932 to 1934.

613. Morris, Mark: "The Mortar-voids Method of Designing Concrete Mixtures," *Proc. ACI,* vol. 29 (1933), pp. 9–26.

614. Portland Cement Association: *Design and Control of Concrete Mixtures,* 10 eds., 1925 to 1952.

615. Swayze, Myron A., and Ernst Gruenwald: "Concrete Mix Design—A Modification of the Fineness Modulus Method," *Proc. ACI,* vol. 43 (1947), pp. 829–843 and (discussion) pp. 844-1 to 844-17.

616. Talbot, A. N., and F. E. Richart: "The Strength of Concrete—Its Relation to the Cement, Aggregate, and Water," *Univ. Illinois Eng. Exp. Sta. Bull.* 137 (1923), 118 pp.

617. U. S. Bureau of Reclamation: *Concrete Manual,* chap. 3, 7th ed., 1963, 642 pp.

618. Waddell, Joseph J.: *Practical Quality Control for Concrete,* chap. 16, McGraw-Hill, New York, 1962, 396 pp.

619. Walker, Stanton, and Delmar L. Bloem: *Estimating Proportions for Concrete,* National Sand and Gravel Association Circular 68 and *National Ready Mixed Concrete Assoc. Pub.* 69 (June, 1962), 47 pp.

620. Weymouth, C. A. G.: "Designing Workable Concrete," *Eng. News-Record,* vol. 121, no. 26 (Dec. 29, 1938), pp. 818–820.

SECTION 700 MANUFACTURE OF CONCRETE

General

700. "Recommended Practice for Measuring, Mixing, and Placing Concrete," report of ACI Committee 614, *J. ACI* (November, 1958), pp. 535–565.

701. Rippon, C. S.: "Methods of Handling and Placing Concrete at Shasta Dam," *Proc. ACI,* vol. 39 (1943), pp. 1–8.

702. Myers, B.: "The Design and Control of Paving Concrete in Iowa," *Proc. ACI,* vol. 37 (1941), pp. 577–588.

703. Cape, E. B.: "Design and Control of Concrete Paving Mixtures: Texas," *Proc. ACI,* vol. 37 (1941), pp. 413–432.

704. Tyler, I. L.: "Concrete Control on the Pennsylvania Turnpike," *Proc. ACI,* vol. 37 (1941), pp. 361–376.

See also Refs. 100–107, 1504, 1800–1804, ASTM C94, C172.

Mixing

730. Hawkins, M. J.: "Concrete Retempering Studies," *J. ACI* (January, 1962), pp. 63–71.

731. Walter, L.: "Automatic Weighing and Handling of Materials," *Cement, Lime, Gravel,* vol. 30, no. 1 (1962), pp. 3–9; no. 2, pp. 49–54.

732. Van Alstine, C. B.: "Mixing Water Control by Use of a Moisture Meter," *J. ACI* (November, 1955) (*Proc. ACI,* vol. 52), pp. 341–348; pt. 2 (December, 1956), pp. 1209–1213.

733. Bray, L. S., and O. Keifer, Jr.: "Check List for Batch Plant Inspection," *J. ACI* (June, 1964), pp. 625–642.

734. Timms, A. G.: "Performance Tests of Concrete Truck Mixers," *Proc. ASTM,* vol. 57 (1957), pp. 1012–1028.

735. Cook, G. C.: "Effect of Time of Haul on Strength and Consistency of Ready-mixed Concrete," *Proc. ACI,* vol. 39 (1943), pp. 413–428.

736. Waugh, W. R.: "The Effect of Grinding in the Large Mixers on Aggregate Grading at Hiwassee Dam," *Proc. ACI,* vol. 39 (1943), pp. 9–20.

737. Wing, S. P., V. Jones, and R. E. Kennedy: "Simplified Test for Evaluating the Effectiveness of Concrete Mixers," *Proc. ASTM,* vol. 43 (1943), pp. 1001–1013.

738. Fitzpatrick, F. L., and W. Sarkin: "Effect of Mixing Sequence on the Properties of Concrete," *Proc. ACI,* vol. 46 (1950), pp. 137–140.

See also Refs. 100–107, 700, 1504; ASTM C94.

Conveying

750. Bell, Charles F.: "Concrete by Pump and Pipeline," *Proc. ACI,* vol. 32 (1936), pp. 343–349.

751. Gray, J. E.: "Laboratory Procedure for Comparing Pumpability of Concrete Mixtures," *Proc. ASTM,* vol. 62 (1962), pp. 964–971.

752. "Pumping Concrete Solves 'Problem' Pours," *Construction Equipment and Materials,* vol. 31, no. 5 (May, 1965), pp. 60–66.

753. Problems and Practices: "Pumping Concrete," *J. ACI* (February, 1966), pp. 291–292.

754. Tuthill, L. H.: "Tunnel Lining Methods for Concrete Compared," *Proc. ACI,* vol. 37 (1941), pp. 29–48.

SECTION 800 MANUFACTURE OF CONCRETE

Placing

800. Clemmer, H. F.: "Placing and Finishing Pavement Concrete," *Proc. ACI,* vol. 37 (1941), pp. 657–664.

801. Rippon, C. S.: "Construction Joint Clean-up Method at Shasta Dam," *Proc. ACI,* vol. 40 (1944), pp. 293–304.

804. Davis, R. E., and H. E. Davis: "Bonding of New Concrete to Old at Horizontal Construction Joints," *Proc. ACI,* vol. 30 (1934), pp. 422–436.

See also Refs. 100–107, 700, 1504.

Compacting

840. "Consolidation of Concrete," report of ACI Committee 609, *J. ACI* (April, 1960), pp. 985–1012.

841. Sawyer, D. H., and S. F. Lee: "Effect of Revibration on Properties of Portland Cement Concrete," *Proc. ASTM,* vol. 56 (1956), pp. 1215–1228.

842. Vollick, C. A.: "Effects of Revibrating Concrete," *J. ACI* (March, 1958), pp. 721–732.

843. Tuthill, L. H., and H. E. Davis: "Overvibration and Revibration of Concrete," *Proc. ACI,* vol. 35 (1939), pp. 41–48.

844. Washa, G. W.: "Comparison of the Physical and Mechanical Properties of Hand Rodded and Vibrated Concrete Made with Different Cements," *Proc. ACI,* vol. 36 (1940), pp. 617–648.

845. Powers, T. C.: "Observations on the Use of Vibration in the Field," *Proc. ACI,* vol. 32 (1936), pp. 74–79.

846. "Compacting Concrete by Vibration," *Proc. ACI,* vol. 49 (1953), pp. 885–956. A symposium.

See also Refs. 100–107, 700, 1504.

Curing

850. "Curing Concrete," report of ACI Committee 612, *J. ACI* (August, 1958), pp. 161–172.

851. Gilkey, H. J.: "The Moist Curing of Concrete," *Eng. News-Record,* vol. 119 (Oct. 14, 1937), pp. 630–633.

852. "Curing of Concrete," *Proc. ACI,* vol. 48 (1952), pp. 701–724. A symposium.

854. Blanks, R. F., H. S. Meissner, and L. H. Tuthill: "Curing Concrete with Sealing Compounds," *Proc. ACI,* vol. 42 (1946), pp. 493–512.

855. Proudley, C. E.: "Evaluation of Curing Compounds for Portland Cement Concrete," *Proc. ASTM,* vol. 53 (1953), pp. 1069–1078.

856. Rhodes, C. C.: "Curing Concrete Pavements with Membranes," *Proc. ACI,* vol. 47 (1951), pp. 277–295.

857. Stanton, T. E.: "Concrete Performance in an Arid Climate," *Proc. ACI,* vol. 37 (1941), pp. 141–156.

858. Burnett, G. E., and M. R. Spindler: "Effect of Time of Application of Sealing Compounds on the Quality of Concrete," *Proc. ACI,* vol. 49 (1953), pp. 193–200.

859. Hanson, J. A.: "Optimum Steam Curing Procedures for Structural Lightweight Concrete," *J. ACI* (June, 1965), pp. 661–672.

860. "Low Pressure Steam Curing," report of ACI Committee 517, *J. ACI* (August, 1963), pp. 953–986; and (March, 1964), pp. 1957–1964.

861. Hanson, J. A.: "Optimum Steam-curing Procedure in Precasting Plants," *J. ACI* (January, 1963), pp. 75–100.

862. Higginson, E. C.: "Effect of Steam Curing on the Important Properties of Concrete," *J. ACI* (September, 1961), pp. 281–298.

863. Ludwig, N. C., and S. A. Pence: "Properties of Portland Cement Pastes Cured at Elevated Temperatures and Pressures," *J. ACI* (February, 1956), pp. 673–688.

864. "High Pressure Steam Curing: Modern Practice, and Properties of Autoclaved Products," report of ACI Committee 516, *J. ACI* (August, 1965), pp. 869–908.

865. Nurse, R. W.: "Steam Curing of Concrete," *Mag. Concrete Res.,* no. 2 (June, 1949), pp. 79–88.

866. Saul, A. G. A.: "Principles Underlying the Steam Curing of Concrete at Atmospheric Pressures," *Mag. Concrete Res.,* no. 6 (March, 1951), pp. 127–135.

867. Hansen, W. C.: "Chemical Reactions in High-pressure Steam Curing of Portland Cement Products," *Proc. ACI,* vol. 49 (1953), pp. 841–856.

868. Thorvaldson, T.: "Effect of Chemical Nature of Aggregate on Strength of Steam-cured Portland Cement Mortars," *J. ACI* (March, 1956), pp. 771–780.

869. Klieger, P.: "Effect of Mixing and Curing Temperature on Concrete Strength," *J. ACI* (June, 1958), pp. 1063–1082.

870. Itakura, C.: "Electric Heating of Concrete in Winter Construction," *Proc. ACI,* vol. 48 (1952), pp. 753–767.

871. Shideler, J. J., et al.: "Entrained Air Simplifies Winter Curing," *Proc. ACI,* vol. 47 (1951), pp. 449–460.

872. "Recommended Practice for Cold Weather Concreting," report of ACI Committee 306, *J. ACI* (September, 1965), pp. 1009–1035.

873. Tuthill, L. H., et al.: "Insulation for Protection of New Concrete in Winter," *Proc. ACI,* vol. 48 (1952), pp. 253–272.

874. Scofield, H. H.: "Effect of Freezing on Permeability, Strength and Elasticity of Concretes and Mortars," *Proc. ASTM,* vol. 37, pt. II (1937), pp. 316.

897. Cordon, W. A., and J. D. Thorpe: "Control of Rapid Drying of Fresh Concrete by Evaporation Control," *J. ACI* (August, 1965), pp. 977–985.

898. "Recommended Practice for Hot Weather Concreting," report of ACI Committee 605, *J. ACI* (November, 1958), pp. 525–534.

See also Refs. 100–107, 110, 1504; ASTM C156, C171.

SECTION 900 FORMS

900. Hurd, M. K., and ACI Committee 622: "Formwork for Concrete," *ACI Spec. Pub.* 4 (1963), 339 pp.

900a. "Recommended Practice for Concrete Formwork," *ACI Standard* 347–63, 52 pp. (1963).

901. Fleming, D. E., and W. H. Wolf: "Testing Program for Lateral Pressure of Concrete," *J. ACI* (May, 1963), pp. 567–573.

902. "Pressures on Formwork," report of ACI Committee 622, *J. ACI* (August, 1958), pp. 173–190.

903. Peurifoy, R. L.: "Lateral Pressure of Concrete on Formwork," *Civil Eng.* (December, 1965), pp. 60–62.

904. Schjödt, R.: "Calculation of Pressure of Concrete on Forms," *Proc. ASCE,* vol. 81 (1955), pp. 680-1–680-16.

906. *Forms for Architectural Concrete,* 2d ed., Portland Cement Association, Chicago, 1942, 55 pp.

910. Johnson, W. R.: "The Use of Absorptive Wall Boards for Concrete Forms," *Proc. ACI,* vol. 37 (1941), pp. 621–632.

911. Vidal, E. N., and R. F. Blanks: "Absorptive Form Lining," *Proc. ACI,* vol. 38 (1942), pp. 253–268.

912. Pittman, H. V.: "Form Linings for Concrete Surfaces," *Eng. News-Record,* vol. 135, no. 18 (November, 1945), pp. 584–588.

913. Anonymous: "Let's Look at Form Coatings," *Concrete Construction* (September, 1966), vol. 11, no. 9, pp. 345–348.

See also Ref. 1504.

SECTION 1000 STRENGTH

1001. Bloem, D. L., and R. D. Gaynor: "Effects of Aggregate Properties on Strength of Concrete," *Proc. ACI,* vol. 60 (1963), pp. 1429–1456.

1002. Burmeister, R. A.: "Tests of Paper Molds for Concrete Cylinders," *Proc. ACI,* vol. 47 (1951), pp. 17–24.

1003. Clark, Arthur P.: "Bond of Concrete Reinforcing Bars," *Proc. ACI,* vol. 46 (1950), pp. 161–184.

1004. Davis, Raymond E., Elwood H. Brown, and J. W. Kelly: "Some Factors Influencing the Bond between Concrete and Reinforcing Steel," *Proc. ASTM,* vol. 38, pt. II (1938), pp. 394–409.

1005. Duke, C. Martin, and Harmer E. Davis: "Some Properties of Concrete under Sustained Combined Stresses," *Proc. ASTM,* vol. 44 (1944), pp. 888–896.

1006. Gilkey, H. J.: "Water-Cement Ratio Versus Strength—Another Look," *Proc. ACI,* vol. 57 (1961), pp. 1287–1312.

1007. Gonnerman, H. F.: "Effect of End Condition of Cylinder on Compressive Strength of Concrete," *Proc. ASTM,* vol. 24, pt. II (1924), pp. 1036–1065.

1008. Gonnerman, H. F.: "Effect of Size and Shape of Test Specimen on Compressive Strength of Concrete," *Proc. ASTM,* vol. 25, pt. II (1925), pp. 237–250.

1008a. Neville, Adam M.: "A General Relation for Strengths of Concrete Specimens of Different Shapes and Sizes," *J. ACI* (October, 1966), pp. 1095–1109.

1009. Gonnerman, H. F., and E. C. Shuman: "Compression, Flexure, and Tension Tests of Plain Concrete," *Proc. ASTM,* vol. 28, pt. II (1928), pp. 527–573.

1010. Greene, Gordon, W.: "Test Hammer Provides New Method of Evaluating Hardened Concrete." *Proc. ACI,* vol. 51 (1955), pp. 249–256.

1011. Grieb, W. E., and George Werner: "Comparison of Splitting Tensile Strength of Concrete with Flexural and Compressive Strengths," *Proc. ASTM,* vol. 62 (1962), pp. 972–995.

1012. Hatt, W. K., and R. E. Mills: "Physical and Mechanical Properties of Portland Cements and Concretes," *Purdue Univ. Eng. Expt. Sta. Bull.* 34 (November, 1928), 97 pp.

1012a. Ople, F. S. Jr., and C. L. Hulsbos: "Probable Fatigue Life of Plain Concrete with Stress Gradient," *J. ACI* (January, 1966), pp. 59–81.

1013. Higginson, Elmo C., George B. Wallace, and Elwood L. Ore: "Effect of Maximum Size Aggregate on Compressive Strength of Mass Concrete," in "Symposium on Mass Concrete," *ACI Spec. Pub.* 6 (1963), pp. 219–256.

1014. Jones, P. G., and F. E. Richart: "The Effect of Testing Speed on Strength and Elastic Properties of Concrete," *Proc. ASTM,* vol. 36, pt. II (1936), pp. 380–391.

1015. Jones, R.: "Non-destructive Testing of Concrete," Cambridge, New York, 1962, 101 pp.

1016. Kesler, C. E.: "Effect of Length to Diameter Ratio on Compressive Strength—An ASTM Cooperative Investigation," *Proc. ASTM,* vol. 59 (1959), pp. 1216–1230.

1017. Kesler, Clyde E., and Y. Higuchi: "Determination of Compressive Strength of Concrete by Using Its Sonic Properties," *Proc. ASTM,* vol. 53 (1953), pp. 1044–1052.

1018. Kluge, R. W.: "Impact Resistance of Reinforced Concrete Slabs," *Proc. ACI,* vol. 39 (1943), pp. 397–412.

1019. Lindner, C. P., and J. C. Sprague: "Effect of Depth of Beam upon the Modulus of Rupture of Plain Concrete," *Proc. ASTM,* vol. 55 (1955), pp. 1062–1084.

1021. Mather, Bryant: "Effect of Type of Test Specimen on Apparent Compressive Strength of Concrete," *Proc. ASTM,* vol. 45 (1945), pp. 802–809.

1022. Mather, Bryant, and W. O. Tynes: "Investigation of Compressive Strength of Molded Cylinders and Drilled Cores of Concrete," *Proc. ACI,* vol. 57 (1961), pp. 767–778.

1023. McHenry, Douglas, and J. J. Shideler: "Review of Data on Effect of Speed in Mechanical Testing of Concrete," *ASTM Spec. Tech. Pub.* 185 (1956), pp. 72–82; reprinted as *Portland Cement Assoc. Bull. D*9 (1956).

1024. Menzel, C. A.: "Some Factors Influencing Results of Pull-out Bond Tests," *Proc. ACI,* vol. 35 (1939), pp. 517–544.

1025. Parkinson, G. A., S. P. Finch, and J. E. Hoff: "Preliminary Report on Strength of Portland Cement Mortar and Its Temperature at Time of Test," *Univ. Texas Bull.* 2825, *Eng. Research Ser.* 26 (1928).

1026. Portland Cement Association: "Cold-weather Concreting," ST94-2 (1962), 6 pp.

1027. Portland Cement Association: "Hot-weather Concreting," ST93 (1963), 4 pp.

1028. Price, W. H.: "Factors Influencing Concrete Strength," *Proc. ACI,* vol. 47 (1951), pp. 417–432.

1029. Richart, F. E., Anton Brantzaeg, and R. L. Brown: "A Study of the Failure of Concrete under Combined Compressive Stress," *Univ. Illinois Eng. Exp. Sta. Bull.* 185 (1928).

1030. Saemann, J. C., and G. W. Washa: "Variation of Mortar and Concrete Properties with Temperature," *Proc. ACI,* vol. 54 (1958), pp. 385–395.

1031. Schuman, L., and John Tucker, Jr.: "Tensile and Other Properties of Concretes Made with Various Types of Cements," *Natl. Bur. Standards Research Paper* 1552 (August, 1943), 18 pp.

1032. Swenson, J. A., L. A. Wagner, and G. L. Pigman: "Effect of Granulometric Composition of Cement on the Properties of Pastes, Mortars, and Concretes," *Natl. Bur. Standards Research Paper* 777 (April, 1935), 30 pp.

1033. Timms, A. G., and N. H. Withey: "Temperature Effects on Compressive Strength of Concrete," *Proc. ACI,* vol. 30 (1934), pp. 159–180 (see also vol. 31 (1935), pp. 165–180 for later report on these investigations).

1034. Tremper, Bailey: "The Measurement of Concrete Strength by Embedded Pull-out Bars," *Proc. ASTM,* vol. 44 (1944), pp. 880–887.

1035. Troxell, G. E.: "The Effect of Capping Methods and End Conditions upon the Compressive Strength of Concrete Test Cylinders," *Proc. ASTM,* vol. 41 (1941), pp. 1038–1052.

1036. U.S. Bureau of Reclamation: "Concrete Manual," 7th ed. (1963), 642 pp.

1037. Walker, Stanton, and D. L. Bloem: "Effects of Aggregate Size on Properties of Concrete," *Proc. ACI,* vol. 57 (1961), pp. 283–298.

1038. Washa, G. W., and P. G. Fluck: "Effect of Sustained Loading on Compressive Strength and Modulus of Elasticity of Concrete," *Proc. ACI,* vol. 46 (1950), pp. 693–700.

1039. Watstein, David: "Effect of Straining Rate on the Compressive Strength and Elastic Properties of Concrete," *Proc. ACI,* vol. 49 (1953), pp. 729–744.

1040. Watstein, David: "Properties of Concrete at High Rates of Loading," in "Symposium on Impact Testing," *ASTM Spec. Tech. Pub.* 176 (1955).

1041. Withey, M. O.: "Some Long-time Tests of Concrete," *Proc. ACI,* vol. 27 (1931), pp. 547–582 (see also vol. 39 (1943), pp. 221–240, for a later report on this investigation).

1042. Zoldners, N. G.: "Effect of High Temperatures on Concretes Incorporating Different Aggregates," *Proc. ASTM,* vol. 60 (1960), pp. 1087–1108.

1043. Zoldners, N. G., V. M. Malhotra, and H. S. Wilson: "High-temperature Behavior of Aluminous Cement Concretes Containing Different Aggregates," *Proc. ASTM,* vol. 63 (1963), pp. 966–995.

1044. Malhotra, H. L.: "The Effect of Temperature on the Compressive Strength of Concrete," *Mag. Concrete Res.* (London) (August, 1956), pp. 85–94.

1045. Saemann, J. C., and G. W. Washa: "Variation of Mortar and Concrete Properties with Temperature," *J. ACI* (November, 1957), pp. 385–396.

1046. Philleo, R.: "Some Physical Properties of Concrete at High Temperature," *J. ACI* (April, 1958), pp. 857–864.

1047. Harmathy, T. Z., and J. E. Berndt: "Hydrated Portland Cement and Lightweight Concrete at Elevated Temperatures," *J. ACI* (January, 1966), pp. 93–112.

1048. Cusens, A. R.: "Strength of Concrete Test Cylinders Cast in Waxed Paper Molds," *J. ACI* (March, 1964), pp. 287–292.

1049. Bloem, D. L., and R. D. Gaynor: "Effects of Aggregate Properties on Strength of Concrete," *J. ACI* (October, 1963), pp. 1429–1456.

1050. Ferguson, Phil M., John E. Breen, and J. Neils Thompson: "Pullout Tests on High Strength Reinforcing Bars," *Proc. ACI,* vol. 62 (1965), pp. 933–950.

1051. Pincus, George, and Hans Gesund: "Evaluating the Tensile Strength of Concrete," *Mater. Res. Std.* (ASTM), vol. 5, no. 9 (1965), pp. 454–458.

1052. McNeely, D. J., and S. D. Lash: "Tensile Strength of Concrete," *J. ACI* (June, 1963), pp. 751–761.

1053. Malhotra, V. M., N. G. Zoldners, and H. M. Woodrooffe: "Ring Test for Tensile Strength of Concrete," *Mater. Res. Std.* (ASTM), (January, 1966), pp. 2–12.

1055. Nordby, G. M.: "Fatigue of Concrete—A Review of Research," *J. ACI* (August, 1958), pp. 191–220.

See also ASTM C495, C496, C513.

SECTION 1100 PERMEABILITY AND DURABILITY

Permeability

1100. McMillan, F. R., and Inge Lyse: "Some Permeability Studies of Concrete," *Proc. ACI,* vol. 26 (1930), pp. 101–142.

1101. Norton, P. T., and D. A. Pletta: "The Permeability of Gravel Concrete," *Proc. ACI,* vol. 27 (1931), pp. 1093–1132.

1102. Glanville, W. H.: "The Permeability of Portland Cement Concrete," *Dept. of Sci. and Ind. Res., Great Britain, Bldg. Res. Tech. Paper* 3 (1931), 62 pp. Bibliography.

1103. Ruettgers, A., E. N. Vidal, and S. P. Wing: "An Investigation of the Permeability of Mass Concrete with Particular Reference to Boulder Dam," *Proc. ACI,* vol. 31 (1935), pp. 382–416, and discussion, vol. 32 (1936), pp. 378–389. Bibliography.

1104. Dunagan, W. M.: "Methods for Measuring the Passage of Water through Concrete," *Proc. ASTM,* vol. 39 (1939), pp. 866–880.

1105. Dunagan, W. M., and G. C. Ernst: "A Study of the Permeability of a Few Integrally Waterproofed Concretes," *Proc. ASTM,* vol. 34 (1934), pp. 383–392.

1106. Wig, R. J., and P. H. Bates: "Tests of Absorptive and Permeable Properties of Portland Cement Mortars and Concretes Together with Tests of Dampproofing and Weather-proofing Compounds and Materials," *Natl. Bur. Standards Technol. Paper* 3 (1911).

1107. Washa, G. W.: "The Efficiency of Surface Treatments on the Permeability of Concrete," *Proc. ACI,* vol. 30 (1934), pp. 1–8.

1108. Wlison, Raymond: "The Limitations of the Absorption Tests for Concrete Products," *Proc. ACI,* vol. 25 (1929), pp. 522–537.

1109. Cook, H. K.: "Permeability Tests of Lean Mass Concrete," *Proc. ASTM,* vol. 51 (1951), pp. 1156–1165.

See also Refs. 101, 102, 106, 110.

Durability, freeze-thaw tests and air-entrained concrete

1120. "Durability of Concrete in Service," report of ACI Committee 201, *J. ACI* (December, 1962), pp. 1771–1820.

1121. Cordon, W. A.: "Freezing and Thawing of Concrete: Mechanisms and Control," *ACI Monograph* 3, 100 pp.

1122. Scholer, C. H.: "Significant Factors Affecting Concrete Durability." *Proc. ASTM,* vol. 52 (1952), pp. 1145–1158.

1125. McNeese, D. C.: "Early Freezing of Non-air-entraining Concrete," *Proc. ACI,* vol. 49 (1953), pp. 293–300.

1127. Tremper, Bailey: "Freeze-Thaw Resistance of Concrete as Affected by Alkalies in Cement," *Proc. ASTM,* vol. 51 (1951), pp. 1097–1107.

1128. Callan, E. J.: "Thermal Expansion of Aggregates and Concrete Durability," *Proc. ACI,* vol. 48 (1952), pp. 485–504.

1129. Verbeck, G., and R. Landgren: "Influence of Physical Characteristics of Aggregates on Frost Resistance of Concrete," *Proc. ASTM,* vol. 60 (1960), pp. 1063–1079.

1130. Klieger, P., and J. A. Hanson: "Freezing and Thawing Tests of Lightweight Aggregate Concrete," *J. ACI* (January, 1961), pp. 779–796.

1131. Long, B. G. and H. J. Kurtz: "Effect of Curing Methods upon the Durability of Concrete as Measured by Changes in the Dynamic Modulus of Elasticity," *Proc. ASTM,* vol. 43 (1943), pp. 1051–1065.

1132. Munger, H. H.: "The Influence of the Durability of Aggregate upon the Durability of the Resulting Concrete," *Proc. ASTM,* vol. 42 (1942), pp. 787–803.

1134. Young, R. B.: "Frost Resistant Concrete," *Proc. ACI,* vol. 36 (1940), pp. 477–493.

1135. Hansen, W. C.: "Influence of Sands, Cements, and Manipulation upon the Resistance of Concrete to Freezing and Thawing," *Proc. ACI,* vol. 39 (1943), pp. 105–124.

1136. Powers, T. C.: "A Working Hypothesis for Further Studies of Frost Resistance of Concrete," *Proc. ACI,* vol. 41 (1945), pp. 245–272.

1137. Flack, H. L.: "Freezing-and-Thawing Resistance of Concrete as Affected by the Method of Test," *Proc. ASTM,* vol. 57 (1957), pp. 1077–1095.

1138. Arni, H. T., B. E. Foster, and R. A. Clevenger: "Automatic Equipment and Comparative Test Results for the Four ASTM Freezing-and-Thawing Methods for Concrete," *Proc. ASTM,* vol. 56 (1956), pp. 1229–1256.

1139. Davis, R. E., H. E. Davis, and J. W. Kelly: "Weathering Resistance of Concretes Containing Fly-ash Cements," *Proc. ACI,* vol. 37 (1941), pp. 281–296.

1140. Wuerpel, C. E., and H. K. Cook: "Automatic Accelerated Freezing-and-Thawing Apparatus for Concrete," *Proc. ASTM,* vol. 45 (1945), pp. 813–823.

1141. "Freezing-and-Thawing Tests of Concrete," *Proc. ASTM,* vol. 46 (1946), pp. 1198–1251. A symposium.

1142. Brewer, H. W., and R. W. Burrows: "Coarse-ground Cement Makes More Durable Concrete," *Proc. ACI,* vol. 47 (1951), pp. 353–360.

1143. Lerch, William: *Basic Principles of Air-entrained Concrete,* Portland Cement Association, 1953, 36 pp.

1144. Blanks, R. F., and W. A. Cordon: "Practices, Experiences and Tests with Air-entraining Agents in Making Durable Concrete," *Proc. ACI,* vol. 45 (1949), pp. 469–487.

1147. Gonnerman, H. F.: "Tests of Concretes Containing Air-entraining Portland Cements or Air-entraining Materials Added to Batch at Mixer," *Proc. ACI,* vol. 40 (1944), pp. 477–507.

1148. Wuerpel, C. E.: "Laboratory Studies of Concrete Containing Air-entraining Admixtures," *Proc. ACI,* vol. 42 (1946), pp. 305–360.

See also Refs. 110, 319–321, 522, 523, 1120; ASTM C290, C291.

Durability—reactive aggregates; alkali and acid resistance

1152. Swenson, E. G., and V. Chaly: "Basis for Classifying Deleterious Characteristics of Concrete Aggregate Materials," *J. ACI* (May, 1956), pp. 987–1002.

1153. Hansen, W. C.: "Expansion and Cracking Studied in Relation to Aggregate and the Magnesia and Alkali Content of Cement," *J. ACI* (February, 1959), pp. 867–878.

1154. Pepper, L., and B. Mather: "Effectiveness of Mineral Admixtures in Preventing Excessive Expansion of Concrete Due to Alkali-Aggregate Reaction," *Proc. ASTM,* vol. 59 (1959), pp. 1178–1203.

1155. Pearson, J. C.: "A Concrete Failure Attributed to Aggregate of Low Thermal Coefficient," *Proc. ACI,* vol. 38 (1942), pp. 29–36.

1157. Brown, L. S., and C. U. Pierson: "Linear Traverse Technique for Measurement of Air in Hardened Concrete," *Proc. ACI,* vol. 47 (1951), pp. 117–124.

1159. Berkey, C. P.: "The Nature of the Processes Leading to the Disintegration of Concrete, with Special Reference to Excess Alkalies," *Proc. ACI,* vol. 37 (1941), pp. 689–692.

1161. Stanton, T. E., et al.: "California Experience with the Expansion of Concrete through Reaction between Cement and Aggregate," *Proc. ACI,* vol. 38 (1942), pp. 209–236; see also *Trans. ASCE,* vol. 107 (1942), pp. 54–84, and vol. 111 (1946), pp. 768–776.

1162. Carlson, R. W.: "Accelerated Tests of Concrete Expansion Due to Alkali-Aggregate Reaction," *Proc. ACI,* vol. 40 (1944), pp. 205–212.

1163. Hansen, W. C.: "Studies Relating to the Mechanism by Which the Alkali-Aggregate Reaction Produces Expansion in Concrete," *Proc. ACI,* vol. 40 (1944), pp. 213–228.

1164. Tremper, B.: "The Effect of Alkalies in Portland Cement on the Durability of Concrete," *Proc. ACI,* vol. 41 (1945), pp. 89–104.

1165. "Effect of Alkalies in Portland Cement on the Durability of Concrete," report of ASTM Committee C1, *ASTM Bull.* 142 (October, 1946), pp. 28–34.

1166. Blanks, R. F., and H. S. Meissner: "The Expansion Test as a Measure of Alkali-Aggregate Reaction," *Proc. ACI,* vol. 42 (1946), pp. 517–540.

1167. Blanks, R. F., and H. S. Meissner: "Deterioration of Concrete Dams Due to Alkali-Aggregate Reaction," *Proc. ASCE,* vol. 71, no. 1 (January, 1945), pp. 3–18, 1089–1110. See also *Trans. ASCE,* vol. 111 (1946), pp. 743–804.

1168. McConnell, D., R. C. Mielenz, W. Y. Holland, and K. T. Greene: "Cement-Aggregate Reaction in Concrete," *Proc. ACI,* vol. 44 (1948), pp. 93–128.

1169. Hanna, W. C.: "Unfavorable Chemical Reactions of Aggregates in Concrete and a Suggested Corrective," *Proc. ASTM,* vol. 47 (1947), pp. 986–1009.

1170. McCoy, W. J., and A. G. Caldwell: "New Approach to Inhibiting Alkali-Aggregate Expansion," *Proc. ACI,* vol. 47 (1951), pp. 693–708.

1171. Powers, T. C., and H. H. Steinour: "An Interpretation of Some Published Researches on the Alkali-Aggregate Reaction," *Proc. ACI,* vol. 51 (1955), pp. 497–516.

1172. Tuthill, L. H.: "Resistance of Cement to the Corrosive Action of Sodium Sulphate Solutions," *Proc. ACI,* vol. 33 (1937), pp. 83–106.

1173. Bloem, Delmar L.: "Soundness and Deleterious Substances," *ASTM Spec. Tech. Pub.* 169-*A* (1966), pp. 497–512.

1174. "Guide for the Protection of Concrete against Chemical Attack by Means of Coatings and Other Corrosion-resistant Materials," report of ACI Committee 515, *J. ACI* (December, 1966), pp. 1305–1392.

1174a. Stanton, T. E., and L. C. Meder: "Resistance of Cements to Attack by Sea Water and by Alkali Soils," *ACI* (March–April, 1938), pp. 433–464.

1174b. Stanton, T. E.: "Durability of Concrete Exposed to Sea Water and Alkali Soils: California Experience," *Proc. ACI,* vol. 44 (1948), pp. 821–847.

1175. Miller, D. G., and P. W. Manson: "Tests of 106 Commercial Cements for Sulfate Resistance," *Proc. ASTM,* vol. 40 (1940), pp. 988–1001.

1176. Hadley, H. M.: "Concrete in Sea Water: A Revised Viewpoint Needed," *Trans. ASCE,* vol. 107 (1942), pp. 345–358.

1177. Dahl, L. A.: "Cement Performance in Concrete Exposed to Sulfate Soils," *Proc. ACI,* vol. 46 (1950), pp. 257–272.

1178. Polivka, M., and E. H. Brown: "Influence of Various Factors on Sulfate Resistance of Concretes Containing Pozzolan," *Proc. ASTM,* vol. 58 (1958), pp. 1077–1100.

1179. Arber, M. G., and H. E. Vivian: "Protection of Mortar from Sulphate Attack by Steam Curing and Carbonation," *Australian J. Appl. Sci.,* vol. 12, no. 4 (1961) pp. 440–445.

1180. Molony, B., M. A. Hashem, and E. Hoffman: "Efflorescence on Concrete Masonry," *Construct. Rev.,* Sydney, vol. 35, no. 9 (September, 1962), pp. 35–36.

1185. Spraul, J. R.: "Acid-resistant Concrete Coatings," *Agr. Eng.,* vol. 22 (1941), pp. 209–210.

See also Refs. 101–107, 110, 322–325, 360–365, 1120; ASTM C227, C289, C342, C441, C452.

Durability—metal corrosion; wear; restoration

1187. Blenkinsop, J. C.: "The Effect on Normal ⅜-in. Reinforcement of Adding Calcium Chloride to Dense and Porous Concretes," *Concrete Research,* vol. 15, no. 43 (March, 1963), pp. 33–38.

1188. Griffin, D. F., and R. L. Henry: "Effect of Salt in Concrete on Compressive Strength, Water Vapor Transmission, and Corrosion of Reinforcing Steel," *Proc. ASTM,* vol. 63 (1963), pp. 1046–1078.

1189. Shalon, R., and M. Raphael: "Influence of Sea Water on Corrosion of Reinforcement," *J. ACI* (June, 1959), pp. 1251–1268.

1190. Monfore, G. E., and G. J. Verbeck: "Corrosion of Prestressed Wire in Concrete," *J. ACI* (November, 1960), pp. 491–516.

1191. Steinour, H. H.: "Influence of the Cement on the Corrosion Behavior of Steel in Concrete," *Res. Develop. Lab. Portland Cement Assoc. Res. Dep. Bull.* 168 (1964), 14 pp.

1191a. Mozer, J. D., A. C. Bianchini, and C. E. Kesler: "Corrosion of Reinforcing Bars in Concrete," *J. ACI* (August, 1965), pp. 909–931.

1191b. McGeary, Frank L.: "Performance of Aluminum in Concrete Containing Chlorides," *J. ACI* (Feburary, 1966), pp. 247–265.

1191c. Monfore, G. E., and Borje Ost: "Corrosion of Aluminum Conduit in Concrete," *J. Portland Cement Assoc. Res. Develop. Labs.* (January, 1965), pp. 10–22.

1192. Smith, F. L.: "Effect of Aggregate Quality on Resistance of Concrete to Abrasion," *ASTM Spec. Tech. Pub.* 205 (1956), pp. 91–106.

1192a. Li, Shu-Tien: "Wear Resistant Concrete Construction," *J. ACI* (February, 1959), pp. 879–892.

1192b. Kennedy, H. L., and G. E. Prior: "Wear Tests of Concrete," *Proc. ASTM,* vol. 53 (1953), pp. 1021–1032.

1192c. Price, W. H.: "Erosion of Concrete by Cavitation and Solids in Flowing Water," *Proc. ACI,* vol. 43 (1947), pp. 1009–1023.

1193. Witte, L. P., and J. E. Backstrom: "Some Properties Affecting the Abrasion Resistance of Air-entrained Concrete," *Proc. ASTM,* vol. 51 (1951), pp. 1141–1155.

1194. Gliddon, C.: "Repairing Concrete Hydraulic Structures," *Proc. ACI,* vol. 44 (1948), pp. 513–518.

1195. Lamprecht, J.: "Concrete Restoration in Water Impounding Structures," *Proc. ACI,* vol. 32 (1936), pp. 533–569.

1196. Capp, F. W.: "Maintaining Concrete Structures," *Proc. ACI,* vol. 32 (1936), pp. 579–592.

1197. Kelly, J. W., and B. D. Keatts: "Two Special Methods of Restoring and Strengthening Masonry Structures," *Proc. ACI,* vol. 42 (1946), pp. 289–304.

1198. Chadwick, W. L.: "Hydraulic Structure Maintenance Using Pneumatically Placed Mortar," *Proc. ACI,* vol. 43 (1947), pp. 533–547.

1199. Keatts, B. D.: "The Maintenance and Reconstruction of Concrete Tunnel Linings with Treated Mortar and Special Concrete," *Proc. ACI,* vol. 43 (1947), pp. 813–826.

1199b. "Guide for Use of Epoxy Compounds with Concrete," report of ACI Committee 403, *J. ACI* (September, 1962), pp. 1121–1142.

1199c. "Guide for Painting Concrete," report of ACI Committee 616, *J. ACI* (March, 1957), pp. 817–832.

See also Refs. 311–313, 1120; ASTM C131, C418.

SECTION 1200 VOLUME CHANGES

Volume changes due to moisture and hydration

1200. Davis, R. E., and G. E. Troxell: "Volumetric Changes in Portland Cement Mortars and Concretes," *Proc. ACI,* vol. 25 (1929), pp. 210–260.

1201. Davis, R. E.: "A Summary of Investigations of Volume Changes in Cements, Mortars and Concretes Produced by Causes Other than Stress," *Proc. ASTM,* vol. 30, pt. I (1930), pp. 668–685. See also *Proc. ACI,* vol. 26 (1930), pp. 407–443.

1202. White, A. H.: "Destruction of Cement Mortars and Concrete through Expansion and Contraction," *Proc. ASTM,* vol. 11 (1911), pp. 531–562; vol. 14, pt. II (1914), pp. 203–241; vol. 42 (1942), pp. 727–741.

1203. Carlson, R. W.: "The Chemistry and Physics of Concrete Shrinkage," *Proc. ASTM,* vol. 35, pt. II (1935), pp. 370–379.

1204. Carlson, R. W.: "Drying Shrinkage of Concrete as Affected by Many Factors," *Proc. ASTM,* vol. 38, pt. II (1938), pp. 419–437.

1205. Carlson, R. W.: "Drying Shrinkage of Large Concrete Members," *Proc. ACI,* vol. 33 (1937), pp. 327–336.

1208. Troxell, G. E.: "A short-time Test for Effect of Type of Cement on Concrete Shrinkage," *Proc. ACI,* vol. 35 (1939), pp. 73–80.

1209. Davis, H. E.: "Autogenous Volume Changes of Concrete," *Proc. ASTM,* vol. 40 (1940), pp. 1103–1110.

1210. Staley, H. R., and D. Peabody, Jr.: "Shrinkage and Plastic Flow of Prestressed Concrete," *Proc. ACI,* vol. 42 (1946), pp. 229–244.

1211. Pickett, G.: "Shrinkage Stresses in Concrete," *Proc. ACI,* vol. 42 (1946), pp. 165–204, 361–398.

1212. Kalousek, G. L., et al.: "Relation of Shrinkage to Moisture Content in Concrete Block," *Proc. ACI,* vol. 50 (1954), pp. 225–240.

1213. Glanville, W. H.: "Studies in Reinforced Concrete. II—Shrinkage Stresses," *Dept. of Sci. and Ind. Res., Great Britain, Bldg. Res. Tech. Paper No.* 11 (1930), 49 pp.

1214. Matsumoto, T.: "Study of the Effect of Moisture Content upon the Expansion and Contraction of Plain and Reinforced Concrete," *Univ. of Ill. Eng. Expt. Sta., Bull.* 126 (Dec. 5, 1921).

1215. Goldbeck, A. T., and F. H. Jackson: "Expansion and Contraction of Concrete and Concrete Roads," *U.S. Dept. Agr. Bull.* 532 (Oct. 13, 1917).

1216. Vetter, C. P.: "Stresses in Reinforced Concrete Due to Volume Changes," *Trans. ASCE,* vol. 98 (1933), pp. 1039–1053.

1217. "The Uses of Expanding Cement Concrete," *Concrete and Constr. Eng.,* vol. 42, no. 1 (January, 1947).

1218. Lynam, C. G.: *Growth and Movement in Portland Cement Concrete,* Oxford, London, 1934, 139 pp.

1219. Hansen, T. C., and K. E. C. Nielsen: "Influence of Aggregate Properties on Concrete Shrinkage," *J. ACI* (July, 1965), pp. 781–794.

1220. Menzel, C. A.: "A Method for Determining the Moisture Condition of Hardened Concrete in Terms of Relative Humidity," *Proc. ASTM,* vol. 55 (1955), pp. 1085–1109.

1221. Powers, T. C.: "Causes and Control of Volume Changes," *J. Portland Cement Assoc. Res. Develop. Labs.* (January, 1959), pp. 29–39.

1222. Lerch, W.: "Plastic Shrinkage," *J. ACI* (February, 1957), pp. 797–802.

1223. Lyse, I.: "Shrinkage and Creep of Concrete," *J. ACI* (February, 1960), pp. 775–782.

1224. Troxell, G. E., J. M. Raphael, and R. E. Davis: "Long-time Creep and Shrinkage Tests of Plain and Reinforced Concrete," *Proc. ASTM,* vol. 58 (1958), pp. 1101–1120.

1225. Hveem, F. N., and B. Tremper: "Some Factors Influencing Shrinkage of Concrete Pavements," *J. ACI* (February, 1957), pp. 781–790.

1226. Pickett, G.: "Effect of Aggregate on Shrinkage of Concrete and a Hypothesis Concerning Shrinkage," *J. ACI* (January, 1956), pp. 581–590.

1227. Bryson, J. O., and D. Watstein: "Comparison of Four Different Methods of Determining Drying Shrinkage of Concrete Masonry Units," *J. ACI,* (August, 1961), pp. 163–184.

1227a. Valore, R. C., Jr., and W. H. Kuenning: "A Cooperative Laboratory Study of the Effect of Testing Environment and Specimen Type on Shrinkage of Masonry Unit Concrete," *J. ACI* (October, 1962), pp. 1391–1433.

1228. Saxer, E. L., and H. T. Toennies: "Measuring Shrinkage of Concrete Block—A Comparison of Test Methods," *Proc. ASTM,* vol. 57 (1957), pp. 988–1011.

1230. Joisel, A.: *Les fissures du ciment. Causes et remédes* ("The cracking of cement. Causes and remedies"), Société d'Editions Scientifiques Techniques et Artistiques, Paris, 1961, 176 pp.

1231. Hsu, Thomas T. C.: "Mathematical Analysis of Shrinkage Stresses in a Model of Hardened Concrete," *Proc. ACI,* vol. 60 (1963), pp. 371–390.

1234. Verbeck, G.: "Carbonation of Hydrated Portland Cement," *ASTM Spec. Tech. Pub.* 205 (1958), pp. 17–36.

1235. Toennies, H. T., and J. J. Shideler: "Plant Drying and Carbonation of Concrete Block—NCMA-PCA Cooperative Program," *J. ACI* (May, 1963), pp. 617–632.

1236. Shideler, J. J.: "Carbonation Shrinkage of Concrete Masonry Units,"*J. Portland Cement Assoc. Res. Develop. Labs.* (September, 1963), pp. 36–51.

1237. Leber, I., and F. A. Blakey: "Some Effects of Carbon Dioxide on Mortars and Concrete," *J. ACI* (September, 1956), pp. 295–308; also (December, 1959), pp. 497–510.

1238. Toennies, H.: "Artificial Carbonation of Concrete Masonry Units," *J. ACI* (February, 1960), pp. 737–756.

1239. Washa, G. W., and R. L. Fedell: "Carbonation and Shrinkage Studies of Nonplastic Expanded Slag Concrete Containing Fly Ash," *J. ACI* (September, 1964), pp. 1109, 1124.

See also Refs. 110, 1273; ASTM C151, C157, C341, C342, C426, C427, C490.

Thermal expansion

1240. Willis, T. F., and M. E. De Reus: "Thermal Volume Change and Elasticity of Aggregates and Their Effect on Concrete," *Proc. ASTM,* vol. 39 (1939), pp. 919–928.

1241. Meyers, S. L.: "Thermal Coefficient of Expansion of Portland Cement," *Ind. and Eng. Chem.*, vol. 32 (1940), pp. 1107–1112.

1242. Mitchell, L. J.: "Thermal Expansion Tests of Aggregates, Neat Cements, and Concretes," *Proc. ASTM,* vol. 53 (1953), pp. 963–977.

See also Refs. 110, 1201.

Creep

1250. Glanville, W. H.: "Studies in Reinforced Concrete. III—The Creep or Flow of Concrete Under Load," *Dept. of Sci. and Ind. Res., Great Britain, Bldg. Res. Tech. Paper No.* 12 (1930), 39 pp.

1251. Davis, R. E., H. E. Davis, and J. S. Hamilton: "Plastic Flow of Concrete under Sustained Stress," *Proc. ASTM,* vol. 34, pt. II (1934), pp. 354–386.

1252. Davis, R. E., H. E. Davis, and E. H. Brown: "Plastic Flow and Volume Changes in Concrete," *Proc. ASTM,* vol. 37, pt. II (1937), pp. 317–330.

1253. Jensen, R. S., and F. E. Richart: "Short-time Creep Tests of Concrete in Compression," *Proc. ASTM,* vol. 38, pt. II (1938), pp. 410–417.

1254. Lorman, W. R.: "The Theory of Concrete Creep," *Proc. ASTM,* vol. 40 (1940), pp. 1082–1102.

1255. Maney, G. A.: "Concrete under Sustained Working Loads: Evidence that Shrinkage Dominates Time Yield," *Proc. ASTM,* vol. 41 (1941), pp. 1021–1030.

1256. Pickett, G.: "The Effect of Change in Moisture Content on the Creep of Concrete under a Sustained Load," *Proc. ACI,* vol. 38 (1942), pp. 333–356.

1257. McHenry, Douglas: "A New Aspect of Creep in Concrete and Its Application to Design," *Proc. ASTM,* vol. 43 (1943), pp. 1069–1084.

1258. Duke, C. M., and H. E. Davis: "Some Properties of Concrete under Sustained Combined Stress," *Proc. ASTM,* vol. 44 (1944), pp. 888–896.

1259. Washa, G. W.: "Plastic Flow of Thin Reinforced Concrete Slabs," *Proc. ACI,* vol. 44 (1948), pp. 237–260.

1260. Davis, R. E., and H. E. Davis: "Flow of Concrete under the Action of Sustained Loads," *Proc. ACI,* vol. 27 (1931), pp. 837–901.

1270. "Symposium on Creep of Concrete," *ACI Spec. Pub.* 9 (1964), 160 pp.

1271. Ali, I., and C. E. Kesler: "Creep in Concrete with and without Exchange of Moisture with the Environment," *University of Illinois, Department of Theoretical and Applied Mechanics Rept.* 641 (1963), 106 pp.

1272. Glucklich, J.: "Creep Mechanism in Cement Mortar," *J. ACI* (July, 1962), pp. 923–948.

1273. Reichard, T. W.: "Creep and Drying Shrinkage of Lightweight and Normal-weight Concretes," *Nat. Bur. Standards Monograph* 74 (1964), 30 pp.

1274. Freudenthal, A. M., and F. Roll: "Creep and Creep-recovery of Concrete under High Compressive Stress," *J. ACI* (June, 1958), pp. 1111–1142.

1275. Ross, A. D.: "Creep of Concrete under Variable Stress," *J. ACI* (March, 1958), pp. 739–758.

1276. Washa, G. W., and P. G. Fluck: "The Effect of Compressive Reinforcement on the Plastic Flow of Reinforced Concrete Beams," *J. ACI* (October, 1952), pp. 89–108.

1277. Polivka, M., D. Pirtz, and R. F. Adams: "Studies of Creep in Mass Concrete," Symposium on Mass Concrete, *ACI Pub.* SP-6 (1963), pp. 257–285.

1278. Fluck, P. G., and G. W. Washa: "Creep of Plain and Reinforced Concrete," *J. ACI* (April, 1958), pp. 879–896.

1279. Erzen, C. Z.: "An Expression for Creep and Its Application to Prestressed Concrete," *J. ACI* (August, 1956), pp. 205–214.

1280. Davis, R. E., and G. E. Troxell: "Properties of Concrete and Their Influence on Prestress Design," *J. ACI* (January, 1954), pp. 381–391.

See also Refs. 110, 1210, 1223, 1224, 1322; ASTM C512.

SECTION 1300 OTHER PROPERTIES

Elastic properties

1300. Walker, S.: "Modulus of Elasticity of Concrete," *Proc. ASTM,* vol. 19, pt. II (1919), pp. 510–606; published also as *Structural Materials Research Lab. Lewis Inst. Bull.* 5, 2d ed., Chicago, 1923.

1301. Davis, R. E., and G. E. Troxell: "Modulus of Elasticity and Poisson's Ratio for Concrete and the Influence of Age and Other Factors upon These Values," *Proc. ASTM,* vol. 29, pt. II (1929), pp. 678–710.

1302. Noble, P. M.: "The Effect of Aggregate and Other Variables on the Elastic Properties of Concrete," *Proc. ASTM,* vol. 31, pt. I (1931), pp. 399–426; published also as *Kansas State Coll. Eng. Expt. Sta. Bull.* 29 (1931).

1303. La Rue, H. A.: "Modulus of Elasticity of Aggregates and Its Effect on Concrete," *Proc. ASTM,* vol. 46 (1946), pp. 1298–1310.

1304. Counto, U. J.: "The Effect of the Elastic Modulus of the Aggregate on the Elastic Modulus, Creep and Creep Recovery of Concrete," *Mag. Concrete Res.* (September, 1964), pp. 129–138.

1305. Richart, F. E., and N. H. Roy: "Digest of Test Data on Poisson's Ratio for Concrete," *Proc. ASTM,* vol. 30, pt. I (1930), pp. 661–667.

1306. Johnson, J. W.: "Relationship between Strength and Elasticity of Concrete in Tension and in Compression," *Iowa State Coll. Eng. Expt. Sta. Bull.* 90 (1920), 43 pp.

1307. Spinner, S., and W. E. Tefft: "A Method for Determining Mechanical Resonance Frequencies and for Calculating Elastic Moduli from These Frequencies," *Proc. ASTM,* vol. 61 (1961), pp. 1221–1238.

1308. Power, T. C.: "Measuring Young's Modulus of Elasticity by Means of Sonic Vibrations," *Proc, ASTM,* vol. 38, pt. II (1938), pp. 460–467.

1309. Thomson, W. T.: "Measuring Changes in Physical Properties of Concrete by the Dynamic Method," *Proc. ASTM,* vol. 40 (1940), pp. 1113–1121.

1310. Obert, L., and W. I. Duvall: "Discussion of Dynamic Methods of Testing Concrete with Suggestions for Standardization," *Proc. ASTM,* vol. 41 (1941), pp. 1053–1070.

1311. Stanton, T. E.; "Tests Comparing the Modulus of Elasticity of Portland Cement Concrete as Determined by the Dynamic (Sonic) and Compression (Secant at 1000 psi.) Methods," *ASTM Bull.* 131 (December, 1944), pp. 17–20.

1312. Long, B. G., H. J. Kurtz, and T. A. Sandenaw: "An Instrument and a Technic for Field Determination of the Modulus of Elasticity, and Flexural Strength, of Concrete (Pavements)," *Proc. ACI,* vol. 41 (1945), pp. 217–232.

1313. Leslie, J. R., and W. J. Cheesman: "An Ultrasonic Method of Studying Deterioration and Cracking in Concrete Structures," *Proc. ACI,* vol. 46 (1950), pp. 17–36.

1314. "Bibliographies on Modulus of Elasticity, Poisson's Ratio, Inelastic Deformation and Volume Changes of Concrete," report of ASTM Committee C-9, *Proc. ASTM,* vol. 28, pt. I (1928), pp. 377–393.

1315. Philleo, R. E.: "Comparison of Results of Three Methods for Determining Young's Modulus of Elasticity of Concrete," *Proc. ACI,* vol. 51 (1955), pp. 461–469.

1316. Pauw, Adrian: "Static Modulus of Elasticity of Concrete as Affected by Density," *J. ACI* (December, 1960), pp. 679–687.

1317. Hirsch, T. J.: "Modulus of Elasticity of Concrete Affected by Elastic Moduli of Cement-paste Matrix and Aggregate," *Proc. ACI,* vol. 59 (1962), pp. 427–451.

1318. Andersen, Paul: "Experiments with Concrete in Torsion," *Trans. ASCE,* vol. 100 (1935), pp. 949–996.

1319. Whitehurst, E. A.: Evaluation of Concrete Properties from Sonic Tests," *ACI Monograph* M-2, 94 pp.

1320. Shideler, J. J.: "Lightweight-aggregate Concrete for Structural Use," *J. ACI* (October, 1957), pp. 299–328. (Figure 4 shows compressive strengths up to 12,000 psi and E up to 6,000 psi, with large differences as between different series of tests.)

1321. Klieger, Paul: "Long-time Study of Cement Performance in Concrete—Chapter 10: Progress Report on Strength and Elastic Properties of Concrete," *J. ACI* (December, 1957), pp. 481–504. (Figures 5-7 show moduli up to 7,500,000 psi for compressive strengths up to 10,000 psi.)

1322. Freudenthal, A. M., and Frederic Roll, "Creep and Creep Recovery of Concrete under High Compressive Stress," *J. ACI* (June, 1958), pp. 1111–1142. (Tables 3–5 show strength and modulus of regular concrete for strengths of 4,510 to 10,340 psi at ages of 28 to 399 days.)

1323. Teller, L. W.: "Elastic Properties," pp. 94–103 in "Significance of Tests and Properties of Concrete and Concrete Aggregates," *ASTM Spec. Tech. Pub.* 169 (1956).

See also Refs. 110, 1238, 1239.

Thermal properties—specific heat and conductivity

1340. Rippon, C. S., and L. J. Snyder: "Themal Properties of Mass Concrete," *Proc. ACI,* vol. 30 (1934), pp. 35–40.

1341. Carman, A. P., and R. A. Nelson: "Thermal Conductivity and Diffusivity of Concrete," *Univ. Ill. Eng. Expt. Sta. Bull.* 122 (1921), 32 pp.

1342. Thomson, W. T.: "A Method of Measuring Thermal Diffusivity and Conductivity of Stone and Concrete," *Proc. ASTM,* vol. 40 (1940), pp. 1073–1080.

1343. Glover, R. E.: "Flow of Heat in Dams," *Proc. ACI,* vol. 31 (1935), pp. 113–124.

1344. Rowley, F. B., A. B. Algren, and C. Carlson: "Thermal Properties of Concrete Construction," *J. Am. Soc. Heating Ventilating Engrs.* (January, 1936), pp. 53–64.

1345. "Furring for Monolithic Concrete Walls—Thermal Insulation," *Portland Cement Assoc. Bull.* AC16 (October, 1936), 4 pp.

1346. "How to Calculate Heat Transmission Coefficients and Vapor Condensation Temperatures of Concrete Masonry Walls," *Portland Cement Assoc. Bull.* CP68 (October, 1946), 8 pp.

1347. "Thermal Insulation of Bldgs.," *Am. Arch. Ref. Data* 11 (May, 1934).

1348. Stone, A.: "Thermal Insulation of Concrete Homes," *Proc. ACI,* vol. 44 (1948), pp. 849–874.

See also *ASTM* C186.

Temperature rise in mass concrete

1360. Davis, R. E., and G. E. Troxell: "Temperatures Developed in Mass Concrete and Their Effect upon Compressive Strength," *Proc. ASTM,* vol. 31, pt. II (1931), pp. 576–594.

1361. Kelly, J. W.: "Some Time-Temperature Effects in Mass Concrete," *Proc. ACI,* vol. 34 (1938), pp. 573–588.

1362. Carlson, Roy W.: "Temperatures and Stresses in Mass Concrete," *Proc. ACI,* vol. 34 (1938), pp. 497–516.

1363. Glover, R. E.: "Calculation of Temperature Distribution in a Succession of Lifts Due to Release of Chemical Heat," *Proc. ACI,* vol. 34 (1938), pp. 105–116.

1364. Carlson, Roy W.: "A Simple Method for the Computation of Temperatures in Concrete Structures," *Proc. ACI,* vol. 34 (1938), pp. 89–104.

1365. Forbrich, L. R.: "The Effect of Various Reagents on the Heat Liberation Characteristics of Portland Cement," *Proc. ACI,* vol. 37 (1941), pp. 161–184.

1366. Forbrich, L. R.: "Temperature Effects Near Concrete Surfaces as Affected by Heat Liberation of Cement," *Proc. ACI,* vol. 38 (1942), pp. 53–64.

1367. Shartsis, Leo, and E. S. Newman: "A Study of the Heat of Solution Procedure for Determining the Heat of Hydration of Portland Cement," *Proc. ASTM,* vol. 43 (1943), pp. 905–916.

1368. Rawhouser, C.: "Cracking and Temperature Control of Mass Concrete," *Proc. ACI,* vol. 41 (1945), pp. 305–348.

See also Refs. 106, 263; ASTM C186.

Extensibility and cracking

1375. Mercer, L. B.: "Classification of Concrete Cracks," *Commonwealth Engr.,* vol. 34, no. 2 (September, 1946).

1376. Watstein, David, and D. E. Parsons: "Width and Spacing of Tensile Cracks in Axially Reinforced Concrete Cylinders," *Natl. Bur. Standards Research Paper* 1545 (July, 1943).

1377. Blanks, R. F., H. S. Meisner, and C. Rawhouser: "Cracking in Mass Concrete," *Proc. ACI,* vol. 34 (1938), pp. 477–495.

1378. Carlson, Roy W.: "Attempts to Measure the Cracking Tendency of Concrete," *Proc. ACI,* vol. 36 (1940), pp. 533–540.

See also Refs. 1361, 1368.

Miscellaneous properties

1385. Menzel, C. A.: "Tests of the Fire Resistance and Stability of Walls of Concrete Masonry Units," *Proc. ASTM,* vol. 31, pt. II (1931), pp. 607–660.

1386. Menzel, C. A.: "Tests of the Fire Resistance and Thermal Properties of Solid Concrete Slabs and Their Significance," *Proc. ASTM,* vol. 43 (1943), pp. 1099–1153.

1387. ACI Committee 216: "Symposium on Fire Resistance of Concrete," *ACI Spec. Pub.* SP-5 (1962), 88 pp.

1398. Nordby, Gene M.: "Fatigue of Concrete—A Review of Research," *J. ACI* (August, 1958), pp. 191–220.

See also Refs. 1042–1047.

SECTION 1400 SPECIAL CONCRETE

Architectural concrete; floors; pneumatically applied mortar

1400. "Concrete as an Architectural Material," *Proc. ACI,* vol. 35 (1939), pp. 349–392. A symposium.

1401. Fischer, H. C.: "Architectural Concrete on the New Naval Medical Center," *Proc. ACI,* vol. 38 (1942), pp. 289–312.

1402. Phillips, R. S.: "Mixtures, Placing and Curing for Architectural Concrete," *Proc. ACI,* vol. 35 (1939), pp. 277–284.

1403. Hogan, J. J.: "Design Details for Architectural Concrete," *Proc. ACI,* vol. 45 (1949), pp. 529–540.

1404. Oberly, E. B.: "Construction Practices for Architectural Concrete," *Proc. ACI,* vol. 45 (1949), pp. 541–552.

1405. Early, J. J.: "The Characteristics of Concrete for Architectural Use," *Proc. ACI,* vol. 35 (1939), pp. 385–389.

1406. "Guide to Cement Plastering," report of ACI Committee 524, *J. ACI* (July, 1963), pp. 817–834.

1407. "Precast Concrete Wall Panels," report of ACI Committee 533, *ACI Spec. Pub.* 11, 143 pp.

1408. Wilson, J. G.: *Exposed Concrete Finishes,* C. R. Books, London, 1962, 144 pp.

1409. "Recommended Practice for Concrete Floor and Slab Construction," report of ACI Committee 302, *J. ACI,* (January, 1966), pp. 1–58.

1410. Boase, A. J.: "Inspection of Concrete Floor Finish Construction," *Proc. ACI,* vol. 39 (1943), pp. 97–104.

1411. Klock, M. B.: "Monolithic and Bonded Floor Finishes," *Proc. ACI,* vol. 45 (1949), pp. 725–730.

1412. Eckert, E. E.: "Heavy Duty Concrete Floors," *Proc. ACI,* vol. 49 (1953), pp. 109–116.

1413. Anonymous: "Non-slip Finishes for Concrete," *Concrete Construction,* vol. 11, no. 9 (September, 1966), pp. 351–353.

1414. "Guide for Painting Concrete," report of ACI Committee 616, *J. ACI* (March, 1957), pp. 817–832.

1416. "ACI Standard Recommended Practice for Shotcreting," report of ACI Committee 506, *J. ACI* (February, 1966), pp. 219–246; also (July, 1966), p. 732.

1416a. "Shotcreting,"*ACI Spec. Pub.* 14 (1966), 224 pp. A symposium. For abstracts, see *J. ACI* (January, 1967), pp. 49–56.

1417. Lorman, W. R.: "Engineering Properties of Shotcrete," *U.S. Naval Civil Eng. Lab. Tech. Report* R429 (May, 1966), 49 pp.

See also Refs. 483–485, 1198, 1504.

Concrete placed under water

1422. Angas, W. M., E. M. Shanley, and J. A. Erickson: "Concrete Problems in the Construction of Graving Docks by the Tremie Method," *Proc. ACI,* vol. 40 (1944), pp. 249–280.

1423. Halloran, P. J., and K. H. Talbot: "The Properties and Behavior Underwater of Plastic Concrete," *Proc. ACI,* vol. 39 (1943), pp. 461–492.

Mass concrete and vacuum concrete

1430. McMillan, F. R.: "Field Survey of Mass Concrete," *Proc. ACI,* vol. 34 (1938), pp. 561–572.

1431. Roberts, H. H.: "Cooling Materials for Mass Concrete," *Proc. ACI,* vol. 47 (1951), pp. 821–832.

1432. Symposium on Mass Concrete, *ACI Spec. Pub.* 6 (1962), 424 pp.

1450. Lockhardt, William F.: "Vacuum Concrete," *Proc. ACI,* vol. 34 (1938), pp. 305–320.

1451. Rippon, C. S.: "Vacuum Processing of Shasta Dam Spillway," *Eng. News-Record,* vol. 134, no. 24 (June 14, 1945), pp. 93–96.

1452. Billner, K. P., and B. M. Thorud: "Vacuum Processes Applied to Precast Concrete Houses," *Proc. ACI,* vol. 46 (1950), pp. 121–128.

1454. Ruud, F. O.: "Prediction and Control of Cooling Stresses in Concrete Blocks," *J. ACI,* (January, 1965), pp. 95–104.

See also Refs. 1340–1343, 1360–1368, 1377, 1378.

Lightweight and heavy weight concrete

1460. Richart, F. E., and V. P. Jensen: "Tests of Plain and Reinforced Haydite Concrete," *Proc. ASTM,* vol. 30, pt. II (1930), pp. 674–706.

1461. Short, A., and W. Kinniburgh: *Lightweight Concrete,* Wiley, New York, 1963, 368 pp.

1462. Gray, W. H., J. F. McLaughlin, and J. D. Antrim: "Fatigue Properties of Lightweight Aggregate Concrete," *J. ACI* (August, 1961), pp. 149–162.

1463. Grimm, C. T.: "Vermiculite Insulating Concrete, Pumped and Sprayed," *Civil Eng.,* vol. 33, no. 11 (November, 1963), pp. 66–69.

1464. Valore, R. C., Jr.: "Insulating Concretes," *J. ACI* (November, 1956), pp. 509–532.

1465. Hanson, E. B., Jr., and N. T. Neelands: "The Effect of Curing Conditions on Compressive, Tensile and Flexural Strength of Concrete Containing Haydite Aggregate," *Proc. ACI,* vol. 41 (1945), pp. 105–116.

1466. "Light Aggregate from Volcanic (Obsidian) Rock," *Western Construction News,* vol. 20 (1945), p. 101.

1467. Petersen, P. H.: "Burned Shale and Expanded Slag Concretes with and without Air-entraining Admixture," *Proc. ACI,* vol. 45 (1949), pp. 165–175.

1468. Kluge, R. W., et al.: "Lightweight-aggregate Concrete," *Proc. ACI,* vol. 45 (1949), pp. 625–642.

1470. "High-strength, High-density Concrete," U.S. Army Corps of Engineers, Waterways Experiment Station *Tech. Report* 6-635 (November, 1963), 65 pp. Also see *J. ACI* (August, 1965), pp. 951–962.

1471. Davis, H. S., and O. E. Borge: "High-density Concrete Made with Hydrous-iron Aggregates," *J. ACI* (April, 1959), pp. 1141–1148.

1472. Davis, H. S.: "High-density Concrete for Shielding Atomic Energy Plants," *J. ACI* (May, 1958), pp. 965–978.

1473. Shirayama, K.: "Properties of Radiation Sheildng Concrete," *J. ACI* (February, 1963), pp. 261–280.

1474. Henrie, J. O.: "Properties of Nuclear Shielding Concrete,"*J. ACI* (July, 1959), pp. 37–46.

1475. ASCE Task Committee: "Concrete for Shielding Nuclear Radiations," *J. Structural Division ASCE* (February, 1962), pp. 123–134.

1476. Thorne, C. P.: "Concrete Properties Relevant to Reactor Shield Behavior," *J. ACI* (May, 1961), pp. 1491–1508.

1477. Witte, L. P., and J. E. Backstrom: "Properties of Heavy Concrete Made with Barite Aggregates," *Proc. ACI,* vol. 51 (1955), pp. 65–88.

1478. Henri, J. O.: "Magnetic Iron Ore Concrete for Nuclear Shielding," *Proc. ACI,* vol. 51 (1955), pp. 541–552.

1479. Callan, E. J.: "Concrete for Radiation Shielding," *Proc. ACI,* vol. 50 (1954), pp. 17–44.

See also Ref. 1504; ASTM C330–C332.

Grouting

1480. Creaghan, T. C.: "The Grouting of Concrete Structures," *Proc. ACI,* vol. 37 (1941), pp. 641–648.

1481. Simonds, A. W.: "Contraction Joint Grouting of Large Dams," *Proc. ACI,* vol. 43 (1947), pp. 637–652.

1482. Minear, V. L.: "Notes on the Theory and Practice of Foundation Grouting," *Proc. ACI,* vol. 43 (1947), pp. 917–931.

1483. Brooks, B. S.: "Dry Mortar as a Bearing and Grouting Material," *Proc. ACI,* vol. 45 (1949), pp. 369–380.

See also Ref. 1504.

Concrete masonry

1486. Wendt, K. F., and P. M. Woodworth: "Tests on Concrete Masonry Units Using Tamping and Vibration Molding Methods," *Proc. ACI,* vol. 36 (1940), pp. 121–164.

1487. Converse, F. J.: "Tests on Reinforced Concrete Masonry," report to the Pacific Coast Building Officials Conference, *Building Standards Monthly,* (February, 1946), pp. 4–11.

See also Ref. 1385.

SECTION 1500 INSPECTION

1500. "The Economic Significance of Specifications," a symposium by ASTM and Western Society of Engineers, 1931, *Proc. ASTM,* vol. 31 (1931), pp. 955–987; see especially articles by J. P. H. Perry and A. R. Lord on specifications for concrete.

1501. *Concrete Pavement Inspectors Manual,* Portland Cement Association Serial R12-30, 1959, 51 pp.

1502. Cohen, A. B.: "Supervision and Inspection of Concrete," *Proc. ACI,* vol. 32 (1936), pp. 40–45.

1503. Young, R. B.: "Inspection," *Proc. ACI,* vol. 32 (1936), pp. 46–50.

1504. *ACI Manual of Concrete Inspection,* American Concrete Institute, 5th ed., 1967, 270 pp.

1505. Springes, G. P.: "Specifications and Inspection," in *Proc. 16th Annual Road School, Purdue Univ. Eng. Ext. Ser. Bull.* 23 (1930), pp. 130–142.

1506. Young, R. B., and W. Schnarr: "A System of Concrete Control for Scattered Small Jobs, as Used by a Large Organization," *Proc. ACI,* vol. 35 (1939), pp. 337–348.

1507. Nichols, J. R.: "Tolerances in Building Construction," *Proc. ACI,* vol. 36 (1940), pp. 493–496.

1508. "Recommended Practice for Concrete Inspection," report of ACI Committee 311, *J. ACI* (November, 1963), pp. 1525–1534.

1509. Bloodgood, G., and L. H. Tuthill: "Responsibilities of an Inspector," *J. ACI* (March, 1957), pp. 899–904.

1510. "Recommended Practice for Concrete Inspection," *ACI Standard* 311–64 (1964), 9 pp.

1511. Tuthill, L. H.: "Inspection of Mass and Related Concrete Construction," *Proc. ACI,* vol. 46 (1950), pp. 349–359.

1512. Clair, M. N.: "The Inspector," *Proc. ACI,* vol. 46 (1950), pp. 709–712.

1520. Stadtfeld, N. T. F.: "Inspection and Testing of Materials," *Proc. ACI,* vol. 46 (1950), pp. 237–247.

SECTION 1600 INSPECTION RECORDS AND REPORTS

See Refs. 106, 1504.

SECTION 1700 PRESENTATION AND ANALYSIS OF TEST DATA

1700. *Manual on Presentation of Data,* American Society for Testing Materials, August, 1940, 73 pp.

1701. Simon, L. E.: *An Engineers' Manual of Statistical Methods,* Wiley, New York, 1941, 231 pp.

1702. Shewhart, W. A.: *Economic Control of Quality of Manufactured Products,* Van Nostrand, New York, 1931, 501 pp.

1703. Arkin, H., and R. R. Colton: *An Outline of Statistical Methods,* Barnes & Noble, New York, 3d ed., 1938, 228 pp.

1704. Shewhart, W. A.: *Statistical Method from the Viewpoint of Quality Control,* U.S. Department of Agriculture Graduate School, 1939, 155 pp.

1705. Worthing, A. G., and J. Geffner: *Treatment of Experimental Data,* Wiley, New York, 1943, 342 pp.

1706. *The Preparation of Statistical Tables, A Handbook,* U.S. Department of Agriculture, Bureau of Agricultural Economics, 1937.

1707. Waddell, J. J.: *Practical Quality Control for Concrete,* McGraw-Hill, New York, 1962, 396 pp.

1708. "Recommended Practice for Evaluation of Compression Test Results of Field Concrete," report of ACI Committee 214, *J. ACI* (December, 1956), pp. 561–579. For revisions, see (September, 1964), pp. 1057–1072.

1709. Lauer, L. R.: "Evaluation of Concrete Compression Test Results," *J. ACI* (April, 1965), pp. 467–478.

1710. Crum, R. W., and H. W. Leavitt: "The Numbers of Specimens or Tests of Concrete and Concrete Aggregates Required for Reasonable Accuracy of the Average," in *Report on Significance of Tests of Concrete and Concrete Aggregates,* 2d ed., ASTM Committee C9, American Society for Testing Materials, Philadelphia, 1943, pp. 163–170.

1711. Humes, C. H., R. F. Passano, and Anson Hayes: "A Study of the Error of Averages and the Application to Corrosion Tests," *Proc. ASTM,* vol. 30, pt. II (1930), pp. 448–455.

1712. Leavitt, H. W., and H. A. Pratt: "A Statistical Analysis of Compression Tests of Mortar Cylinders, Cubes and Prisms," *Proc. ASTM,* vol. 39 (1939), pp. 851–859.

1713. Creskoff, J. J.: "Estimating 28-Day Strength of Concrete from Earlier Strengths—Including the Probable Error of the Estimate," *Proc. ACI,* vol. 41 (1945), pp. 493–512.

1714. Bloem, D. L.: "Studies of Uniformity of Compressive Strength Tests of Ready Mixed Concrete," *ASTM Bull.* (May, 1955), pp. 65–70.

See also Refs. 106, 110.

SECTION 1800 SPECIFICATIONS

General

1800. *Report of the Joint Committee on Standard Specifications for Concrete and Reinforced Concrete,* American Concrete Institute, Detroit, Mich., and American Society for Testing Materials, Philadelphia, June, 1940.

1801. "Specifications for Structural Concrete for Buildings," *ACI Standard* 301–66 (1966), 56 pp.

1802. *Standard Specifications for ighway Materials and Methods of Sampling and Testing,* 6th ed., American Association of State Highway Officials, Washington, 1950, pt. I, Specifications, 231 pp., pt. II, Methods, 414 pp.

1804. "Construction Specifications for Concrete Work on the Small Job," *Proc. ACI,* vol. 27 (1931), pp. 65–96.

1806. Perry, J. P. H.: "The Economic Significance of Specifications for Materials from the Point of View of a Producer of Concrete," *Proc. West. Soc. Engrs.*, vol. 36 (1931), pp. 284–291; see also *Eng. and Cont.*, vol. 70 (November, 1931), pp. 285–288.

1807. "Specifications for Concrete Pavements and Bases (ACI 617)," *Proc. ACI*, vol. 47 (1951), pp. 721–744.

1808. Abbott, R. W.: *Engineering Contracts and Specifications*, 3d ed., Wiley, New York, 1954, 429 pp.

ASTM SPECIFICATIONS AND METHODS OF TESTING

Below is given a list of selected ASTM Specifications and methods of testing pertaining to concrete and materials for making concrete.

C10 Specifications for Natural Cement
C14 Specifications for Concrete Sewer Storm-drain and Culvert Pipe
C29 Method of Test for Unit Weight of Aggregate
C30 Method of Test for Voids in Aggregate for Concrete
C31 Method of Making and Storing Concrete-compression and Flexure-test Specimens in the Field
C33 Specifications for Concrete Aggregates
C39 Method of Test for Compressive Strength of Molded Concrete Cylinders
C40 Method of Test for Organic Impurities in Sands for Concrete
C42 Method of Obtaining and Testing Drilled Cores and Sawed Beams of Concrete
C55 Spcifications for Concrete Building Brick
C70 Method of Test for Surface Moisture in Fine Aggregate
C76 Specifications for Reinforced Concrete Culvert, Storm Drain, and Sewer Pipe
C78 Method of Test for Flexural Strength of Concrete (Using Simple Beam with Third-point Loading)
C85 Method of Test for Cement Content of Hardened Portland Cement Concrete
C87 Method of Test for Effect of Organic Impurities in Fine Aggregate Strength of Mortars
C88 Method of Test for Soundness of Aggregates by Use of Sodium Sulfate or Magnesium Sulfate
C90 Specifications for Hollow Load-bearing Concrete Masonry Units
C91 Specifications for Masonry Cement
C94 Specifications for Ready-mixed Concrete
C109 Method of Test for Compressive Strength of Hydraulic Cement Mortars
C114 Methods of Chemical Analysis of Hydraulic Cement
C115 Method of Test for Fineness of Portland Cement by the Turbidimeter
C116 Method of Test for Compressive Strength of Concrete Using Portions of Beams Broken in Flexure
C117 Method of Test for Amount of Material Finer than No. 200 Sieve in Mineral Aggregates by Washing
C118 Specifications for Concrete Pipe for Irrigation or Drainage
C123 Method of Test for Lightweight Pieces in Aggregate
C124 Method of Test for Flow of Portland Cement Concrete by Use of the Flow Table
C125 Definition of Terms Relating to Concrete and Concrete Aggregates
C127 Method of Test for Specific Gravity and Absorption of Coarse Aggregate
C128 Method of Test for Specific Gravity and Absorption of Fine Aggregate
C129 Specifications for Hollow Non-load-bearing Concrete Masonry Units
C131 Method of Test for Resistance to Abrasion of Small Size Coarse Aggregate by Use of the Los Angeles Machine

C136 Method of Test for Sieve or Screen Analysis of Fine and Coarse Aggregates
C138 Method of Test for Weight per Cubic Foot, Yield, and Air Content (Gravimetric) of Concrete (also see C173 and C231)
C139 Specifications for Concrete Masonry Units for Construction of Catch Basins and Manholes
C140 Methods of Sampling and Testing Concrete Masonry Units
C142 Method of Test for Clay Lumps in Natural Aggregates
C143 Method of Test for Slump of Portland Cement Concrete
C144 Specifications for Aggregate for Masonry Mortar
C145 Specification for Solid Load-bearing Concrete Masonry Units
C150 Specifications for Portland Cement
C151 Method of Test for Autoclave Expansion of Portland Cement
C156 Method of Test for Water Retention Efficiency of Liquid Membrane-forming Compounds and Impermeable Sheet Materials for Curing Concrete
C157 Method of Test for Length Change of Cement Mortar and Concrete
C171 Specifications for Waterproof Paper for Curing Concrete
C172 Method of Sampling Fresh Concrete
C173 Method of Test for Air Content (Volumetric) of Freshly Mixed Concrete by the Volumetric Method (also see C138 and C231)
C174 Method of Measuring Length of Drilled Concrete Cores
C175 Specifications for Air-entraining Portland Cement
C183 Methods of Sampling Hydraulic Cement
C184 Method of Test for Fineness of Hydraulic Cement by the No. 200 Sieve
C185 Method of Test for Air Content of Hydraulic Cement Mortar
C186 Method of Test for Heat of Hydration of Portland Cement
C187 Method of Test for Normal Consistency of Hydraulic Cement
C188 Method of Test for Specific Gravity of Hydraulic Cement
C189 Method of Test for Soundness of Hydraulic Cement over Boiling Water (Pat Test)
C190 Method of Test for Tensile Strength of Hydraulic Cement Mortars
C191 Methods of Test for Time of Setting of Hydraulic Cement by Vicat Needle
C192 Method of Making and Curing Concrete Compression and Flexure Test Specimens in the Laboratory
C204 Method of Test for Fineness of Portland Cement by Air-permeability Apparatus
C205 Specifications for Portland Blast-furnace Slag Cement
C219 Definition of Terms Relating to Hydraulic Cement
C226 Specification for Air-entraining Additions for Use in the Manufacture of Air-entraining Portland Cement
C227 Method of Test for Potential Alkali Reactivity of Cement-Aggregate Combinations
C230 Specification for Flow Table for Use in Tests of Hydraulic Cement
C231 Method of Test for Air Content of Freshly Mixed Concrete by the Presssure Method (also see C138 and C173)
C232 Method of Test for Bleeding of Concrete
C233 Method of Testing Air-entraining Admixtures for Concrete
C234 Method of Test for Comparing Concretes on the Basis of the Bond Developed with Reinforcing Steel
C235 Method of Test for Scratch Hardness of Coarse Aggregate Particles
C243 Method of Test for Bleeding of Cement Pastes and Mortars
C260 Specification for Air-entraining Admixtures for Concrete
C265 Method of Test for Calcium Sulfate in Hydrated Portland Cement Mortar
C266 Method of Test for Time of Setting of Hydraulic Cement by Gillmore Needles
C267 Method of Test for Chemical Resistance of Mortars

C270 Specification for Mortar for Unit Masonry

C289 Method of Test for Potential Reactivity of Aggregates (Chemical Method)

C290 Method of Test for Resistance of Concrete Specimens to Rapid Freezing and Thawing in Water

C291 Method of Test for Resistance of Concrete Specimens to Rapid Freezing in Air and Thawing in Water

C293 Method of Test for Flexural Strength of Concrete (Using Simple Beam with Center-point Loading)

C294 Descriptive Nomenclature of Constituents of Natural Mineral Aggregates

C295 Recommended Practice for Petrographic Examination of Aggregates for Concrete

C305 Method for Mechanical Mixing of Hydraulic Cement Pastes and Mortars of Plastic Consistency

C309 Specification for Liquid Membrane-forming Compounds for Curing Concrete

C311 Methods of Sampling and Testing Fly Ash for Use as an Admixture in Portland Cement Concrete

C330 Specification for Lightweight Aggregates for Structural Concrete

C331 Specifications for Lightweight Aggregates for Concrete Masonry Units

C332 Specification for Lightweight Aggregates for Insulating Concrete

C340 Specification for Portland-Pozzolan Cement

C341 Method of Test for Volume Change of Concrete Products

C342 Method of Test for Potential Volume Change of Cement-Aggregate Combinations

C348 Method of Test for Flexural Strength of Hydraulic Cement Mortars

C349 Method of Test for Compressive Strength of Hydraulic Cement Mortars Using Portions of Prisms Broken in Flexure

C350 Specification for Fly Ash for Use as an Admixture in Portland Cement Concrete

C358 Specifications for Slag Cement

C359 Method of Test for False Set of Portland Cement (Mortar Method)

C360 Method of Test for Ball Penetration in Fresh Portland Cement Concrete

C402 Specifications for Raw or Calcined Natural Pozzolans for Use as Admixtures in Portland Cement Concrete

C403 Method of Test for Time of Setting of Concrete Mixtures by Penetration Resistance

C418 Method of Test for Abrasion Resistance of Concrete

C426 Method of Test for Drying Shrinkage of Concrete Block

C427 Method of Test for Moisture Condition of Hardened Concrete by the Relative Humidity Method

C430 Method of Test for Fineness of Hydraulic Cement by the No. 325 Sieve

C440 Specification for Cotton Mats for Curing Concrete

C441 Method of Test for Effectiveness of Mineral Admixtures in Preventing Excessive Expansion of Concrete Due to the Alkali-Aggregate Reaction

C451 Method of Test for False Set of Portland Cement (Paste Method)

C452 Method of Test for Potential Expansion of Portland Cement Mortars Exposed to Sulfate

C465 Specification for Processing Additions for Use in the Manufacture of Portland Cement

C470 Specification for Single-use Molds for Forming 6 by 12-inch Concrete Compression-test Cylinders

C490 Specification for Apparatus for Use in Measurement of Volume Change of Cement Paste, Mortar, and Concrete

C494 Specification for Chemical Admixtures for Concrete

C495 Method of Test for Compressive Strength of Lightweight Insulating Concrete

C496 Method of Test for Splitting Tensile Strength of Molded Concrete Cylinders

C497 Methods of Test for Determining Physical Properties of Concrete Pipe or Tile

C506 Specifications for Reinforced-concrete Arch Culvert, Storm-drain, and Sewer Pipe

C507 Specifications for Reinforced-concrete Elliptical Culvert, Storm-drain, and Sewer Pipe

C511 Specification for Moist Cabinets and Rooms Used in the Testing of Hydraulic Cements and Concretes

C512 Method of Test for Creep of Concrete in Compression

C513 Method for Securing, Preparing, and Testing Specimens from Hardened Lightweight Insulating Concrete for Compressive Strength

C535 Method of Test for Resistance to Abrasion of Large-size Coarse Aggregate by Use of the Los Angeles Machine

C566 Method of Test for Total Moisture Content of Aggregate by Drying

C586 Method of Test for Potential Alkali Reactivity of Carbonate Rocks for Concrete Aggregates (Rock Cylinder Method)

D98 Specifications for Calcium Chloride

D2419 Method of Test for Sand-equivalent Value of Soils and Fine Aggregate

E11 Specifications for Sieves for Testing Purposes (Wire Cloth Sieves, Round-hole and Square-hole Screens or Sieves)

E12 Definition of Terms Relating to Density and Specific Gravity of Solids, Liquids and Gases

E13 Definition of the Term Screen (Sieve)

E20 Recommended Practice for Analysis by Microscopical Methods for Particle Size Distribution of Particulate Substances of Subsieve Sizes

E119 Methods of Fire Tests of Building Construction and Materials

INDEX

INDEX

Abram's method of proportioning, 157, 158
Abrasion test, 283
Absorption by aggregates (*see* Aggregates)
Absorption capacity, 75, 76, 426–428
Absorptive form linings, 283
ACI (American Concrete Institute), 475
ACI proportioning method, 142–150, 445
  for no-slump concrete, 154
  for small jobs, 150, 151
  for structural lightweight concrete, 151–154
Adjustment of mix (*see* Mix adjustments)
Admixtures, 95–105
  accelerators, 98, 102, 103
  air-entraining agents, 4, 98–101 (*See also* specific air entries)
  bonding, 105
  Darex, 99
  early strength, 98, 102, 103
  expansion-producing, 103
  gas-forming, 99, 101
  lignosulfonates, 98
  pozzolans, 55, 103–105, 201, 277, 302
  retarders, 97, 98
  uses, 99
  waterproofing, 106
  water-reducing, 97–98
  workability, 97
Aggregate-cement ratio, 133
Aggregate content of concrete, 3
Aggregate-paste relationships, 133–136
Aggregates, absorption by, 74, 429, 430
Aggregates, absorption by, effective, 75
    test for, 426–430
  absorption capacity, 75, 76, 426–428
  artificial, 61, 368–370
  bulking of, 72
  bulking factor, 73
  burned-clay, 61, 368
  clay and silt in, 87, 88
  coarse, 4, 148
  crushed-stone, 61, 143
  deleterious substances, 87–90
  effect of, on properties of concrete, 4, 5
  fine, 3, 62
    recommended percentages, 138, 139
  fineness modulus, 79, 80
  function of, 4, 5
  gap grading, 77
  general characteristics, 62
  gradation, 60, 62, 76–86
  grading requirements, 84–86
  Haydite, 368, 370
  heavy-weight, 61, 62, 367, 368
  lightweight, 61, 62, 74, 368–370
  maximum size, 82, 83, 145
    recommended values, 145
  moisture content, 64, 74–76, 428, 429
  natural, 61
  particle interference, 77
  pit-run, 62
  proportioning, 60, 64, 80
  pumice, 62, 74, 368
  quality requirements, 62, 86, 87
  quartering, 65
  radiation shielding, 61, 62
  reactive, 90, 91, 104, 275–278
    correctives for, 277, 278

Aggregates, reactive, tests for, 277
  sampling of, 65, 66
  sand in, percentage, 137–140
  sand equivalent test, 88, 89
  separated, 78
  shape, 87, 89
  sieve analyses, 77–81, 424, 425
  slag, 62
  solid volume, 63, 269
  soundness, 88
  source, 61
  space occupied by, 3, 60
  specific gravity, 66–68, 426–429
    bulk, 66, 427, 428
  stone sand, 61
  storage, 92, 93
  surface areas, 156
  surface moisture, 64, 74–76, 428, 429
  unit weight, 68–70, 74, 430
  voids in, 68–72
  water for washing, 95
  wear resistance, 89
  weathering test, 88
Aggressive waters, 278–281
Air, entrapped, 120
Air-content tests of concrete, 121–123
  gravimetric, 123
  pressure, 121
  volumetric, 122
Air contents, of cement mortars, 52
  recommended for concrete, 99
Air-entraining agents, 4, 98–101
Air-entraining portland cement (*see* Portland cement)
Air entrainment, demonstration, 458, 459
  effect of, on concrete, 98, 117, 120, 272, 273
    on design of mix, 99
  measurement, 121–123
Air meter, 121
Alite, 24, 25
Alkali in cements, 16, 22
Alkali in cements, reaction with aggregates, 16, 22, 48
Alkali soils, 278
Alumina, 14, 15, 18
Aluminous cement, 58, 102
Aluminum powder, 101, 370
Architectural concrete, 352
Aspdin, Joseph, 10
ASTM (American Society for Testing Materials), 475
ASTM specifications, list of, 510–513
  for portland cement, 49–54
Autoclave test for soundness, 51, 52

Ball test, 466
  comparison with slump test, 113, 114
Batch mixer, 169, 170
Batch quantities, 140
Batch timer, 171
Batching, cement, 167
  irregularities, 167
  for test cylinders, 470, 471
  uniformity requirement, 162, 163
Batching equipment, 163, 169
  admixtures, 167–169
  calibration of, 164
  volumetric batching, 166
  water batching, 167, 169
  weight batching, 163–166
    checking accuracy, 164, 165
Bearing conditions, 255–257
Belite, 24, 35
Belt conveyors, 183, 184
Bituminous coatings, for curing, 198, 199
  for waterproofing, 266
Blast-furnace slag, 56
Bleeding of concrete, 41, 116, 117, 194
Bond strength, 234, 236
Boyle's law, 121, 459
Briquet, 51, 52
  testing machine, 463
Bulk specific gravity, 66, 427–429

Bulk volume, 64
Bulking of sand, 72, 73
Burned-clay aggregate, 61, 368

Calcium chloride, 96, 102, 103, 105, 205–207
  for acceleration of strength, 102, 103, 205–207
  effect on shrinkage, 302
Calcium hydroxide, 281
Calcium sulfate (*see* Gypsum)
Calcium sulfoaluminate, 33
Calculations for mix proportions, 143–150
Capping of cylinders, 256, 257, 450, 451
  with cement, 256, 257
  with gypsum, 256, 257, 472
  with sulfur, 256, 473, 474
Capping materials, test of, 450, 451
Cardboard cylinder molds, 254
Carlson, R. W., 304, 305, 335
Cavitation, 282, 283
Celite, 24
Cement, expanding, 103
  free lime in, 23, 25
  high-alkali, 16, 22
  high-early-strength (*see* High-early-strength cement)
  hydraulic, 10
  low-heat, 12, 27–30, 50, 52, 53
  modified (*see* Modified cement)
  natural, 18*n*.
  pozzolan, 103–105
  setting of, 32, 33, 36–40, 43, 52, 53, 118, 119, 415–417
  slag, 56
  specifications, 50, 52, 53
  sulfate-resisting, 12, 27, 28, 104
  surface areas, 29
  unit weight, 11
  (*See also* Portland cement; Portland-pozzolan cement)
Cement-aggregate ratio, 132
Cement-aggregate reaction, 90, 91, 275–278
Cement content in concrete, 5
Cement factor, 123–125, 140
Cement fineness (*see* Fineness of cement)
Cement paste, function of, 4
Cement-strength curves, 238, 241, 245
Cement-voids ratio, 158–160
Cement-water paste, functions of, 4
Central-mixed concrete, 173
Chapman flask, 429
Chemical attack, 281
Chemical composition of cement, 13–18, 21–28
Chert, 88, 90, 103
Chutes, 177, 183, 184
Cinders, 3, 62
Clay, 18, 56, 87–90
  and silt in aggregates, 87–90
Clinker, 19
  glass component, 25
Clinkering temperature, 19
Coal, 88, 90
Coarse aggregate, 3, 4, 148
Coefficient of variation, 400–402
Cohesiveness, test for, 433
Cold-weather concreting, 55, 202–205
Colorimetric test, 88
Compaction of concrete, by hand, 192
  by vibrators (*see* Vibratory compaction)
Composition of concrete, 3–6
  hardened solid portion, 4
Compression test, of cement mortar, 418–421
  of concrete, 436–442
  specimens, 242, 251, 384
    preparation of, 470, 471
Compressive strength (*see* Strength)
Condensation on walls, 332, 333
Consistency, 108–116
  apparatus, 112–116

Consistency, controlling factors, 108–110
desirable, 111
exchanging aggregate for cement, effect of, 5
normal, 37
tests for, 110–116
ball, 112–114
flow, 110–112
slump, 110, 112–114
values, 110
Control of operations, 8
Conversion factors, 461
Conveying of concrete, 175–184
batch containers, 175–178
belts, 183, 184
buckets, 175, 177
buggies, 176
chutes, 177, 183, 184
pneumatic method, 183
pump and pipeline, 178–183
cleaning of pump, 182, 183
pump sizes, 181, 182
Corrosion of metal, 206, 281, 282
Cracking, 49, 196, 207, 339–345
causes, 49, 196, 340–345
plastic shrinkage, 196, 207
by thermal stresses, 343–345
Creep, effect on, of age, 311, 312
of aggregate, 314
of cement, 312, 313
of moisture, 315
of prestressing, 319
of reinforcement, 318, 319
of size of mass, 315, 316
of stress, amount, 311, 312
kind, 316
of water-cement ratio, 312, 313
estimation of, 316, 317
factors affecting, 309, 310
recovery, 310, 317
significance, 291
Crushed stone, 61, 143
(*See also* Aggregates)
Curing, 4, 105, 106, 136, 195–209
compounds, 198, 199
Curing, conditions, test for effect on strength, 422, 423
definition, 4, 105, 136, 195, 196
effect of freezing, 200
mass, 209
methods, 105, 106, 196–199, 209
standard, 209
steam, 201, 202
temperatures, 199, 200
Curing period, length of, 195, 196
Cylinders, capping of (*see* Capping of cylinders)
care of, 385
testing of, 255–259

Damage prevention, 211
Darex, 99
Data, analysis, 385–406
central tendency, 398
deviation, 399–401
standard, 399–401
dispersion, 398–402
frequency diagram, 397
frequency distribution, 397, 398
grouping, 396–398
number of tests required, 404, 405
probable error, 402
raw, 385
significant figures, 405, 406
summary, 385, 406
uncertainty limits of average, 402–404
variation, 396
variation coefficient, 401, 402
Degree of saturation, 74, 75
Deleterious substances, 87–90
Delivery of concrete, 175–181, 183
Density, 66
(*See also* Voids)
Diary, 392
Diatomaceous earth, 55, 103
Dicalcium silicate, 11, 24–28, 34, 35
abbreviation for, 17
composition, 17

Dicalcium silicate, computation of, 26
  effect on cement, 28
Diffusivity, 334
Distribution of concrete, 175–184
  belt conveyor, 181, 184
  buggies, 176
  chuting, 177, 183, 184
  pipe line, 178–183
  pneumatic method, 183
Dunagan method, 125
Durability, 7, 268–284
  effect on, of aggregate, 269, 270, 275–277
    of air-entrainment, 98, 99, 271–273
    of cement, 11, 272, 273, 278–281
    of chemical attack, 281
    of leaching, 281
    of quality of paste, 7
    of sea water, 279
    of sulfate waters, 12, 269, 278–281
    of temperature changes, 270
    of water-cement ratio, 270
    of water content, 270
  factors affecting, 268
  freeze-thaw tests, 273–275
    sonic-modulus determinations, 274, 275
  weathering, 268
  (*See also* Soundness)
Durability factor, 275

Earth curing, 197
Edwards, L. N., 156
Effective absorption, 75
Efflorescence, 281
Elasticel, 101
Elasticity (*see* Modulus, of elasticity)
End conditions, effect on strength, 255–257
  test for, 450, 451
Entrained air (*see* Air entrainment)
Ettringite, 33, 48

Expansive cement, 57
  for prestressing, 373
Extensibility, 342, 343

Factors affecting concrete strengths, 238–259
  (*See also* Strength)
Failures of test specimens, typical, 228
  Mohr rupture diagram, 229
False set, 39, 53
Fatigue strength, 348–350
Felite, 24
Field batch weights, 148
Figures, 407–409
Final set, 37, 38
Fine aggregate (*see* Aggregates)
Fineness of cement, 29–31, 49, 52
  by Blaine apparatus, 30–31
  effect on strength, 135
  by hydrometer, 31
  by Roller air analyzer, 31
  size of particles, 29, 30
  specific surface, 39–41
  specification requirement, 52
  surface area, 30
  by turbidimeter, 30, 31
Fineness modulus, 79, 80
Finishing of concrete, 354, 355
Fire resistance, 345–349
Flash set, 38, 206
Flexural strength, 233, 234
Float finish, 354, 355
Floors, base preparation, 353, 354
  curing, 355, 356
  finishing, 354, 355
  mix, 354
  placing, 354
  protection, 355, 356
  requirements, 353
  surface hardness, 356
  types, 353
Flow test, 54, 467
Fly ash, 100, 103
Forms, absorptive linings for, 283

Forms, coatings for, 224, 225
  construction of, 220–222
  design for, 219
  failure of, 225
  materials for, 217–219
  pressures on, 119, 213–216
  removal of, 209, 210
  requirements of, 213
  slip type of, 222–224
  ties for, 218–219
  uplift on, 216
Free lime, 23, 35, 49
Freezing, effect of, 200
Freezing-thawing, resistance to, 172, 273
Freezing-and-thawing test, 273–275
Frequency distribution, 397, 398
Fresh concrete, 108–128
  characteristics, 432–435
  composition, 125, 126
  pressure on form, 119, 213–216
  set of, 118, 119
  temperature, 127, 128

Gas-forming agents, 101
Gel, 34–36, 38, 41–45
Gillmore test, 38, 52
Go-devil, 182
Graded aggregate, 60, 62, 76–86
Gravel, 61
  (*See also* Aggregates)
Gravimetric method, 123
Grouted concrete, 373
Grouting, 371, 372
Gunite, 356–360
  aggregate, 358, 359
  base preparation, 358
  equipment, 357
  limitations, 357
  mix, 359
  procedure, 359, 360
  rebound, 359
  sand, 358
  uses, 356
Gypsum, 15
  for control of time of setting, 20, 33, 34

Hammer test of concrete, 236–238
Hand mixing of concrete, 173
Hardeners, surface, 356
Hardening process, 36, 39
Haydite, 368, 370
Heat of hydration, 35, 36, 53, 54
  ASTM method for, 36
Heat generation, 335, 336
Heat transmission, 332
Heating of materials, 202
Heavy-weight concrete, 367, 368
High-alkali cement, 22
  reaction with aggregates, 16, 22, 48
High-alumina cement, 58
High-early-strength cement, 13, 54, 55, 135
  advantages, 13
  ASTM specification requirements, 50, 52, 53
  calcium chloride as accelerator for, 102, 103, 205–207
High-lime cement, 27
Hoover Dam, 36
Hot-weather concreting, 207–209
Hydration, of cement, 17, 31–35
  conditions favoring, 4
  heat of, 35, 36

Igneous rock, 61
Ignition, loss on, 22
Initial set, 37, 38
Insoluble residue, 21, 22
Inspection, after concreting, 384
  before concreting, 383
  of concreting, 383
    specimens molded, 384, 385
      required strength, 385, 386
  field laboratory, 386, 387
  organization, 376, 377

Inspection, records (*see* Records)
reports (*see* Reports)
scope of, 375, 376
Inspector, authority, 382
qualifications, 377, 378
relations of, with contractor, 379–381
with superiors, 379
responsibility, 378
training, 378, 379
Instructions for laboratory work, 413–460
general, 413
reports, 414
Inundation, 73
Iron, 15, 18

Job test specimens, 250, 251, 385, 386
Joint cleanup, 185, 186

Kelly ball, 113, 114, 466
Kiln, 19
Klein turbidimeter, 31

Laboratory, in field, 386, 387
work instructions, 413–460
Laitance, 8, 41*n.*, 117, 185
Leaching, 281
Lightweight concrete, 350, 368–370
Lime, 13, 14, 18, 27, 34, 35
Limestone, 18
Los Angeles rattler, 89
Loss on ignition, 21, 22
Low-heat cement, 27, 29
ASTM specification requirements, 50, 52, 53
compound composition, 27

Magnesia, 16, 18, 22, 35
Magnesium sulfate, 12
Making of concrete, 7–9
Making of concrete, control problems, 8, 9
desired properties, 8
factors affecting production, 7
Masonry cement, 56, 57
Mass concrete, 36, 185, 203, 360–364
characteristics, 360, 361
minimum temperature permitted, 203
properties, 362–364
temperature rise, 335–339
thermal stresses, 343–345
wet-screened, 361, 362
Mass curing, 209
Maximum size of aggregate (*see* Aggregates)
Measurement, of aggregates, 162–164, 166, 167
of cement, 167
of water, 167–169
Measuring cylinder, 430
Metamorphic rocks, 61
Mica, 62
Mineral aggregates (*see* Aggregates)
Mix adjustments, 5, 149
test, 446–449
Mixers, 169, 170
efficiency of, 171, 172
paving, 170
Mixes (*see* Proportioning)
Mixing, 169–174
effect of, on strength, 171, 172
hand, 173
for test cylinders, 470
Mixing plant, 169, 170, 173, 174
Mixing time, 171, 174
Mixing water, 94, 145
Modified cement, 12, 13, 27
ASTM specification requirements, 50, 52, 53
compound composition, 27
Modulus, of elasticity, 321, 330
effect on, of age, 326, 327
of aggregate, 327, 328
of mix, 326, 327

Modulus, of elasticity, effect on, of moisture, 328
of stress, 330
of test method, 325, 326
of type of load, 330
of unit weight, 329
of water-cement ratio, 326, 327
methods of determination, 321–323
in relation to strength, 329
sonic, 323, 324
sustained, 321, 326, 330
of rigidity, 324, 325
of rupture, 231–233, 241
Moisture, in aggregates, 64, 74–76, 428, 429
surface, 74
test for, 428, 429
Molding, of briquets, 418, 419
of cubes, 419, 420
of cylinders, 470, 471
Mortar, standard, 51–53
strength test, 51–53, 418, 420

Natural cement, 18*n*.
Normal consistency, 37, 38
test for, 415, 416
Number of specimens, 242, 251, 384

Opaline silica, 22
Organic impurities, 62, 67
Ottawa sand, 51, 53

Paper covering, 196, 199
Patching, 210, 211
Paving mixers, 170
Percentage of sand in aggregate, 137–140
Periclase, 35
Perlite, 369
Permeability, 47, 261–268
Permeability, effect on, of absorption, 267, 268
of admixtures, 265, 266
of aggregates, 265
of cement fineness, 263
of coatings, 265
of curing, 263, 265, 266
of placement method, 266, 267
of pore structure, 47, 261
of uniformity, 266, 267
of vibratory compaction, 267
of water-cement ratio, 263–265
factors affecting, 263–267
significance of, 262
tests, 262, 263
Physical requirements, portland cement, 52, 53
Placing of concrete, 185–195
in cold weather, 202–205
calcium chloride for, 205, 206
in hot weather, 207–209
preparations for, 185–187
on slope, 188, 190
under water, 365
Pneumatically placed concrete, 183
Pneumatically placed mortar, 286, 356–360
Poisson's ratio, 330, 331
Ponding, 197
Pore space, 45–47
Portland blast-furnace-slag cement, 56
Portland cement, acceptance tests, 50, 52, 53
air-entraining, 13, 98–101
advantages, 98, 99
voids developed by, 99–101
alkalies in, 16, 22
ASTM specifications for, 50–54
clinker, 19
composition, 21
chemical, 13–18, 21–28
compound, 23–27
influence of, on characteristics, 28
fineness (*see* Fineness of cement)

Portland cement, heat liberated by, 35, 36, 335–338
high-early-strength (*see* High-early-strength cement)
hydration of, 31–35
low-heat, 27, 29, 50, 52, 53
magnesia in, 16, 18, 22, 35
manufacture of, 18–20
modified, 12, 13, 27, 50, 52, 53
origin of name, 10
paste, 4, 6, 7, 40–48
physical requirements, 52, 53
raw materials, 18
set, 32, 33, 36–40, 43
size of particles, 29
soundness, 23, 35, 49, 52
specific gravity, 11, 461
specifications for, 50–53
sulfate-resisting, 12, 13, 27, 28, 50, 52, 53, 278–281
types, 12, 13–27
unit weight, 11
unsoundness, 23, 35, 49, 52
(*See also* Cement)
Portland Cement Association, 476
Portland-pozzolan cement, 55, 104, 335, 336, 363
comparison with portland cement, 55
Power float, 355
Powers, T. C., 115, 468
Pozzolan, 55, 103–105, 201, 277, 302
amount used, 55, 104
as corrective for reactive aggregate, 277
definition, 55, 103
Prepacked concrete, 286–288, 373
Pressures on forms, 119, 213–216
Properties of concrete, 454, 455
Proportioning, 5–7, 131–160
adjustments, 5, 141, 142, 149
arbitrary, 155
general, 130
by maximum density of aggregate, 155, 156
Proportioning, methods, 142–160
Abrams', 157, 158
ACI, 142–150, 445
for no-slump concrete, 154
for small jobs, 150–151
for structural lightweight concrete, 151–154
Edwards', 156
fineness modulus, 157, 158
mortar-voids, 158–160
surface area of aggregate, 156
Talbot and Richart, 158–160
trial, 137–141
voids in coarse aggregate, 160
voids-cement ratio, 158–160
variable factors in, 136, 137
Proportions, general, 5, 130, 131
methods of expressing, 131–133
mix, calculations for, 142–150
wet field mix, 125
Protection of concrete, from chemicals, 281
from cold, 199, 202–205
from drying, 196–199
from heat and wind, 207–209
Pumicite, 55, 103, 277, 278
Pumping of concrete, 178–183
Pycnometer, 76, 426

Quality of concrete, 8, 9
Quantities of materials, 139, 148, 149
Quartering method, 65, 66

Radiation shielding, 367, 368
Rate of loading, 257–259
Reactive aggregates, 90, 91, 275–278
Ready-mixed concrete, 173, 174, 203
Records, batching, 390
curing, 391
diary, 392
general comments, 389, 390

Records, materials, 390, 391
  mixing, 390
  photographic, 392
  placing, 391
Reference material, 475–513
Reinforcement, 106
Remolding test, 115, 468, 469
Removal of forms, 209, 210
Repair methods, 284–288
Reports, daily, 391, 392
  monthly, 393, 394
  summary, 393, 394
Residue, insoluble, 21, 22
Retempering, 173
Revibration, 195
Richart and Talbot voids-cement ratio, 158–160
Rock pockets, cause, 8, 176, 177, 187, 188
  prevention, 176, 177, 188, 189
Rodding procedure, 470
Rotary kiln, 19, 20

Salamander, 204
Sampling of aggregates, 65, 66
Sand, 3, 61, 138
  bulking of, 72
  equivalent, 88, 89
  standard, 51, 53
  stone, 61
  streaks, 194
  (*See also* Aggregates)
Sand splitter, 65
Sandstone, 301, 308, 309, 314
Saturation, degree of, 75
Schmidt concrete test hammer, 236–238
Sea water, 95, 279
Sealing compounds, 196, 198, 199, 204, 209
Sedimentary rocks, 61
Segregation, cause, 8, 176, 177, 187, 188
  prevention, 176, 177, 188, 189
Setting, of cement, 32, 33, 36–40, 43, 52, 118, 119, 415–417
    false, 38, 53, 119
    flash, 38, 119
  of concrete, 118, 119
Shale, 88
Shape, effect of, on strength, 251–254, 452, 453
Shear strength, 234
Shipping samples, 385
Shotcrete (*see* Gunite)
Shrink-mixed concrete, 173
Shrinkage (*see* Volume changes)
Sieve analysis, 77–81, 424, 425
  computations for combinations of aggregates, 80, 81
  curves, 81, 82
  method of test, 77–79, 424, 425
Sieves, specifications for, 77, 79
Silica, 14, 18
  diatomaceous, 55
  opaline, 22
Silicates, 10, 11
Silt, 62, 88, 90
Size, maximum, of aggregates, 82–84, 145
    effect of, on concrete, 83, 84
  of mixers, 169, 170
  of specimen, effect on strength, 251–254
Slag, 56
  cement, 56
Slump cone, 112
Slump-consistency relationship, 110, 112
Slump test, 112–114, 465
Slump values, 143
  comparison of, with ball values, 113, 114
    with flow values, 110
  effect of temperature, 112, 113
  recommended limits, 143
Slump-water content relationship, 110
Slurry, 19
Sodium silicate, 103

Sodium sulfate, 278
Soft fragments of aggregate, 87–90
Solid unit weight, 67
Solid volume, 63, 64, 67, 124, 144
Sonic modulus, 323, 324
Soundness, of aggregate, 88
  of cement, 23, 49, 52
    autoclave test for, 51, 52
    specification requirements, 49, 52
  of concrete, 268–284
  of rock, 88, 269, 270
Specific gravity, of aggregates, 66–68, 426–429
  bulk, 66, 427, 428
  of portland cement, 11, 461
  test for, 426–428
Specific heat, 334
Specific surface, 29–31
Specifications for portland cement, 49–53
Specimen shape, effect of, on strength, 251–254, 452, 453
Specimens of concrete, molding, 384, 385, 470, 471
  number recommended for test, 242, 251, 384
  shipping, 385
  storing, 385
Sprayed mortar (*see* Gunite)
Sprinkling, 197
Standard deviation, 399–401
Standard sand, 51
Steam curing, 201, 202
Steel-trowel finish, 355
Stock piles, 92, 93
Stone sand, 61
Storage, of aggregates, 92, 93
  of test specimens, 385
Straw covering, 197
Strength, bond, 234–236
  compressive, 47, 48, 52, 230–231
    relation of, to modulus of rupture, 231, 232, 241
      to tensile strength, 231, 232
    specimen, 251, 470, 471
Strength, effect on, of admixtures, 240, 273
    of age, 241, 245
    of aggregate, 239, 240
    of bearing conditions, 255–257
    of caps, 256, 257
    of casting conditions, 254
    of cement, amount, 240, 241
      type, 238, 239
    of curing, mass, 244
      moisture, 47, 242, 243, 422, 423
      steam, 201, 202
      temperature, 243–246
    of high temperatures, 248–250
    of lateral compression, 230
    of loading conditions, 247, 248, 257–259
    of moisture content, 254, 255
    of proportions, 240, 241
    of rate of loading, 257–259
    of size and shape of specimen, 251–254, 452, 453
    of temperature at test, 255
    of water-cement ratio, 240, 241, 436–442
    of wet screening, 253
  fatigue, 348–350
  flexural, 233, 234
  kinds, 230–236
  measure of quality, 228
  nature of, 228
  nondestructive indications, 236–238
  required, 242, 406
  shear, 234
  significance of, 227–230
  specimens recommended, 250, 251
  tensile, 51–53, 231–233, 456, 457
  test of cement mortar, 51, 418–423
    effect of curing conditions, 422, 423
  water-cement ratio, effect on, 7, 147, 241

Strength-age relationship, 238, 241
Sulfate-resisting cement, 27, 28, 104
ASTM specification requirements, 50, 52, 53
compound composition, 27
Sulfate waters, 12, 48
Sulfoaluminates, 33, 34, 103
Surface-area method of proportioning, 156
Surface areas, of aggregates, 156
of cement, 29
Surface moisture, 74–76
method of test for, 76, 428, 429
Surface treatments, 281

Tables, 407
Talbot and Richart voids-cement ratio, 158–160
Tamping rod, 465, 467
Temperature, effect of, on strength, 243–246, 248–250
of wet mix, 127
Temperature rise, 335–339
effect of precooling, 339
Temperatures for curing, 199, 202
Tensile strength of concrete, 231–233, 456, 457
Tension tests of cement mortars, 51–53, 418
Test specimens, 242, 251, 384, 385
required strength, 385, 386
storage, 385
Testing, aggregates, 65–83, 88–91
briquets, 53, 418, 463
cement, 49–54, 415–421
in compression, 51, 52, 230, 231, 251–259, 436–442, 450–453
cubes, 52, 419, 420
in flexure, 233, 253
in tension, 232, 418, 456, 457
Testing-machine instructions, 462–464
Tetracalcium aluminoferrite, 17, 24–28
Tetracalcium aluminoferrite, abbreviation for, 17
composition, 17
effect of, on cement, 28
Thermal conductivity, 331–334
Thermal expansion, 308, 309
Thermal properties, 331–339
Thermal stress, 343–345
Time, of mixing, 171
of setting, 37–39, 415–417
Time limit between mixing and placing, 174
Timing device, 171
Tobermorite, 35
gel, 42, 43, 45
Transit-mixed concrete, 173, 174
Transportation of concrete, 175–184
Tremie, 365
Trial batches, 140
Trial method of proportioning, 137–141, 443, 444
Tricalcium aluminate, 17, 24–26, 28, 36
abbreviation for, 17
composition, 17
computation of, 26
effect on cement, 28, 36
resistance to aggressive waters, 12
Tricalcium silicate, 11, 24–28, 34–36
abbreviation for, 17
composition, 17
computation of, 26
effect on cement, 28, 36
Triethanolamine, 102
Troweling workability, test for, 433
Truck agitators, 173
Truck mixers, 173, 174
Turbidimeter, 31

Underwater concrete, 365, 366
Unit weight, of aggregate, 64, 67
of cement, 11
of concrete, 123, 350, 351
test for, 430

Unsoundness, 23, 49, 52

Vacuum concrete, 116, 283, 366, 367
Vermiculite, 369
Vibrators, 192–195
efficiency of, 193
Vibratory compaction, concrete mix for, 193, 194
proper use of, 194, 195
revibration, 195
Vicat test, 38, 52
Vinsol resin, 100
Voids, 45
Volume changes, 7, 48, 201, 290–309
autogenous, 292–294
carbonation shrinkage, 48
drying shrinkage, 291–308
effect on, of admixtures, 296–298, 302
of age at first observation, 302
of aggregates, 298–301
of carbonation, 48, 306, 307
of cement, amount, 295, 296
composition, 297, 298
expansive cement, 294
fineness, 297, 298
of drying period, 304
of forms, 305
of gel structure, 292
of moisture, 295, 296, 303
of pozzolan, 297, 298
of reinforcement, 307, 308
of specimen size, 304, 305
of temperature, 303
of water content, 295, 296, 303
factors affecting, 294, 295
fresh concrete, 292
prepacked concrete, 308
setting shrinkage, 117
significance, 291
thermal, 308, 309

Wagner turbidimeter, 31
Washing of aggregates, 95
Water, absorbed, 64, 74–76, 426–430
for curing, 95
effect of, on strength, 7, 147, 241, 436–442
on workability, 109–111, 436, 437
free, in concrete, 64
measurement of, 167–169
mixing, 94, 145
sea, 95
for standard mortar, 51
Water-cement ratio, effect of, mix on, 6, 7
on properties of concrete, 6, 133, 134, 141, 145, 147
test for, 436–442
law, 134
recommended values, 146
strength curves, 147
Water content of concrete, 6, 145, 146
Water gain, 41, 116, 117, 194
Waterproof paper, 105, 196, 199
Waterproofing admixtures, 106
Wear, 282–284
Weathering (*see* Durability)
Weighing of materials, 163–166
Weight, 350, 351
Wet process, 19
Wet-screened concrete for specimens, 361, 362
Wheelbarrows, 167
Workability, 8, 108–110, 115, 116, 120
admixtures, 97
apparatus, 115, 116
factors affecting, 108–110
measurement of, 115, 116

Yield, 123, 124, 140